Design and Analysis of Biomolecular Circuits

Heinz Koeppl • Douglas Densmore
Gianluca Setti • Mario di Bernardo
Editors

Design and Analysis of Biomolecular Circuits

Engineering Approaches to Systems and Synthetic Biology

Springer

Editors
Heinz Koeppl
ETH Zurich
Automatic Control Laboratory
Physikstrasse 3
8092 Zurich
Switzerland
koeppl@ethz.ch

Douglas Densmore
Department of Electrical
and Computer Engineering
Boston University
8 Saint Mary's St, Boston Massachusetts
USA
dougd@bu.edu

Gianluca Setti
Università di Bologna
ARCES
Via Toffano 2/2
40125 Bologna
Italy
gianluca.setti@unife.it

Mario di Bernardo
Università di Napoli Federico II
Via Claudio 21
80125 Napoli
Italy
mario.dibernardo@unina.it

ISBN 978-1-4419-6765-7 e-ISBN 978-1-4419-6766-4
DOI 10.1007/978-1-4419-6766-4
Springer New York Dordrecht Heidelberg London

Library of Congress Control Number: 2011926002

Printed on acid-free paper

Springer is part of Springer Science+Business Media (www.springer.com)

Preface

The book is devoted to the design and analysis of biomolecular circuits as considered in systems biology and synthetic biology – two very dynamic and promising fields of research. Combining expertise and know-how from the biological and physical sciences with computer science, mathematics and engineering, their potential to impact society is only limited by the imagination of those working in the fields.

Synthetic biology promises to introduce new bio-therapeutic, bio-remediation, and bio-sensing applications. For example, synthesizing bacteria to seek out and destroy cancer cells, grass that glows red if planted on top of a land mine, cells that perform arithmetic operations, and small organisms that detect and remove heavy metals from the world's most dangerous drinking water. These systems are all possible by introducing key concepts in the way we abstract and standardize the process by which biological systems are developed. One of the current goals in synthetic biology is showing that, starting with well characterized biological primitives, complex systems can be composed using rules for system composition and automated with algorithms, biophysical models, and liquid handling robotics. A prime example of the synthetic biology community is the exciting International Genetically Engineered Machine competition (iGEM) held every year at MIT in Boston (USA). This event is just the tip of the iceberg. The field is in its infancy much the way the semiconductor industry was in the 1940's. Tremendous advances can be gained by not only furthering our knowledge of the biological phenomenon underlying these systems but also making sure the overall design process is formal, rigorous, and standardized.

Underpinning the advance and application of systems and synthetic biology is the development of appropriate modeling and computational tools for analysis and design purposes. This is an important ongoing research area. The idea is to model biological processes and reactions so as to allow experiments to be carried out *in-silico* before moving to the wet-lab. This is strongly reminiscent of the early days of electronics, where mathematical models had to be formulated to allow, for example, computer aided design (CAD) of integrated circuits and the efficient testing and design of complex devices.

The idea for this book arose during a successful special session on "Design of Biological Circuits and Systems" held at the IEEE International Symposium on Circuits and Systems (ISCAS) in 2009 in Taipei (Taiwan) organized by co-editors Heinz Koeppl and Gianluca Setti. The aim of the book is to present in a coherent framework some of the most recent work on the analysis, simulation and design of biomolecular circuits and systems reflecting the interdisciplinary and collaborative nature of the field. The results discussed in the book range from how these systems should be modeled and analyzed to how they should be physically designed and implemented.

The book is organized around four general thematic areas:

A. Analysis and Simulation
B. Modularity and Abstraction
C. Design and Standardization
D. Enabling Technologies

Drawing parallels to electronic circuit design the chosen organization of the book indicate – what the editors believe – are the important necessary steps to build complex synthetic circuits. Based on an appropriate mathematical formalism of how to describe, analyze and simulate basic cellular processes one can start to abstract away part of this overwhelming complexity (Part A). Abstraction and the clear definition of functional entities or *modules* that can be composed is the crucial step toward large-scale integration (Part B). If well-defined building blocks with well-defined interfaces are in place, standards can be created and the *in silico* design process can be automated (Part C). The concluding theme of the book discusses the experimental feasibility of the corresponding *in vivo* design and analysis process (Part D).

Each of these themes is organized in different chapters that are self-contained so that they can be read individually by experts but also read sequentially by someone wanting to get an overview of the field. The book is intended for computational scientists, e.g. mathematicians, physicists, computer scientist or engineers as well as for researchers from the life sciences. Special efforts have been made to make the chapters accessible for a broad, multi-disciplinary readership. For instance, in the experimental chapters of Part D, care has been taken so that computational researchers can follow the otherwise rather technical expositions on the technologies applied in experimental systems and synthetic biology.

We would like to thank Springer for their help and support in assembling this book together. In particular, we want to thank Charles Glaser for his constant drive behind this book project. The commitment taken and the effort invested by all contributors to deliver the chapter on time is gratefully acknowledged.

We hope the reader will find this book enjoyable and motivating. Systems and synthetic biology are exciting emerging research areas where skills and know-how from science and engineering are required. We believe this book offers a balanced overview of the many open problems and research challenges in the *design and analysis of biomolecular circuits*.

ETH Zurich, Switzerland — Heinz Koeppl
Boston University, USA — Douglas Densmore
University of Ferrara, Italy — Gianluca Setti
University of Bristol, UK and University of Naples Federico II, Italy — Mario di Bernardo
April 29, 2011

Contents

Contributors

David F. Anderson Departments of Mathematics, University of Wisconsin-Madison, 480 Lincoln Drive, Madison, WI 53706-1388, anderson@math.wisc.edu

Murad Banaji Department of Mathematics, University of Portsmouth, Lion Gate Building, Lion Terrace, Portsmouth, Hampshire PO1 3HF, UK, murad.banaji@port.ac.uk

Jacob Beal BBN Technologies, Cambridge, Massachusetts, jakebeal@bbn.com

Frank T. Bergmann California Institute of Technology, Pasadena, California
and
University of Washington, Seattle, Washington, fbergman@caltech.edu

Kevin Burrage Computing Laboratory, Wolfson Building, Parks Road, Oxford, OX1 3QD, UK
and
Department of Mathematical Sciences, Queensland University of Technology, Brisbane 4001, Australia, kevin.burrage@qut.edu.au

Pamela M. Burrage Queensland University of Technology, Brisbane 4001, Australia

Yizhi Cai Johns Hopkins University, Baltimore, Maryland, cai@jhmi.edu

Deepak Chandran University of Washington, Seattle, Washington, deepakc@u.washington.edu

Austin Che Ginkgo Bioworks, Boston, Massachusetts, austin@csail.mit.edu

Holger Conzelmann Max-Planck-Institute for Dynamics of Complex Technical Systems, Magdeburg, Germany
and
Institute for System Dynamics, University of Stuttgart, Germany

Gheorghe Craciun Department of Mathematics and Department of Biomolecular Chemistry, University of Wisconsin, craciun@math.wisc.edu

Domitilla Del Vecchio Department of Mechanical Engineering, Massachusetts Institute of Technology, 77 Massachusetts Ave. Cambridge, 02139, ddv@mit.edu

Douglas Densmore Department of Electrical and Computer Engineering, Boston University, Boston, MA, dougd@bu.edu

Mario di Bernardo Department of Systems and Computer Engineering, University of Bristol
and
Department of Engineering Mathematics, University of Naples Federico II, mario.dibernardo@unina.it

Michael Ederer Institute for System Dynamics, University of Stuttgart, Germany

Michal Galdzicki University of Washington, Seattle, Washington, mgaldzic@u.washington.edu

John H. Gennari University of Washington, Seattle, Washington, gennari@u.washington.edu

Ernst Dieter Gilles Max-Planck-Institute for Dynamics of Complex Technical Systems, Magdeburg, Germany

Marc Hafner Ecole Polytechnique Fédérale de Lausanne (EPFL), Lausanne, Switzerland

Nathan J. Hillson Joint BioEnergy Institute, Emeryville, California, njhillson@lbl.gov

Kyung Hyuk Kim Department of Bioengineering, University of Washington, William H. Foege Building, Box 355061, Seattle, WA 98195-5061, U.S.A., kkim@uw.edu

Heinz Koeppl Swiss Federal Institute of Technology Zurich (ETHZ), Zurich, Switzerland, koeppl@ethz.ch

Thomas G. Kurtz Departments of Mathematics, University of Wisconsin-Madison, 480 Lincoln Drive, Madison, WI 53706-1388, kurtz@math.wisc.edu

André Leier Department of Biosystems Science and Engineering, Swiss Federal Institute of Technology (ETH) Zurich, Universitätsstrasse 6, 8092 Zurich, Switzerland

Michael Liss GENEART AG, Josef-Engert-Str. 11, Regensburg 93053, Germany

James Lu Swiss Federal Institute of Technology Zurich (ETHZ), Zurich, Switzerland

Pier Luigi Luisi Biology Department, University of Roma Tre; V.le G. Marconi 446, 00146 Rome, Italy, luisi@mat.ethz.ch

Tatiana Marquez-Lago Department of Biosystems Science and Engineering, Swiss Federal Institute of Technology (ETH) Zurich, Universitätsstrasse 6, 8092 Zurich, Switzerland

Dan V. Nicolau, Jr Department of Integrative Biology, University of California at Berkeley, Berkeley CA 94704, United States

Casian Pantea Department of Mathematics, University of Wisconsin, pantea@math.wisc.edu

Serge Pelet Institute of Biochemistry, Department of Biology, Schaffmattstr 18, CH 8093-Zürich, ETH Zürich, serge.pelet@bc.biol.ethz.ch

Matthias Peter Department of Biology, Institute of Biochemistry, ETH Zürich

Tatjana Petrov Swiss Federal Institute of Technology Zurich (ETHZ), Zurich, Switzerland

Andrew Phillips Microsoft Research, Cambridge, UK, andrew.phillips@microsoft.com

Giovanni Russo Department of Systems and Computer Engineering, University of Naples Federico II, Via Claudio 21, 80125 Napoli, Italy, giovanni.russo2@unina.it

Julio Saez-Rodriguez European Bioinformatics Institute (EMBL-EBI) and EMBL-Heidelberg, Genome Biology Unit, Wellcome Trust Genome Campus, Cambridge CB10 1SD, UK, saezrodriguez@ebi.ac.uk

Herbert M. Sauro Department of Bioengineering, University of Washington, Seattle, Washington, hsauro@u.washington.edu

Jean Jacques Slotine Massachusetts Institute of Technology, Nonlinear Systems Laboratory, jjs@mit.edu

Eduardo D. Sontag Department of Mathematics, Rutgers University, Piscataway, NJ, sontag@math.rutgers.edu

Pasquale Stano Biology Department, University of Roma Tre, V.le G. Marconi 446, 00146 Rome, Italy

Ralf Wagner Life Technologies Inc. Geneart AG, Im Gewerbepark B35, Regensburg 93059, Germany
and
University of Regensburg, Molecular Microbiology and Gene Therapy, Franz-Josef-Strauss-Allee 11, Regensburg 93053, Germany, ralf.wagner@lifetech.com

Part I
Analysis and Simulation

Chapter 1
Continuous Time Markov Chain Models for Chemical Reaction Networks

David F. Anderson* and Thomas G. Kurtz†

Abstract A reaction network is a chemical system involving multiple reactions and chemical species. The simplest stochastic models of such networks treat the system as a continuous time Markov chain with the state being the number of molecules of each species and with reactions modeled as possible transitions of the chain. This chapter is devoted to the mathematical study of such stochastic models. We begin by developing much of the mathematical machinery we need to describe the stochastic models we are most interested in. We show how one can represent counting processes of the type we need in terms of Poisson processes. This *random time-change representation* gives a stochastic equation for continuous-time Markov chain models. We include a discussion on the relationship between this stochastic equation and the corresponding martingale problem and Kolmogorov forward (master) equation. Next, we exploit the representation of the stochastic equation for chemical reaction networks and, under what we will refer to as the classical scaling, show how to derive the deterministic law of mass action from the Markov chain model. We also review the diffusion, or Langevin, approximation, include a discussion of first order reaction networks, and present a large class of networks, those that are weakly reversible and have a deficiency of zero, that induce product-form stationary distributions. Finally, we discuss models in which the numbers of molecules and/or the reaction rate constants of the system vary over several orders of magnitude. We show that one consequence of this wide variation in scales is that different subsystems may evolve on different time scales and this time-scale variation can be exploited to identify reduced models that capture the behavior of parts of the system. We will discuss systematic ways of identifying the different time scales and deriving the reduced models.

*Research supported in part by NSF grant DMS 05-53687
†Research supported in part by NSF grants DMS 05-53687 and DMS 08-05793

T.G. Kurtz (✉)
Departments of Mathematics, University of Wisconsin-Madison, 480 Lincoln Drive, Madison, WI 53706-1388
e-mail: kurtz@math.wisc.edu

H. Koeppl et al. (eds.), *Design and Analysis of Biomolecular Circuits: Engineering Approaches to Systems and Synthetic Biology*, DOI 10.1007/978-1-4419-6766-4_1,

Keywords Reaction network · Markov chain · Law of mass action · Law of large numbers · Central limit theorem · Diffusion approximation · Langevin approximation · Stochastic equations · Multiscale analysis · Stationary distributions

MSC 2010 60J27, 60J28, 60J80, 60F17, 80A30, 92C40

Introduction

The idea of modeling chemical reactions as a stochastic process at the molecular level dates back at least to [12] with a rapid development beginning in the 1950s and 1960s. (See, for example, [6, 7, 39].) For the reaction

$$A + B \rightharpoonup C$$

in which one molecule of A and one molecule of B are consumed to produce one molecule of C, the intuition for the model for the reaction is that the probability of the reaction occurring in a small time interval $(t, t + \Delta t]$ should be proportional to the product of the numbers of molecules of each of the reactants and to the length of the time interval. In other words, since for the reaction to occur a molecule of A and a molecule of B must be close to each other, the probability should be proportional to the number of pairs of molecules that can react. A more systematic approach to this conclusion might be to consider the following probability problem: Suppose k red balls (molecules of A) and l black balls (molecules of B) are placed uniformly at random in n boxes, where n is much larger than k and l. What is the probability that at least one red ball ends up in the same box as a black ball? We leave it to the reader to figure that out. For a more physically based argument, see [21].

Our more immediate concern is that the calculation, however justified, assumes that the numbers of molecules of the chemical species are known. That assumption means that what is to be computed is a *conditional probability*, that is, a computation that uses information that might not (or could not) have been known when the experiment was first set up.

Assuming that at time t there are $X_A(t)$ molecules of A and $X_B(t)$ molecules of B in our system, we express our assumption about the probability of the reaction occurring by

$$P\{\text{reaction occurs in } (t, t + \Delta t] | \mathcal{F}_t\} \approx \kappa X_A(t) X_B(t) \Delta t \qquad (1.1)$$

where $\mathcal{F}_t$ represents the information about the system that is available at time t and κ is a positive constant, the *reaction rate constant*. Since Kolmogorov's fundamental work [28], probabilists have modeled information as a σ-algebra (a collection of sets with particular properties) of events (subsets of possible outcomes) in the sample space (the set of all possible outcomes). Consequently, mathematically, $\mathcal{F}_t$ is a σ-algebra, but readers unfamiliar with this terminology should just keep the idea of information in mind when we write expressions like this, that is, $\mathcal{F}_t$ just represents the information available at time t.

One of our first goals will be to show how to make the intuitive assumption in (1.1) into a precise mathematical model. Our model will be formulated in terms of X_A, X_B, and X_C which will be *stochastic processes,* that is, random functions of time. The triple $X(t) = (X_A(t), X_B(t), X_C(t))$ gives the *state* of the process at time t. Simple bookkeeping implies

$$X(t) = X(0) + R(t)\begin{pmatrix} -1 \\ -1 \\ 1 \end{pmatrix}, \tag{1.2}$$

where $R(t)$ is the number of times the reaction has occurred by time t and $X(0)$ is the vector giving the numbers of molecules of each of the chemical species in the system at time zero. We will assume that two reactions cannot occur at exactly the same time, so R is a counting process, that is, $R(0) = 0$ and R is constant except for jumps of plus one.

Our first task, in section "Counting Processes and Continuous Time Markov Chains", will be to show how one can represent counting processes of the type we need in terms of the most elementary counting process, namely, the Poisson process. Implicit in the fact that the right side of (1.1) depends only on the current values of X_A and X_B is the assumption that the model satisfies the Markov property, that is, the future of the process only depends on the current value, not on values at earlier times. The representation of counting processes in terms of Poisson processes then gives a stochastic equation for a general continuous-time Markov chain. There are, of course, other ways of specifying a continuous-time Markov chain model, and section "Counting Processes and Continuous Time Markov Chains" includes a discussion of the relationship between the stochastic equation and the corresponding martingale problem and Kolmogorov forward (master) equation. We also include a brief description of the common methods of simulating the models.

Exploiting the representation as a solution of a stochastic equation, in section "Reaction Networks" we discuss stochastic models for chemical reaction networks. Under what we will refer to as the classical scaling, we show how to derive the deterministic law of mass action from the Markov chain model and introduce the diffusion or Langevin approximation. We also discuss the simple class of networks in which all reactions are unary and indicate how the large literature on branching processes and queueing networks provides useful information about this class of networks. Many of these networks have what is known in the queueing literature as product form stationary distributions, which makes the stationary distributions easy to compute. The class of networks that have stationary distributions of this form is not restricted to unary networks, however. In particular, all networks that satisfy the conditions of the zero-deficiency theorem of Feinberg [15, 16], well-known in deterministic reaction network theory, have product-form stationary distributions. There is also a brief discussion of models of reaction networks with delays.

The biological systems that motivate the current discussion may involve reaction networks in which the numbers of molecules of the chemical species present in the system vary over several orders of magnitude. The reaction rates may also vary widely. One consequence of this wide variation in scales is that different subsystems may evolve on different time scales and this time-scale variation can be exploited to identify reduced models that capture the behavior of parts of the system. Section "Multiple Scales" discusses systematic ways of identifying the different time scales and deriving the reduced models.

Although much of the discussion that follows is informal and is intended to motivate rather than rigorously demonstrate the ideas and methods we present, any lemma or theorem explicitly identified as such is rigorously justifiable, or at least we intend that to be the case. Our intention is to prepare an extended version of this paper that includes detailed proofs of most or all of the theorems included.

Counting Processes and Continuous Time Markov Chains

The simplest counting process is a Poisson process, and Poisson processes will be the basic building blocks that we use to obtain more complex models.

Poisson Processes

A Poisson process is a model for a series of random observations occurring in time.

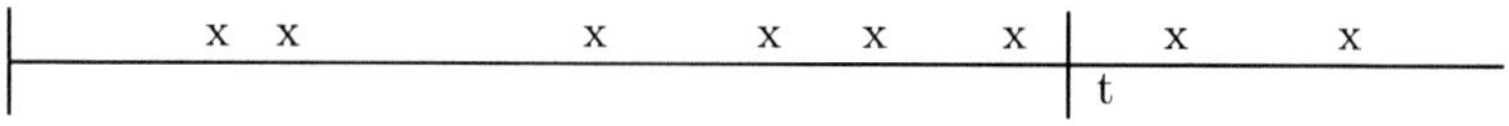

Let $Y(t)$ denote the number of observations by time t. In the figure above, $Y(t) = 6$. Note that for $t < s$, $Y(s) - Y(t)$ is the number of observations in the time interval $(t, s]$. We make the following assumptions about the model.

1. Observations occur one at a time.
2. Numbers of observations in disjoint time intervals are independent random variables, i.e., if $t_0 < t_1 < \cdots < t_m$, then $Y(t_k) - Y(t_{k-1})$, $k = 1, \ldots, m$ are independent random variables.
3. The distribution of $Y(t + a) - Y(t)$ does not depend on t.

The following result can be found in many elementary books on probability and stochastic processes. See, for example, Ross [41].

Theorem 1.1. *Under assumptions (1), (2), and (3), there is a constant $\lambda > 0$ such that, for $t < s$, $Y(s) - Y(t)$ is Poisson distributed with parameter $\lambda(s - t)$, that is,*

$$P\{Y(s) - Y(t) = k\} = \frac{(\lambda(s-t))^k}{k!} e^{-\lambda(s-t)}. \tag{1.3}$$

If $\lambda = 1$, then Y is a *unit* (or rate one) Poisson process. If Y is a unit Poisson process and $Y_\lambda(t) \equiv Y(\lambda t)$, then Y_λ is a Poisson process with parameter λ. Suppose $Y_\lambda(t) = Y(\lambda t)$ and $\mathcal{F}_t$ represents the information obtained by observing $Y_\lambda(s)$, for $s \leq t$. Then by the independence assumption and (1.3)

$$\begin{aligned} P\{Y_\lambda(t+\Delta t) - Y_\lambda(t) > 0 | \mathcal{F}_t\} &= P\{Y_\lambda(t+\Delta t) - Y_\lambda(t) > 0\} \\ &= 1 - e^{-\lambda \Delta t} \approx \lambda \Delta t. \end{aligned} \tag{1.4}$$

The following facts about Poisson processes play a significant role in our analysis of the models we will discuss.

Theorem 1.2. *If Y is a unit Poisson process, then for each $u_0 > 0$,*

$$\lim_{n\to\infty} \sup_{u \leq u_0} \left| \frac{Y(nu)}{n} - u \right| = 0 \quad a.s.$$

Proof. For fixed u, by the independent increments assumption, the result is just the ordinary law of large numbers. The uniformity follows by monotonicity. □

The classical central limit theorem implies

$$\lim_{n\to\infty} P\left\{ \frac{Y(nu) - nu}{\sqrt{n}} \leq x \right\} = \int_{-\infty}^{x} \frac{1}{\sqrt{2\pi}} e^{-y^2/2} dy = P\{W(u) \leq x\},$$

where W is a *standard Brownian motion*. In fact, the approximation is uniform on bounded time intervals in much the same sense that the limit in Theorem 1.2 is uniform. This result is essentially Donsker's functional central limit theorem [13]. It suggests that for large n

$$\frac{Y(nu) - nu}{\sqrt{n}} \approx W(u), \qquad \frac{Y(nu)}{n} \approx u + \frac{1}{\sqrt{n}} W(u)$$

where the approximation is uniform on bounded time intervals. One way to make this approximation precise is through the strong approximation theorem of Komlós, Major, and Tusńady [29, 30], which implies the following.

Lemma 1.3. *A unit Poisson process Y and a standard Brownian motion W can be constructed so that*

$$\Gamma \equiv \sup_{t \geq 0} \frac{|Y(t) - t - W(t)|}{\log(2 \vee t)} < \infty \quad a.s.$$

and there exists $c > 0$ such that $E[e^{c\Gamma}] < \infty$.

Proof. See Corollary 7.5.5 of [14]. □

Note that

$$\left| \frac{Y(nt) - nt}{\sqrt{n}} - \frac{1}{\sqrt{n}} W(nt) \right| \le \frac{\log(nt \vee 2)}{\sqrt{n}} \Gamma, \tag{1.5}$$

and that $\frac{1}{\sqrt{n}} W(nt)$ is a standard Brownian motion.

Continuous Time Markov Chains

The calculation in (1.4) and the time-change representation $Y_\lambda(t) = Y(\lambda t)$ suggest the possibility of writing R in (1.2) as

$$R(t) = Y\left(\int_0^t \kappa X_A(s) X_B(s) ds\right)$$

and hence

$$\begin{pmatrix} X_A(t) \\ X_B(t) \\ X_C(t) \end{pmatrix} \equiv X(t) = X(0) + \begin{pmatrix} -1 \\ -1 \\ 1 \end{pmatrix} Y\left(\int_0^t \kappa X_A(s) X_B(s) ds\right). \tag{1.6}$$

Given Y and the initial state $X(0)$ (which we assume is independent of Y), (1.6) is an equation that uniquely determines X for all $t > 0$. To see that this assertion is correct, let τ_k be the kth jump time of Y. Then letting

$$\zeta = \begin{pmatrix} -1 \\ -1 \\ 1 \end{pmatrix},$$

(1.6) implies $X(t) = X(0)$ for $0 \le t < \tau_1$, $X(t) = X(0) + \zeta$ for $\tau_1 \le t < \tau_2$, and so forth. To see that the solution of this equation has the properties suggested by (1.1), let $\lambda(X(t)) = \kappa X_A(t) X_B(t)$ and observe that occurrence of the reaction in $(t, t + \Delta t]$ is equivalent to $R(t + \Delta t) > R(t)$, so the left side of (1.1) becomes

$$\begin{aligned} &P\{R(t + \Delta t) > R(t) | \mathcal{F}_t\} \\ &\quad = 1 - P\{R(t + \Delta t) = R(t) | \mathcal{F}_t\} \\ &\quad = 1 - P\left\{Y\left(\int_0^t \lambda(X(s))ds + \lambda(X(t))\Delta t\right) = Y\left(\int_0^t \lambda(X(s))ds\right)\right\} \\ &\quad = 1 - e^{-\lambda(X(t))\Delta t} \approx \lambda(X(t))\Delta t, \end{aligned}$$

where the third equality follows from the fact that $Y(\int_0^t \lambda(X(s))ds)$ and $X(t)$ are part of the information in $\mathcal{F}_t$ (are $\mathcal{F}_t$-measurable in the mathematical terminology) and the independence properties of Y.

More generally, a continuous time Markov chain X taking values in $\mathbb{Z}^d$ is specified by giving its *transition intensities* (*propensities* in much of the chemical physics literature) λ_l that determine

$$P\left\{X(t+\Delta t)-X(t)=\zeta_l|\mathcal{F}_t^X\right\}\approx\lambda_l(X(t))\Delta t, \tag{1.7}$$

for the different possible jumps $\zeta_l\in\mathbb{Z}^d$, where $\mathcal{F}_t^X$ is the $\sigma-$algebra generated by X (all the information available from the observation of the process up to time t). If we write

$$X(t)=X(0)+\sum_l\zeta_l R_l(t)$$

where $R_l(t)$ is the number of jumps of ζ_l at or before time t, then (1.7) implies

$$P\left\{R_l(t+\Delta t)-R_l(t)=1\middle|\mathcal{F}_t^X\right\}\approx\lambda_l(X(t))\Delta t,\quad l\in\mathbb{Z}^d.$$

R_l is a *counting process* with intensity $\lambda_l(X(t))$, and by analogy with (1.6), we write

$$X(t)=X(0)+\sum\zeta_l Y_l\left(\int_0^t\lambda_l(X(s))ds\right), \tag{1.8}$$

where the Y_l are independent unit Poisson processes. This equation has a unique solution by the same jump by jump argument used above provided $\sum_l\lambda_l(x)<\infty$ for all x. Unless we add additional assumptions, we cannot rule out the possibility that the solution only exists up to some finite time. For example, if $d=1$ and $\lambda_1(k)=(1+k)^2$, the solution of

$$X(t)=Y_1\left(\int_0^t(1+X(s))^2ds\right)$$

hits infinity in finite time. To see why this is the case, compare the above equation to the ordinary differential equation

$$\dot{x}(t)=(1+x(t))^2,\quad x(0)=0.$$

Equivalence of Stochastic Equations and Martingale Problems

There are many ways of relating the intensities λ_l to the stochastic process X, and we will review some of these in later sections, but the stochastic equation (1.8) has the advantage of being intuitive (λ_l has a natural interpretation as a 'rate') and easily generalized to take into account such properties as external noise, in which (1.8) becomes

$$X(t)=X(0)+\sum\zeta_l Y_l\left(\int_0^t\lambda_l(X(s),Z(s))ds\right)$$

where Z is a stochastic process independent of $X(0)$ and the Y_l, or delays, in which (1.8) becomes

$$X(t) = X(0) + \sum \zeta_l Y_l \left(\int_0^t \lambda_l(X(s), X(s-\delta))ds \right),$$

or perhaps the λ_l become even more complicated functions of the past of X. We will also see that these stochastic equations let us exploit well-known properties of the Poisson processes Y_l to study the properties of X.

The basic building blocks of our models remain the counting processes R_l and their intensities expressed as functions of the past of the R_l and possibly some additional stochastic input independent of the Y_l (for example, the initial condition $X(0)$ or the environmental noise Z).

For the moment, we focus on a finite system of counting processes $R = (R_1, \ldots, R_m)$ given as the solution of a system of equations

$$R_l(t) = Y_l \left(\int_0^t \gamma_l(s, R)ds \right), \tag{1.9}$$

where the γ_l are *nonanticipating* in the sense that

$$\gamma_l(t, R) = \gamma_l(t, R(\cdot \wedge t)), \quad t \geq 0,$$

that is, at time t, $\gamma_l(t, R)$ depends only on the past of R up to time t, and the Y_l are independent, unit Poisson processes. The independence of the Y_l ensures that only one of the R_l jumps at a time. Let τ_k be the kth jump time of R. Then any system of this form has the property that for all l and k,

$$M_l^k(t) \equiv R_l(t \wedge \tau_k) - \int_0^{t \wedge \tau_k} \gamma_l(s, R)ds$$

is a *martingale*, that is, there exists a filtration $\{\mathcal{F}_t\}$ such that

$$E[M_l^k(t+s)|\mathcal{F}_t] = M_l^k(t), \quad t, s \geq 0.$$

Note that

$$\lim_{k\to\infty} E[R_l(t \wedge \tau_k)] = \lim_{k\to\infty} E\left[\int_0^{t\wedge\tau_k} \gamma_l(s, R)ds \right],$$

allowing $\infty = \infty$, and if the limit is finite for all l and t, then $\tau_\infty = \infty$ and for each l,

$$M_l(t) = R_l(t) - \int_0^t \gamma_l(s, R)ds$$

is a martingale.

There is a converse to these assertions. If $(R_1, \ldots, R_m)$ are counting processes adapted to a filtration $\{\mathcal{F}_t\}$ and $(\lambda_1, \ldots, \lambda_m)$ are nonnegative stochastic processes adapted to $\{\mathcal{F}_t\}$ such that for each k and l,

$$R_l(t \wedge \tau_k) - \int_0^{t \wedge \tau_k} \lambda_l(s) ds$$

is a $\{\mathcal{F}_t\}$-martingale, we say that λ_l is the $\{\mathcal{F}_t\}$-intensity for R_l.

Lemma 1.4. *Assume that $R = (R_1, \ldots, R_m)$ is a system of counting processes with no common jumps and λ_l is the $\{\mathcal{F}_t\}$-intensity for R_l. Then there exist independent unit Poisson processes $Y_1, \ldots, Y_m$ (perhaps on an enlarged sample space) such that*

$$R_l(t) = Y_l\left(\int_0^t \lambda_l(s) ds\right).$$

Proof. See Meyer [40] and Kurtz [35]. □

This lemma suggests the following alternative approach to relating the intensity of a counting process to the corresponding counting process. Again, given nonnegative, nonanticipating functions γ_l, the intuitive problem is to find counting processes R_l such that

$$P\{R_l(t + \Delta t) > R_l(t) | \mathcal{F}_t\} \approx \gamma_l(t, R)\Delta t,$$

which we now translate into the following *martingale problem.* In the following definition $\mathbb{J}_m[0, \infty)$ denotes the set of m–dimensional cadlag (right continuous with left limits at each $t > 0$) counting paths.

Definition 1.5. Let γ_l, $l = 1, \ldots, m$, be nonnegative, nonanticipating functions defined on $\mathbb{J}_m[0, \infty)$. Then a family of counting processes $R = (R_1, \ldots, R_m)$ is a solution of the *martingale problem* for $(\gamma_1, \ldots, \gamma_m)$ if the R_l have no simultaneous jumps and there exists a filtration $\{\mathcal{F}_t\}$ such that R is adapted to $\{\mathcal{F}_t\}$ and for each l and k,

$$R_l(t \wedge \tau_k) - \int_0^{t \wedge \tau_k} \gamma_l(s, R) ds$$

is a $\{\mathcal{F}_t\}$-martingale.

Of course, the solution of (1.9) is a solution of the martingale problem and Lemma 1.4 implies that every solution of the martingale problem can be written as a solution of the stochastic equation. Consequently, the stochastic equation and the martingale problem are equivalent ways of specifying the system of counting processes that corresponds to the γ_l. The fact that the martingale problem uniquely characterizes the system of counting processes is a special case of a theorem of Jacod [23].

Thinning of Counting Processes

Consider a single counting process R_0 with $\{\mathcal{F}_t\}$-intensity λ_0, and let $p(t, R_0)$ be a cadlag, nonanticipating function with values in $[0, 1]$. For simplicity, assume

$$E[R_0(t)] = E\left[\int_0^t \lambda_0(s) ds\right] < \infty.$$

We want to construct a new counting process R_1 such that at each jump of R_0, R_1 jumps with probability $p(t-, R_0)$. Perhaps the simplest construction is to let $\xi_0, \xi_1, \ldots$ be independent, uniform [0, 1] random variables that are independent of R_0 and to define

$$R_1(t) = \int_0^t \mathbf{1}_{[0,p(s-,R_0)]}(\xi_{R_0(s-)})dR_0(s).$$

Since with probability one,

$$R_1(t) = \lim_{n\to\infty} \sum_{k=0}^{\lfloor nt \rfloor} \mathbf{1}_{[0,p(\frac{k}{n},R_0)]}\left(\xi_{R_0(\frac{k}{n})}\right)\left(R_0\left(\frac{k+1}{n}\right) - R_0\left(\frac{k}{n}\right)\right),$$

where $\lfloor z \rfloor$ is the integer part of z, setting $\widetilde{R}_0(t) = R_0(t) - \int_0^t \lambda_0(s)ds$, we see that

$$\begin{aligned} R_1(t) &- \int_0^t \lambda_0(s)p(s, R_0)ds \\ &= \int_0^t (\mathbf{1}_{[0,p(s-,R_0)]}(\xi_{R_0(s-)}) - p(s-, R_0))dR_0(s) \\ &\quad + \int_0^t p(s-, R_0)d\widetilde{R}_0(s) \end{aligned}$$

is a martingale (because both terms on the right are martingales). Hence, R_1 is a counting process with intensity $\lambda_0(t)p(t, R_0)$. We could also define

$$R_2(t) = \int_0^t \mathbf{1}_{(p(s-,R_0),1]}(\xi_{R_0(s-)})dR_0(s),$$

so that R_1 and R_2 would be counting processes without simultaneous jumps having intensities $\lambda_0(t)p(t, R_0)$ and $\lambda_0(t)(1 - p(t, R_0))$.

Note that we could let p be a nonanticipating function of both R_0 and R_1, or equivalently, R_1 and R_2. With that observation in mind, let $\gamma_0(t, R)$ be a nonnegative, nonanticipating function of $R = (R_1, \ldots, R_m)$, and let $p_l(t, R)$, $l = 1, \ldots, m$, be cadlag nonnegative, nonanticipating functions satisfying $\sum_{l=1}^m p_l(t, R) \equiv 1$. Let Y be a unit Poisson process and $\xi_0, \xi_1, \ldots$ be independent, uniform [0, 1] random variables that are independent of Y, and set $q_0 = 0$ and for $1 \le l \le m$ set $q_l(t, R) = \sum_{i=1}^l p_i(t, R)$. Now consider the system

$$R_0(t) = Y\left(\int_0^t \gamma_0(s, R)ds\right) \tag{1.10}$$

$$R_l(t) = \int_0^t \mathbf{1}_{(q_{l-1}(s-,R),q_l(s-,R)]}(\xi_{R_0(s-)})dR_0(s). \tag{1.11}$$

Then $R = (R_1, \ldots, R_m)$ is a system of counting processes with intensities $\lambda_l(t) = \gamma_0(t, R)p_l(t, R)$.

If, as in the time-change equation (1.9) and the equivalent martingale problem described in Definition 1.5, we start with intensities $\gamma_1, \ldots, \gamma_m$, we can define

$$\gamma_0(t, R) = \sum_{l=1}^{m} \gamma_l(t, R), \quad p_l(t, R) = \frac{\gamma_l(t, R)}{\gamma_0(t, R)},$$

and the solution of the system (1.10) and (1.11) will give a system of counting processes with the same distribution as the solution of the time-change equation or the martingale problem. Specializing to continuous-time Markov chains and defining

$$\lambda_0(x) = \sum_l \lambda_l(x), \quad q_l(x) = \sum_{i=1}^{l} \lambda_i(x)/\lambda_0(x),$$

the equations become

$$R_0(t) = Y\left(\int_0^t \lambda_0(X(s))ds\right) \tag{1.12}$$
$$X(t) = X(0) + \sum_l \zeta_l \int_0^t \mathbf{1}_{(q_{l-1}(X(s-)),q_l(X(s-)))}(\xi_{R_0(s-)})dR_0(s).$$

This representation is commonly used for simulation, see section "Simulation".

The Martingale Problem and Forward Equation for Markov Chains

Let X satisfy (1.8), and for simplicity, assume that $\tau_\infty = \infty$, that only finitely many of the λ_l are not identically zero, and that

$$E[R_l(t)] = E\left[\int_0^t \lambda_l(X(s))ds\right] < \infty, \quad l = 1, \ldots, m.$$

Then for f a bounded function on $\mathbb{Z}^d$,

$$f(X(t)) = f(X(0)) + \sum_l \int_0^t (f(X(s-) + \zeta_l) - f(X(s-)))dR_l(t)$$

and defining

$$\widetilde{R}_l(t) = R_l(t) - \int_0^t \lambda_l(X(s))ds,$$

we see that

$$\begin{aligned} f(X(t)) - f(X(0)) - \int_0^t \sum_l \lambda_l(X(s))(f(X(s) + \zeta_l) - f(X(s)))ds \\ = \sum_l \int_0^t (f(X(s-) + \zeta_l) - f(X(s-)))d\widetilde{R}_l(s) \end{aligned}$$

is a martingale.

Define

$$Af(x) = \sum_l \lambda_l(x)(f(x + \zeta_l) - f(x)).$$

Allowing $\tau_\infty < \infty$, define $X(t) = \infty$ for $t \geq \tau_\infty$. If $\tau_\infty < \infty$,

$$\lim_{k \to \infty} |X(\tau_k)| = \infty,$$

and this definition gives a 'continuous' extension of X to the time interval $[0, \infty)$. Let f satisfy $f(x) = 0$ for $|x|$ sufficiently large, and define $f(\infty) = 0$. Then for any solution of (1.8),

$$f(X(t)) - f(X(0)) - \int_0^t Af(X(s))ds \tag{1.13}$$

is a martingale.

Definition 1.6. A right continuous, $\mathbb{Z}^d \cup \{\infty\}$-valued stochastic process X is a solution of the *martingale problem* for A if there exists a filtration $\{\mathcal{F}_t\}$ such that for each f satisfying $f(x) = 0$ for $|x|$ sufficiently large, (1.13) is a $\{\mathcal{F}_t\}$-martingale. X is a *minimal solution*, if in addition, $X(t) = \infty$ for $t \geq \tau_\infty$.

The following lemma follows from Lemma 1.4.

Lemma 1.7. *If X is a minimal solution of the martingale problem for A, then there exist independent unit Poisson processes Y_l (perhaps on an enlarged sample space) such that*

$$R_l(t) = Y_l\left(\int_0^t \lambda_l(X(s))ds\right).$$

The martingale property implies

$$E[f(X(t))] = E[f(X(0))] + \int_0^t E[Af(X(s))]ds$$

and taking $f(x) = \mathbf{1}_{\{y\}}(x)$, we have

$$P\{X(t) = y\} = P\{X(0) = y\} + \int_0^t \left(\sum_l \lambda_l(y - \zeta_l) P\{X(s) = y - \zeta_l\} - \sum_l \lambda_l(y) P\{X(s) = y\} \right) ds$$

giving the Kolmogorov forward or *master equation* for the distribution of X. In particular, defining $p_y(t) = P\{X(t) = y\}$ and $\nu_y = P\{X(0) = y\}$, $\{p_y\}$ satisfies the system of differential equations

$$\dot{p}_y(t) = \sum_l \lambda_l(y - \zeta_l) p_{y-\zeta_l}(t) - \left(\sum_l \lambda_l(y) \right) p_y(t), \tag{1.14}$$

with initial condition $p_y(0) = \nu_y$.

Lemma 1.8. *Let $\{\nu_y\}$ be a probability distribution on $\mathbb{Z}^d$, and let $X(0)$ satisfy $P\{X(0) = y\} = \nu_y$. The system of differential equations (1.14) has a unique solution satisfying $p_y(0) = \nu_y$ and $\sum_y p_y(t) \equiv 1$ if and only if the solution of (1.8) satisfies $\tau_\infty = \infty$.*

Simulation

The stochastic equations (1.8) and (1.12) suggest methods of simulating continuous-time Markov chains, and these methods are, in fact, well known. Equation (1.8) corresponds to the *next reaction* (next jump) method as defined by Gibson and Bruck [18].

The algorithm obtained by simulating (1.12) is known variously as the *embedded chain method* or Gillespie's [19, 20] *direct method* or the *stochastic simulation algorithm* (SSA).

If we define an Euler-type approximation for (1.8), that is, for $0 = t_0 < t_1 < \cdots$, recursively define

$$\widehat{X}(t_n) = X(0) + \sum_l \zeta_l Y_l \left(\sum_{k=0}^{n-1} \lambda_l \left(\widehat{X}(t_k) \right) (t_{k+1} - t_k) \right),$$

we obtain Gillespie's [22] τ-leap method.

Stationary Distributions

We restrict our attention to continuous-time Markov chains for which $\tau_\infty = \infty$ for all initial values and hence, given $X(0)$, the process is uniquely determined as a solution of (1.8), (1.12), or the martingale problem given by Definition 1.6, and the one-dimensional distributions are uniquely determined by (1.14). A probability distribution π is called a stationary distribution for the Markov chain if $X(0)$ having distribution π implies X is a stationary process, that is, for each choice of $0 \leq t_1 < \cdots < t_k$, the joint distribution of

$$(X(t + t_1), \ldots, X(t + t_k))$$

does not depend on t.

If $X(0)$ has distribution π, then since $E[f(X(0))]=E[f(X(t))] =\sum_x f(x)\pi(x)$, the martingale property for (1.13) implies

$$0 = E\left[\int_0^t Af(X(s))ds\right] = t\sum_x Af(x)\pi(x),$$

and as in the derivation of (1.14),

$$\sum_l \lambda_l(y-\zeta_l)\pi(y-\zeta_l) - \left(\sum_l \lambda_l(y)\right)\pi(y) = 0.$$

Reaction Networks

We consider a *network* of r_0 chemical reactions involving s_0 chemical species, $S_1, \ldots, S_{s_0}$,

$$\sum_{i=1}^{s_0} \nu_{ik} S_i \rightharpoonup \sum_{i=1}^{s_0} \nu'_{ik} S_i, \quad k = 1, \ldots, r_0,$$

where the ν_{ik} and ν'_{ik} are nonnegative integers. Let the components of $X(t)$ give the numbers of molecules of each species in the system at time t. Let ν_k be the vector whose ith component is ν_{ik}, the number of molecules of the ith chemical species consumed in the kth reaction, and let ν'_k be the vector whose ith component is ν'_{ik}, the number of molecules of the ith species produced by the kth reaction. Let $\lambda_k(x)$ be the rate at which the kth reaction occurs, that is, it gives the propensity/intensity of the kth reaction as a function of the numbers of molecules of the chemical species.

If the kth reaction occurs at time t, the new state becomes

$$X(t) = X(t-) + \nu'_k - \nu_k.$$

The number of times that the kth reaction occurs by time t is given by the counting process satisfying

$$R_k(t) = Y_k\left(\int_0^t \lambda_k(X(s))ds\right),$$

where the Y_k are independent unit Poisson processes. The state of the system then satisfies

$$\begin{aligned}X(t) &= X(0) + \sum_k R_k(t)(\nu_k' - \nu_k)\\ &= X(0) + \sum_k Y_k\left(\int_0^t \lambda_k(X(s))ds\right)(\nu_k' - \nu_k).\end{aligned}$$

To simplify notation, we will write

$$\zeta_k = \nu_k' - \nu_k.$$

Rates for the Law of Mass Action

The stochastic form of the law of mass action says that the rate at which a reaction occurs should be proportional to the number of distinct subsets of the molecules present that can form the inputs for the reaction. Intuitively, the mass action assumption reflects the idea that the system is well-stirred in the sense that all molecules are equally likely to be at any location at any time. For example, for a binary reaction $S_1 + S_2 \rightharpoonup S_3$ or $S_1 + S_2 \rightharpoonup S_3 + S_4$,

$$\lambda_k(x) = \kappa_k x_1 x_2,$$

where κ_k is a rate constant. For a unary reaction $S_1 \rightharpoonup S_2$ or $S_1 \rightharpoonup S_2 + S_3$, $\lambda_k(x) = \kappa_k x_1$. For $2S_1 \rightharpoonup S_2$, $\lambda_k(x) = \kappa_k x_1(x_1 - 1)$.

For a binary reaction $S_1 + S_2 \rightharpoonup S_3$, the rate should vary inversely with volume, so it would be better to write

$$\lambda_k^N(x) = \kappa_k N^{-1} x_1 x_2 = N\kappa_k z_1 z_2,$$

where classically, N is taken to be the volume of the system times Avogadro's number and $z_i = N^{-1}x_i$ is the concentration in moles per unit volume. For $2S_1 \rightarrow S_2$, since N is very large,

$$\frac{1}{N}\kappa_k x_1(x_1 - 1) = N\kappa_k z_1\left(z_1 - \frac{1}{N}\right) \approx N\kappa_k z_1^2.$$

Note that unary reaction rates also satisfy

$$\lambda_k(x) = \kappa_k x_i = N\kappa_k z_i.$$

Although, reactions of order higher than binary may not be physical, if they were, the analogous form for the intensity would be

$$\lambda_k^N(x) = \kappa_k \frac{\prod_i \nu_{ik}!}{N^{|\nu_k|-1}} \prod_i \binom{x_i}{\nu_{ik}} = N\kappa_k \frac{\prod_i \nu_{ik}!}{N^{|\nu_k|}} \prod \binom{x_i}{\nu_{ik}},$$

where $|\nu_k| = \sum_i \nu_{ik}$. Again $z = N^{-1}x$ gives the concentrations in moles per unit volume, and

$$\lambda_k^N(x) \approx N\kappa_k \prod_i z_i^{\nu_{ik}} \equiv N\widetilde{\lambda}_k(z), \tag{1.15}$$

where $\widetilde{\lambda}_k$ is the usual deterministic form of mass action kinetics.

General Form for the Classical Scaling

Setting $C^N(t) = N^{-1}X(t)$ and using (1.15)

$$\begin{aligned} C^N(t) &= C^N(0) + \sum_k N^{-1} Y_k\left(\int_0^t \lambda_k^N(X(s))ds\right)\zeta_k \\ &\approx C^N(0) + \sum_k N^{-1} Y_k\left(N\int_0^t \widetilde{\lambda}_k(C^N(s))ds\right)\zeta_k \\ &= C^N(0) + \sum_k N^{-1} \widetilde{Y}_k\left(N\int_0^t \widetilde{\lambda}_k(C^N(s))ds\right)\zeta_k + \int_0^t F(C^N(s))ds, \end{aligned}$$

where $\widetilde{Y}_k(u) = Y_k(u) - u$ is the centered process and

$$F(z) \equiv \sum_k \kappa_k \prod_i z_i^{\nu_{ik}} \zeta_k.$$

The law of large numbers for the Poisson process, Lemma 1.2, implies $N^{-1}\widetilde{Y}(Nu) \approx 0$, so

$$C^N(t) \approx C^N(0) + \sum_k \int_0^t \kappa_k \prod_i C_i^N(s)^{\nu_{ik}} \zeta_k ds = C^N(0) + \int_0^t F(C^N(s))ds,$$

which in the limit as $N \to \infty$ gives the classical deterministic law of mass action

$$\dot{C}(t) = \sum_k \kappa_k \prod_i C_i(t)^{\nu_{ik}} \zeta_k = F(C(t)). \tag{1.16}$$

(See [31, 33, 34] for precise statements about this limit.)

Since by (1.5),

$$\frac{1}{\sqrt{N}}\widetilde{Y}_k(Nu) = \frac{Y_k(Nu) - Nu}{\sqrt{N}}$$

is approximately a Brownian motion,

$$\begin{aligned}
V^N(t) &\equiv \sqrt{N}(C^N(t) - C(t)) \\
&\approx V^N(0) + \sqrt{N}\left(\sum_k \frac{1}{N} Y_k\left(N\int_0^t \widetilde{\lambda}_k(C^N(s))ds\right)\zeta_k - \int_0^t F(C(s))ds\right) \\
&= V^N(0) + \sum_k \frac{1}{\sqrt{N}}\widetilde{Y}_k\left(N\int_0^t \widetilde{\lambda}_k(C^N(s))ds\right)\zeta_k \\
&\quad + \int_0^t \sqrt{N}(F(C^N(s)) - F(C(s)))ds \\
&\approx V^N(0) + \sum_k W_k\left(\int_0^t \widetilde{\lambda}_k(C(s))ds\right)\zeta_k + \int_0^t \nabla F(C(s))V^N(s)ds,
\end{aligned}$$

where the second approximation follows from (1.15). The limit as N goes to infinity gives $V^N \Rightarrow V$ where

$$V(t) = V(0) + \sum_k W_k\left(\int_0^t \widetilde{\lambda}_k(C(s))ds\right)\zeta_k + \int_0^t \nabla F(C(s))V(s)ds. \quad (1.17)$$

(See [32, 34, 42] and Chap. 11 of [14].) This limit suggests the approximation

$$C^N(t) \approx \widehat{C}_N(t) \equiv C(t) + \frac{1}{\sqrt{N}}V(t). \quad (1.18)$$

Since (1.17) is a linear equation driven by a Gaussian process, V is Gaussian as is $\widehat{C}_N$.

Diffusion/Langevin Approximations

The first steps in the argument in the previous section suggest simply replacing the rescaled centered Poisson processes $\frac{1}{\sqrt{N}}\widetilde{Y}_k(N\cdot)$ by independent Brownian motions and considering a solution of

$$D^N(t) = D^N(0) + \sum_k \frac{1}{\sqrt{N}} W_k\left(\int_0^t \widetilde{\lambda}_k(D^N(s))ds\right)\zeta_k + \int_0^t F(D^N(s))ds \quad (1.19)$$

as a possible approximation for C^N. Unfortunately, even though only ordinary integrals appear in this equation, the theory of the equation is not quite as simple as it looks. Unlike (1.8) where uniqueness of solutions is immediate, no general uniqueness theorem is known for (1.19) without an additional requirement on the solution. In particular, setting

$$\tau_k^N(t) = \int_0^t \widetilde{\lambda}_k \left(D^N(s)\right) ds,$$

we must require that the solution D^N is *compatible* with the Brownian motions W_k in the sense that $W_k(\tau_k^N(t) + u) - W_k(\tau_k^N(t))$ is independent of $\mathcal{F}_t^{D^N}$ for all k, $t \geq 0$, and $u \geq 0$. This requirement is intuitively natural and is analogous to the requirement that a solution of an Itô equation be *nonanticipating*. In fact, we have the following relationship between (1.19) and a corresponding Itô equation.

Lemma 1.9. *If D^N is a compatible solution of (1.19), then there exist independent standard Brownian motions B_k (perhaps on an enlarged sample space) such that D^N is a solution of the Itô equation*

$$D^N(t) = D^N(0) + \sum_k \frac{1}{\sqrt{N}} \int_0^t \sqrt{\widetilde{\lambda}(D^N(s))} dB_k(s)\zeta_k + \int_0^t F(D^N(s))ds. \tag{1.20}$$

Proof. See [34, 35] and Chap. 11 of [14]. For a general discussion of compatibility, see [36], in particular, Example 3.20. □

In the chemical physics literature, D^N is known as the Langevin approximation for the continuous-time Markov chain model determined by the master equation. Just as there are alternative ways of determining the continuous-time Markov chain model, there are alternative approaches to deriving the Langevin approximation. For example, C^N is a solution of the martingale problem corresponding to

$$A_N f(x) = \sum_k N\lambda_k(x)(f(x + N^{-1}\zeta_k) - f(x)),$$

and if f is three times continuously differentiable with compact support,

$$A_N f(x) = L_N f(x) + O(N^{-2}),$$

where

$$L_N f(x) = \frac{1}{2N} \sum_k \lambda_k(x)\zeta_k^\top \partial^2 f(x)\zeta_k + F(x) \cdot \nabla f(x),$$

and any compatible solution of (1.19) is a solution of the martingale problem for L_N, that is, there is a filtration $\{\mathcal{F}_t^N\}$ such that

$$f\left(D^N(t)\right) - f\left(D^N(0)\right) - \int_0^t L_N f\left(D^N(s)\right) ds$$

is a $\{\mathcal{F}_t^N\}$-martingale for each twice continuously differentiable function having compact support. The converse also holds, that is, any solution of the martingale problem for L_N that does not hit infinity in finite time can be obtained as a compatible solution of (1.19) or equivalently, as a solution of (1.20).

Finally, the Langevin approximation can be derived starting with the master equation. First rewrite (1.14) as

$$\dot{p}^N(y,t) = \sum_l N\lambda_l\left(y - N^{-1}\zeta_l\right) p^N\left(y - N^{-1}\zeta_l, t\right) - \left(\sum_l N\lambda_l(y)\right) p^N(y,t), \tag{1.21}$$

where now

$$p^N(y,t) = P\{C^N(t) = y\}.$$

Expanding $\lambda_l(y - N^{-1}\zeta_l)p^N(y - N^{-1}\zeta_l)$ in a Taylor series (the Kramer-Moyal expansion, or in this context, the system-size expansion of van Kampen; see [42]) and discarding higher order terms gives

$$\dot{p}^N(y,t) \approx \frac{1}{2N}\sum_l \zeta_l^\top \partial^2\left(\lambda_l(y)p^N(y,t)\right)\zeta_k - \sum_l \zeta_l \cdot \nabla\left(\lambda_l(y)p^N(y,t)\right).$$

Replacing $\approx$ by $=$ gives the Fokker-Planck equation

$$\dot{q}^N(y,t) = \frac{1}{2N}\sum_l \zeta_l^\top \partial^2\left(\lambda_l(y)q^N(y,t)\right)\zeta_k - \sum_l \zeta_l \cdot \nabla\left(\lambda_l(y)q^N(y,t)\right)$$

corresponding to (1.20). These three derivations are equivalent in the sense that any solution of the Fokker-Planck equation for which $q^N(\cdot,t)$ is a probability density for all t gives the one-dimensional distributions of a solution of the martingale problem for L_N, and as noted before, any solution of the martingale problem that does not hit infinity in finite time can be obtained as a solution of (1.20) or (1.19). See [37] for a more detailed discussion.

The approximation (1.18) is justified by the convergence of V^N to V, but the justification for taking D^N as an approximation of C^N is less clear. One can, however, apply the strong approximation result, Lemma 1.3, to construct D^N and C^N in such a way that in a precise sense, for each $T > 0$,

$$\sup_{t \le T} |D^N(t) - C^N(t)| = \mathcal{O}\left(\frac{\log N}{N}\right).$$

First Order Reaction Networks

If all reactions in the network are unary, for example,

$$S_1 \rightharpoonup S_2$$

$$S_1 \rightharpoonup S_2 + S_3$$
$$S_1 \rightharpoonup S_1 + S_2$$
$$S_1 \rightharpoonup \emptyset,$$

then the resulting process is a multitype branching process, and if reactions of the form

$$\emptyset \rightharpoonup S_1$$

are included, the process is a branching process with immigration. Networks that only include the above reaction types are termed first order reaction networks. For simplicity, first consider the system

$$\emptyset \rightharpoonup S_1$$
$$S_1 \rightharpoonup S_2$$
$$S_2 \rightharpoonup 2S_1 \ .$$

The stochastic equation for the model becomes

$$X(t) = X(0) + Y_1(\kappa_1 t)\begin{pmatrix}1\\0\end{pmatrix} + Y_2\left(\kappa_2 \int_0^t X_1(s)ds\right)\begin{pmatrix}-1\\1\end{pmatrix}$$
$$+ Y_3\left(\kappa_3 \int_0^t X_2(s)ds\right)\begin{pmatrix}2\\-1\end{pmatrix},$$

for some choice of $\kappa_1, \kappa_2, \kappa_3 > 0$. Using the fact that $E[Y_k(\int_0^t \lambda_k(s)ds)] = E[\int_0^t \lambda_k(s)ds]$, we have

$$E[X(t)] = E[X(0)] + \begin{pmatrix}\kappa_1\\0\end{pmatrix} t + \int_0^t \kappa_2 E[X_1(s)]ds \begin{pmatrix}-1\\1\end{pmatrix}$$
$$+ \kappa_3 \int_0^t E[X_2(s)]ds \begin{pmatrix}2\\-1\end{pmatrix}$$
$$= E[X(0)] + \begin{pmatrix}\kappa_1\\0\end{pmatrix} t + \int_0^t \begin{pmatrix}-\kappa_2 & 2\kappa_3\\ \kappa_2 & -\kappa_3\end{pmatrix} E[X(s)]ds$$

giving a simple linear system for the first moments, $E[X(t)]$. For the second moments, note that

$$X(t)X(t)^\top = X(0)X(0)^\top + \int_0^t X(s-)dX(s)^\top + \int_0^t dX(s)X(s-)^\top + [X]_t,$$

where $[X]_t$ is the matrix of quadratic variations which in this case is simply

$$[X]_t = Y_1(\kappa_1 t)\begin{pmatrix}1 & 0\\ 0 & 0\end{pmatrix} + Y_2\left(\kappa_2\int_0^t X_1(s)ds\right)\begin{pmatrix}1 & -1\\ -1 & 1\end{pmatrix}$$
$$+ Y_3\left(\kappa_3\int_0^t X_2(s)ds\right)\begin{pmatrix}4 & -2\\ -2 & 1\end{pmatrix}.$$

Since

$$X(t) - X(0) - \kappa_1 t\begin{pmatrix}1\\ 0\end{pmatrix} - \kappa_2\int_0^t X_1(s)ds\begin{pmatrix}-1\\ 1\end{pmatrix} - \kappa_3\int_0^t X_2(s)ds\begin{pmatrix}2\\ -1\end{pmatrix}$$

is a martingale,

$$\begin{aligned}
E[X(t)X(t)^\top] &= E[X(0)X(0)^\top] \\
&\quad + \int_0^t E\left[X(s)\left(\begin{pmatrix}\kappa_1 & 0\end{pmatrix} + X(s)^\top\begin{pmatrix}-\kappa_2 & 2\kappa_3\\ \kappa_2 & -\kappa_3\end{pmatrix}^\top\right)\right]ds \\
&\quad + \int_0^t E\left[\left(\begin{pmatrix}\kappa_1\\ 0\end{pmatrix} + \begin{pmatrix}-\kappa_2 & 2\kappa_3\\ \kappa_2 & -\kappa_3\end{pmatrix}X(s)\right)X(s)^\top\right]ds \\
&\quad + \begin{pmatrix}\kappa_1 & 0\\ 0 & 0\end{pmatrix}t + \int_0^t \left(\kappa_2 E[X_1(s)]\begin{pmatrix}1 & -1\\ -1 & 1\end{pmatrix}\right. \\
&\qquad\qquad \left. + \kappa_3 E[X_2(s)]\begin{pmatrix}4 & -2\\ -2 & 1\end{pmatrix}\right)ds \\
&= E[X(0)X(0)^\top] + \int_0^t \kappa_1\begin{pmatrix}2E[X_1(s)] & E[X_2(s)]\\ E[X_2(s)] & 0\end{pmatrix}ds \\
&\quad + \int_0^t \left(E[X(s)X(s)^\top]\begin{pmatrix}-\kappa_2 & 2\kappa_3\\ \kappa_2 & -\kappa_3\end{pmatrix}^\top\right. \\
&\qquad \left. + \begin{pmatrix}-\kappa_2 & 2\kappa_3\\ \kappa_2 & -\kappa_3\end{pmatrix}E\left[X(s)X(s)^\top\right]\right)ds + \begin{pmatrix}\kappa_1 & 0\\ 0 & 0\end{pmatrix}t \\
&\quad + \int_0^t \left(\kappa_2 E[X_1(s)]\begin{pmatrix}1 & -1\\ -1 & 1\end{pmatrix}\right. \\
&\qquad \left. + \kappa_3 E[X_2(s)]\begin{pmatrix}4 & -2\\ -2 & 1\end{pmatrix}\right)ds.
\end{aligned}$$

In general, the stochastic equation for first order networks will be of the form

$$X(t) = X(0) + \sum_k Y_k^0(\alpha_k^0 t)\zeta_k^0 + \sum_{l=1}^{s_0}\sum_k Y_k^l\left(\alpha_k^l\int_0^t X_l(s)ds\right)\zeta_k^l,$$

where all components of ζ_k^0 are nonnegative and all components of ζ_k^l are nonnegative except for the possibility that the lth component of ζ_k^l may be -1. The martingale properties of the Y_k^l imply that the expectation of X satisfies

$$E[X(t)] = E[X(0)] + at + \int_0^t AE[X(s)]ds, \tag{1.22}$$

where $a = \sum_k \alpha_k^0 \zeta_k^0$ and A is the matrix whose lth column is $A_l = \sum_k \alpha_k^l \zeta_k^l$. Note that the solution of (1.22) is given by

$$E[X(t)] = e^{At} E[X(0)] + \int_0^t e^{A(t-s)} a\, ds,$$

and if A is invertible

$$E[X(t)] = e^{At} E[X(0)] + A^{-1}(e^{At} - I)a, \tag{1.23}$$

where I is the identity matrix.

Similarly to before, the matrix of second moments satisfies

$$\begin{aligned} E\left[X(t)X(t)^\top\right] = E\left[X(0)X(0)^\top\right] &+ \int_0^t \left(E[X(s)]a^\top + aE[X(s)]^\top\right) ds \\ &+ \int_0^t \left(AE\left[X(s)X(s)^\top\right] + E\left[X(s)X(s)^\top\right]A^\top\right) ds \\ &+ B_0 t + \sum_l \int_0^t E[X_l(s)]B_l ds\ , \end{aligned}$$

where

$$B_0 = \sum_k \alpha_k^0 \zeta_k^0 \zeta_k^{0\top}, \quad B_l = \sum_k \alpha_k^l \zeta_k^l \zeta_k^{l\top}.$$

See [3], Sect. V.7.

A system that only includes reactions of the form

$$\begin{aligned} \emptyset &\rightharpoonup S_i \\ S_i &\rightharpoonup S_j \\ S_i &\rightharpoonup \emptyset \end{aligned}$$

can be interpreted as an infinite server queueing network, with $\emptyset \rightharpoonup S_i$ corresponding to an 'arrival', $S_i \rightharpoonup \emptyset$, a 'departure', and $S_i \rightharpoonup S_j$ the movement of a 'customer' from station i to station j. Customers (molecules) that start in or enter the system move (change type) independently until they leave the system. This independence implies that if $\{X_i(0)\}$ are independent Poisson distributed random variables, then $\{X_i(t)\}$ are independent Poisson distributed random variables for all

$t \geq 0$. Since the Poisson distribution is determined by its expectation, under the assumption of an independent Poisson initial distribution, the distribution of $X(t)$ is determined by $E[X(t)]$, that is, by the solution of (1.22).

Suppose that for each pair of species S_i and S_j, it is possible for a molecule of S_i to be converted, perhaps through a sequence of intermediate steps, to a molecule of S_j. In addition, assume that the system is open in the sense that there is at least one reaction of the form $\emptyset \rightharpoonup S_i$ and one reaction of the form $S_j \rightharpoonup \emptyset$. Then A is invertible, so $E[X(t)]$ is given by (1.23), and as $t \to \infty$, $e^{At} \to 0$ so $E[X(t)] \to -A^{-1}a$. It follows that the stationary distribution for X is given by a vector $\overline{X}$ of independent Poisson distributed random variables with $E[\overline{X}] = -A^{-1}a$.

If the system is closed so that the only reactions are of the form $S_i \rightharpoonup S_j$ and the initial distribution is multinomial with parameters $(n, p_1(0), \ldots, p_{s_0}(0))$, that is, for $k = (k_1, \ldots, k_{s_0})$ with $\sum_i k_i = n$,

$$P\{X(0) = k\} = \binom{n}{k_1, \ldots, k_{s_0}} \prod p_i(0)^{k_i},$$

then $X(t)$ is multinomial $(n, p_1(t), \ldots, p_{s_0}(t))$, where $p(t) = (p_1(t), \ldots, p_{s_0}(t))$ is given by

$$p(t) = e^{At} p(0).$$

Note that if the intensity for the reaction $S_i \rightharpoonup S_j$ is $\kappa_{ij} X_i(t)$, then the model is equivalent to n independent continuous-time Markov chains with state space $\{1, \ldots, s_0\}$ and transition intensities given by the κ_{ij}. Consequently, if the independent chains have the same initial distribution, $p(0) = (p_1(0), \ldots, p_{s_0}(0))$, then they have the same distribution at time t, namely $p(t)$. The multinomial distribution with parameters $(n, \overline{p})$ with $\overline{p} = \lim_{t\to\infty} p(t)$ will be a stationary distribution, but $\overline{p}$ is not unique unless the assumption that every chemical species S_i can be converted into every other chemical species S_j holds.

See [17] for additional material on first order networks.

Product Form Stationary Distributions

The Poisson and multinomial stationary distributions discussed above for unary systems are special cases of what are known as *product form stationary distributions* in the queueing literature. As noted in Chap. 8 of [27] and discussed in detail in [2], a much larger class of reaction networks also has product form stationary distributions. In fact, stochastic models of reaction networks that satisfy the conditions of the zero deficiency theorem of Feinberg [15] from deterministic reaction network theory have this property.

Let $\mathcal{S} = \{S_i : i = 1, \ldots, s_0\}$ denote the collection of chemical species, $\mathcal{C} = \{\nu_k, \nu'_k : k = 1, \ldots, r_0\}$ the collection of *complexes*, that is, the vectors that give either the inputs or the outputs of a reaction, and $\mathcal{R} = \{\nu_k \to \nu'_k : k = 1, \ldots, r_0\}$ the collection of reactions. The triple, $\{\mathcal{S}, \mathcal{C}, \mathcal{R}\}$ determines the reaction network.

Definition 1.10. A chemical reaction network, $\{\mathcal{S}, \mathcal{C}, \mathcal{R}\}$, is called *weakly reversible* if for any reaction $\nu_k \to \nu_k'$, there is a sequence of directed reactions beginning with ν_k' as a source complex and ending with ν_k as a product complex. That is, there exist complexes $\nu_1, \dots, \nu_r$ such that $\nu_k' \to \nu_1, \nu_1 \to \nu_2, \dots, \nu_r \to \nu_k \in \mathcal{R}$. A network is called *reversible* if $\nu_k' \to \nu_k \in \mathcal{R}$ whenever $\nu_k \to \nu_k' \in \mathcal{R}$.

Let $\mathcal{G}$ be the directed graph with nodes given by the complexes $\mathcal{C}$ and directed edges given by the reactions $\mathcal{R} = \{\nu_k \to \nu_k'\}$, and let $\mathcal{G}_1, \dots, \mathcal{G}_\ell$ denote the connected components of $\mathcal{G}$. $\{\mathcal{G}_j\}$ are the *linkage classes* of the reaction network. Note that a reaction network is weakly reversible if and only if the linkage classes are strongly connected.

Definition 1.11. $S = \text{span}_{\{\nu_k \to \nu_k' \in \mathcal{R}\}}\{\nu_k' - \nu_k\}$ is the *stoichiometric subspace* of the network. For $c \in \mathbb{R}^{s_0}$, we say $c + S$ and $(c + S) \cap \mathbb{R}^{s_0}_{>0}$ are the *stoichiometric compatibility classes* and *positive stoichiometric compatibility classes* of the network, respectively. Denote $\dim(S) = s$.

Definition 1.12. The *deficiency* of a chemical reaction network, $\{\mathcal{S}, \mathcal{C}, \mathcal{R}\}$, is $\delta = |\mathcal{C}| - \ell - s$, where $|\mathcal{C}|$ is the number of complexes, ℓ is the number of linkage classes, and s is the dimension of the stoichiometric subspace.

For $x, c \in \mathbb{Z}^{s_0}_{\geq 0}$, we define $c^x \equiv \prod_{i=1}^{s_0} c_i^{x_i}$, where we interpret $0^0 = 1$, and $x! \equiv \prod_{i=1}^{s_0} x_i!$. If for each complex $\eta \in \mathcal{C}$, $c \in \mathbb{R}^{s_0}_{>0}$ satisfies

$$\sum_{k:\nu_k=\eta} \kappa_k c^{\nu_k} = \sum_{k:\nu_k'=\eta} \kappa_k c^{\nu_k}, \tag{1.24}$$

where the sum on the left is over reactions for which η is the source complex and the sum on the right is over those for which η is the product complex, then c is a special type of equilibrium of the system (you can see this by summing each side of (1.24) over the complexes), and the network is called *complex balanced*. The following is the Deficiency Zero Theorem of Feinberg [15].

Theorem 1.13. *Let $\{\mathcal{S}, \mathcal{C}, \mathcal{R}\}$ be a weakly reversible, deficiency zero chemical reaction network governed by deterministic mass action kinetics,* (1.16). *Then, for any choice of rate constants κ_k, within each positive stoichiometric compatibility class there is precisely one equilibrium value c, that is $\sum_k \kappa_k c^{\nu_k}(\nu_k' - \nu_k) = 0$, and that equilibrium value is locally asymptotically stable relative to its compatibility class. Moreover, for each $\eta \in \mathcal{C}$,*

$$\sum_{k:\nu_k=\eta} \kappa_k c^{\nu_k} = \sum_{k:\nu_k'=\eta} \kappa_k c^{\nu_k}. \tag{1.25}$$

For stochastically modeled systems we have the following theorem.

Theorem 1.14. *Let $\{\mathcal{S}, \mathcal{C}, \mathcal{R}\}$ be a chemical reaction network with rate constants κ_k. Suppose that the deterministically modeled system is complex balanced with* *equilibrium* $\overline{c} \in \mathbb{R}^m_{>0}$. *Then, for any irreducible communicating equivalence class,* Γ, *the stochastic system has a product form stationary measure*

$$\pi(x) = M\frac{\overline{c}^x}{x!}, \quad x \in \Gamma, \tag{1.26}$$

where M *is a normalizing constant.*

Theorem 1.13 then shows that the conclusion of Theorem 1.14 holds, regardless of the choice of rate constants, for all stochastically modeled systems with a reaction network that is weakly reversible and has a deficiency of zero.

Models with Delay

Modeling chemical reaction networks as continuous-time Markov chains is intuitively appealing and, as noted, consistent with the classical deterministic law of mass action. Cellular reaction networks, however, include reactions for which the exponential timing of the simple Markov chain model is almost certainly wrong. These networks typically involve assembly processes (transcription or translation), referred to as elongation, in which an enzyme or ribosome follows a DNA or RNA template to create a new DNA, RNA, or protein molecule. The exponential holding times in the Markov chain model reflect an assumption that once the molecules come together in the right configuration, the time it takes to complete the reaction is negligible. That is not, in general, the case for elongation. While each step of the assembly process might reasonably be assumed to take an exponentially distributed time, the total time is a sum of such steps with the number of summands equal to the number of nucleotides or amino acids. Since this number is large and essentially fixed, if the individual steps have small expectations, the total time that the reaction takes once the assembly is initiated may be closer to deterministic than exponential. See [5, 8] for examples of stochastic models of cellular reaction networks with delays.

One reasonable (though by no means only) way to incorporate delays into the models is to assume that for a reaction with deterministic delay ξ_k that initiates at time t^* the input molecules are lost at time t^* and the product molecules are produced at time $t^* + \xi_k$. Noting that the number of initiations of a reaction by time t can still be modeled by the counting process $Y_k(\int_0^t \lambda_k(X(s))ds)$, we may let Γ_1 denote those reactions with no delay and Γ_2 those with a delay, and conclude that the system should satisfy the equation

$$\begin{aligned} X(t) = X(0) &+ \sum_{k\in\Gamma_1} Y_{k,1}\left(\int_0^t \lambda_k(X(s))ds\right)(\nu_k' - \nu_k) \\ &- \sum_{k\in\Gamma_2} Y_{k,2}\left(\int_0^t \lambda_k(X(s))ds\right)\nu_k + \sum_{k\in\Gamma_2} Y_{k,2}\left(\int_0^{t-\xi_k} \lambda_k(X(s))ds\right)\nu_k', \end{aligned}$$

where we take $X(s) \equiv 0$, and hence $\lambda_k(X(s)) \equiv 0$, for $s < 0$. Existence and uniqueness of solutions to this equation follow by the same jump by jump argument used in Section "Continuous Time Markov Chains".

Simulation of reaction networks modeled with delay is no more difficult than simulating those without delay. For example, the above equation suggests a simulation strategy equivalent to the next reaction method [1, 18]. There are also analogues of the stochastic simulation algorithm, or Gillespie's algorithm [8].

Multiple Scales

The classical scaling that leads to the deterministic law of mass action assumes that all chemical species are present in numbers of the same order of magnitude. For reaction networks in biological cells, this assumption is usually clearly violated. Consequently, models derived by the classical scaling may not be appropriate. For these networks some species are present in such small numbers that they should be modeled by discrete variables while others are present in large enough numbers to reasonably be modeled by continuous variables. These large numbers may still differ by several orders of magnitude, so normalizing all 'large' quantities in the same way may still be inappropriate. Consequently, methods are developed in [4, 25, 26] for deriving simplified models in which different species numbers are normalized in different ways appropriate to their numbers in the system.

Derivation of the Michaelis-Menten Equation

Perhaps the best known examples of reaction networks in which multiple scales play a role are models that lead to the Michaelis-Menten equation. Darden [9, 10] gave a derivation starting from a stochastic model, and we prove his result using our methodology.

Consider the reaction system

$$S_1 + S_2 \underset{\kappa_2'}{\overset{\kappa_1'}{\rightleftharpoons}} S_3 \overset{\kappa_3'}{\rightharpoonup} S_4 + S_2,$$

where S_1 is the substrate, S_2 the enzyme, S_3 the enzyme-substrate complex, and S_4 the product. Assume that the parameters scale so that

$$\begin{aligned} Z_1^N(t) = Z_1^N(0) - N^{-1}Y_1\left(N\int_0^t \kappa_1 Z_1^N(s)Z_2^N(s)ds\right) \\ + N^{-1}Y_2\left(N\int_0^t \kappa_2 Z_3^N(s)ds\right) \end{aligned}$$

$$Z_2^N(t) = Z_2^N(0) - Y_1\left(N\int_0^t \kappa_1 Z_1^N(s)Z_2^N(s)ds\right) + Y_2\left(N\int_0^t \kappa_2 Z_3^N(s)ds\right)$$
$$+ Y_3\left(N\int_0^t \kappa_3 Z_3^N(s)ds\right)$$
$$Z_3^N(t) = Z_2^N(0) + Y_1\left(N\int_0^t \kappa_1 Z_1^N(s)Z_2^N(s)ds\right) - Y_2\left(N\int_0^t \kappa_2 Z_3^N(s)ds\right)$$
$$- Y_3\left(N\int_0^t \kappa_3 Z_3^N(s)ds\right)$$
$$Z_4^N(t) = N^{-1}Y_3\left(N\int_0^t \kappa_3 Z_3^N(s)ds\right),$$

where $\kappa_1, \kappa_2, \kappa_3$ do not depend upon N. Note that we scale the numbers of molecules of the substrate and the product as in the previous section, but we leave the enzyme and enzyme-substrate variables discrete. Note also that $M = Z_3^N(t) + Z_2^N(t)$ is constant, and define

$$\widehat{Z}_2^N(t) = \int_0^t Z_2^N(s)ds = Mt - \int_0^t Z_3^N(s)ds.$$

Theorem 1.15. *Assume that $Z_1^N(0) \to Z_1(0)$ and that M does not depend on N. Then $(Z_1^N, \widehat{Z}_2^N)$ converges to $(Z_1(t), \widehat{Z}_2(t))$ satisfying*

$$Z_1(t) = Z_1(0) - \int_0^t \kappa_1 Z_1(s)\dot{\widehat{Z}}_2(s)ds + \int_0^t \kappa_2(M - \dot{\widehat{Z}}_2(s))ds \tag{1.27}$$
$$0 = -\int_0^t \kappa_1 Z_1(s)\dot{\widehat{Z}}_2(s)ds + \int_0^t (\kappa_2+\kappa_3)(M - \dot{\widehat{Z}}_2(s))ds,$$

and hence $\dot{\widehat{Z}}_2(s) = \frac{(\kappa_2+\kappa_3)M}{\kappa_2+\kappa_3+\kappa_1 Z_1(s)}$ and

$$\dot{Z}_1(t) = -\frac{M\kappa_1\kappa_3 Z_1(t)}{\kappa_2+\kappa_3+\kappa_1 Z_1(s)}. \tag{1.28}$$

Proof. Relative compactness of the sequence $(Z_1^N, \widehat{Z}_2^N)$ is straightforward, that is, at least along a subsequence, we can assume that $(Z_1^N, \widehat{Z}_2^N)$ converges in distribution to a continuous process $(Z_1, \widehat{Z}_2)$ (which turns out to be deterministic). Dividing the second equation by N and passing to the limit, we see $(Z_1, \widehat{Z}_2)$ must satisfy

$$0 = -\int_0^t \kappa_1 Z_1(s)d\widehat{Z}_2(s) + (\kappa_2+\kappa_3)Mt - \int_0^t (\kappa_2+\kappa_3)d\widehat{Z}_2(s). \tag{1.29}$$

Since $\widehat{Z}_2$ is Lipschitz, it is absolutely continuous, and rewriting (1.29) in terms of the derivative gives the second equation in (1.27). The first equation follows by a similar argument. □

Of course, (1.28) is the Michaelis-Menten equation.

Scaling Species Numbers and Rate Constants

Assume that we are given a model of the form

$$X(t) = X(0) + \sum_k Y_k \left(\int_0^t \lambda'_k(X(s))ds \right) (\nu'_k - \nu_k)$$

where the λ'_k are of the form

$$\lambda'_k(x) = \kappa'_k \prod_i \nu_{ik}! \prod_i \binom{x_i}{\nu_{ik}}.$$

Let $N_0 \gg 1$. For each species i, define the *normalized abundance* (or simply, the abundance) by

$$Z_i(t) = N_0^{-\alpha_i} X_i(t),$$

where $\alpha_i \geq 0$ should be selected so that $Z_i = O(1)$. The abundance may be the species number ($\alpha_i = 0$) or the species concentration or something else.

Since the rate constants may also vary over several orders of magnitude, we write $\kappa'_k = \kappa_k N_0^{\beta_k}$ where the β_k are selected so that $\kappa_k = O(1)$. For a binary reaction

$$\kappa'_k x_i x_j = N_0^{\beta_k + \alpha_i + \alpha_j} \kappa_k z_i z_j,$$

and we can write

$$\beta_k + \alpha_i + \alpha_j = \beta_k + \nu_k \cdot \alpha.$$

We also have,

$$\kappa'_k x_i = N_0^{\beta_k + \nu_k \cdot \alpha} z_i, \quad \kappa'_k x_i (x_i - 1) = N_0^{\beta_k + \nu_k \cdot \alpha} z_i \left(z_i - N_0^{-\alpha_i} \right),$$

with similar expressions for intensities involving higher order reactions.

We replace N_0 by N in the above expressions and consider a family of models,

$$Z_i^N(t) = Z_i^N(0) + \sum_k N^{-\alpha_i} Y_k \left(\int_0^t N^{\beta_k + \nu_k \cdot \alpha} \lambda_k(Z^N(s))ds \right) \left(\nu'_{ik} - \nu_{ik} \right),$$

where the original model is $Z = Z^{N_0}$. Note that for reactions of the form $2S_i \rightharpoonup *$, where $*$ represents an arbitrary linear combination of the species, the rate is $N^{\beta_k + 2\alpha_i} Z_i^N(t)(Z_i^N(t) - N^{-\alpha_i})$, so if $\alpha_i > 0$, we should write λ_k^N instead of λ_k, but to simplify notation, we will simply write λ_k.

We have a family of models indexed by N for which $N = N_0$ gives the 'correct' or original model. Other values of N and any limits as $N \to \infty$ (perhaps with a change of time-scale) give approximate models. The challenge is to select the α_i and the β_k in a reasonable way, but once that is done, the initial condition for index N is given by

$$Z_i^N(0) = N^{-\alpha_i} \left\lfloor N^{\alpha_i} \frac{X_i(0)}{N_0^{\alpha_i}} \right\rfloor,$$

where $\lfloor z \rfloor$ is the integer part of z and the $X_i(0)$ are the initial species numbers in the original model.

Allowing a change of time-scale, where t is replaced by tN^γ, suppose $\lim_{N\to\infty} Z_i^N(\cdot N^\gamma) = Z_i^\infty$. Then we should have

$$X_i(t) \approx N_0^{\alpha_i} Z_i^\infty(t N_0^{-\gamma}).$$

Determining the Scaling Exponents

There are, of course, many ways of selecting the α_i and β_k, but we want to make this selection so that there are limiting models that give reasonable approximations for the original model. Consequently, we look for natural constraints on the α_i and β_k.

For example, suppose that the rate constants satisfy

$$\kappa_1' \geq \kappa_2' \geq \cdots \geq \kappa_{r_0}'.$$

Then it seems natural to select

$$\beta_1 \geq \cdots \geq \beta_{r_0},$$

although it may be reasonable to impose this constraint separately for the binary reactions and the unary reactions.

To get a sense of the issues involved in selecting exponents that lead to reasonable limits, consider a reaction network in which the reactions involving S_3 are

$$S_1 + S_2 \rightharpoonup S_3 + S_4 \qquad S_3 + S_5 \rightharpoonup S_6.$$

Then

$$\begin{aligned} Z_3^N(t) = Z_3^N(0) &+ N^{-\alpha_3} Y_1\left(N^{\beta_1+\alpha_1+\alpha_2} \int_0^t \kappa_1 Z_1^N(s) Z_2^N(s) ds\right) \\ &- N^{-\alpha_3} Y_2\left(N^{\beta_2+\alpha_3+\alpha_5} \int_0^t \kappa_2 Z_3^N(s) Z_5^N(s) ds\right), \end{aligned}$$

or scaling time

$$Z_3^N(tN^\gamma) = Z_3^N(0) + N^{-\alpha_3} Y_1\left(N^{\beta_1+\alpha_1+\alpha_2+\gamma}\int_0^t \kappa_1 Z_1^N(sN^\gamma)Z_2^N(sN^\gamma)ds\right)$$
$$-N^{-\alpha_3} Y_2\left(N^{\beta_2+\alpha_3+\alpha_5+\gamma}\int_0^t \kappa_2 Z_3^N(sN^\gamma)Z_5^N(sN^\gamma)ds\right).$$

Assuming that for the other species in the system $Z_i^N = O(1)$, we see that $Z_3^N = O(1)$ if

$$(\beta_1+\alpha_1+\alpha_2+\gamma)\vee(\beta_2+\alpha_3+\alpha_5+\gamma)\le\alpha_3$$

or if

$$\beta_1+\alpha_1+\alpha_2 = \beta_2+\alpha_3+\alpha_5 > \alpha_3.$$

Note that in the latter case, we would expect $Z_3^N(t) \approx \frac{\kappa_1 Z_1^N(t)Z_2^N(t)}{\kappa_2 Z_5^N(t)}$. If these conditions both fail, then either Z_3^N will blow up as $N\to\infty$ or will be driven to zero.

With this example in mind, define $Z_i^{N,\gamma}(t) = Z_i^N(tN^\gamma)$ so

$$Z_i^{N,\gamma}(t) = Z_i^N(0) + \sum_k N^{-\alpha_i} Y_k\left(\int_0^t N^{\gamma+\beta_k+\nu_k\cdot\alpha}\lambda_k(Z^{N,\gamma}(s))ds\right)(\nu'_{ik}-\nu_{ik}).$$

Recalling that $\zeta_k = \nu'_k - \nu_k$, for $\theta_i \ge 0$, consider

$$\sum_i \theta_i N^{\alpha_i} Z_i^{N,\gamma}(t)$$
$$= \sum_i \theta_i N^{\alpha_i} Z_i^N(0) + \sum_k Y_k\left(\int_0^t N^{\gamma+\beta_k+\nu_k\cdot\alpha}\lambda_k(Z^{N,\gamma}(s))ds\right)\langle\theta,\zeta_k\rangle,$$

where $\langle\theta,\zeta_k\rangle = \sum_i \theta_i\zeta_{ik}$, and define $\alpha_\theta = \max\{\alpha_i : \theta_i > 0\}$. If all $Z_i^{N,\gamma} = O(1)$, then the left side is $O(N^{\alpha_\theta})$, and as in the single species example above, we must have

$$\gamma + \max\{\beta_k + \nu_k\cdot\alpha : \langle\theta,\zeta_k\rangle \ne 0\} \le \alpha_\theta. \tag{1.30}$$

or

$$\max\{\beta_k+\nu_k\cdot\alpha : \langle\theta,\zeta_k\rangle > 0\} = \max\{\beta_k+\nu_k\cdot\alpha : \langle\theta,\zeta_k\rangle < 0\}. \tag{1.31}$$

Note that (1.30) is really a constraint on the time-scale determined by γ saying that if (1.31) fails for some θ, then γ must satisfy

$$\gamma \le \alpha_\theta - \max\{\beta_k+\nu_k\cdot\alpha : \langle\theta,\zeta_k\rangle \ne 0\}.$$

The value of γ given by

$$\gamma_i = \alpha_i - \max\{\beta_k + \nu_k \cdot \alpha : \zeta_{ik} \neq 0\}$$

gives the natural time-scale for S_i in the sense that $Z_i^{N,\gamma}$ is neither asymptotically constant nor too rapidly oscillating to have a limit. The γ_i are values of γ for which interesting limits may hold. Linear combinations $\langle \theta, Z^{N,\gamma} \rangle$ may have time-scales

$$\gamma_\theta = \alpha_\theta - \max\{\beta_k + \nu_k \cdot \alpha : \langle \theta, \zeta_k \rangle \neq 0\}$$

that are different from all of the species time-scales and may give *auxiliary variables* (see, for example, [38]) whose limits capture interesting properties of the system.

The equation (1.31) is called the *balance equation*, and together, the alternative (1.31) and (1.30) is referred to as the *balance condition*. To employ this approach to the identification of simplified models, it is not necessary to solve the balance equations for every choice of θ. The equations that fail simply place restrictions on the time-scales γ that can be used without something blowing up. The goal is to find α_i and β_k that give useful limiting models, and solving some subset of the balance equations can be a useful first step. Natural choices of θ in selecting the subset of balance equations to solve include those for which $\langle \theta, \zeta_k \rangle = 0$ for one or more of the ζ_k. See section "First Order Reaction Networks" of [26] for a more detailed discussion.

In the next subsection, we apply the balance conditions to identify exponents useful in deriving a reduced model for a simple reaction network. For an application to a much more complex model of the heat shock response in E. coli, see [24].

An Application of the Balance Conditions

Consider the simple example

$$\emptyset \overset{\kappa_1'}{\rightharpoonup} S_1 \underset{\kappa_3'}{\overset{\kappa_2'}{\rightleftharpoons}} S_2, \qquad S_1 + S_2 \overset{\kappa_4'}{\rightharpoonup} S_3$$

Assume $\kappa_k' = \kappa_k N_0^{\beta_k}$. Then a useful subset of the balance equations is

$$\begin{array}{ll} S_2 & \beta_2 + \alpha_1 = (\beta_3 + \alpha_2) \vee (\beta_4 + \alpha_1 + \alpha_2) \\ S_1 & \beta_1 \vee (\beta_3 + \alpha_2) = (\beta_2 + \alpha_1) \vee (\beta_4 + \alpha_1 + \alpha_2) \\ S_3 & \beta_4 + \alpha_1 + \alpha_2 = -\infty \\ S_1 + S_2 & \beta_1 = \beta_4 + \alpha_1 + \alpha_2 \end{array}$$

where we take the maximum of the empty set to be $-\infty$. Of course, it is not possible to select parameters satisfying the balance equation for S_3, so we must restrict γ by

$$\gamma \leq \alpha_3 - (\beta_4 + \alpha_1 + \alpha_2). \tag{1.32}$$

Let $\alpha_1 = 0$ and $\beta_1 = \beta_2 > \beta_3 = \beta_4$, so balance for S_1, S_2, and $S_1 + S_2$ is satisfied if $\alpha_2 = \beta_2 - \beta_3$, which we assume. Taking $\alpha_3 = \alpha_2$, (1.32) becomes

$$\gamma \le -\beta_4 = -\beta_3.$$

The system of equations becomes

$$\begin{aligned}
Z_1^N(t) &= Z_1^N(0) + Y_1\left(\kappa_1 N^{\beta_1} t\right) - Y_2\left(\kappa_2 N^{\beta_2} \int_0^t Z_1^N(s)ds\right)\\
&\quad + Y_3\left(\kappa_3 N^{\beta_3+\alpha_2} \int_0^t Z_2^N(s)ds\right)\\
&\quad - Y_4\left(\kappa_4 N^{\beta_4+\alpha_2} \int_0^t Z_1^N(s)Z_2^N(s)ds\right)\\
Z_2^N(t) &= Z_2^N(0) + N^{-\alpha_2} Y_2\left(\kappa_2 N^{\beta_2} \int_0^t Z_1^N(s)ds\right)\\
&\quad - N^{-\alpha_2} Y_3\left(\kappa_3 N^{\beta_3+\alpha_2} \int_0^t Z_2^N(s)ds\right)\\
&\quad - N^{-\alpha_2} Y_4\left(\kappa_4 N^{\beta_4+\alpha_2} \int_0^t Z_1^N(s)Z_2^N(s)ds\right)\\
Z_3^N(t) &= Z_3^N(0) + N^{-\alpha_3} Y_4\left(\kappa_4 N^{\beta_4+\alpha_2} \int_0^t Z_1^N(s)Z_2^N(s)ds\right).
\end{aligned}$$

There are two time-scales of interest in this model, $\gamma = -\beta_1$, the time-scale of S_1, and $\gamma = -\beta_3$, the time-scale of S_2 and S_3. Recalling that $\alpha_2 + \beta_3 = \alpha_2 + \beta_4 = \beta_1 = \beta_2$, for $\gamma = -\beta_1$,

$$\begin{aligned}
Z_1^{N,-\beta_1}(t) &= Z_1^N(0) + Y_1(\kappa_1 t) - Y_2\left(\kappa_2 \int_0^t Z_1^{N,-\beta_1}(s)ds\right)\\
&\quad + Y_3\left(\kappa_3 \int_0^t Z_2^{N,-\beta_1}(s)ds\right)\\
&\quad - Y_4\left(\kappa_4 \int_0^t Z_1^{N,-\beta_1}(s)Z_2^{N,-\beta_1}(s)ds\right)\\
Z_2^{N,-\beta_1}(t) &= Z_2^N(0) + N^{-\alpha_2} Y_2\left(\kappa_2 \int_0^t Z_1^{N,-\beta_1}(s)ds\right)\\
&\quad - N^{-\alpha_2} Y_3\left(\kappa_3 \int_0^t Z_2^{N,-\beta_1}(s)ds\right)\\
&\quad - N^{-\alpha_2} Y_4\left(\kappa_4 \int_0^t Z_1^{N,-\beta_1}(s)Z_2^{N,-\beta_1}(s)ds\right)\\
Z_3^{N,-\beta_1}(t) &= Z_3^N(0) + N^{-\alpha_3} Y_4\left(\kappa_4 \int_0^t Z_1^{N,-\beta_1}(s)Z_2^{N,-\beta_1}(s)ds\right),
\end{aligned}$$

and the limit of $Z^{N,-\beta_1}$ satisfies

$$\begin{aligned}
Z_1(t) &= Z_1(0) + Y_1(\kappa_1 t) - Y_2\left(\kappa_2 \int_0^t Z_1(s)ds\right) + Y_3\left(\kappa_3 \int_0^t Z_2(s)ds\right) \\
&\quad - Y_4\left(\kappa_4 \int_0^t Z_1(s)Z_2(s)ds\right) \\
Z_2(t) &= Z_2(0) \\
Z_3(t) &= Z_3(0).
\end{aligned}$$

Note that the stationary distribution for Z_1 is Poisson with $E[Z_1] = \frac{\kappa_1+\kappa_3 Z_2(0)}{\kappa_2+\kappa_4 Z_2(0)}$.

For $\gamma = -\beta_3$,

$$\begin{aligned}
Z_1^{N,-\beta_3}(t) &= Z_1^N(0) + Y_1\left(\kappa_1 N^{\beta_1-\beta_3} t\right) - Y_2\left(\kappa_2 N^{\beta_2-\beta_3} \int_0^t Z_1^{N,-\beta_3}(s)ds\right) \\
&\quad + Y_3\left(\kappa_3 N^{\alpha_2} \int_0^t Z_2^{N,-\beta_3}(s)ds\right) \\
&\quad - Y_4\left(\kappa_4 N^{\alpha_2} \int_0^t Z_1^{N,-\beta_3}(s) Z_2^{N,-\beta_3}(s)ds\right) \\
Z_2^{N,-\beta_3}(t) &= Z_2^N(0) + N^{-\alpha_2} Y_2\left(\kappa_2 N^{\beta_2-\beta_3} \int_0^t Z_1^{N,-\beta_3}(s)ds\right) \\
&\quad - N^{-\alpha_2} Y_3\left(\kappa_3 N^{\alpha_2} \int_0^t Z_2^{N,-\beta_3}(s)ds\right) \\
&\quad - N^{-\alpha_2} Y_4\left(\kappa_4 N^{\alpha_2} \int_0^t Z_1^{N,-\beta_3}(s) Z_2^{N,-\beta_3}(s)ds\right) \\
Z_3^{N,-\beta_3}(t) &= Z_3^N(0) + N^{-\alpha_3} Y_4\left(\kappa_4 N^{\alpha_2} \int_0^t Z_1^{N,-\beta_3}(s) Z_2^{N,-\beta_3}(s)ds\right),
\end{aligned}$$

and dividing the first equation by $N^{\beta_1-\beta_3} = N^{\beta_2-\beta_3} = N^{\alpha_2}$, we see that

$$\int_0^t Z_1^{N,-\beta_3}(s)\left(\kappa_2 + \kappa_4 Z_2^{N,-\beta_3}(s)\right)ds - \int_0^t \left(\kappa_1 + \kappa_3 Z_2^{N,-\beta_3}(s)\right)ds \to 0.$$

Since $Z_2^{N,-\beta_3}$ is well-behaved, this limit can be shown to imply

$$\int_0^t Z_1^{N,-\beta_3}(s)ds - \int_0^t \frac{\kappa_1 + \kappa_3 Z_2^{N,-\beta_3}(s)}{\kappa_2 + \kappa_4 Z_2^{N,-\beta_3}(s)} ds \to 0. \tag{1.33}$$

We emphasize that ${Z_1}^{N,-\beta_3}$ is not converging, but it is oscillating rapidly and averages locally so that this limit holds. It follows that the other components $(Z_2^{N,-\beta_3}, Z_3^{N,-\beta_3})$ converge to the solution of

$$\begin{aligned}
Z_2(t) &= Z_2(0) + \int_0^t \left((\kappa_2 - \kappa_4 Z_2(s)) \frac{\kappa_1 + \kappa_3 Z_2(s)}{\kappa_2 + \kappa_4 Z_2(s)} - \kappa_3 Z_2(s) \right) ds \\
&= Z_2(0) + \int_0^t \left(\kappa_1 - \frac{2\kappa_4 Z_2(s)(\kappa_1 + \kappa_3 Z_2(s))}{\kappa_2 + \kappa_4 Z_2(s)} \right) \\
Z_3(t) &= Z_3(0) + \int_0^t \kappa_4 Z_2(s) \frac{\kappa_1 + \kappa_3 Z_2(s)}{\kappa_2 + \kappa_4 Z_2(s)} ds. \qquad (1.34)
\end{aligned}$$

Hybrid Limits

Suppose that for some choice of γ, $Z_i^\gamma = \lim_{N\to\infty} Z_i^{N,\gamma}$ exists and is a well-behaved process. Then if $\alpha_i = 0$, $Z_i^{\infty,\gamma}$ will be an integer-valued, pure-jump process, and if $\alpha_i > 0$, Z_i^γ will have continuous sample paths. In fact, if $\alpha_i > 0$, typically Z_i^γ will satisfy an equation of the form

$$Z_i^\gamma(t) = Z_i(0) + \int_0^t F_i(Z^\gamma(s))ds.$$

Consequently, the natural class of limits will by *hybrid* or *piecewise deterministic* (in the sense of Davis [11]) *models* in which some components are discrete and some are absolutely continuous. See section "Reaction Networks" of [4] and Sect. 6.3 of [26] for examples.

It is possible to obtain diffusion processes as limits, but these are not typical for reaction networks. (Note that the diffusion approximations discussed in section "Diffusion/Langevin Approximations" do not arise as *limits* of a sequence of processes.) One example that is more naturally interpreted as a model in population genetics (a Moran model) but can be interpreted as a reaction network would be

$$S_1 + S_2 \rightharpoonup 2S_1, \quad S_1 + S_2 \rightharpoonup 2S_2,$$

where both reactions have the same rate constant. Suppose the normalized system has the form

$$\begin{aligned}
Z_1^N(t) &= Z_1^N(0) + N^{-1/2} Y_1\left(\kappa N \int_0^t Z_1^N(s) Z_2^N(s) ds\right) \\
&\quad - N^{-1/2} Y_2\left(\kappa N \int_0^t Z_1^N(s) Z_2^N(s) ds\right) \\
Z_2^N(t) &= Z_2^N(0) + N^{-1/2} Y_2\left(\kappa N \int_0^t Z_1^N(s) Z_2^N(s) ds\right) \\
&\quad - N^{-1/2} Y_1\left(\kappa N \int_0^t Z_1^N(s) Z_2^N(s) ds\right).
\end{aligned}$$

If we center Y_1 and Y_2, the centerings cancel, and assuming

$$\left(Z_1^N(0), Z_2^N(0)\right) \Rightarrow \left(Z_1^\infty(0), Z_2^\infty(0)\right),$$

(Z_1^N, Z_2^N) converges to a solution of

$$\begin{aligned} Z_1(t) &= Z_1(0) + W_1\left(\kappa \int_0^t Z_1(s)Z_2(s)ds\right) - W_2\left(\kappa \int_0^t Z_1(s)Z_2(s)ds\right) \\ Z_2(t) &= Z_2(0) + W_2\left(\kappa \int_0^t Z_1(s)Z_2(s)ds\right) - W_1\left(\kappa \int_0^t Z_1(s)Z_2(s)ds\right). \end{aligned}$$

Central Limit Theorems and Diffusion Approximations

In section "Derivation of the Michaelis-Menten Equation", Z_2^N and Z_3^N do not converge, but $\int_0^t Z_2^N(s)ds$ and $\int_0^t Z_3^N(s)ds$ do, that is, the rapid fluctuations in Z_2^N and Z_3^N average out. Similarly, to obtain (1.34), we used the fact that for $\gamma = -\beta_3$, the rapid fluctuations in $Z_1^{N,\gamma} = Z_1^N(\cdot N^\gamma)$ average to something well-behaved.

Both of these examples have deterministic limits, and it is natural to seek the same kind of central limit theorem that holds under the classical scaling. Define

$$F(z_2) = \frac{\kappa_1 + \kappa_3 z_2}{\kappa_2 + \kappa_4 z_2}(\kappa_2 - \kappa_4 z_2),$$

and recall that we are assuming $\gamma = -\beta_3$ and $\beta_1 - \beta_3 = \beta_2 - \beta_3 = \alpha_2$. For fluctuations around (1.34), we have

$$\begin{aligned} V^N(t) &= N^{\alpha_2/2}\left(Z_2^{N,\gamma}(t) - Z_2(t)\right) \\ &= V^N(0) + N^{-\alpha_2/2}\widetilde{Y}_2\left(\kappa_2 N^{\alpha_2}\int_0^t Z_1^{N,\gamma}(s)ds\right) \\ &\quad - N^{-\alpha_2/2}\widetilde{Y}_3\left(\kappa_3 N^{\alpha_2}\int_0^t Z_2^{N,\gamma}(s)ds\right) \\ &\quad - N^{-\alpha_2/2}\widetilde{Y}_4\left(\kappa_4 N^{\alpha_2}\int_0^t Z_1^{N,\gamma}(s)Z_2^{N,\gamma}(s)ds\right) \\ &\quad + N^{\alpha_2/2}\int_0^t \left(Z_1^{N,\gamma}(s)\left(\kappa_2 - \kappa_4 Z_2^{N,\gamma}(s)\right) - F(Z_2(s))\right) \\ &\quad - \kappa_3\int_0^t V^N(s)ds. \qquad (1.35) \end{aligned}$$

Assuming $V^N(0)$ converges, the convergence of $Z_2^{N,\gamma}$ and $\int Z_1^{N,\gamma} ds$ and the functional central limit theorem for the renormalized Poisson processes imply the convergence of the first four terms on the right and we would have a central limit theorem similar to that described in section "General Form for the Classical Scaling" if it were not for the fifth term on the right.

To treat the fifth term, we exploit the martingale properties discussed in section "The Martingale Problem and Forward Equation for Markov Chains". In particular, if

$$f_N(z_1, z_2) = N^{-\alpha_2/2} z_1 \frac{\kappa_2 - \kappa_4 z_2}{\kappa_2 + \kappa_4 z_2},$$

as in (1.13),

$$\begin{aligned} M_N(t) &= f_N\left(Z_1^{N,\gamma}(t), Z_2^{N,\gamma}(t)\right) - f_N\left(Z_1^{N,\gamma}(0), Z_2^{N,\gamma}(0)\right) \\ &\quad - \int_0^t A_N f_N\left(Z_1^{N,\gamma}(s), Z_2^{N,\gamma}(s)\right) ds \\ &\approx N^{\alpha_2/2} \int_0^t \left(Z_1^{N,\gamma}(s)\left(\kappa_2 - \kappa_4 Z_2^{N,\gamma}(s)\right) - F\left(Z_2^{N,\gamma}(s)\right)\right) ds \end{aligned}$$

is a martingale, and (1.35) becomes

$$\begin{aligned} V^N(t) &= V^N(0) + N^{-\alpha_2/2}\widetilde{Y}_2\left(\kappa_2 N^{\alpha_2} \int_0^t Z_1^{N,\gamma}(s) ds\right) \\ &\quad - N^{-\alpha_2/2}\widetilde{Y}_3\left(\kappa_3 N^{\alpha_2} \int_0^t Z_2^{N,\gamma}(s) ds\right) \\ &\quad - N^{-\alpha_2/2}\widetilde{Y}_4\left(\kappa_4 N^{\alpha_2} \int_0^t Z_1^{N,\gamma}(s) Z_2^{N,\gamma}(s) ds\right) \\ &\quad + M_N(t) + N^{\alpha_2/2} \int_0^t \left(F\left(Z_2^{N,\gamma}(s)\right) - F(Z_2(s))\right) ds \\ &\quad - \kappa_3 \int_0^t V^N(s) ds + O\left(N^{-\alpha_2/2}\right) \\ &= V^N(0) + \widehat{M}_N(t) + N^{\alpha_2/2} \int_0^t \left(F(Z_2^{N,\gamma}(s)) - F(Z_2(s))\right) ds \\ &\quad - \kappa_3 \int_0^t V^N(s) ds + O(N^{-\alpha_2/2}), \end{aligned}$$

where $\widehat{M}_N$ is defined by the above equality.

Define

$$\Delta f(z_1, z_2, \delta_1, \delta_2) = f(z_1 + \delta_1, z_2 + \delta_2) - f(z_1, z_2).$$

Then the quadratic variation of M_N is

$$\begin{aligned}[M_N]_t &= \int_0^t \Delta f_N\left(Z_1^{N,\gamma}(s-), Z_2^{N,\gamma}(s-), 1, 0\right)^2 dR_1^N(s) \\ &+ \int_0^t \Delta f_N\left(Z_1^{N,\gamma}(s-), Z_2^{N,\gamma}(s-), -1, N^{-\alpha_2}\right)^2 dR_2^N(s) \\ &+ \int_0^t \Delta f_N\left(Z_1^{N,\gamma}(s-), Z_2^{N,\gamma}(s-), 1, -N^{-\alpha_2}\right)^2 dR_3^N(s) \\ &+ \int_0^t \Delta f_N\left(Z_1^{N,\gamma}(s-), Z_2^{N,\gamma}(s-), -1, -N^{-\alpha_2}\right)^2 dR_4^N(s).\end{aligned}$$

Observing that each of the integrands is asymptotically

$$N^{-\alpha_2}\left(\frac{\kappa_2 - \kappa_4 Z_2^{N,\gamma}(s)}{\kappa_2 + \kappa_4 Z_2^{N,\gamma}(s)}\right)^2$$

and that, for example by (1.33),

$$N^{-\alpha_2} R_2^N(t) \to \int_0^t \kappa_2 \frac{\kappa_1 + \kappa_3 Z_2(s)}{\kappa_2 + \kappa_4 Z_2(s)} ds,$$

we have $[M_N]_t \to C(t)$ where

$$\begin{aligned}C(t) = \int_0^t \left(\frac{\kappa_2 - \kappa_4 Z_2(s)}{\kappa_2 + \kappa_4 Z_2(s)}\right)^2 \Bigg(\kappa_1 + \kappa_2 \frac{\kappa_1 + \kappa_3 Z_2(s)}{\kappa_2 + \kappa_4 Z_2(s)} + \kappa_3 Z_2(s) \\ + \kappa_4 Z_2(s) \frac{\kappa_1 + \kappa_3 Z_2(s)}{\kappa_2 + \kappa_4 Z_2(s)}\Bigg) ds,\end{aligned}$$

which, by the martingale central limit theorem (see, for example, Theorem 7.1.4 of [14]), implies $M_N \Rightarrow M$ where M can be written as the time change of a Brownian motion, that is, $M(t) = W(C(t))$.

Unfortunately, M is not independent of the limits of the three renormalized Poisson processes, so rather than applying the martingale central limit theorem to M_N, we need to apply it to $\widehat{M}_N$. The quadratic variation for $\widehat{M}_N$ is

$$\begin{aligned}[\widehat{M}_N]_t &= \int_0^t \Delta f_N\left(Z_1^{N,\gamma}(s-), Z_2^{N,\gamma}(s-), 1, 0\right)^2 dR_1^N(s) \\ &+ \int_0^t \left(N^{-\alpha_2/2} + \Delta f_N\left(Z_1^{N,\gamma}(s-), Z_2^{N,\gamma}(s-), -1, N^{-\alpha_2}\right)\right)^2 dR_2^N(s) \\ &+ \int_0^t \left(-N^{-\alpha_2/2} + \Delta f_N\left(Z_1^{N,\gamma}(s-), Z_2^{N,\gamma}(s-), 1, -N^{-\alpha_2}\right)\right)^2 dR_3^N(s) \\ &+ \int_0^t \left(-N^{-\alpha_2/2} + \Delta f_N\left(Z_1^{N,\gamma}(s-), Z_2^{N,\gamma}(s-), -1, -N^{-\alpha_2}\right)\right)^2 dR_4^N(s),\end{aligned}$$

and $[\widehat{M}_N]_t$ converges to

$$\widehat{C}(t) = \int_0^t \left(\kappa_1 \left(\frac{\kappa_2 - \kappa_4 Z_2(s)}{\kappa_2 + \kappa_4 Z_2(s)} \right)^2 \right.$$
$$+ \left(\frac{\kappa_2 - \kappa_4 Z_2(s)}{\kappa_2 + \kappa_4 Z_2(s)} - 1 \right)^2 \left(\kappa_2 \frac{\kappa_1 + \kappa_3 Z_2(s)}{\kappa_2 + \kappa_4 Z_2(s)} + \kappa_3 Z_2(s) \right)$$
$$\left. + \left(\frac{\kappa_2 - \kappa_4 Z_2(s)}{\kappa_2 + \kappa_4 Z_2(s)} + 1 \right)^2 \kappa_4 Z_2(s) \frac{\kappa_1 + \kappa_3 Z_2(s)}{\kappa_2 + \kappa_4 Z_2(s)} \right) ds.$$

Consequently, $\widehat{M}_N \Rightarrow W(\widehat{C}(t))$ and $V^N \Rightarrow V$ satisfying

$$V(t) = V(0) + W(\widehat{C}(t)) + \int_0^t (F'(Z_2(s)) - \kappa_3)V(s)ds,$$

which, as in (1.17) is a Gaussian process.

Let

$$G(z_2) = \left(\kappa_1 \left(\frac{\kappa_2 - \kappa_4 z_2}{\kappa_2 + \kappa_4 z_2} \right)^2 + \left(1 + \frac{\kappa_2 - \kappa_4 z_2}{\kappa_2 + \kappa_4 z_2} \right)^2 \left(\kappa_2 \frac{\kappa_1 + \kappa_3 z_2}{\kappa_2 + \kappa_4 z_2} + \kappa_3 z_2 \right) \right.$$
$$\left. + \left(\frac{\kappa_2 - \kappa_4 z_2}{\kappa_2 + \kappa_4 z_2} - 1 \right)^2 \kappa_4 z_2 \frac{\kappa_1 + \kappa_3 z_2}{\kappa_2 + \kappa_4 z_2} \right).$$

Then the analysis above suggests the following diffusion or Langevin approximation for $Z_2^{N,\gamma}$:

$$D^N(t) = D^N(0) + N^{-\alpha_2/2} W \left(\int_0^t G(D^N(s))ds \right)$$
$$+ \int_0^t (F(D^N(s)) - \kappa_3 D^N(s))ds.$$

See [26] for a detailed discussion of the central limit theorem and diffusion approximations for multiscaled models. In particular, that paper contains a systematic discussion of the treatment of integral terms with rapidly oscillating integrands.

References

1. Anderson DF (2007) A modified next reaction method for simulating chemical systems with time dependent propensities and delays. J Chem Phys 127(21):214107
2. Anderson DF, Craciun G, Kurtz TG (2010) Product-form stationary distributions for deficiency zero chemical reaction networks. Bull Math Biol 72(8):1947–1970

3. Athreya KB, Ney PE (1972) Branching processes. Springer-Verlag, New York. Die Grundlehren der mathematischen Wissenschaften, Band 196
4. Ball K, Kurtz TG, Popovic L, Rempala G (2006) Asymptotic analysis of multiscale approximations to reaction networks. Ann Appl Probab 16(4):1925–1961
5. Barrio M, Burrage K, Leier A, Tian T (2006) Oscillatory regulation of Hes1: discrete stochastic delay modelling and simulation. PLoS Comp Biol 2:1017–1030
6. Bartholomay AF (1958) Stochastic models for chemical reactions. I. Theory of the unimolecular reaction process. Bull Math Biophys 20:175–190
7. Bartholomay AF (1959) Stochastic models for chemical reactions. II. The unimolecular rate constant. Bull Math Biophys 21:363–373
8. Bratsun D, Volfson D, Tsimring LS, Hasty J (2005) Delay-induced stochastic oscillations in gene regulation. PNAS 102:14593–14598
9. Darden T (1979) A pseudo-steady state approximation for stochastic chemical kinetics. Rocky Mt J Math 9(1):51–71. Conference on Deterministic Differential Equations and Stochastic Processes Models for Biological Systems, San Cristobal, N.M., 1977
10. Darden TA (1982) Enzyme kinetics: stochastic vs. deterministic models. In: Reichl LE, Schieve WC (eds) Instabilities, bifurcations, and fluctuations in chemical systems (Austin, Tex., 1980). University of Texas Press, Austin, TX, pp 248–272
11. Davis MHA (1993) Markov models and optimization. Monographs on statistics and applied probability, vol 49. Chapman & Hall, London
12. Delbrück M (1940) Statistical fluctuations in autocatalytic reactions. J Chem Phys 8(1): 120–124
13. Donsker MD (1951) An invariance principle for certain probability limit theorems. Mem Amer Math Soc 1951(6):12
14. Ethier SN, Kurtz TG (1986) Markov processes. Wiley series in probability and mathematical statistics: probability and mathematical statistics. John Wiley & Sons Inc, New York. Characterization and convergence
15. Feinberg M (1987) Chemical reaction network structure and the stability of complex isothermal reactors i. the deficiency zero and deficiency one theorems. Chem Engr Sci 42(10):2229–2268
16. Feinberg M (1988) Chemical reaction network structure and the stability of complex isothermal reactors ii. multiple steady states for networks of deficiency one. Chem Engr Sci 43(1):1–25
17. Gadgil C, Lee CH, Othmer HG (2005) A stochastic analysis of first-order reaction networks. Bull Math Biol 67(5):901–946
18. Gibson MA, Bruck J (2000) Efficient exact simulation of chemical systems with many species and many channels. J Phys Chem A 104(9):1876–1889
19. Gillespie DT (1976) A general method for numerically simulating the stochastic time evolution of coupled chemical reactions. J Comput Phys 22(4):403–434
20. Gillespie DT (1977) Exact stochastic simulation of coupled chemical reactions. J Phys Chem 81:2340–61
21. Gillespie DT (1992). A rigorous derivation of the chemical master equation. Physica A 188:404–425
22. Gillespie DT (2001) Approximate accelerated stochastic simulation of chemically reacting systems. J Chem Phys 115(4):1716–1733
23. Jacod J (1974/75) Multivariate point processes: predictable projection, Radon-Nikodým derivatives, representation of martingales. Z Wahrscheinlichkeit und Verw Gebiete 31:235–253
24. Kang HW (2009) The multiple scaling approximation in the heat shock model of e. coli. In Preparation
25. Kang HW, Kurtz TG (2010) Separation of time-scales and model reduction for stochastic reaction networks. Ann Appl Probab (to appear)
26. Kang HW, Kurtz TG, Popovic L (2010) Diffusion approximations for multiscale chemical reaction models. In Preparation
27. Kelly FP (1979) Reversibility and stochastic networks. Wiley series in probability and mathematical statistics. John Wiley & Sons Ltd, Chichester

28. Kolmogorov AN (1956) Foundations of the theory of probability. Chelsea Publishing Co, New York. Translation edited by Nathan Morrison, with an added bibliography by A. T. Bharucha-Reid
29. Komlós J, Major P, Tusnády G (1975) An approximation of partial sums of independent RV's and the sample DF. I. Z Wahrscheinlichkeit und Verw Gebiete 32:111–131
30. Komlós J, Major P, Tusnády G (1976) An approximation of partial sums of independent RV's, and the sample DF. II. Z Wahrscheinlichkeit und Verw Gebiete 34(1):33–58
31. Kurtz TG (1970) Solutions of ordinary differential equations as limits of pure jump Markov processes. J Appl Probab 7:49–58
32. Kurtz TG (1971) Limit theorems for sequences of jump Markov processes approximating ordinary differential processes. J Appl Probab 8:344–356
33. Kurtz TG (1972) The relationship between stochastic and deterministic models for chemical reactions. J Chem Phys 57(7):2976–2978
34. Kurtz TG (1977/78) Strong approximation theorems for density dependent Markov chains. Stoch Proc Appl 6(3):223–240
35. Kurtz TG (1980) Representations of Markov processes as multiparameter time changes. Ann Probab 8(4):682–715
36. Kurtz TG (2007) The Yamada-Watanabe-Engelbert theorem for general stochastic equations and inequalities. Electron J Probab 12:951–965
37. Kurtz TG (2010) Equivalence of stochastic equations and martingale problems. In: Dan Crisan (ed) Stochastic analysis 2010. Springer, Heidelberg
38. E W, Liu D, Vanden-Eijnden E (2005) Nested stochastic simulation algorithm for chemical kinetic systems with disparate rates. J Chem Phys 123(19):194107
39. McQuarrie DA (1967) Stochastic approach to chemical kinetics. J Appl Probab 4:413–478
40. Meyer PA (1971) Démonstration simplifiée d'un théorème de Knight. In: Dellacherie C, Meyer PA (eds) Séminaire de Probabilités, V (Univ. Strasbourg, année universitaire 1969–1970). Lecture Notes in Math, vol 191. Springer, Berlin, pp 191–195
41. Ross S (1984) A first course in probability, 2ed edn Macmillan Co, New York
42. van Kampen NG (1961) A power series expansion of the master equation. Canad J Phys 39:551–567

Chapter 2
Stochastic Simulation for Spatial Modelling of Dynamic Processes in a Living Cell

Kevin Burrage, Pamela M. Burrage, André Leier, Tatiana Marquez-Lago, and Dan V. Nicolau, Jr

Abstract One of the fundamental motivations underlying computational cell biology is to gain insight into the complicated dynamical processes taking place, for example, on the plasma membrane or in the cytosol of a cell. These processes are often so complicated that purely temporal mathematical models cannot adequately capture the complex chemical kinetics and transport processes of, for example, proteins or vesicles. On the other hand, spatial models such as Monte Carlo approaches can have very large computational overheads. This chapter gives an overview of the state of the art in the development of stochastic simulation techniques for the spatial modelling of dynamic processes in a living cell.

Keywords Plasma membrane · Chemical kinetics · Gene regulation · Stochastic simulation algorithm · Multiscale stochastic modelling · Diffusion · Delayed reactions · Stochastic simulators

Introduction

Why Do We Need Spatial Models of a Cell?

Recent progress in genetic sequencing, microscopy and other experimental methods has shed light on membrane structures and phenomena, including the discovery that the plasma membrane of a cell may possess significant lateral structure (microdomains). Similar progress has been made in the understanding of transport phenomena on the membrane, of ion channel function and of transport across

K. Burrage (✉)
Computing Laboratory, Wolfson Building, Parks Road, Oxford, OX1 3QD, UK
and
Department of Mathematical Sciences, Queensland University of Technology, Brisbane 4001, Australia
e-mail: kevin.burrage@qut.edu.au

H. Koeppl et al. (eds.), *Design and Analysis of Biomolecular Circuits: Engineering Approaches to Systems and Synthetic Biology*, DOI 10.1007/978-1-4419-6766-4_2,

the membrane [20, 58, 63]. Indeed, a number of research groups are now building dynamic maps of all the ultrastructure within a living cell. The Visible Cell [47] is one such project using electron tomography and 3D rendering to build a complete view of the ultrastructure within a pancreatic Beta cell. However, the integration of this information into comprehensive and coherent models of cellular transport and kinetics, for example linking the plasma membrane with transport processes in the cell to complex genetic regulatory processes, has been slow. In order to produce a coherent picture of cellular dynamics, mathematical modelling and simulation provide an indispensable tool.

On the other hand, the modelling of cellular processes poses mathematical challenges of its own. The main modelling challenges are due to the essential multiscale nature of the processes we are trying to understand. For example, the classical view of the plasma membrane lipid bilayer as a two-dimensional fluid acting as a neutral solvent for membrane proteins in which all particles diffuse freely [65] has been substantially modified in recent years. The plasma membrane is in fact a highly complex structure that is compartmentalized on multiple length and time scales. This compartmentalization is driven by a variety of lipid-lipid, lipid-protein and actin-cytoskeleton interactions [2, 20, 39, 50]. In addition, an important role that has been ascribed to all plasma membrane microdomains is that of selectively concentrating proteins to facilitate the assembly of signalling complexes [64]. However, little quantitative analysis has been attempted to explore the basic mechanics of how microdomains might drive protein-protein interactions as demanded of their role in supporting the assembly of signalling platforms. For example, if microdomains do aggregate proteins, are there any constraints on size and dynamics that need to be imposed for them to achieve this function? If so, are these constraints realistic and how do the predictions compare with recent estimates of microdomain size and dynamics?

Thus, the building of mathematical models and innovative simulation techniques provide a vital component when we attempt to understand the complex feedbacks between dynamic processes on the membrane and, for example, genetic regulation. In building these models we must address two fundamental questions, namely (1) are the processes well-described by assuming homogenization and deterministic principles, or (2) do we need to explicitly model the underlying heterogeneity and stochasticity of these cellular processes?

Why Do We Need Stochastic Models?

There has been a long and successful history in computational cell biology of using rate kinetic ordinary differential equations to model chemical kinetics within a living cell. These techniques have been applied on the plasma membrane, in the cytosol, and in the nucleus of eukaryotic cells to understand gene regulation. Modifications via delay differential equations were first considered as far back as [31], in order to represent the fact that the complex regulatory processes of transcription and translation were not immediate but were in fact examples of delayed processes.

It was the pioneering work of Gillespie [29] and Kurtz [38] who challenged this deterministic view of cellular kinetics. They argued that when the cellular environment contained small to moderate numbers of proteins, that the Law of Mass Action is not an adequate description of the underlying chemical kinetics because it only describes the average behaviour. In this regard, the fundamental principle is that of intrinsic noise. Intrinsic noise is associated with the inherent uncertainty in knowing when a reaction occurs and what that reaction is. The variance associated with this uncertainty increases as the number of proteins in the cellular environment becomes small. Gillespie [29] and Kurtz [38] showed how to model intrinsic noise through the concept of nonlinear discrete Markov processes, and Poisson processes, respectively. These two approaches both model the same processes and are now lumped together under the title the Stochastic Simulation Algorithm (SSA). The essential observation underlying the SSA is that the waiting time between reactions is exponentially distributed and that the most likely reaction to occur in this time interval is based on the relative sizes of the propensity functions. However, the need for a time step small enough to capture one reaction at each step can lead to prohibitive computational costs.

The SSA describes the evolution of a nonlinear discrete Markov process and as such this stochastic process has a probability density function whose solution is described by the Chemical Master Equation (CME). The solution of the CME can be reduced to the computation of the evolution of the exponential of a matrix times an initial probability vector. As there is one equation for each possible configuration of the state space this can be very computationally challenging, although recently developed methods can cope with some of these computational costs [23,35,43,44,55].

There is in fact an intermediate regime that can still capture the inherent stochastic effects but reduce the computational complexity associated with the SSA. This intermediate framework is called the Chemical Langevin Equation. It is described by an Itô stochastic differential equation (SDE) driven by a set of Wiener processes that describes the fluctuations in concentrations of the molecular species. Various numerical methods can then be applied to this equation – the simplest method being the Euler-Maruyama method [37]. These temporal approaches are applied under the principle of homogeneity. It is well known, however, that diffusion on the cell membrane is not only highly anomalous but the diffusion rate of proteins on live cell membranes is between one and two orders of magnitude slower than in reconstituted artificial membranes with the same composition [50]. Furthermore, diffusion is dependent on the dimensions of the medium so that diffusion on the highly disordered cell membrane is not a perfectly mixing process and therefore the assumptions underlying the classical theory of chemical kinetics fail, requiring new approaches to modelling chemistry on a spatially crowded membrane [53].

Simulation Toolkits

Increasingly, scientists are using a mix of experimental, mathematical modelling and simulation to extract deep biological insights at subcellular and tissue levels.

Table 2.1 Spatial stochastic simulators ordered by their underlying simulation method

Simulation method		Software	Spatial mode	Spatial scale	Time	Features
Lattice-based	RDME-based	MesoRD	Lattice	Meso	E	RDS
		SmartCell	Lattice	Meso	E	RDS
		GMP	Lattice	Meso	E+F	RDS
		STEPS	Tetrahedral Mesh	Meso	E	RDS
	Micro-scopic lattice	GridCell	Lattice	Micro	F	EV
		E-Cell (Spatiocyte)	Lattice	Micro+ Meso	E+F	RDS, EV
Off-lattice particle		ChemCell	Continuum[a]	Micro	F	RDS
		MCell	Continuum[a]	Micro	A	RDS
		Smoldyn	Continuum[a]	Micro	F	RDS, EV
		Cell++	Continuum+ Gradient	Micro+ Macro	F	RDS
		CyberCell	Continuum	Micro	F	EV
	GFRD	E-Cell (eGFRD)	Continuum	Micro	E	EV

E event-based, *F* fixed time steps, *A* adaptive, *EV* excluded volume effect is reproducible, *RDS* reaction and diffusion on surfaces and between surfaces and volumes is supported (at different degrees of accuracy)

[a]These programs represent surfaces as lists of primitive objects. Depending on the program this can be triangles, spheres, boxes, or others elements

This not only enriches biology but also enriches computer science and computational and applied mathematics through new methodologies operating at a variety of spatial and temporal scales. Some of these ideas have been incorporated into software toolkits, for example [42]. See Table 2.1 for a list of spatial stochastic simulators. At the same time markup language environments such as CellML [33] and FieldML [17] are being developed which act as both a repository and an environment for simulating a variety of biological models.

Temporal Models of Chemical Kinetics

In a purely temporal homogeneous setting and when there are large numbers of molecules present, chemical reactions are modelled by ordinary differential equations that are based on the laws of Mass Action and that estimate reaction rates on the basis of average values of the reactant density. Any set of m chemical reactions can be characterised by two sets of quantities: the stoichiometric vectors (update rules for each reaction) $\nu_1, \ldots, \nu_m$ and the propensity functions $a_1(X(t)), \ldots, a_m(X(t))$. The propensity functions represent the relative probabilities of each of the m reactions occurring. They are formed by multiplying the rate constant and the product of the reactants on the left-hand-side of each reaction. Here $X(t)$ is the vector of concentrations at time t of the N species involved in the reactions. The ODE that describes this chemical system is given by

$$X'(t) = \sum_{j=1}^{m} \nu_j a_j(X(t)).$$

In the case of small numbers of molecules the appropriate formulation is the Stochastic Simulation Algorithm (SSA) [29], as ODEs can only describe a mean behaviour. The SSA is an exact procedure that describes the evolution of a discrete nonlinear Markov process. It accounts for the inherent stochasticity (internal noise) of the m reacting channels and only assigns integer numbers of molecules to the state vector. At each step, the SSA samples two random numbers from the uniform distribution U[0,1] to evaluate an exponential waiting time, τ, for the next reaction to occur and an integer j between 1 and m that indicates which reaction occurs. The state vector is updated at the new time point by the addition of the jth stoichiometric vector to the previous value of the state vector, that is

$$X(t+\tau) = X(t) + \nu_j.$$

The main limiting feature of SSA is that the time step can become very small, especially if there are large numbers of molecules or widely varying rate constants. τ-leap methods have been suggested in which the sampling of likely reactions is taken from either Poisson [30] or Binomial [72] distributions. In these approaches a much larger time step can be used at the loss of a relatively small amount of accuracy.

A different approach is to compute the probability density function associated with the SSA, which is the solution of the Chemical Master Equation (CME). The CME is a discrete parabolic partial differential equation in which there is one equation for each configuration of the 'state space'. When the state space is enumerated, the CME becomes a linear ODE and the probability density function takes the form

$$p(t) = e^{At} p(0)$$

where A is the state-space matrix. Even for relatively small systems, the dimension of A can be in the millions, but a variety of techniques have been proposed [23, 35, 43, 44, 55] to make this a very feasible technique.

The regime intermediate to the discrete stochastic regime and the continuous deterministic ODE regime is described by the Chemical Langevin Equation (CLE). The CLE attempts to preserve the correct dynamics for the first two moments of the SSA and takes the form

$$dX = \sum_{j=1}^{m} \nu_j a_j(X(t)) + B(X(t))dW(t).$$

Here $W(t) = (W_1(t), \ldots, W_N(t))^{\mathrm{T}}$ is a vector of N independent Wiener processes whose increments $\Delta W_j = W_j(t+h) - W_j(t)$ are $N(0, h)$ and where

$$B(x) = \sqrt{C}, \quad C = (\nu_1, \ldots, \nu_m) Diag(a_1(X), \ldots, a_m(X))(\nu_1, \ldots, \nu_m)^T.$$

Here h is the time discretisation step. Effective methods designed for the numerical solution of SDEs [11, 13, 37] can be used to simulate the chemical kinetics in this intermediate regime. Mélykúti et al. [48] have shown how to construct the CLE so that it minimizes the number of Wiener processes. Furthermore, adaptive multiscale methods have been developed that attempt to move back and forth between these three regimes as the numbers of molecules change [10].

Sometimes temporal models are not sufficiently rich to capture complicated spatial effects. But rather than abandoning temporality, it is possible to capture important spatial aspects and incorporate them into temporal models. This can be done in a number of ways. For example, compartmental models have been developed that couple together the plasma membrane, cytosol and nucleus – see for example Tian et al. [73], in which an SSA implementation of Ras nanoclusters on the plasma membrane is coupled with an ODE model for the MAPK pathway in the cytosol. Diffusion and translocation can be captured through the use of distributed delays that can then be incorporated into mathematical frameworks through the use of delay differential equations or delay variants of the Stochastic Simulation Algorithm (see [8], for example). Very recently, Marquez-Lago et al. [46] have explored a number of spatial scenarios, run detailed spatial simulations to capture diffusion and translocation processes and then incorporated this information into purely temporal models through distributed delays. Another way in which spatial information can be captured and then incorporated into purely temporal models is the area of anomalous diffusion, where spatial crowding and molecular binding can affect chemical kinetics. In this setting the mean square deviation of a diffusing molecule is no longer linear but sublinear in time t and of the form

$$E\left[X^2(t)\right] = 2Dt^{\alpha}, \quad \alpha \in (0, 1].$$

Here, α is called the anomalous diffusion parameter. If the value of α can be estimated, either experimentally or from detailed Monte Carlo simulations, then the SSA can be modified so that the waiting time between reactions is no longer exponentially distributed but has a heavy tail [53].

Monte-Carlo Approaches

In many cases, the heterogeneous nature of a living cell means that spatial models are mandatory. The fundamental transport process within a cell is either diffusion or the motion of proteins or vesicles along microtubules by molecular motors. To capture these processes, we can use continuum models based on partial differential equations in which diffusion is represented by a Laplacian operator and directed transport by a convective term. If chemical kinetics are involved, then this leads to the framework of the reaction-diffusion partial differential equation. However as spatial structures become more and more complex, so that the homogenisation process breaks down, Monte Carlo simulations become more appropriate. One such environment is the plasma membrane.

The plasma membrane is an extremely complex and crowded environment that has many roles including signalling, cell-cell communication, cell feeding and excretion and protection of the interior of a cell. It is heterogeneous – the cytoskeletal structure just inside the plasma membrane can corral and compartmentalize membrane proteins. Chemically inert objects can form barriers to protein diffusion on the plasma membrane, and this can lead to anomalous diffusion rather than pure diffusion. Trying to capture such complexity using higher-level mathematical frameworks such as partial differential equations is extremely challenging, so instead a stochastic spatial model using the Monte-Carlo technique becomes appropriate, especially as the domain of interest is essentially two dimensional. In such a simulation the plasma membrane can be mapped to a two dimensional lattice, usually regular but not necessarily so. The size of each computational cell "voxel" depends on what biological questions are being asked, but taking into account volume-exclusion effects, usually the voxel is such that at most one protein per voxel is allowed. Given the dimensions of typical membrane protein anchors, a typical voxel size is thus of the order of 1–3 nm. Assuming that we wish to model the dynamics on the plasma membrane of a typical cell, say $10 \times 10\,\mu$m, then the computational lattice has $5{,}000 \times 5{,}000$ voxels. In a spatial Monte Carlo simulation of this type, a random walk is carried out by each protein on the membrane; a protein is selected at random, and a movement direction (north, south, east or west, in the case of a rectangular lattice arrangement) is randomly determined. The distance moved depends on the diffusion rates for each species. Chemical reactions can be simulated by checking the chemical reaction rules and then replacing that protein and/or creating a new protein at that location whenever a collision (volume exclusion event) occurs. In one step, each protein must statistically (but not necessarily deterministically) be selected to move, so if we are interested in the dynamics over large regions or long time scales, then this approach is computationally demanding. In addition, since the approach is stochastic, a number of simulations must be run in order to collect appropriate statistics. Nicolau et al. [51, 52] have used this approach to model the effects of compartmentalisation, anomalous diffusion and the motion of lipid rafts on the chemical kinetics taking place on the membrane. However, only relatively small sections of the membrane on short time scales are considered due to the slow computational performance. There are a number of approaches to address this issue, one of which we discuss below. As computational speeds and memory sizes increase, longer times and larger systems will be able to be modelled.

Recently, in [12], a plasma membrane stochastic spatial model was implemented in a parallel setting, so that both long-range diffusion rates could be investigated and an entire cell membrane could be modelled with the simulation running for several real-time seconds. Domain decomposition was applied to allocate portions of the membrane to different parallel processes; each membrane portion was populated with molecules of various species, inert obstacles, cytoskeletal fences and mobile or immobile lipid rafts. OpenMPI was used for message-passing between processes, and MPI's Cartesian topology management commands ensured the physical proximity of processes with adjacent membrane portions. By allowing each parallel process to have a "ghost" copy of its right-hand neighbour's leftmost column and its left-hand neighbour's rightmost column of membrane data, along with its own membrane portion, communication costs between processes can be minimised while maintaining an accurate view of the current membrane state. Data moving between processes was first moved into these "ghost" columns and these were then merged with the neighbours' membrane portions at every time step of the simulation. Slave to slave implementations gained excellent speed-up [12] so a cross-cluster implementation can allow a simulation on the entire cell membrane to be run for several real-time seconds and hence may provide valuable data and insights for biologists.

In a spatial Monte Carlo simulation of proteins diffusing on a lattice, diffusion can be considered as a unimolecular reaction. Thus if we order the voxel elements within the lattice into a vector, we can consider this approach in the SSA framework and apply the same tools that have been developed in the purely temporal setting. In particular, there is a spatial CME associated with this approach [19]. Finally, we note that although we have emphasised Monte Carlo simulations for heterogeneous models, there is a recent approach where a continuum finite element partial differential equation method has been used to understand the effect of lipid rafts on chemical kinetics on the plasma membrane [62].

Multi-scale Modelling

Lattice Versus Off-Lattice Methods

Spatially resolved simulations are computationally expensive, a fact that becomes more evident when compared with their solely temporal counterparts. Nevertheless, adequate modelling of biological systems often requires spatially resolved simulations. By consequence, one should always keep in mind the trade-off between simulation time and necessary level of resolution.

Of all spatial methods, the option with the lowest computational cost consists of solving reaction-diffusion partial differential equations, each of which represents the concentration of a molecular species in the system. This approach is only valid if and when: (1) all molecular species in the system have large molecular concentrations, and (2) noise is not amplified throughout the system. If at least one of these conditions fails to hold, we must rely on spatial stochastic simulators, which can be of discrete or continuous nature. In turn, the highly resolved end of the discrete spatial stochastic simulators spectrum is represented by lattice and off-lattice particle based methods.

In off-lattice methods, all particles in the system have explicit spatial coordinates, at all times. At each time step, molecules with non-zero diffusion coefficients are able to move, in a random walk fashion, to new positions. In many cases, reaction bins whose size depends on the particular diffusion rates are drawn around each particle. If one or more molecules happen to be inside such a bin, appropriate chemical reactions can take place with a certain probability, and if a reaction is readily performed, the reactant particles are flagged, to avoid repetition of chemical events. Noticeably, in off-lattice methods, the domains and/or compartments can still be discretized, to aid the localization of particles within the simulation domain. Particle methods can provide very detailed simulations of highly complex systems at the cost of exceedingly large amounts of computational time and, possibly, restrictions on the size of the simulation domain. Hence, such detailed simulations can often only yield short simulation time spans that, in many cases, are of no interest to the experimentalists.

For lattice methods, and within the context of Molecular Biology, a computational mesh (generally two dimensional or three dimensional) is used to represent a cellular compartment, such as a membrane or the interior of some part of a cell [49, 51, 75]. The lattice is then "populated" with particles of the different molecular species that comprise the system, either at random or at chosen spatial locations, depending on the theoretical question at hand. All particles with non-zero diffusion coefficient are able to diffuse throughout the simulation domain by jumping to empty neighbouring sites and, depending on user-specified reaction rules, appropriate chemical reactions can take place with a certain probability. It is worth noting the mesh can represent microscopic or mesoscopic domains. In the former, each lattice site is allowed to host at most one molecule. These microscopic lattice-based simulators are often called Kinetic Monte Carlo Methods, which might create some confusion as this name is shared by general stochastic simulation strategies.

A less computationally intensive alternative, albeit still costly in many scenarios, is to consider molecular interactions in the mesoscopic realm. Here, the discretization of the Reaction-Diffusion Master Equation (RDME) results in reactive neighbouring sub-volumes within which several particles can coexist, while well-mixedness is assumed in each subvolume. Following this line of thought, there are a few algorithms in the literature extending discrete stochastic simulators to approximate solutions of the RDME by introducing diffusion steps as first order reactions, with a reaction rate constant proportional to the diffusion coefficient. For instance, in [7, 67] the authors provide the specific outline for extending discrete stochastic simulators to the RDME regime, while the algorithms in [1, 21] provide clever extensions of the 'next reaction method' [28], commonly known as the 'next subvolume method'. Additionally, a review on the construction of such methods can be found in [25]. Figure 2.1 attempts to provide an illustration of the spatio-temporal advancement scheme of lattice and off-lattice algorithms.

A few additional aspects are worth considering. First, in mesoscopic lattice methods, as well as inefficiently posed off-lattice methods, problems may arise due to neglecting the 'volume exclusion' effect (for example, whenever a (sub)domain

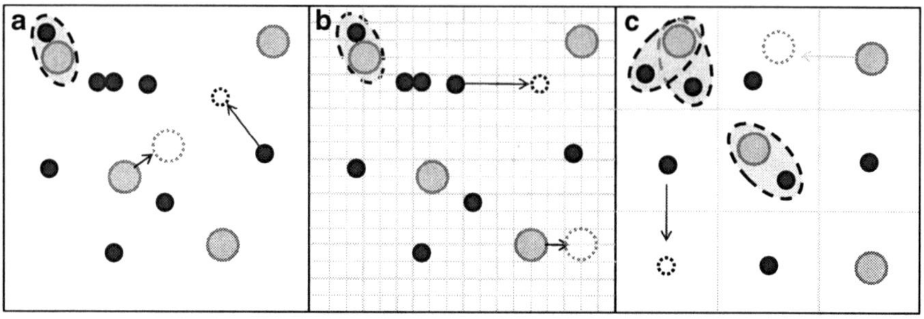

Fig. 2.1 Illustration of the advancement schemes for idealized **(a)** off-lattice methods, **(b)** microscopic lattice methods and **(c)** mesoscopic lattice methods. *Arrows* in *black* represent diffusion over the same time scale, while *grey arrows* represent diffusion over a considerable larger time step. *Dashed black lines* represent potential reaction partners, over a single time step (before or after diffusion changes, not simultaneously)

Table 2.2 Software and corresponding websites

MesoRD	http://mesord.sourceforge.net/
SmartCell	http://smartcell.crg.es/
GMP	http://www.science.uva.nl/research/scs/CellMath/GMP
STEPS	http://sourceforge.net/projects/steps
GridCell	http://iml.ece.mcgill.ca/GridCell/
ChemCell	http://www.sandia.gov/~sjplimp/chemcell.html
MCell	http://www.mcell.cnl.salk.edu/
Smoldyn	http://www.smoldyn.org/
Cell++	http://www.compsysbio.org/lab/cell_simulations
E-cell	http://www.e-cell.org/

is populated by a large number of molecules that would not physically fit). The same would hold for inefficiently posed microscopic lattice methods, where each molecule is set to occupy a single site, irrespective of its physical size.

Secondly, molecular crowding can prevent reacting molecules from reaching regions of the domain, due to the high concentration of macromolecules impeding their passage [5]. While this effect can be explicitly treated by microscopic lattice methods (as well as some off-lattice methods), mesoscopic lattice methods are in a great disadvantage, their expected accuracy being low when treating these cases. Lastly, the artificial nature of the lattice may not only limit the spatial resolution of the simulation, but also introduce lattice anisotropy [59].

In the following section, we describe the Next Subvolume Method in more detail, as well as a coarse-grained version that accelerates simulations by 2–3 orders of magnitude. We additionally report recent extensions and corrections to these algorithms. Lastly, we refer to Table 2.2 for a list of publicly available lattice and off-lattice simulators.

The Next Subvolume Method and Its Coarse-Grained Version, Bτ-SSSA

The Next Subvolume Method (NSM) [1, 21, 22, 32] is a generalization of the SSA [29], where the simulation domain is divided into uniform separate subvolumes that are small enough to be considered homogeneous by diffusion over the time scale of the reaction. At each step, the state of the system is updated by performing an appropriate reaction within a single subvolume, or by allowing a molecule to jump to a randomly selected neighbouring subvolume. Diffusion is then modelled as a unary reaction, with rate proportional to the two-dimensional molecular diffusion coefficient divided by the length of a side of the subvolume. In this way, diffusion inside the algorithm becomes another possible event with a regular propensity function, and follows the same update procedure as any chemical reaction. The expected time for the next event in a subvolume is calculated in a similar way to the SSA algorithm, including the reaction and diffusion propensities of all molecules contained in that subvolume, at that particular time. However, times for subsequent events will only be recalculated for those SVs that were involved in the current time step, and they are subsequently re-ordered in an event queue.

Even though mesoscopic simulations are much faster than their microscopic counterparts, often we will need to coarse-grain the simulation to provide for a spatially resolved model that yields accurate chemical kinetics in meaningful simulation times that are of actual biological interest to the experimentalists. With this in mind, a very natural extension of the NSM comes down to performing τ-leaps (slightly larger time steps) that account for one or more diffusion and reaction events, without significantly compromising spatio-temporal accuracy. This is the idea behind the Binomial τ-leap Spatial Simulation Algorithm, Bτ-SSSA [45] where, at each iteration, the subvolume with shortest reaction-diffusion τ-leap is selected. Then, the algorithm performs a number of possible, yet randomly chosen, events inside each subvolume, re-calculates a new τ-leap for all subvolumes that were affected by current reactive or diffusive events, re-orders the time event queue in increasing time, and consequently chooses the subvolume indicated by the top of the time event queue.

A few additional notes and issues are worth considering. First, in order to calculate the expected time for the next event in any subvolume, the sum of diffusion propensities needs to be multiplied by the number of directions in which the molecules can diffuse, that is, the number of neighbours contained in the connectivity matrix for that subvolume. While this is explicitly stated in [22], it is implicitly assumed in [21]. The same holds for the coarse-grained version presented in [45], which implicitly considers the number of neighbouring subvolumes when calculating τ-leaps. We have noticed that, neglecting the number of neighbours for the calculation of the diffusion propensity yields radically low accuracy, whenever the system contains zero-order reactions (namely, when one of the molecular species is constitutively created).

Secondly, and perhaps more importantly, it has been readily noticed that accuracy is lost when considering commonly used implementations of bimolecular reactions. Early on, [7] reported that: "Too large a cell size violates the cell statistical homogeneity assumptions, whereas too small a cell size may compromise the separability of reaction and diffusion viewed as independent elementary processes in significantly small subvolumes". Furthermore, the authors in [34] reported their concern for the dependence of solutions to the RDME on mesh spacing. In response to these open issues, Erban and Chapman suggested clever corrections for the propensity values in lattice based models, hence homogenizing results with varying degrees of mesh finesse [26]. Extrapolations of these considerations are current work in progress, and include several implications in: (1) two-dimensional domains, (2) coarse-grained scenarios, and (3) unstructured meshes [41].

Hybrid Discrete-Continuous

There are probably many ways to create hybrid reaction-diffusion algorithms. For example, we can group species according to size, treat small molecules deterministically, and treat large particles and their interactions with large and small molecules stochastically. The latter implies molecules exist in high numbers and/or diffuse rapidly.

In another approach, Engblom et al. [24] split the time integration of the RDME into a macroscopic diffusion (for species with large numbers of molecules) and a stochastic mesoscopic reaction/diffusion part (for species with small numbers of molecules) obtaining the mesoscopic diffusion coefficients from proper FEM discretizations. At the same time, as with purely temporal hybrid algorithms, there are a number of options for designing a spatial hybrid algorithm, in particular with respect to the methods being combined and the partitioning criteria.

Software and Other Spatial Approaches

Capturing Spatial Attributes

Despite increasingly available computational resources, simulating highly-resolved spatial models of cellular processes and pathways can still be computationally demanding, if not prohibitively expensive. This motivates the search for alternative, indirect ways to incorporate spatial information in purely temporal models, while aiming at reasonable accuracy when compared to its fully spatial counterparts.

Recently, such an alternative methodology based on reactions with associated distributed time-delay was presented in Marquez-Lago et al. [46]. The method consists of two steps: (1) distribution fitting and (2) stochastic simulation. The delay distributions stem from diffusion profiles and can be directly obtained from in silico

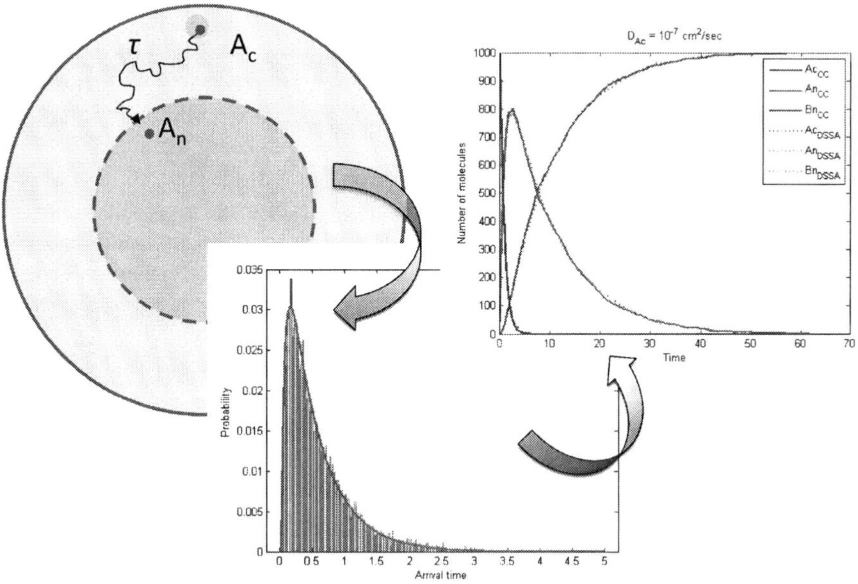

Fig. 2.2 Simplification of the idea behind distributed delays. For a translocation process, we can obtain first arrival (first passage) times either directly from experiments, from analytical or numerical solutions of a corresponding PDE, or from particle simulators such as ChemCell (illustrated in the figure on the *left*). We calculate the delay distribution (*middle*) and draw from it random delays, to be used in the delay stochastic simulation algorithm whenever a translocation reaction is occurring (*right*)

(particle-based) simulations, in vitro experiments, or by solving the corresponding PDEs. Once these tailored distributions are calculated they are used with their associated reactions in a modified version of the delay stochastic simulation algorithm (DSSA) (Fig. 2.2).

The methodology captures some spatial processes with accuracy that is unmatched by any other purely temporal method.

Spatial Simulation Software

In this section, we give a brief and not exhaustive overview of software packages for spatial modelling of chemical reaction networks and cellular processes. Spatial stochastic simulation software can be classified according to its underlying methodology, that is, the choice of spatial representation and the temporal evolution scheme. It is customary to distinguish between lattice and off-lattice (particle) methods. As explained above, the former can be subdivided into spatial SSA implementations, leading to trajectories that are exact or approximate realizations of the reaction-diffusion master equation (RDME), and methods based on microscopic discretizations of the spatial domain, such as kinetic Monte Carlo methods. Off-lattice methods can be subdivided into particle based methods with fixed time-stepping, and Green's Function Reaction Dynamics (GFRD) methods. GFRD effectively reduce the multi-body reaction-diffusion problem into multiple single-body and two-body problems that are then solved using Green's functions. The simulation advances from the time of one reaction to the time of the next reaction. Table 2.1 summarizes the method and features of each approach. Table 2.2 lists the software packages and the corresponding webpages for additional information and download links. More detailed summaries and recent comparisons between particle-based simulators can be found in [4, 6, 70].

RDME-Based Methods

Stundzia and Lumsden [67] were the first to introduce diffusion reactions between sub-volumes to the SSA framework [29, 30], followed by [7]. Essentially, Stundzia and Lumsden implemented a spatial version of Gillespie's Direct Method (DM) and used it to simulate the propagation of an ionic reaction-diffusion calcium wave through a cell. Later, Ander et al. [1] and Elf et al. [21] independently developed spatial schemes of the Next Reaction Method (NRM) [28] the latter being coined the Next-Subvolume Method (NSM). This method is now included in two software packages, namely in **SmartCell** (version 4.2) and **MesoRD** [32] (version 0.3). A recent extension of MesoRD allows also for a correct description of reaction rates when the subvolume size is close to the reaction radii of the molecules. The MesoRDToolBox is a MatLab toolbox for visualizing MesoRD simulation data. SmartCell provides a Java GUI to process its outputs. Both SmartCell and MesoRD allow the option of choosing other methods for simulating a chemical diffusion-reaction system. MesoRD has been used to study spatial oscillation patterns displayed by the Min system of Escherichia coli [27]. To our knowledge, coarse-grained versions of NSM have not been implemented in any publicly available software package.

Other algorithms that have been recently proposed for simulating reaction-diffusion systems include the **Gillespie-Multi-Particle (GMP)** method [60] and the multinomial simulation algorithm (MSA) [40]. In GMP, the reaction and diffusion processes are executed independently of each other (operator splitting scheme). The method uses the multi-particle method (a.k.a. Lattice Gas Automata algorithm) [14, 15] to simulate diffusion: at each diffusion step, molecules from one subvolume are uniformly distributed among its adjacent subvolumes. While diffusion steps occur at predetermined times, reactions are simulated in between those steps using standard SSA. The fixed diffusion time step corresponds to the average time between diffusion events in the RDME. In Dobrzyński et al. [18], GMP has been compared to MesoRD, Smoldyn, GFRD, and (non-spatial) SSA, using a model of regulated gene expression and diffusion of phosphorylated CheY in the E.coli chemotaxis pathway as case studies.

Unlike GMP, the MSA does not decouple reaction and diffusion events, instead diffusive steps are taken until time to the next reaction is reached. The probabilities for diffusion events are taken from a multinomial distribution and diffusion is not limited to directly neighbouring subvolumes. However, the coupling of diffusive steps and reactions causes a problem, namely when particles are chosen for diffusion that have been chosen before as reactants in the next reaction event. The MSA deals with this issue by removing reactants immediately once a reaction is chosen (such that they are no longer available for diffusion) and adding the product(s) after the reaction occurs. There is no source code available for MSA.

Finally, **STEPS** [78] (v1.1.1) is a platform for simulating reaction-diffusion processes in 3D, using irregular, tetrahedral meshes for volume discretization. STEPS has the spatial version of the direct reaction method implemented, but also allows non-spatial SSA implementations on individual compartments ignoring the domain discretization. Volumes are bounded by membranes that can contain stationary reaction molecules such as channel proteins. It is intended to extend STEPS allowing diffusion also on membranes.

Methods Based on Microscopic Lattices

GridCell [9] (v1.2) subdivides the 3D domain into microscopic compartments (voxels), each having 26 surrounding neighbours plus itself (a so-called D3Q27 model) for diffusion and reaction. Each voxel may contain at most one molecule. Molecules diffuse by hopping with a species-dependent moving probability to a randomly chosen neighbour and can react with other molecules in neighbouring voxels. Particle moves and reactions are independent events. The simulation evolves in fixed time steps in which each particle may move and/or react only once. As with all other methods based on microscopic lattices, GridCell is able to capture the effects of volume exclusion and molecular crowding by introducing inert particles.

Spatiocyte [5, 6] has been designed to simulate reaction and diffusion in 3D and 2D volumes and between 3D and 2D compartments, and to reproduce implications of molecular crowding. In Spatiocyte, the domain is discretized into a hexagonal close-packed lattice. Each compartment is a voxel with a radius equal to that of the simulated molecules and has 12 adjoining neighbours. Aside from individual molecules that are simulated on the microscopic scale, Spatiocyte also supports the simulation of homogeneously distributed (HD) species on a compartmental scale. Diffusion-influenced reactions are modelled using a discretized version of Collins and Kimball [16] for obtaining a reaction probability. Temporal evolution is based on hybrid time-driven and event-driven methods [69]. Diffusion steps occur at predefined time steps while diffusion-independent reactions are performed according to the NRM [28].

The spatiocyte algorithm has been implemented as a plug-in module to **ECell** [5, 6]. ECell [5, 68, 74] (v3.2.0) is a simulation platform for modelling and analyzing chemical reaction networks. Originally, it only supported temporal simulation algorithms ranging from stochastic simulation algorithms (such as SSA or explicit/implicit tau-leap methods) to ODE and DAE solvers. Recently, it has been extended by two spatial algorithms, one of which is Spatiocyte. This has been recently used to investigate the MinDE system. Simulation results showed a link between transient membrane attachment of *E. coli* MinE and the formation of E-rings [5].

Off-Lattice Particle Methods

ChemCell [56, 57], **MCell** [36, 66], and **Smoldyn** [3, 4] are the most popular and widely used off-lattice stochastic particle-based reaction-diffusion simulators. While ChemCell (last version 10 Sep. 2008) also allows for non-spatial simulations using ODE solvers or standard SSA, MCell (v3.1.846) and Smoldyn (v2.1) offer particle simulations only. Smoldyn is the only simulator that supports reaction and diffusion of particles in 1D, 2D, and 3D. ChemCell and Smoldyn use a fixed time step, whereas MCell uses an adaptive time-stepping, but allows the user to specify an upper limit. These particle-based simulators vary in many more features, such as system boundaries, geometric primitives, support of 0-order reactions, and surface interactions, to name a few. We refer for a more detailed comparison of these methods to the paper by Andrews et al. [4].

Another particle simulator is **Cell++** [61]. This spatial modelling and simulation platform combines a cellular automata engine with Brownian dynamics. Cell++ allows the simulation of large numbers of small molecules (such as calcium ions, pyruvate, ATP), while simultaneously treating larger molecules, such as enzymes, as entities. The continuum domain is superimposed by a 3D lattice. Each cubic subvolume describes the cellular environment (cytosol, nucleus, or membrane) and stores the local concentration of small molecules. The simulation evolves in discrete time steps. At each step, the relative flux of small molecule (metabolite) concentrations between two lattice cells, the diffusion step of all large molecules (enzymes), reactions among large molecules and those between small and large molecules are calculated. Cell++ was designed to study the impact of spatial organization on several biochemical systems including metabolism, signalling pathways, calcium waves and lipid raft mediated signalling.

CyberCell [59] is a particle-based simulator that was built to study the volumetric impact of macromolecular crowding on cellular reaction-diffusion systems. The simulation advances in discrete time steps of fixed size. Particles diffuse by a fixed length in a random direction, uniformly distributed over the surface of a sphere. The moving probability depends on the diffusion constant. After the diffusion step, particles that moved are checked for potential collisions with other particles. If a collision is observed, it is determined if a reaction occurred. If no reaction happened, the move is rejected. This approach is used to enforce volume exclusion.

Greens-Function Reaction Dynamics (GFRD) Methods

The idea of Green's Function Reaction Dynamics (GFRD) [76, 77] is to choose a maximum time step that is still small enough such that only single particles or pairs of particles have to be considered, and no particle can collide with more than one other particle during this step. The corresponding Einstein equation (for propagating single particles) and Smoluchowski equation (for the two-body problem) can be solved analytically, using Green's functions. From these solutions, probabilities for the next unimolecular and bimolecular reactions to happen are derived, and these probabilities can be subsequently used in an event-driven GFRD algorithm. This algorithm iteratively (1) determines the maximum time step, the next reaction to occur and the time when it occurs (within the maximum time step), (2) propagates all particles, and (3) updates particles according to the reaction.

The GFRD algorithm is very accurate and much faster than other particle-based methods for systems that are diffusion-dominated. However, the GFRD algorithm is not exact since the decomposition into one-body and two-body problems involves cut-off distances [71]. Also, the original GFRD can be computationally very intensive due to the synchronized updates of all particles at each step, and the fact that the system evolves according to the smallest tentative reaction time. Recently, Takahashi et al. [71] introduced an asynchronous version of the GFRD, called eGFRD, which was inspired by the work of Oppelstrup et al. [54]. In eGFRD, spherical protective domains are placed around single particles and pairs of particles, making the new scheme exact.

In the upcoming **E-Cell** version 4 a particle-based simulator based on the eGFRD algorithm will be included. In Takahashi et al. [71] eGFRD has been used to demonstrate how spatiotemporal correlations can change the response of the MAPK pathway.

Conclusions

The last few years have seen a rapid development in spatial algorithms and toolkits. The remaining challenges are to make these approaches robust and efficient in multiscale heterogeneous environments and to allow them to be integrated in, for example, whole organ models.

References

1. Ander M, Beltrao P, Di Ventura B et al (2004) SmartCell, a framework to simulate cellular processes that combines stochastic approximation with diffusion and localisation: analysis of simple networks. Syst Biol 1:129–138
2. Anderson RG, Jacobson K (2002) A role for lipid shells in targeting proteins to caveolae, rafts, and other lipid domains. Science 296:1821–1825

3. Andrews SS, Bray D (2004) Stochastic simulation of chemical reactions with spatial resolution and single molecule detail. Phys Biol 1:137–151
4. Andrews SS, Addy NJ, Brent R, Arkin AP (2010) Detailed simulations of cell biology with Smoldyn 2.1. PLoS Comput Biol 6:e1000705
5. Arjunan SNV, Tomita M (2009) Modeling reaction-diffusion of molecules on surface and in volume spaces with the E-cell system. IJCSIS 3:10060913
6. Arjunan SNV, Tomita M (2010) A new multicompartmental reaction-diffusion modeling method links transient membrane attachment of E. coli MinE to E-ring formation. Syst Synth Biol 4:35–53
7. Baras F, Mansour MM (1996) Reaction-diffusion master equation: a comparison with microscopic simulations. Phys Rev E 54(6):6139–6148
8. Barrio M, Burrage K, Leier A, Tian T (2006) Oscillatory regulation of Hes1: discrete stochastic delay modelling and simulation. PLoS Comput Biol 2(9):e117
9. Boulianne L, Al Assaad S, Dumontier M, Gross WJ (2008) GridCell: a stochastic particle-based biological system simulator. BMC Syst Biol 2:66
10. Burrage K, Tian T, Burrage PM (2004) A multi-scaled approach for simulating chemical reaction systems. Prog Biophys Mol Biol 85:217–234
11. Burrage PM, Burrage K (2002) A variable stepsize implementation for stochastic differential equations. SIAM J Sci Comput 24(3):848–864
12. Burrage PM, Burrage K, Kurowski K, Lorenc M, Nicolau DV, Swain M, Ragan M (2009) A parallel plasma membrane simulation, In: Guerrero J (ed) Proceedings of 1st international workshop on high performance computational systems biology (HiBi2009), Conference Publishing Services, IEEE Computer Society, Trento, Italy, 14–16 October 2009, pp 105–112, ISBN: 978-0-7695-3809-9
13. Burrage PM, Herdiana R, Burrage K (2004) Adaptive stepsize based on control theory for SDEs. J Comput Appl Math 170:317–336
14. Chopard B, Frachebourg L, Droz M (1994) Multiparticle lattice gas automata for reaction diffusion systems. Int J Mod Phys C 5:47–63
15. Chopard B, Droz M (1998) Cellular automata modeling of physical systems. Cambridge University Press, Cambridge, UK
16. Collins FC, Kimball GE (1949) Diffusion-controlled reaction rates. J Colloid Sci 4:425–437
17. Crampin E, Smith N, Hunter P (2004) Multi-scale modelling and the IUPS Physiome Project. J Mol Histol 35(7):707–714
18. Dobrzyński M, Rodríguez JV, Kaandorp JA, Blom JG (2007) Computational methods for diffusion-influenced biochemical reactions. Bioinformatics 23:1969–1977
19. Drawert B, Lawson MJ, Petzold L, Khammash M (2010) The diffusive finite state projection algorithm for efficient simulation of the stochastic reaction-diffusion master equation. J Chem Phys 132:074101. doi:10.1063/1.3310809
20. Edidin M (2003) The state of lipid rafts: from model membranes to cells. Annu Rev Biophys Biomol Struct 32:257–283
21. Elf J, Doncic A, Ehrenberg M (2003) Mesoscopic reaction-diffusion in intracellular signaling. In: Bezrukov SM, Frauenfelder H, Moss F (eds) Fluctuations and noise in biological, biophysical, and biomedical systems, Proceedings of the SPIE 5110, pp 114–125
22. Elf J, Ehrenberg M (2004) Spontaneous separation of bi-stable biochemical systems into spatial domains of opposite phases. Syst Biol 1:230–236
23. Engblom S (2009) Galerkin spectral method applied to the chemical master equation. Commun Comput Phys v5(i5):871–896
24. Engblom S, Ferm L, Hellander A, Loetstedt P (2009) Simulation of stochastic reaction-diffusion processes on unstructured meshes. SIAM J Sci Comput 31:1774–1797
25. Erban R, Chapman SJ, Maini PK (2007) A practical guide to stochastic simulations of reaction–diffusion processes. arXiv:0704.1908
26. Erban R, Chapman SJ (2009) Stochastic modelling of reaction-diffusion processes: algorithms for bimolecular reactions. Phys Biol 6:046001
27. Fange D, Elf J (2006) Noise-induced Min phenotypes in E. coli. PLoS Comput Biol 2:e80

28. Gibson MA, Bruck J (2000) Efficient exact atochastic simulation of chemical systems with many species and many channels. J Phys Chem 104(9):1876–1889
29. Gillespie DT (1977) Exact stochastic simulation of coupled chemical reactions. J Phys Chem 81(25):2340–2361
30. Gillespie DT (2001) Approximate accelerated stochastic simulation of chemically reacting systems. J Chem Phys 115(4):1716–1733
31. Goodwin BC (1965) Oscillatory behavior in enzymatic control processes. Adv Enzyme Regul 3:425–438
32. Hattne J, Fange D, Elf J (2005) Stochastic reaction-diffusion simulation with MesoRD. Bioinformatics 21:2923–2924
33. Hedley W, Nelson MR, Bullivant DP, Nielsen PF (2001) A short introduction to CellML. Philos Trans R Soc Lond A 359:1073–1089
34. Isaacson SA (2009) The reaction-diffusion master equation as an asymptotic approximation of diffusion to a small target. SIAM J Appl Math 70:77–111
35. Jahnke T, Galan S (2008) Solving chemical master equations by an adaptive wavelet method. In: Simos TE, Psihoyios G, Tsitouras C (eds) Numerical analysis and applied mathematics: international conference on numerical analysis and applied mathematics 2008, vol. 1048 of AIP Conference Proceedings, Psalidi, Kos, Greece, 16–20 September 2008, pp. 290–293
36. Kerr RA, Bartol TM, Kaminski B et al (2008) Fast Monte Carlo simulation methods for biological reaction-diffusion systems in solution and on surfaces. SIAM J Sci Comput 30:3126
37. Kloeden PE, Platen E (1992) Numerical solution of stochastic differential equations. Springer-Verlag, Berlin
38. Kurtz TG (1972) The relationship between stochastic and deterministic models for chemical reactions. J Chem Phys 57(7):2976–2978
39. Kusumi A, Koyama-Honda I, Suzuki K (2004) Molecular dynamics and interactions for creation of stimulation-induced stabilized rafts from small unstable steady-state rafts. Traffic 5:213–230
40. Lampoudi S, Gillespie DT, Petzold L (2009) The multinomial simulation algorithm for discrete stochastic simulation of reaction-diffusion systems. J Chem Phys 130:094104
41. Leier A, Marquez-Lago TT (2011) Correction factors for boundary diffusion and bimolecular reactions in reaction-diffusion master equations. To be submitted
42. Loew LM, Schaff JC (2001) The virtual cell: a software environment for computational cell biology. Trends Biotechnol 19(10):401–406
43. MacNamara S, Bersani AM, Burrage K, Sidje RB (2008) Stochastic chemical kinetics and the total quasi-steady-state assumption: application to the stochastic simulation algorithm and chemical master equation. J Chem Phys 129(9):095105
44. MacNamara S, Burrage K, Sidje RB (2008) Multiscale modeling of chemical kinetics via the master equation. SIAM J Multiscale Model Simul 6(4):1146–1168
45. Marquez-Lago TT, Burrage K (2007) Binomial tau-leap spatial stochastic simulation algorithm for applications in chemical kinetics. J Chem Phys 127:104101
46. Marquez-Lago TT, Leier A, Burrage K (2010) Probability distributed time delays: integrating spatial effects into temporal models. BMC Syst Biol 4:19
47. Marsh BJ (2006) Toward a 'visible cell'... and beyond. Aust Biochemist 37:5–10
48. Mélykúti B, Burrage K, Zygalakis KC (2010) Fast stochastic simulation of biochemical reaction systems by alternative formulations of the chemical Langevin equation. J Chem Phys 132:1
49. Morton-Firth CJ, Bray D (1998) Predicting temporal fluctuations in an intracellular signalling pathway. J Theor Biol 192:117–128
50. Murase K, Fujiwara T, Umemura TY (2004) Ultrafine membrane compartments for molecular diffusion as revealed by single molecule techniques. Biophys J 86:4075–4093
51. Nicolau Jr, DV, Burrage K, Parton RG et al (2006) Identifying optimal lipid raft characteristics required to promote nanoscale protein-protein interactions on the plasma membrane. Mol Cell Biol 26(1):313–323
52. Nicolau Jr, DV, Hancock JF, Burrage K (2007) Sources of anomalous diffusion on cell membranes: a Monte Carlo study. Biophys J 92:1975–1987

53. Nicolau Jr, DV, Burrage K (2008) Stochastic simulation of chemical reactions in spatially complex media. Comput Math Appl 55(5):1007–1018
54. Oppelstrup T, Bulatov VV, Donev A et al (2006) First-passage kinetic Monte Carlo method. Phys Rev Lett 97:230602
55. Peleš S, Munsky B, Khammash M (2006) Reduction and solution of the chemical master equation using time scale separation and finite state projection. J Chem Phys 125:204104–1–13
56. Plimpton SJ, Slepoy A (2003) ChemCell: a particle-based model of protein chemistry and diffusion in microbial cells. Sandia National Laboratory Technical Report 2003, Albuquerque, NM
57. Plimpton SJ, Slepoy A (2005) Microbial cell modeling via reacting diffusing particles. J Physiol 16:305
58. Prior IA, Muncke C, Parton RG et al (2003) Direct visualization of Ras proteins in spatially distinct cell surface microdomains. J Cell Biol 160:165–170
59. Ridgway D, Broderick G, Lopez-Campistrous A et al (2008) Coarse-grained molecular simulation of diffusion and reaction kinetics in a crowded virtual cytoplasm. Biophys J 94:3748–3759
60. Rodríguez JV, Kaandorp JA, Dobrzyński M, Blom JG (2006) Spatial stochastic modelling of the phosphoenolpyruvate-dependent phosphotransferase (PTS) pathway in Escherichia coli. Bioinformatics 22:1895–1901
61. Sanford C, Yip MLK, White C, Parkinson J (2006) Cell++–simulating biochemical pathways. Bioinformatics 22:2918–2925
62. Séguis J-C, Burrage K, Erban R, Kay D (2010) Efficient numerical model for lipid rafts and protein interactions on a cell membrane, in preparation
63. Sharma P, Varma R, Sarasij RC et al (2004) Nanoscale organization of multiple GPI-anchored proteins in living cell membranes. Cell 116:577–589
64. Simons K, Toomre D (2000) Lipid rafts and signal transduction. Nat Rev Mol Cell Biol 1:31–39
65. Singer SJ, Nicolson GL (1972) The fluid mosaic model of the structure of cell membranes. Science 175:720–731
66. Stiles JR, Bartol TM (2001) Monte Carlo methods for simulating realistic synaptic microphysiology using MCell. CRC Press, Boca Raton, FL
67. Stundzia AB, Lumsden CJ (1996) Stochastic simulation of coupled reaction-diffusion processes. J Comp Physiol 127:196–207
68. Takahashi K, Ishikawa N, Sadamoto Y et al (2003) E-Cell 2: multi-platform E-Cell simulation system. Bioinformatics 19:1727–1729
69. Takahashi K, Kaizu K, Hu B, Tomita M (2004) A multi-algorithm, multi-timescale method for cell simulation. Bioinformatics 20:538–546
70. Takahashi K, Arjunan SNV, Tomita M (2005) Space in systems biology of signaling pathways – towards intracellular molecular crowding in silico. FEBS Lett 579:1783–1788
71. Takahashi K, Tănase-Nicola S, ten Wolde PR (2010) Spatio-temporal correlations can drastically change the response of a MAPK pathway. PNAS 107(6):2473–2478
72. Tian T, Burrage K (2004) Binomial leap methods for simulating stochastic chemical kinetics. J Chem Phys 121:10356–10364
73. Tian T, Harding A, Westbury E, Hancock J (2007) Plasma membrane nano-switches generate robust high-fidelity Ras signal transduction. Nat Cell Biol 9:905–914
74. Tomita M, Hashimoto K, Takahashi K et al (1999) E-CELL: software environment for whole-cell simulation. Bioinformatics 15:72–84
75. Turner TE, Schnell S, Burrage K (2004) Stochastic approaches for modelling in vivo reactions. Comput Biol Chem 28:165–178
76. van Zon JS, ten Wolde PR (2005) Green's-function reaction dynamics: a particle-based approach for simulating biochemical networks in time and space. J Chem Phys 123: 1–16
77. van Zon JS, ten Wolde PR (2005) Simulating biochemical networks at the particle level and in time and space: Green's function reaction dynamics. Phys Rev Lett 94:128103
78. Wils S, De Schutter E (2009) STEPS: modeling and simulating complex reaction-diffusion systems with Python. Front Neuroinform 3:15

Chapter 3
Graph-Theoretic Analysis of Multistability and Monotonicity for Biochemical Reaction Networks

Gheorghe Craciun, Casian Pantea, and Eduardo D. Sontag

Abstract Mathematical models of biochemical reaction networks are usually high dimensional, nonlinear, and have many unknown parameters, such as reaction rate constants, or unspecified types of chemical kinetics (such as mass-action, Michaelis-Menten, or Hill kinetics). On the other hand, important properties of these dynamical systems are often determined by the network structure, and do not depend on the unknown parameter values or kinetics. For example, some reaction networks may give rise to multiple equilibria (i.e., they may function as a biochemical switch) while other networks have unique equilibria for any parameter values. Or, some reaction networks may give rise to monotone systems, which renders their dynamics especially stable. We describe how the species-reaction graph (SR graph) can be used to analyze both multistability and monotonicity of networks.

Keywords Biochemical reaction networks · Multistability · Monotonicity · SR graph

Introduction

There is great interest in methods that draw conclusions about the dynamical properties of a chemical reaction network based only on the network structure, i.e., with limited or absent knowledge about many kinetic details [7, 26]. Here we will concentrate on the properties of *multistability* and *monotonicity*.

Multistability refers to the capacity of a biochemical system to operate at several discrete, alternative steady-states, and plays an important role in cell signaling, division, and differentiation [2, 23, 29].

G. Craciun (✉)
Department of Mathematics and Department of Biomolecular Chemistry,
University of Wisconsin
e-mail: craciun@math.wisc.edu

H. Koeppl et al. (eds.), *Design and Analysis of Biomolecular Circuits: Engineering Approaches to Systems and Synthetic Biology*, DOI 10.1007/978-1-4419-6766-4_3,

Monotone systems display well-ordered behavior that excludes the possibility for chaotic dynamics [4, 20, 27]. Moreover, perturbations of such systems have unambiguous global effects and a predictability characteristic that confers robustness and adaptability [21].

In this chapter we describe some of the main results on the use of the SR graph of a reaction network to analyze its multistability and monotonicity properties, as described especially in [9] and [5] , respectively. Our focus will *not* be on presenting the most powerful results in full generality (for these the reader should consult [5,9], and also [6, 10]). Instead, we will concentrate on simpler versions of these results, and will especially focus on pointing out how these results can be formulated in an unified language based on the notion of SR graph.

Definitions and Notation

Dynamical Systems Derived from Chemical Reaction Networks

A chemical reaction system in which n reactants participate in m reactions has dynamics governed by the system of ordinary differential equations

$$\frac{dx}{dt} = Sv(x) \tag{3.1}$$

where $x = (x_1, \ldots, x_n)^t$ is the nonnegative n-vector of species concentrations, $v = (v_1, \ldots, v_m)^t$ is the m-vector of reaction rates, and S is the $n \times m$ stoichiometric matrix.

Arbitrary orderings can be chosen on the sets of substrates and reactions. Further, S is only defined up to an arbitrary re-signing of its columns, equivalent to a switching of the left and right-hand sides of a reaction. The equation (3.1) defines a dynamical system on the nonnegative orthant of $\mathbb{R}^n$. If we also assume that all species may have some inflow (which is allowed to be zero) and some outflow which increases strictly with concentration, we obtain the related system

$$\frac{dx}{dt} = F + Sv(x) - Q(x) \tag{3.2}$$

Here F is a constant nonnegative vector representing the inflow, and the diagonal function $Q(x) = (Q_1(x_1), \ldots, Q_n(x_n))^t$ represents the outflow or degradation, and we assume that $\frac{\partial Q_i}{\partial x_i} > 0$ for each i.

For example, for the reaction network

$$X_1 + X_2 \rightleftharpoons X_3, \; 2X_1 \rightleftharpoons X_2, \; X_2 + X_3 \rightleftharpoons X_4 \tag{3.3}$$

we can choose

$$S = \begin{pmatrix} -1 & -2 & 0 \\ -1 & 1 & -1 \\ 1 & 0 & -1 \\ 0 & 0 & 1 \end{pmatrix} \tag{3.4}$$

where each column of S corresponds to one reaction in the network. Sometimes the inflow and outflow terms F and $Q(x)$ are included in the reaction network as 'inflow reactions' $X_i \to 0$ and 'outflow reactions' $0 \to X_j$. Here we choose to associate to a reaction network (such as (3.3)) either the closed system (3.1), or the open system (3.2). Note that the dynamical properties of these two types of systems may be very different from each other, and some theorems might apply to only one or the other of them.

We assume that for each reversible reaction its reaction rate v_i can be decomposed as

$$v_i(x) = v_i^+(x) - v_i^-(x),$$

where v_i^+ is the rate of the forward reaction, and v_i^- is the rate of the reverse reaction.

In biochemical applications, the most common types of reaction rates are *mass-action, Michaelis-Menten*, or *Hill kinetics*. For example, for the reaction $X_1 + X_2 \to X_3$, we could have

$$v_1^+(x) = k_1 x_1 x_2$$

or

$$v_1^+(x) = \frac{k_1 x_1 x_2}{1 + k_2 x_1 x_2}$$

or

$$v_1^+(x) = \frac{k_1 x_1 x_2^2}{k_2 + x_2^2}.$$

for some positive constants k_1 and k_2.

Given a reaction network we define its *SR graph* as follows. The SR graph is a bipartite undirected graph, where the nodes are partitioned into *species nodes* and *reaction nodes*. We draw an *edge* from a species node to a reaction node if that species appears in the reaction, i.e., we draw an edge from species node i to reaction node j if the s_{ij} entry of the stoichiometric matrix S is not zero. Moreover, if $s_{ij} > 0$ we say that it is a *positive edge* (and will draw it with a *solid line*), and if $s_{ij} < 0$ we say that it is a *negative edge* (and will draw it with a *dashed line*). Finally, if the stoichiometric coefficient of a species within a reaction is two or more, then we label the corresponding edge with this stoichiometric coefficient (so if an edge does not have a numeric label, it will follow that the corresponding stoichiometric coefficient is 1). The SR graph of reaction network (3.3) is shown in Fig. 3.1.

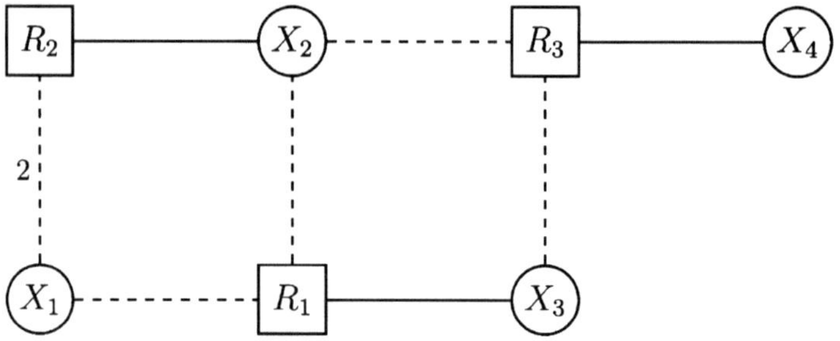

Fig. 3.1 The SR graph of reaction network (3.3). Positive edges are shown as *solid lines*, and negative edges are shown as *dashed lines*. Note that the graph contains three cycles, and any two of them have S-to-R intersection. Also, all cycles are o-cycles

Note that the SR graph in Fig. 3.1 contains several cycles.[1] We will show that multistability and monotonicity of a network is strongly related to the *types* of cycles present in its SR graph. For this we need to be able to distinguish among several types of cycles.

Consider a cycle that has p edges, and q of them are negative edges. We say that this cycle is an *e-cycle* if $q \equiv \frac{p}{2} \pmod 2$, i.e., the number of negative edges along the cycle has the same parity as the total number of edges along the cycle divided by 2. (Note that the total number of edges along any cycle must be even, because the SR graph is a bipartite graph.) Otherwise, i.e., if the number of negative edges has different parity from the total number of edges divided by 2, we say that the cycle is an *o-cycle*. For example, the cycle $R_2 - X_2 - R_3 - X_3 - R_1 - X_1 - R_2$ in Fig. 3.1 is an o-cycle, since it has $p = 6$ edges and $q = 4$ negative edges, and the numbers q and $\frac{p}{2}$ have different parities.[2]

Another relevant type of cycle is called *s-cycle*. A cycle $\mathcal{C}$ is called an s-cycle if we have

$$\prod_{i=1}^{p/2} \sigma_{2i-1} = \prod_{i=1}^{p/2} \sigma_{2i},$$

where p is the number of edges of $\mathcal{C}$, and $\sigma_1, \sigma_2, \ldots, \sigma_p$ are the stoichiometric coefficients of the edges of $\mathcal{C}$, in the order in which they occur along $\mathcal{C}$ (it is easy to see that it does not matter where we start along $\mathcal{C}$). In other words, $\mathcal{C}$ is an s-cycle if the two possible ways of multiplying the stoichiometric coefficients of every other edge of $\mathcal{C}$ give rise to the same result. Obviously, if all the stoichiometric coefficients along a cycle are 1, then that cycle is an s-cycle, and if exactly one stoichiometric

[1] In [5] cycles are called 'loops'.

[2] The original definition of e-cycles and o-cycles in [11] describes these types of cycles in terms of 'c-pairs': e-cycles have an even number of c-pairs and o-cycles have an odd number of c-pairs. The two definitions are equivalent for networks that do not have one-step catalysis, which are our main focus here. Compare also with Lemma 4.4 in [5].

coefficient along a cycle is $\neq 1$, then that cycle is not an s-cycle. For example, for the SR graph in Fig. 3.1, the cycle $X_2 - R_3 - X_3 - R_1 - X_2$ is an s-cycle, and the cycles $R_2 - X_2 - R_1 - X_1 - R_2$ and $R_2 - X_2 - R_3 - X_3 - R_1 - X_1 - R_2$ are not s-cycles.

Sometimes not only the types of cycles are important, but also the way cycles intersect within the SR graph. We say that two cycles *have an S-to-R intersection* if the connected components of their intersection are paths that go from a species node to a reaction node (and not from a species node to another species node, or from a reaction node to another reaction node). For example, consider the cycles $R_2 - X_2 - R_3 - X_3 - R_1 - X_1 - R_2$ and $R_2 - X_2 - R_1 - X_1 - R_2$ in Fig. 3.1. Their intersection has a single connected component, which is the path $R_1 - X_1 - R_2 - X_2$. Therefore, these two cycles have an S-to-R intersection.

Consider some closed pointed convex cone $K \subset \mathbb{R}^n$. We say that an autonomous dynamical system

$$\dot{x} = f(x) \tag{3.5}$$

is *monotone with respect to* K if for any two solutions $x_1(t)$ and $x_2(t)$ of (3.5), such that $x_1(0) - x_2(0) \in K$, it follows that $x_1(t) - x_2(t) \in K$ for all $t > 0$. (Note that we assume that solutions $x(t)$ exist for all times $t > 0$.)

A property relevant to monotonicity is *persistence*. A dynamical system defined on a domain contained within the nonnegative orthant of $\mathbb{R}^n$ is called *persistent* if any trajectory with positive initial condition does not have any ω-limit points on the boundary of the nonnegative orthant. In other words, the system is persistent if for any solution $x(t)$ with positive initial condition such that $x(t_n) \to L$ for some sequence $t_n \to \infty$, it follows that all the coordinates of L are positive.

The Main Results

Throughout this chapter we assume that the following properties are satisfied by the reaction network and its reaction rate functions:

Assumption 1. The reaction network does not have one-step catalysis, i.e., if a species appears on one side of a reaction then it does not appear on the other side of that reaction.

Assumption 2. For each irreversible reaction (and also separately for the forward and reverse reactions of a reversible reaction), its reaction rate depends only on the concentrations of the reactants, which are the species that are being consumed by the reaction. Moreover, the partial derivatives of the rate function with respect to the concentrations of the reactants are nonnegative.

Neither one of these two assumptions are very restrictive; on the other hand, neither one of them is truly necessary for analyzing multistability (see [10] for details).

In this section we formulate two theorems that use the SR graph of a reaction network to analyze its multistability and monotonicity properties.

Theorem 3.1 (Banaji and Craciun [9]). *Consider a reaction network such that its SR graph satisfies the following two conditions:*

(1) all cycles are o-cycles or s-cycles (or both),
(2) no two e-cycles have an S-to-R intersection.

Then the system (3.2) does not have multiple positive equilibria, and the system (3.1) does not have multiple positive nondegenerate equilibria within any affine invariant subspace.

Note that, in the presence of any conservation laws, the *relevant* multistability question is *not* whether there exists a unique equilibrium, but whether there exists a unique equilibrium within any affine invariant subspace, since for well-behaved systems we expect that one equilibrium should exist in every such invariant subspace (see also [14, 18]). Note also that for the system (3.2) there can be no conservation law, due to the presence of nondegenerate outflow or degradation terms.

Theorem 3.1 does apply for reaction network (3.3) because all cycles in Fig. 3.1 are o-cycles. For more examples see [9, 10].

If there exist conserved quantities, additional analysis is needed to rule out degenerate equilibria. For mass-action systems, conditions that exclude the possibility of degenerate equilibria are described in [15]. For non-mass-action systems such conditions are described in [17, 19].

Consider now the system

$$\frac{dr}{dt} = v(x_0 + Sr(t)), \tag{3.6}$$

where r_j is called the *extent* of the jth reaction, $j = 1, \ldots, m$. The following theorem allows us to analyze the monotonicity of this system (i.e., monotonicity in reaction coordinates [5]), and also provides information on the dynamics of the related system (3.1).

Theorem 3.2 (Angeli, DeLeenheer and Sontag [5]). *Consider a reaction network such that its SR graph satisfies the following two conditions:*

(1) each species node is adjacent to at most two edges,
(2) each cycle is an e-cycle.

Assume in addition that all stoichiometric compatibility classes are compact sets, that all reaction rates vanish if the concentrations of some of their reactants are zero, and that all reaction rates are strictly increasing with respect to the concentrations of their reactant species. Then the system (3.6) is monotone with respect to an order induced by some orthant cone.

Assume moreover that the system (3.1) is persistent, and all reactions are reversible. Then almost all positive solutions of (3.1) converge to the set of equilibria, i.e., the measure of the set of possibly non-converging initial conditions is zero.[3]

[3] Often much more can be said, e.g., under some additional assumptions it follows that all positive solutions converge to an equilibrium. See Theorem 2 in [5] for details.

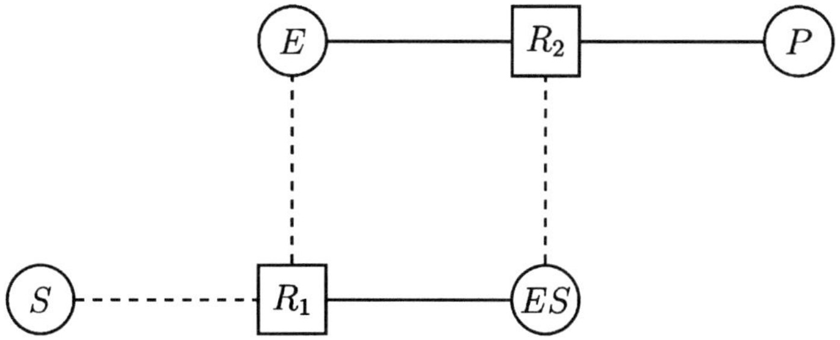

Fig. 3.2 The SR graph of reaction network (3.7). Note that the graph contains one cycle, and it is an e-cycle. Also, note that each species node is adjacent to at most two reaction nodes

A proof of Theorem 3.2 follows from Proposition 4.5, Corollary 1, and Theorem 2 in [5]. In order to apply Corollary 1 in [5] note that, if all reactions are reversible, then the connectivity of the 'directed SR graph' is the same as for the SR graph, and if the SR graph has several connected components then they will generate uncoupled subsystems.[4]

For example, consider the reaction network

$$E + S \rightleftharpoons ES \rightleftharpoons E + P. \tag{3.7}$$

A version of this network was analyzed in detail in [5]. Theorem 3.2 does apply to this network, since its SR graph, shown in Fig. 3.2, has the property that its only cycle is an e-cycle, and each species node is adjacent to at most two edges. Moreover, the network (3.7) is persistent (see [3]), and all its reactions are reversible.

In general, if there is only one cycle in an SR graph then Theorem 3.1 applies if this cycle is o- or s-cycle (or both), while Theorem 3.2 applies if the cycle is an e-cycle and no species node is adjacent to more than two edges. Therefore Theorem 3.1 also applies to network (3.7).

Note also that, if Theorem 3.2 does apply, and in particular if in the SR graph each species has at most two adjacent edges, then no two cycles can have an S-to-R intersection (because if a connected component of the intersection of two cycles is an S-to-R path, then there must be at least three adjacent edges to the species node at one end of the path). Therefore, if Theorem 3.2 does apply, and in addition all stoichiometric coefficients are 1, then the hypotheses (1) and (2) of Theorem 3.1 also hold.

Finally, consider the reaction network

$$A + B \rightleftharpoons P, \;\; B + C \rightleftharpoons Q, \;\; C \rightleftharpoons 2A, \tag{3.8}$$

which was also analyzed in [12] under the assumption of mass-action kinetics.

[4] On the other hand, note that the notion of 'directed SR graph' in [5] is different from the notion of 'DSR graph' (also called directed SR graph) introduced in [10].

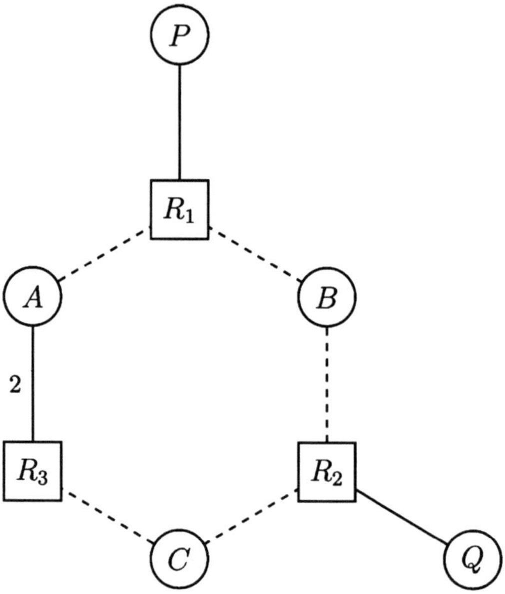

Fig. 3.3 The SR graph of reaction network (3.8). The graph contains only one cycle, which is an e-cycle, and each species node is adjacent to at most two reaction nodes. Moreover, the cycle is *not* an s-cycle

The network (3.8) is persistent because it admits a positive P-semiflow, and every minimal siphon contains the support of a P-semiflow (see [3] for details; see also [1]). Since the only cycle in its SR graph is an e-cycle, and no species node is adjacent to more than two edges, it follows that Theorem 3.2 does apply for this network (see Fig. 3.3). Note that the cycle $A - R_1 - B - R_2 - C - R_3 - A$ in Fig. 3.3 is neither an o-cycle nor an s-cycle, so Theorem 3.1 does not apply. On the other hand, if the kinetics of this network is mass-action, then deficiency theory [18] guarantees that there is an unique equilibrium in each stoichiometric compatibility class, and there also exists a globally defined strict Lyapunov function. This, together with persistence, guarantees global convergence of all positive trajectories within a stoichiometric compatibility class to the unique equilibrium in that class [28].

Discussion

The SR graph was first introduced in [11] for the analysis of mass-action systems, inspired by the SCL graph of Schlosser and Feinberg [24,25]; see also [13,16]. The case of networks that may contain one-step catalysis is discussed in [11,13,16] for mass-action kinetics, and in [10] for general kinetics.

The free software package `BioNetX` provides algorithms for examining dynamical properties of biochemical reaction networks [22]. In particular, this software computes the SR graph of a network, and verifies the conditions (1) and (2) from Theorem 3.1.

Monotonicity was also considered in [30], were it was treated in an algebraic fashion. In [8] conditions are determined in order to characterize the set of cones and associated partial orders which make a certain reaction monotone, and it is established that, under some minor assumptions, monotonicity of a network with respect to a given partial order is equivalent to asking that each individual reaction be monotone with respect to that same order. This result is also independent of reaction kinetics.

Acknowledgments We thank David Angeli, Murad Banaji, and Patrick DeLeenheer for very useful discussions about various topics described in this chapter. The authors acknowledge support from NIH grant R01GM086881, and from the DOE BACTER Institute.

References

1. Anderson DF (2008) Global asymptotic stability for a class of nonlinear chemical equations. SIAM J Appl Math 68:1464–1476
2. Angeli D, Ferrell JE, Sontag ED (2004) Detection of multistability, bifurcations, and hysteresis in a large class of biological positive-feedback systems. Proc Natl Acad Sci USA 101:1822–1827
3. Angeli D, DeLeenheer P, Sontag ED (2007) A Petri net approach to the study of persistence in chemical reaction networks. Math Biosci 210:598–618
4. Angeli D, Hirsch MW, Sontag ED (2009) Attractors in coherent systems of differential equations. J Differ Equ 246(8):3058–3076
5. Angeli D, DeLeenheer P, Sontag ED (2010) Graph-theoretic characterizations of monotonicity of chemical networks in reaction coordinates. J Math Biol 61(4):581–616
6. Angeli D, Sontag ED (2008) Translation-invariant monotone systems, and a global convergence result for enzymatic futile cycles. Nonlinear Anal B 9:128–140
7. Bailey JE (2001) Complex biology with no parameters. Nat Biotechnol 19:503–504
8. Banaji M (2009) Monotonicity in chemical reaction systems. Dyn Syst 24:1–30
9. Banaji M, Craciun G (2010) Graph-theoretic criteria for injectivity and unique equilibria in general chemical reaction systems. Adv Appl Math 44:168–184
10. Banaji M, Craciun G (2009) Graph-theoretic approaches to injectivity and multiple equilibria in systems of interacting elements. Commun Math Sci 7:867–900
11. Craciun G (2002) Systems of nonlinear equations deriving from complex chemical reaction networks. Ph.D. thesis, Department of Mathematics, The Ohio State University, Columbus
12. Craciun G, Feinberg M (2005) Multiple equilibria in complex chemical reaction networks: I. The injectivity property. SIAM J Appl Math 65:1526–1546
13. Craciun G, Feinberg M (2006) Multiple equilibria in complex chemical reaction networks: II. The Species-Reactions graph. SIAM J Appl Math 66:1321–1338
14. Craciun G, Feinberg M (2006) Multiple equilibria in complex chemical reaction networks: extensions to entrapped species models. IEE Proc Syst Biol 153(4):179–186
15. Craciun G, Feinberg M (2010) Multiple equilibria in complex chemical reaction networks: semiopen mass action systems. SIAM J Appl Math 70:1859–1877
16. Craciun G, Tang Y, Feinberg M (2006) Understanding bistability in complex enzyme-driven reaction networks. Proc Natl Acad Sci USA 103:8697–98702
17. Craciun G, Helton JW, Williams RJ (2008) Homotopy methods for counting reaction network equilibria. Math Biosci 216(2):140–149
18. Feinberg M (1979) Lectures on chemical reaction networks. Notes of lectures given at the Mathematics Research Center of the University of Wisconsin. http://www.che.eng.ohio-state.edu/~FEINBERG/LecturesOnReactionNetworks

19. Helton JW, Klep I, Gomez R (2009) Determinant expansions of signed matrices and of certain jacobians. SIAM J Matrix Anal Appl 31(2):732–754
20. Hirsch MW, Smith HL (2005) Monotone dynamical systems. In: Canada A, Drabek P, Fonda A (eds) Handbook of differential equations, ordinary differential equations, vol 2. Elsevier, Amsterdam
21. Ma'ayan A, Lipshtat A, Iyengar R, Sontag ED (2010) Proximity of intracellular regulatory networks to monotone systems. IET Syst Biol 2:103–112
22. Pantea C, Craciun G (2010) Computational methods for analyzing bistability in biochemical reaction networks. Proceedings of the IEEE International Symposium on Circuits and Systems (ISCAS) 549–552
23. Pomerening JR, Sontag ED, Ferrell JE (2003) Building a cell cycle oscillator: hysteresis and bistability in the activation of Cdc2. Nat Cell Biol 5:346–351
24. Schlosser P (1988) A graphical determination of the possibility of multiple steady states in complex isothermal CFSTRs. Ph.D. thesis, Department of Chemical Engineering, University of Rochester, Rochester
25. Schlosser P, Feinberg M (1994) A theory of multiple steady states in isothermal homogeneous CFSTRs with many reactions. Chem Eng Sci 49:1749–1767
26. Shinar G, Feinberg M (2010) Structural sources of robustness in biochemical reaction networks. Sci 327:1389–1391
27. Smith HL (1995) Monotone dynamical systems: an introduction to the theory of competitive and cooperative systems. Mathematical surveys and monographs, vol 41. AMS, Providence
28. Sontag ED (2001) Structure and stability of certain chemical networks and applications to the kinetic proofreading model of T-Cell receptor signal transduction. IEEE Trans Automat Contr 46:1028–1047
29. Tyson JJ, Chen KC, Novak B (2003) Sniffers, buzzers, toggles and blinkers: dynamics of regulatory and signaling pathways in the cell. Curr Opin Cell Biol 15:221–231
30. Volpert AI, Volpert VA, Volpert VA (1994) Traveling wave solutions of parabolic systems. Mathematical monographs, vol 140. AMS, Providence

Chapter 4
From Structure to Dynamics in Biological Networks

Murad Banaji

Abstract Biological systems often display behaviour that is robust to considerable perturbation. In fact, experimental and computational work suggests that some behaviours are 'structural' in that they occur in all systems with particular qualitative features. In this chapter, some relationships between structure and dynamics in biological networks are explored. The emphasis is on chemical reaction networks, regarded as special cases of more general classes of dynamical systems termed interaction networks. The mathematical approaches described involve relating patterns in the Jacobian matrix to the dynamics of a system. Via a series of examples, it is shown how simple computations on matrices and related graphs can lead to strong conclusions about allowed behaviours.

Keywords Biological networks · Chemical reaction networks · Jacobian matrix · SR graph · Stability

Introduction

Crucial to the functioning of biological entities is the fact that they display behaviour robust to internal noise and environmental perturbation. Providing mathematically precise definitions and analyses of this ability to function reliably is, however, not easy. Progress in this direction must involve: (1) Listing qualitative dynamical behaviours which occur in biological systems, and which may be functionally significant; (2) Providing some mathematical meaning to the notion of 'structure' in biological systems, and (3) Elucidating the relationships between structure and behaviour.

M. Banaji (✉)
Department of Mathematics, University of Portsmouth, Lion Gate Building, Lion Terrace, Portsmouth, Hampshire PO1 3HF, UK
e-mail: murad.banaji@port.ac.uk

H. Koeppl et al. (eds.), *Design and Analysis of Biomolecular Circuits: Engineering Approaches to Systems and Synthetic Biology*, DOI 10.1007/978-1-4419-6766-4_4,

Qualitative behaviours include, amongst others, multistability, oscillation and chaos. Asking when these might occur or be forbidden in a biological system is not merely of mathematical interest. There are many examples in the literature of biological systems which permit multistability, with different attracting states having different biological significance. Sometimes a great many attracting states may be allowed [31]. Similarly, a number of important biological oscillators have been documented [27], and there is some evidence for subtle behaviours such as frequency encoding of information [12]. There is also a significant literature on chaos in experimental and *in silico* biological systems. On the other hand, there are also many biological systems which appear to allow only simple behaviour, such as global attraction of all trajectories to a unique steady state.

Here, the emphasis is on defining easily computable conditions guaranteeing that certain behaviours will *not* occur. This approach also provides insight into instability of various kinds. For example, proving that some class of systems can have no more than one equilibrium rules out saddle-node bifurcations in these systems; this, in turn, may guide the search for these bifurcations in systems where they are not structurally forbidden.

The starting point for any rigorous investigation of the structure-dynamics relationship in biological systems must be an attempt to define 'structure' in ways which are both natural, and allow tools from analysis, algebra, and combinatorics to be brought into play. In this context, the characterisation of biological systems as **networks** has proved particularly useful. After introducing some broad ideas, examples of particular strands of theory and model classes to which this theory can be applied will be presented. The common thread is that all the approaches involve examining the Jacobian matrix to make claims about asymptotic behaviour in ordinary differential equation (O.D.E.) models of biological systems.

The reader will notice a particular emphasis on chemical reaction networks (CRNs). This springs not only from their central role in biology, but also because theory developed initially for CRNs can provide insight into processes which are not necessarily chemical in nature. This is true both an abstract level: theory on multistationarity developed for CRNs was later generalised to much wider contexts; but also more concretely: quantities such as membrane potentials may behave mathematically like chemicals, being increased and decreased by chemical processes, except without certain formal requirements such as that they must remain nonnegative, or obey stoichiometric rules.

Qualitative Models from Biology

Consider some biological system, and assume that the state of the system is defined by a set of n quantities whose allowed values define a state-space $X \subseteq \mathbb{R}^n$. Assume that the evolution of the system can be modelled by the autonomous O.D.E.

$$\dot{x} = F(x), \tag{4.1}$$

where $x=(x_1,\ldots,x_n)^T \in X$, $F:X\to\mathbb{R}^n$, and $F\in C^1(X)$, the set of all continuously differentiable functions on X. Only this finite dimensional, continuous-time, autonomous case is discussed here, although many of the fundamental ideas can be extended to other model classes.

Differentiability of F ensures the existence and uniqueness of solutions of (4.1). Invariance of X is an *a priori* requirement, giving rise to certain restrictions on F: for example, if $x_i \geq 0$ (as, say, if x_i is a chemical concentration), then in any reasonable model, we must have that $x_i = 0$ implies $\dot{x}_i \geq 0$. Beyond such automatic restrictions, we can generally expect biological or physical constraints to provide further information about F. Characterising these constraints, and thus the set of allowed models of the system, is a task for the biological modeller. Assume that this task has been carried out and $\mathcal{F} \subset C^1(X)$ has been chosen as the set of all models which may reasonably describe the biological system. $\mathcal{F}$ can be termed a 'qualitative model' of the biological system.

It is often the case that each $F \in \mathcal{F}$ has restrictions on its Jacobian $\mathrm{d}F$. A well-known case is when entries in $\mathrm{d}F$ are of fixed sign over all of X (or sometimes, the interior of X). In this situation, the basic goal becomes to analyse how the sign-pattern of $\mathrm{d}F$, often best studied via the associated signed digraph or 'I-graph' (see Fig. 4.1), restricts the behaviour of the system. The conjectures and results of Kaufman, Soulé and Thomas discussed in [21] provide examples of work in this area. Some results, for example on the relationship between circuits in signed digraphs and periodic attractors, have only recently been proved [2], while others are still open.

The ideas to be presented involve factorisations of the Jacobian, and can be seen as generalisations of work on systems with signed Jacobian. In general, there may be an arbitrary number of matrices in these factorisations, and these may be constant, or have entries of fixed sign. Moreover, the factors may have relationships to each other. Similar to the signed Jacobian case, the techniques aim to make claims about asymptotic dynamics of the systems. It is worth noting that Jacobian

$$\begin{pmatrix} -a & -b & 0 \\ c & -d & e \\ f & 0 & g \end{pmatrix}$$

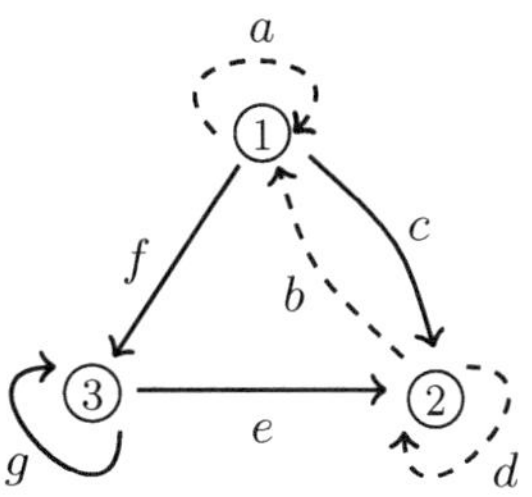

Fig. 4.1 The sign pattern of an $n \times n$ matrix corresponds to a signed digraph on n vertices. *Left*. A 3×3 matrix. Assume that $a,b,c,d,e,f,g > 0$. *Right*. The corresponding signed digraph on three vertices, sometimes termed an interaction graph or I-graph, with vertices and edges labelled for clarity. Each edge corresponds to an entry in the matrix. Negative edges are *dashed lines*, while positive edges are *bold lines*. A directed edge from vertex j to vertex i corresponds to the ijth entry in the matrix

factorisations can also play a role in numerical exploration of complex biological networks for which only partial quantitative data is available [18]. The emphasis here is on analytical approaches, but it is likely that these can also be used to guide such numerical work.

Defining Structure in Biological Systems

Interaction Networks

It is useful to begin with the abstract notion of an interaction network. Consider models where there are 'species' whose evolution we are interested in, and where these species 'interact' with each other. The species may be chemical, biological, or indeed physical quantities. An interaction is any event which affects some nonempty subset of the species, and whose occurrence is affected by some subset of the species (possibly empty). Since a continuous-time description is assumed, events occur at a 'rate' which is a real number dependent on the 'amounts' (i.e., concentrations, populations, etc.) of the species involved.

Assume that there are n species in some biological system. A successful model will aim to describe the evolution of the amounts $x_1, \ldots, x_n$ of these species. Define $x = [x_1, \ldots, x_n]^T$. Assume that there are m interactions between the species, occuring at rates $v_1(x), \ldots, v_m(x)$, and define $v(x) = [v_1(x), \ldots, v_m(x)]^T$. To complete the model, we need to describe how $x_1, \ldots, x_n$ are affected by $v_1(x), \ldots, v_m(x)$. For this we need n interaction functions, where the ith interaction function, $f_i(v(x))$, is just the rate of change of species i as a function of the rates of interaction.

Knowledge of the interaction rates and interaction functions gives the evolution $\dot{x}_i = f_i(v(x)), i = 1, \ldots, n$, or more briefly,

$$\dot{x} = F(x) \equiv f(v(x)), \tag{4.2}$$

where $f(v(x)) = [f_1(v(x)), f_2(v(x)), \ldots, f_n(v(x))]^T$. Heuristically, this equation tells us that to understand the evolution of x_i at some moment in time, we need to know the rates of all the interactions which affect x_i, and how its evolution depends on these rates. We assume continuous differentiability of the scalar functions f_j, v_k.

Any dynamical system which can be constructed in this way can be termed an **interaction network**. As each x_i can be assumed to be a real number lying in some interval (perhaps unbounded), the state-space for an interaction network is quite naturally a **rectangular subset** of $\mathbb{R}^n$, that is, the product of n intervals, defining allowed values of each quantity. Each of these intervals may or may not be closed and/or bounded.

At its most abstract, an interaction network is simply a dynamical system where the right hand side can be written as a composition of two functions. In this formal sense, every dynamical system $\dot{x} = F(x)$ can be written as an interaction network $\dot{x} = F(id(x))$ or $\dot{x} = id(F(x))$, where id refers to the identity on $\mathbb{R}^n$. Perhaps surprisingly, this formal approach can indeed prove useful, and leads to generalisations of results on systems with signed Jacobian [4]. But the power of treating a system as an interaction network is generally most apparent when the decomposition is given by the biology itself.

Jacobian Factorisations and Generalised Graphs

Differentiating $F(x) = f(v(x))$ gives

$$\mathrm{d}F(x) = \mathrm{d}f(v(x))\mathrm{d}v(x).$$

This application of the chain rule from basic calculus tells us that for an interaction network, the Jacobian $\mathrm{d}F(x)$ has a certain factorisation at each point. For the techniques to be presented below, it is this factorisation which proves most important, rather than the original composition structure of F which gave rise to it. Moreover, as will be seen by example below, sometimes one or both of the factors can be further factorised. So, quite generally, assume that at each $x \in X$ we can write $\mathrm{d}F(x) = A^{(1)}(x)A^{(2)}(x)\cdots A^{(k)}(x)$, where the dimensions of matrices $A^{(i)}$ are such that they can be multiplied, and their product is square. Then, particular structures and relationships between the $A^{(i)}$ can be used to make a variety of claims about $\mathrm{d}F$, and thus about F.

Before proceeding to examples of such analysis it should be mentioned that matrices and lists of matrices have a variety of graphical representations. For example, associated with any real square matrix is a signed digraph discussed earlier. Similarly, corresponding to pairs of matrices whose products are square are bipartite graphs termed SR graphs and directed SR graphs (DSR graphs). A starting point for the definitions of these bipartite objects is the association of a simple, signed, labelled graph G_M, also termed an SR graph, with any matrix M (see Fig. 4.2).

$$M = \begin{pmatrix} -a & -b & 0 & c \\ -d & 0 & e & -f \\ 0 & g & h & j \end{pmatrix}$$

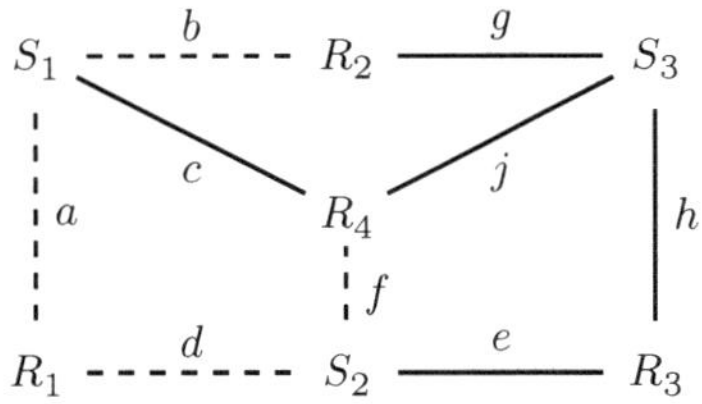

Fig. 4.2 Real rectangular matrices can be represented as signed, labelled, bipartite graphs. *Left.* A rectangular matrix M. Assume that $a, b, c, d, e, f, g, h, j > 0$. *Right.* The corresponding SR graph G_M. Vertices corresponding to rows of M have been labelled S_1, S_2, S_3 while vertices corresponding to columns have been labelled R_1, R_2, R_3, R_4. Each edge corresponds to an entry in the matrix. Positive edges are *bold lines* while negative edges are *dashed lines*

$$\mathrm{E+S0} \rightleftharpoons \mathrm{ES0} \rightarrow \mathrm{E+S1} \rightleftharpoons \mathrm{ES1} \rightarrow \mathrm{E+S2}$$
$$\mathrm{F+S2} \rightleftharpoons \mathrm{FS2} \rightarrow \mathrm{F+S1} \rightleftharpoons \mathrm{FS1} \rightarrow \mathrm{F+S0}$$

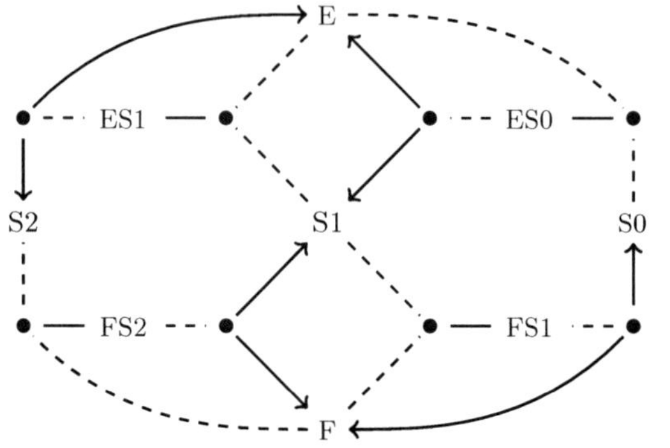

Fig. 4.3 A system of reactions representing the MAPK cascade derived from [24], and the DSR graph derived from factorisation of the associated Jacobian under weak assumptions on the kinetics. S0, S1 and S2 represent, respectively, the unphosphorylated, monophosphorylated, and biphosphorylated forms of MAPK. E represents MAPK kinase, and F represents MAP kinase phosphatase. ES0, ES1, FS1 and FS2 are complexes. In the DSR graph, the 9 reactants are labelled vertices while the 8 reactions are filled circles. Dashed lines represent negative edges, while bold lines represent positive edges. Edge-labels are all 1 and have been omitted. An interesting question is whether this network permits stable oscillation for any kinetics

Associating SR graphs with matrices leads naturally to more complex constructions. An example of a system of chemical reactions, the DSR graph derived from a Jacobian factorisation of the system, and the kind of question we would like to be able to answer using the DSR graph, are shown in Fig. 4.3. Note that the DSR graph encodes certain weak assumptions about the kinetics [6], and is *not* simply a convenient graphical representation of the reaction scheme. The DSR graph for a CRN has certain formal relationships to other graphical objects such as Petri nets, but some care is needed in interpreting this correspondence.

Going beyond SR/DSR graphs, more general graphs can be associated with a list of matrices which can be multiplied, and whose product is square. Graphical objects are useful not only because they provide a visualisation of the Jacobian structure, but also as formal objects to which the tools of computational graph theory can be applied to make claims about the underlying matrices. For example, it is well known that a square matrix is irreducible if and only if its associated digraph is strongly connected [9], and this is often the easiest test for irreducibility. A number of results which involve checking conditions on cycles in graphs to make claims about the associated matrices are known, and some of these will be described and applied below.

Qualitative Classes and Qualitative Rules

Qualitative biological and physical knowledge often translates into restrictions on entries in the Jacobian of the associated dynamical system. Consider some interaction network $\dot{x} = f(v(x))$ with Jacobian $\mathrm{d}F(x) = \mathrm{d}f(v(x))\mathrm{d}v(x)$. Define $\mathcal{I}_k \subseteq \{1,\ldots,n\}$ $(k = 1,\ldots,m)$ to be the indices of species which affect v_k, i.e., $i \in \mathcal{I}_k$ if and only if $\partial v_k/\partial x_i$ is not identically zero. Similarly, let $\mathcal{J}_k \subseteq \{1,\ldots,m\}$ $(k = 1,\ldots,n)$ be the indices of interactions whose rates affect species k, i.e., $i \in \mathcal{J}_k$ if and only if $\partial f_k/\partial v_i$ is not identically zero. Suppose there is a rule that 'a species' value is affected by a particular process only if the species affects the rate of that process'. Mathematically this states that given $k_1 \in \{1,\ldots,m\}$ and $k_2 \in \{1,\ldots,n\}$, then $(k_1 \in \mathcal{J}_{k_2}) \Rightarrow (k_2 \in \mathcal{I}_{k_1})$. This implies a relationship between the zero entries in $\mathrm{d}v(x)$ and $\mathrm{d}f(v(x))$, namely, $([\mathrm{d}v(x)]_{ji} = 0) \Rightarrow ([\mathrm{d}f(v(x))]_{ij} = 0)$. If 'only if' is changed to 'if and only if' in the above statement, then, $(k_1 \in \mathcal{J}_{k_2}) \Leftrightarrow (k_2 \in \mathcal{I}_{k_1})$, or in terms of the matrices, $([\mathrm{d}v(x)]_{ji} = 0) \Leftrightarrow ([\mathrm{d}f(v(x))]_{ij} = 0)$, i.e., $\mathrm{d}f(v(x))$ and $(\mathrm{d}v(x))^T$ have the same pattern of zeros.

In the case of chemical reaction networks, the chemical species interact with fixed stoichiometries. As a consequence $\Gamma = \mathrm{d}f(v(x))$ is a constant matrix, generally termed the stoichiometric matrix, and in fact Eq. 4.2 can be written

$$\dot{x} = \Gamma v(x). \tag{4.3}$$

Constant $\mathrm{d}f(v(x))$ can arise in other models too, where the occurrence of some interaction (e.g., a predation interaction in an ecological model) is assumed to cause fixed changes in the amounts of species involved. A weak assumption about reaction kinetics that is often satisfied is that (1) $(\Gamma_{ij} = 0) \Rightarrow ((\mathrm{d}v)_{ji} = 0)$, and (2) $\Gamma_{ij}(\mathrm{d}v)_{ji} \leq 0$. In words: (1) if a species concentration affects a rate of reaction, then it must participate in the reaction, and (2) if a species occurs on the left (right) of a reaction, then increasing its concentration cannot decrease the rate at which the reaction proceeds to the right (left).

To abbreviate such rules one can employ the notion of a qualitative class of matrices, and generalisations of this idea [10, 25]. A matrix M determines the qualitative class $\mathcal{Q}(M)$ consisting of all matrices with the same sign pattern as M. Explicitly, $\mathcal{Q}(M)$ consists of all matrices X with the same dimensions as M, and satisfying $(M_{ij} > 0) \Rightarrow (X_{ij} > 0)$, $(M_{ij} < 0) \Rightarrow (X_{ij} < 0)$ and $(M_{ij} = 0) \Rightarrow (X_{ij} = 0)$. The closure of $\mathcal{Q}(M)$ is here denoted as $\mathcal{Q}_0(M)$. With this notation, the pair of conditions $\Gamma_{ij}(\mathrm{d}v)_{ji} \leq 0$ and $(\Gamma_{ij} = 0) \Rightarrow ((\mathrm{d}v)_{ji} = 0)$ can be phrased as '$\mathrm{d}v$ lies in the closure of the qualitative class of $-\Gamma^T$', or more succinctly $\mathrm{d}v \in \mathcal{Q}_0(-\Gamma^T)$. Several abbreviated or omitted proofs of results to follow rely fundamentally on manipulations involving qualitative classes.

Mathematical Background

Before proceeding to examples, some background material is needed. Define $\mathbb{R}^n_+$ to be the nonnegative orthant in $\mathbb{R}^n$, i.e.

$$\mathbb{R}^n_+ = \{x \in \mathbb{R}^n \;:\; x_i \geq 0 \;\text{ for }\; i = 1, \dots, n\}\,.$$

From now on, in general, the state-space X is assumed to be some rectangular region, while in the case of CRNs, $X = \mathbb{R}^n_+$.

Chemical reactions: left and right hand sides. Since the examples are drawn mostly from the analysis of systems of chemical reactions, some basic observations and terminology are needed. First, note that in chemical reaction networks, the sets $\mathcal{I}_k$ can generally be partitioned as $\mathcal{I}_k = \mathcal{I}_k^- \cup \mathcal{I}_k^+$, where $\mathcal{I}_k^-$ (resp. $\mathcal{I}_k^+$) are the indices of reactants occurring on the left (resp. right) hand side of reaction k. Thus the signs of entries in the kth column of the stoichiometric matrix Γ serve to define $\mathcal{I}_k^-$ and $\mathcal{I}_k^+$. Note however that Γ is not uniquely defined, as the notions of left and right hand sides are interchangeable for each reaction (provided the reaction rates are suitably redefined), and all theory must be robust to such interchange.

Chemical reactions: stoichiometry classes. Consider any vector $p \in \ker(\Gamma^T)$, i.e., such that $p^T\Gamma = 0$. Then $H_p(x) \equiv p^T x$ is a conserved quantity of the system. This is immediate as

$$\dot{H}_p(x) = p^T\dot{x} = p^T\Gamma v(x) = 0.$$

Since $\mathrm{Im}(\Gamma) = (\ker(\Gamma^T))^{\perp}$, this means that all trajectories of the system lie in cosets of $\mathrm{Im}(\Gamma)$. Since trajectories are also restricted to $\mathbb{R}^n_+$, the intersections between cosets of $\mathrm{Im}(\Gamma)$ and $\mathbb{R}^n_+$ are invariant sets for the system, and are termed stoichiometry classes[1] of the system (see also [14]). A stoichiometry class containing a positive vector is termed a nontrivial stoichiometry class.

Matrices: notation and terminology. For an $n \times m$ matrix A, $A(\alpha|\beta)$ will refer to the submatrix of A with rows indexed by the set $\alpha \subseteq \{1, \dots, n\}$ and columns indexed by the set $\beta \subseteq \{1, \dots, m\}$. A **minor** is the determinant of a square submatrix of A. So, if $A(\alpha|\beta)$ is square, then $A[\alpha|\beta] = \det(A(\alpha|\beta))$. A **principal minor** of a square matrix A is the determinant of a principal submatrix of A, i.e., a submatrix of the form $A[\alpha|\alpha]$. A square matrix A is **sign nonsingular** (SNS) [10] if all matrices in $\mathcal{Q}(A)$ are nonsingular. It is **sign singular** (SS) if all matrices in $\mathcal{Q}(A)$ are singular. A square matrix with nonnegative off-diagonal elements is termed **quasi-positive**. A square matrix, with eigenvalues all having negative real part is **Hurwitz**. A **signature matrix** is a diagonal matrix with diagonal entries ± 1.

[1] Stoichiometry classes are also sometimes referred to as 'stoichiometric compatibility classes'.

P-matrices and injectivity of functions. P-matrices are square matrices all of whose principal minors are positive. P_0-matrices are matrices all of whose principal minors are nonnegative, i.e., matrices in the closure of the P-matrices. $P^{(-)}$-matrices ($P_0^{(-)}$-matrices) are matrices which are P-matrices (P_0-matrices) after a reversal of sign. Differentiable functions on a rectangular domain with P- or $P^{(-)}$-matrix Jacobian at each point are injective [15]. This theoretical result has proved highly applicable to proving injectivity of vector fields arising in biology and chemistry [4, 30]. With some additional restrictions, the requirement of a P-matrix Jacobian can be weakened to that of a P_0-matrix Jacobian [29]. In the context of a CRN, a P_0 or $P_0^{(-)}$-matrix Jacobian guarantees that there cannot be multiple positive nondegenerate equilibria (henceforth abbreviated to **MPNE**) on any stoichiometry class (see Fig. 4.4).[2]

Cones and order preserving dynamical systems. Any nonempty set of vectors in $\mathbb{R}^n$ (finite or infinite) defines a closed, convex cone consisting of nonnegative combinations of these vectors [9]. Suppose a cone $K \subseteq \mathbb{R}^n$ satisfies, in addition, that $K \cap (-K) = \{0\}$, then K defines a partial order on $\mathbb{R}^n$. Cones which do not contain x and $-x$ for any nonzero vector x will be termed **pointed**.[3] Given any cone $K \subseteq \mathbb{R}^n$, an $n \times n$ matrix J is termed **K-quasipositive** if for each $y \in K$, there exists $t \in \mathbb{R}$ such that $Jy + ty \in K$. When J is the Jacobian of a dynamical system, and K is closed, convex, and pointed, K-quasipositivity implies monotonicity of the system with respect to the order defined by K, i.e., ordered initial conditions remain ordered under evolution. When K is also solid, i.e., has nonzero n-dimensional volume, this has important dynamical implications [19]. Closed, convex, pointed and solid cones are termed **proper** cones (see Fig. 4.5).

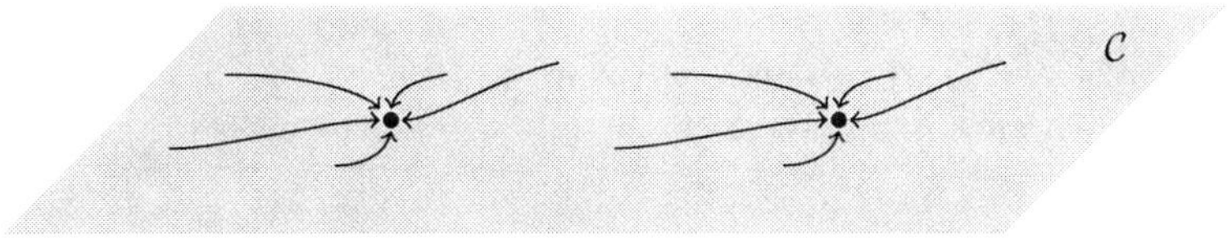

Fig. 4.4 The grey region C represents a portion of the relative interior of some nontrivial stoichiometry class. It can be shown using results in [15] and arguments from degree theory [6] that a CRN with P_0-matrix Jacobian can have no more than one positive nondegenerate equilibrium on each stoichiometry class. Thus the behaviour illustrated, where there are two positive equilibria on C, each locally asymptotically stable on C, cannot occur in a CRN with P_0-matrix Jacobian

[2] Although conclusions for CRNs with P_0 or $P_0^{(-)}$ Jacobian are stated in terms of the absence of multiple positive nondegenerate equilibria, additional structure, for example involving inflow and outflow of substrates, can imply a P or $P^{(-)}$ Jacobian, and thus the existence of no more than one equilibrium on all of state space.

[3] There is some ambiguity in terminology in different strands of the literature. Cones referred to as 'pointed' here and in [9] are termed 'salient' in some references, with the word 'pointed' referring to cones containing the zero vector. Since all cones discussed here are closed, they are all 'pointed' in this other sense too.

Fig. 4.5 Proper cones generate orders which may be preserved by a CRN. Such order preservation has important dynamical consequences, including, for example, ruling out attracting periodic orbits intersecting the interior of $\mathbb{R}^n_+$. Stronger implications including convergence of most (in a measure-theoretic sense), or even all, bounded orbits to equilibria, can follow from further assumptions. However, identifying preserved orders may be a nontrivial task

Graphs and cycles. Consider any signed graph or multigraph which may or may not have directed edges. In the usual way, **cycles** in such graphs are minimal undirected (directed) paths from some vertex to itself. The sign of a cycle is defined as the product of signs of edges in the cycle. The size $|c|$ of a cycle c is the number of edges in c. If the graph is bipartite, then any cycle c has a parity

$$P(c) = (-1)^{|c|/2}\mathrm{sign}(c).$$

c is termed an **e-cycle** if $P(c) = 1$, and an **o-cycle** otherwise. Suppose that, in addition to the graph being bipartite, each edge e in the graph has associated with it a numerical label $l(e)$. Then a cycle c containing edges $e_1, e_2, \ldots, e_{2r}$ such that e_i and $e_{(i \bmod 2r)+1}$ are adjacent for each $i = 1, \ldots, 2r$ is an **s-cycle** if:

$$\prod_{i=1}^{r} l(e_{2i-1}) - \prod_{i=1}^{r} l(e_{2i}) = 0.$$

Two oriented cycles in a graph are compatibly oriented, if each induces the same orientation on every edge in their intersection. Two cycles (possibly unoriented) are compatibly oriented if there is an orientation for each so that this requirement is fulfilled [6]. In a bipartite graph, two cycles have **S-to-R intersection** if they are compatibly oriented and each component of their intersection contains an odd number of edges.

Applying the Theory

Ruling out MPNE in General CRNs

Define the following conditions on a matrix M, and the SR graph G_M associated with M:

C1. G_M contains no e-cycles.
C2. All e-cycles in G_M are s-cycles, and no two e-cycles have S-to-R intersection.
C3. Every square submatrix of M is either SNS or SS.
C4. Every square submatrix of M is either SNS or singular.
C5. $M'N$ is a P_0-matrix for all $M' \in \mathcal{Q}_0(M)$, $N \in \mathcal{Q}_0(M^T)$.
C6. MN is a P_0-matrix for all $N \in \mathcal{Q}_0(M^T)$.

Proposition 4.1. *The following implications hold:*

$$\begin{array}{ccccc} C1 & \Leftrightarrow & C3 & \Leftrightarrow & C5 \\ \Downarrow & & \Downarrow & & \Downarrow \\ C2 & \Rightarrow & C4 & \Leftrightarrow & C6. \end{array}$$

The implications $C1 \Rightarrow C2$, $C3 \Rightarrow C4$, and $C5 \Rightarrow C6$ follow by definition, while the other implications follow from results in [6–8], or minor generalisations of these results.

Proposition 4.1 can be used to show the absence of MPNE in systems of chemical reactions with only weak assumptions on the kinetics:

Theorem 4.2. *Consider a CRN $\dot{x} = \Gamma v(x)$, with $\mathrm{d}v \in \mathcal{Q}_0(-\Gamma^T)$. If G_Γ satisfies Condition C2, or Γ satisfies Condition C4, then the system cannot have MPNE on any stoichiometry class.*

By Proposition 4.1, both Condition C2 and Condition C4 imply Condition C6 on the Jacobian $\Gamma \mathrm{d}v$, which rules out MPNE on any stoichiometry class. Nontrivial examples of applications of these results to biological systems have been previously presented in the references above.

Interestingly, condition C1 can be used to show that the absence of MPNE sometimes follows regardless of the stoichiometries of substrates involved. An example of a family of systems to which this applies is presented in Fig. 4.6.

Generalised Mass-Action Kinetics

Consider a CRN $\dot{x} = \Gamma v(x)$ where reactions are all assumed to be irreversible (i.e., a reversible reaction is treated as two irreversible ones). Assume for definiteness that the system is written with substrates on the left and products on the right. As before, let $\mathcal{I}_k^-$ be the indices of reactants occurring on the left hand side of reaction k.

$$p_i A_i + q_i A_{i+1} \rightleftharpoons r_i B_i,$$
$$i = 1, \ldots, n+1$$
$$s A_{n+2} \rightleftharpoons t A_1$$

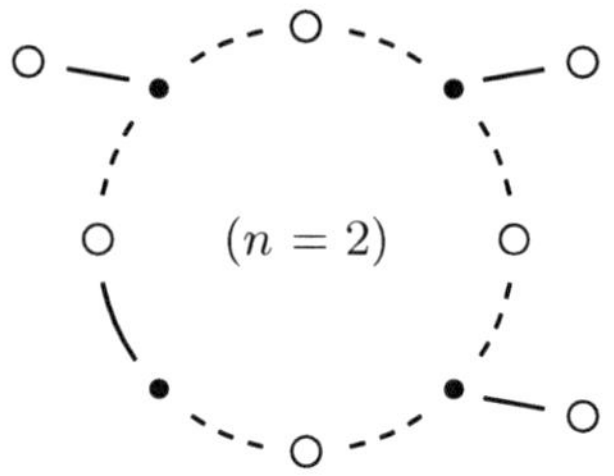

Fig. 4.6 *Left.* A system consisting of $n + 2$ chemical reactions ($n \geq 1$). The stoichiometries p_i, q_i, r_i, s and t are arbitrary. *Right.* The DSR graph for $n = 2$ with edge-labels omitted. Substrates are represented as open circles, while reactions are filled circles. For each n, the DSR graph contains only a single cycle. For even n this can be computed to be an o-cycle, while for odd n it is an e-cycle. Thus, by Proposition 4.1, for even n, this system forbids MPNE for arbitrary stoichiometries

Define 'generalised mass-action kinetics' by the choice of rate functions $v_k(x) = p_k \prod_{j \in \mathcal{I}_k^-} x_j^{a_{jk}}$. The quantities p_k and indices a_{jk} are positive constants. Mass-action kinetics is the special case, $a_{jk} = -\Gamma_{jk}$. Choosing any x in the interior of $\mathbb{R}_+^n$, and differentiating gives

$$\frac{\partial v_k}{\partial x_i} = \begin{cases} \frac{a_{ik}}{x_i} v_k & i \in \mathcal{I}_k^- \\ 0 & \text{otherwise.} \end{cases}$$

More succinctly, let $V = [\partial v_k / \partial x_i]$, D_x be the $n \times n$ positive diagonal matrix with entries $\frac{1}{x_i}$ on the diagonal, D_v the $m \times m$ positive diagonal matrix with entries $v_k(x)$ on the diagonal, and $A = [a_{ik}]$. Then the above equation can be written

$$V = D_v A^T D_x\,, \quad \text{and so} \quad J = \Gamma V = \Gamma D_v A^T D_x\,. \tag{4.4}$$

In the special case of mass-action kinetics, $A = -\Gamma_-^T$, where Γ_- is the matrix Γ with positive entries replaced with zeros. Then

$$V = -D_v \Gamma_-^T D_x\,, \quad \text{and so} \quad J = \Gamma V = -\Gamma D_v \Gamma_-^T D_x\,.$$

A question of interest is when the structure of Γ and A in (4.4) ensure that J is a $P_0^{(-)}$-matrix, ruling out MPNE. A necessary and sufficient condition can be found, which involves only basic computation on Γ and A:

Theorem 4.3. *For fixed Γ and A, $J = \Gamma D_v A^T D_x$ is a $P_0^{(-)}$-matrix for all positive D_v and D_x, if and only if $\Gamma[\alpha|\beta]A[\beta|\alpha] \leq 0$ for every nonempty $\alpha \subseteq \{1, \ldots, n\}$ and $\beta \subseteq \{1, \ldots, m\}$ with $|\alpha| = |\beta|$.*

The proof uses the Cauchy-Binet formula [16] and is a minor generalisation of results in [8]. Although the decomposition $J = \Gamma D_v A^T D_x$ is only defined in the interior of the nonnegative orthant, the result that J is a $P_0^{(-)}$ matrix applies, by closure arguments, on the boundary too. Thus a very basic computation involving only the stoichiometric matrix and the matrix of exponents suffices to rule out MPNE.

Consider the reaction system:

$$A + B \rightleftharpoons C, \quad 2A + B \rightleftharpoons D$$

which has stoichiometric matrix, in reversible and irreversible forms

$$\Gamma_r = \begin{pmatrix} -1 & -2 \\ -1 & -1 \\ 1 & 0 \\ 0 & 1 \end{pmatrix}, \quad \Gamma_{ir} = \begin{pmatrix} -1 & 1 & -2 & 2 \\ -1 & 1 & -1 & 1 \\ 1 & -1 & 0 & 0 \\ 0 & 0 & 1 & -1 \end{pmatrix}.$$

If we assume mass-action kinetics so that the matrix of exponents is

$$A = \begin{pmatrix} 1 & 0 & 2 & 0 \\ 1 & 0 & 1 & 0 \\ 0 & 1 & 0 & 0 \\ 0 & 0 & 0 & 1 \end{pmatrix},$$

then the conditions on Γ_{ir} and A from Theorem 4.3 hold, and so MPNE are ruled out. On the other hand, Condition C4 defined above does not apply (to Γ_r or Γ_{ir}), and so MPNE cannot be ruled out if more general kinetics are allowed.

Structural Conditions for Local Stability of Equilibria

Other functional forms arising in biological (not necessary strictly chemical) models can give other useful factorisations. Here, an example is presented of a class of systems with Jacobian which is everywhere Hurwitz. This guarantees local stability of all equilibria.

Define

$$q(x) = [q_1(x_1), \ldots, q_n(x_n)]^T,$$

where, for each i, $\mathrm{d}q_i/\mathrm{d}x_i > 0$. Such a strictly increasing diagonal function can be termed an outflow function. Consider the system

$$\dot{x} = y + \Gamma v(x) - q(x), \tag{4.5}$$

where y is a constant nonnegative vector, representing inflow/production of the quantities x_i, while $q(x)$ represents their outflow or degradation.

Theorem 4.4. *Let h_k, g_j be differentiable scalar functions with positive derivative, and assume that the rate functions v_k can be written:*

$$v_k(x) = h_k\left(-\sum_j \Gamma_{jk} g_j(x_j)\right).$$

Then all equilibria of (4.5) are locally stable.

Such rate terms have been used, for example, in models of mitochondria [13], where the physical interpretation is, roughly speaking, that the rate of each redox reaction depends on a weighted sum of potentials corresponding to the associated half-reactions.

The result can be seen quite easily, and as usual the process begins by finding a useful form for the Jacobian. Differentiating, and abbreviating $h_k'(\sum_j \Gamma_{jk} g_j(x_j))$ to h_k' and $g_i'(x_i)$ to g_i', gives

$$\frac{\partial v_k}{\partial x_i} = -h_k' \Gamma_{ik} g_i'.$$

More succinctly,

$$[\partial v_k/\partial x_i] \equiv V = -D_h \Gamma^T D_g, \text{ and } J = \Gamma V - \mathrm{d}q = -\Gamma D_h \Gamma^T D_g - \mathrm{d}q,$$

where D_h is the $m \times m$ positive diagonal matrix with entries h_k' on the diagonal, D_g is the $n \times n$ positive diagonal matrix with entries g_i' on the diagonal, and $\mathrm{d}q$ is the derivative of q, again a positive diagonal matrix.

Define $J_0 = \Gamma D_h \Gamma^T$. Applying the Cauchy-Binet formula, it is immediate that J_0 is a P_0-matrix – in fact it is also symmetric and hence positive semidefinite. Now $-J = J_0 D_g + \mathrm{d}q = J_0' D_g$, where $J_0' \equiv J_0 + \mathrm{d}q\, D_g^{-1}$. Since J_0 is a symmetric P_0-matrix, and $\mathrm{d}q\, D_g^{-1}$ is a positive diagonal matrix, by basic properties of P-matrices, J_0' is a symmetric P-matrix. Consequently $J = -J_0' D_g$ is Hurwitz [20], implying local stability of any equilibria of such a system.

Monotonicity in General CRNs

An important and difficult question is to find necessary and sufficient conditions for dynamical systems arising in biology to preserve a partial order on their state space, with the goal of making claims about absence of periodic attractors, or convergence of orbits. The special structure of CRNs makes this question somewhat more tractable, and a variety of examples of order-preserving CRNs have appeared in the literature (see [11,22,23] for example). A general, geometric approach to this question was presented in [3]. Here a closely related, but more direct, algebraic, approach is described.

Any finite set of r vectors in $\mathbb{R}^n$ can be written as the columns of an $n \times r$ matrix, and equally, any $n \times r$ matrix A generates the cone $K(A) \subseteq \mathbb{R}^n$ consisting of nonnegative combinations of its column vectors:

$$K(A) = \{Ay \; : \; y \in \mathbb{R}^r_+\}.$$

Clearly $K(A)$ is closed and convex, but may or may not be pointed and solid. Such a cone, generated by a finite number of vectors is termed polyhedral. The following is a useful starting point for asking when dynamical systems preserve orderings defined by polyhedral cones:

Proposition 4.5. *Consider some $n \times r$ matrix A, some $r \times n$ matrix B, and define $J = AB$.*

1. *J is $K(A)$-quasipositive if and only if given any $y \in \mathbb{R}^r_+$, there exists $\alpha \in \mathbb{R}$ and $z \in \mathbb{R}^r_+$ such that $(BA + \alpha I)y - z \in \ker(A)$. This condition is satisfied if there exists an $r \times r$ matrix B' such that $\mathrm{Im}(B') \subseteq \ker(A)$ (i.e., $AB' = 0$), and $BA + B'$ is quasipositive.*
2. *If A has rank r, then J is $K(A)$-quasipositive if and only if BA is quasipositive.*

The proof of this proposition is presented in the appendix. Three examples of its application are now given. The first two involve applying the (easier) Claim 2, while the third applies Claim 1 directly.

Example A. A first application of Claim 2 in Proposition 4.5 is the following.

Theorem 4.6. *Consider a CRN $\dot{x} = \Gamma v(x)$ with Γ an $n \times m$ matrix, and $\mathrm{d}v \in \mathcal{Q}_0(-\Gamma^T)$. Suppose that (1) Γ has trivial kernel, (2) each row of Γ contains no more than two nonzero entries, and (3) the SR graph G_Γ contains no o-cycles. Then there exists a signature matrix D such that the Jacobian $\Gamma \mathrm{d}v$ is $K(\Gamma D)$-quasipositive.*

Note that Condition (3) can also be stated in terms of the the matrix Γ, but the most brief and elegant statement is in terms of the SR graph. One dynamical implication of Theorem 4.6 is that periodic orbits which include a positive concentration, and are attracting on their stoichiometry classes, cannot occur for such systems. This follows because restricting to $\mathrm{Im}(\Gamma)$, $K(\Gamma D)$ is a proper cone, in that it closed, convex, pointed and has non-empty relative interior in $\mathrm{Im}(\Gamma)$.

To see why Theorem 4.6 holds, consider any $m \times m$ signature matrix D and let $\Gamma' = \Gamma D$ so that $\Gamma = \Gamma' D$. Since Γ has trivial kernel, so does Γ'. By Claim 2 in Proposition 4.5, the Jacobian $\Gamma' D \mathrm{d}v$ is $K(\Gamma')$-quasipositive if and only if $D \mathrm{d}v\, \Gamma D$ is quasipositive. What remains is to show that the requirement $\mathrm{d}v \in \mathcal{Q}_0(-\Gamma^T)$ along with conditions (2) and (3) above imply that there exists some signature matrix D such that $D \mathrm{d}v\, \Gamma D$ is quasipositive. But it is well known that a matrix is similar, via a signature matrix, to a quasipositive matrix if and only if the associated I-graph contains no nontrivial negative cycles ([17] for example). The problem then reduces to showing that all matrices in $\mathrm{d}v\, \Gamma$ belong in the closure of the same qualitative class, and that the I-graph associated with this qualitative class does indeed contain no nontrivial negative cycles. These facts are not proved here, but the proof is not hard.

As a specific example of the use of Theorem 4.6, consider again the systems in Fig. 4.6, this time with n odd. These systems fulfil assumptions (1), (2) and (3) in Theorem 4.6, and thus, for any reaction rates satisfying $\mathrm{d}v \in \mathcal{Q}_0(-\Gamma^T)$, there can be no positive periodic attractors on any stoichiometry class.

Related, but more powerful results than Theorem 4.6 are possible. Suspending requirement (1) in the theorem, it is no longer necessarily possible to find a preserved order for the system restricted to each stoichiometry class; however it is possible to prove that the evolution of the so-called 'extents' of reactions is monotone. With some effort, convergence properties of the original dynamical system (on the space of chemical concentrations), can then be inferred from this conclusion [1].

Example B. A more nontrivial example of the use of Claim 2 in Proposition 4.5, is the following:

Theorem 4.7. *Consider a CRN* $\dot{x} = \Gamma v(x)$ *with* Γ *an* $n \times m$ *matrix, and* $\mathrm{d}v \in \mathcal{Q}_0(-\Gamma^T)$. *Further, suppose we can factorise* $\Gamma = \Lambda\Theta$ *where*

1. Λ *is an* $n \times r$ *matrix with exactly one nonzero entry in each row.*
2. Θ *is an* $r \times m$ *matrix such that each column of* Θ *contains no more than one positive entry and no more than one negative entry.*

Then the Jacobian $\Gamma\mathrm{d}v$ *is everywhere* $K(\Lambda)$*-quasipositive.*

Note first that $\ker(\Lambda^T) \subseteq \ker(\Gamma^T)$, and hence $\mathrm{Im}(\Gamma) \subseteq \mathrm{Im}(\Lambda)$. So all stoichiometry classes and hence all trajectories of the system lie in cosets of $\mathrm{Im}(\Lambda)$, which can be termed Λ-classes of the system. The assumption on Λ implies that it has rank r, and hence $K(\Lambda)$ is a closed, convex, pointed cone. Claim 2 in Proposition 4.5 then implies that $J(x)$ is $K(\Lambda)$-quasipositive if and only if $\Theta\mathrm{d}v(x)\Lambda$ has nonnegative off-diagonal elements. It can be calculated directly with some effort that the assumptions imply that $\Theta\mathrm{d}v(x)\,\Lambda$ has nonnegative off-diagonal elements (and nonpositive diagonal elements) for any x.

There is also a more explicit meaning in terms of recoordinatisation: the arguments imply that on each Λ-class of the system there exist local coordinates y, whose evolution is cooperative. In other words, these coordinates evolve according to $\dot{y} = \tilde{F}(y)$ where $\mathrm{d}\tilde{F}$ has nonnegative off-diagonal elements. Consider any $c \in \mathbb{R}^n_+$ and some vector $x = c + \Lambda y$ lying on the same Λ-class as c. Then $\Lambda\dot{y} = \dot{x} = \Lambda\Theta v(x)$. Since Λ has rank r, there exists a matrix Λ' such that $\Lambda'\Lambda = I$. Multiplying both sides of the equation by Λ' gives $\dot{y} = \Theta v(c + \Lambda y)$. Defining $\tilde{F}(y) = \Theta v(c + \Lambda y)$ and differentiating gives

$$\mathrm{d}\tilde{F}(y) = \Theta\mathrm{d}v(c + \Lambda y)\Lambda\,.$$

A similar interpretation in terms of recoordinatisation also applies to the previous example.

In order to draw out the dynamical implications of Theorem 4.7, further assumptions and theory are needed. This is because the stoichiometry classes will generally be of lower dimension than the Λ-classes, and $K(\Lambda)$ does not, in general, induce an ordering on the stoichiometry classes. In fact it is possible for all points on the stoichiometry classes to be unordered with respect to the ordering induced by $K(\Lambda)$. However, with additional assumptions, ideas from [26], generalised in [5], can be used to infer restrictions on the dynamics.

Example C. Directly applying Claim 1 in Proposition 4.5 is more difficult than applying Claim 2. To illustrate the harder case where $\ker(\Lambda)$ is not trivial, consider the system of chemical reactions

$$A + B \rightleftharpoons C, \quad A \rightleftharpoons B.$$

We can factorise the stoichiometric matrix $\Gamma = \Lambda\Theta$ as follows:

$$\begin{pmatrix} -1 & -1 \\ -1 & 1 \\ 1 & 0 \end{pmatrix} = \begin{pmatrix} -1 & 0 & 0 & 1 \\ 0 & 1 & -1 & 0 \\ 1 & 0 & 1 & 0 \end{pmatrix} \begin{pmatrix} 1 & 0 \\ -1 & 1 \\ 0 & 0 \\ 0 & -1 \end{pmatrix}.$$

Note that $\ker(\Lambda)$ consists of nonnegative multiples of $(-1, 1, 1, -1)^T$, and that $K(\Lambda)$ is pointed and solid in $\mathbb{R}^3$ [3]. Assuming that at each x, $\mathrm{d}v(x) \in \mathcal{Q}_0(-\Gamma^T)$, means that $\mathrm{d}v(x)$ takes the form

$$\begin{pmatrix} a & b & -c \\ d & -e & 0 \end{pmatrix},$$

where $a, b, c, d, e \geq 0$. So $\Theta \mathrm{d}v(x)\Lambda \equiv R$ takes the form

$$\begin{pmatrix} 1 & 0 \\ -1 & 1 \\ 0 & 0 \\ 0 & -1 \end{pmatrix} \begin{pmatrix} a & b & -c \\ d & -e & 0 \end{pmatrix} \begin{pmatrix} -1 & 0 & 0 & 1 \\ 0 & 1 & -1 & 0 \\ 1 & 0 & 1 & 0 \end{pmatrix}$$

$$= \begin{pmatrix} -(c+a) & b & -(c+b) & a \\ a+c-d & -(b+e) & b+c+e & d-a \\ 0 & 0 & 0 & 0 \\ d & e & -e & -d \end{pmatrix}.$$

Clearly, not all off-diagonal elements of R are necessarily nonnegative. However, defining

$$P = \begin{pmatrix} -d & 0 & b+c+e & -a \\ d & 0 & -b-c-e & a \\ d & 0 & -b-c-e & a \\ -d & 0 & b+c+e & -a \end{pmatrix},$$

one can observe that $\mathrm{Im}(P) = \ker(\Lambda)$, and moreover $R + P$ has nonnegative off-diagonal elements. Thus, by Claim 1 in Proposition 4.5, the Jacobian $\Gamma \mathrm{d}v(x)$ is $K(\Lambda)$-quasipositive at each x. Mild additional assumptions giving global convergence of all trajectories to equilibria are provided in [5].

The instructive point about this example is that the apparently redundant third column of Λ and third row of Θ are crucial for the argument to work: removing these gives Λ with trivial kernel, and $\Theta \mathrm{d}v(x)\Lambda$ fails to be quasipositive.

Conclusions

It has been illustrated that structural analysis of biological networks can lead to surprisingly strong conclusions about dynamical behaviour. A variety of useful approaches begin by deriving constraints on the Jacobian of the system from qualitative (biological) knowledge. This idea can sometimes reduce a difficult problem in the qualitative theory of ordinary differential equations to simple calculations on matrices or graphs. Apparently disparate strands of theory, for example on systems with signed Jacobian, and on chemical reaction networks, can sometimes be brought under a single umbrella via these techniques [4].

One important conclusion is that the stoichiometric matrix of a system of chemical reactions contains a considerable volume of information. It is well known that analysis of the stoichiometric matrix of a CRN can provide biologically useful information on conserved quantities [14], or optimal flux distributions [28]; but it is far from obvious that it can also encode information on the type, number, and local/global stability of limit sets.

Despite the successes, there still remains considerable progress to be made in this area. For example, the problem of identifying when systems of chemical reactions (with general or restricted kinetics) give rise to order preserving dynamical systems is far from solved. Seeking structural conditions for local and global stability is a difficult but important task. A starting point is to ask when matrix products are structurally Hurwitz (see [25] for the case of when a qualitative class is Hurwitz). Difficult and subtle questions remain on how best to deal with conservation laws in CRNs.

Finally, as mentioned in the introduction, a number of results in this area potentially have constructive converses, namely theorems which assert that certain behaviours *will* occur in some model class, and provide rules for guaranteeing occurrence of these behaviours. As synthetic biology gains maturity, there is the exciting possibility of using such theory to derive qualitative rules to aid the design of biological systems with novel behaviours.

Appendix: Proof of Proposition 4.5

By definition, J is $K(A)$-quasipositive if and only if for each $y' \in K(A)$, there exists $\alpha \in \mathbb{R}$ such that $Jy' + \alpha y' \in K(A)$. Suppose this is the case. Then for each y such that $Ay \in K(A)$ (which includes $\mathbb{R}^r_+$ by definition), $\exists \alpha \in \mathbb{R}, z \in \mathbb{R}^r_+$ such that $ABAy + \alpha Ay = Az$, i.e., $A(BAy + \alpha y - z) = 0$. Conversely, suppose that for each $y \in \mathbb{R}^r_+$, there exist $\alpha \in \mathbb{R}$ and $z \in \mathbb{R}^r_+$ such that $(BA + \alpha I)y - z \in \ker(A)$. Given $y' \in K(A)$, choose some $y \in \mathbb{R}^r_+$, such that $Ay = y'$, and fix α, z such that $A((BA+\alpha I)y - z) = 0$, i.e., $A(BA+\alpha I)y = Az$. So $AB(Ay)+\alpha(Ay) \in K(A)$, i.e., $Jy' + \alpha y' \in K(A)$.

Suppose there exists an $r \times r$ matrix B' such that $\mathrm{Im}(B') \subseteq \ker(A)$, and $BA + B'$ is quasipositive. By quasipositivity, there exists $\alpha \in \mathbb{R}$, such that $BA + B' + \alpha I$ is a nonnegative matrix, and so, given any $y \in \mathbb{R}^r_+$, $z \equiv BAy + B'y + \alpha y \in \mathbb{R}^r_+$. But then $A((BA + \alpha I + B')y - z) = 0$, implying $A((BA + \alpha I)y - z) = 0$. (The last implication follows because $AB' = 0$). This completes the proof of Claim 1.

If A has rank r, then $\ker(A)$ is trivial, and $A((BA+\alpha I)y-z) = 0$ is equivalent to $(BA+\alpha I)y = z$, so $(BA+\alpha I)y \in \mathbb{R}^r_+$. Since y is an arbitrary vector in $\mathbb{R}^r_+$, α can be found to satisfy this restriction if and only if BA has nonnegative off-diagonal elements, proving the second claim.

References

1. Angeli D, De Leenheer P, Sontag ED (2009a) Graph-theoretic characterizations of monotonicity of chemical reaction networks in reaction coordinates. J Math Biol. doi:10.1007/s00285-009-0309-0
2. Angeli D, Hirsch MW, Sontag E (2009b) Attractors in coherent systems of differential equations. J Diff Eq 246:3058–3076
3. Banaji M (2009) Monotonicity in chemical reaction systems. Dyn Syst 24(1):1–30
4. Banaji M (2010) Graph-theoretic conditions for injectivity of functions on rectangular domains. J Math Anal Appl 370:302–311
5. Banaji M, Angeli D (2010) Convergence in strongly monotone systems with an increasing first integral. SIAM J Math Anal 42(1):334–353
6. Banaji M, Craciun G (2009) Graph-theoretic approaches to injectivity and multiple equilibria in systems of interacting elements. Commun Math Sci 7(4):867–900
7. Banaji M, Craciun G (2010) Graph-theoretic criteria for injectivity and unique equilibria in general chemical reaction systems. Adv Appl Math 44:168–184
8. Banaji M, Donnell P, Baigent S (2007) P matrix properties, injectivity and stability in chemical reaction systems. SIAM J Appl Math 67(6):1523–1547
9. Berman A, Plemmons R (1979) Nonnegative matrices in the mathematical sciences. Academic, New York
10. Brualdi RA, Shader BL (1995) Matrices of sign-solvable linear systems. Number 116 in Cambridge tracts in mathematics. Cambridge University Press, Cambridge, UK
11. De Leenheer P, Angeli D, Sontag ED (2007) Monotone chemical reaction networks. J Math Chem 41(3):295–314
12. Dolmetsch RE, Xu K, Lewis RS (1998) Calcium oscillations increase the efficiency and specificity of gene expression. Nature 392:933–936
13. Donnell P, Banaji M, Baigent S (2009) Stability in generic mitochondrial models. J Math Chem 46(2):322–339
14. Famili I, Palsson BO (2003) The convex basis of the left null space of the stoichiometric matrix leads to the definition of metabolically meaningful pools. Biophys J 85:16–26
15. Gale D, Nikaido H (1965) The Jacobian matrix and global univalence of mappings. Math Ann 159:81–93
16. Gantmacher FR (1959) The theory of matrices. Chelsea Publishing Company, New York
17. Gouzé J-L (1998) Positive and negative circuits in dynamical systems. J Biol Sys 6:11–15

18. Grimbs S, Selbig J, Bulik S, Holzhütter H-G, Steuer R (2007) The stability and robustness of metabolic states: identifying stabilizing sites in metabolic networks. Mol Syst Biol 3(146) doi:10.1038/msb4100186
19. Hirsch MW, Smith H (2005) Chapter monotone dynamical systems. In: Battelli F and Fečkan M (ed.) Handbook of differential equations: ordinary differential equations, vol II. Elsevier BV, Amsterdam, pp 239–357
20. Kafri WS (2002) Robust D-stability. Appl Math Lett 15:7–10
21. Kaufman M, Soulé C, Thomas R (2007) A new necessary condition on interaction graphs for multistationarity. J Theor Biol 248(4):675–685
22. Kunze H, Siegel D (2002a) A graph theoretic approach to strong monotonicity with respect to polyhedral cones. Positivity 6:95–113
23. Kunze H, Siegel D (2002b) Monotonicity properties of chemical reactions with a single initial bimolecular step. J Math Chem 31(4):339–344
24. Markevich NI, Hoek JB, Kholodenko BN (2004) Signaling switches and bistability arising from multisite phosphorylation in protein kinase cascades. J Cell Biol 164(3):353–359
25. Maybee J, Quirk J (1969) Qualitative problems in matrix theory. SIAM Rev 11(1):30–51
26. Mierczyński, J (1987) Strictly cooperative systems with a first integral. SIAM J Math Anal 18(3):642–646
27. Novák B, Tyson JJ (2008) Design principles of biochemical oscillators. Nat Rev Mol Cell Biol 9:981–991
28. Orth JD, Thiele I, Palsson BØ(2010) What is flux balance analysis? Nat Biotechnol 28:245–248
29. Parthasarathy T (1983) On global univalence theorems volume 977 of Lecture notes in mathematics. Springer-Verlag, Berlin, Heidelberg, New York
30. Soulé C (2003) Graphic requirements for multistationarity. Complexus 1:123–133
31. Thomson M, Gunawardena J (2009) Unlimited multistability in multisite phosphorylation systems. Nature 460:274–277

Chapter 5
Contraction Theory for Systems Biology

Giovanni Russo, Mario di Bernardo, and Jean Jacques Slotine

Abstract In this chapter, we present a theoretical framework for the analysis, synchronization and control of biochemical circuits and systems modeled by means of ODEs. The methodology is based on the use of contraction theory, a powerful concept from the theory of dynamical systems, ensuring convergence of all trajectories of a system of interest towards each other. After introducing contraction theory, we present some application to biochemical networks. Specifically, we introduce a graphical approach to verify if a system is contracting and apply it to synthesize networks of self-synchronizing Repressilators. We then present a more general analysis of quorum sensing networks.

Keywords Contraction theory · Graphical algorithm · Entrainment · Synchronization

Introduction

With the increasing number of biological circuits and devices being analyzed and designed in Systems and Synthetic biology, the availability of appropriate mathematical tools for the investigation of their properties is a pressing research problem. A classical example is the study of synchronization in networks of biological oscillators coupled via quorum sensing or the analysis of their entrainment to some external periodic input. Typically, when differential equations are used to model the circuit of interest, Lyapunov-based techniques or methods based on linearization are used to obtain global or local results on the stability of the synchronous evolution, the possible entrainment of the network under investigation and so on. The aim of this chapter is to review recent results on applying a different tool from dynamical system theory, *contraction theory*, to systems and synthetic biology.

G. Russo (✉)
Department of Systems and Computer Engineering, University of Naples Federico II, Via Claudio 21, 80125 Napoli, Italy
e-mail: giovanni.russo2@unina.it

H. Koeppl et al. (eds.), *Design and Analysis of Biomolecular Circuits: Engineering Approaches to Systems and Synthetic Biology*, DOI 10.1007/978-1-4419-6766-4_5,

Essentially, a nonlinear time-varying dynamic system will be called *contracting* if initial conditions or temporary disturbances are forgotten exponentially fast, i.e. if trajectories of the system converge towards each other with an exponential convergence rate. It turns out that relatively simple algebraic conditions can be given for this stability-like property to be verified, and that this property is preserved through basic system combinations and aggregations.

A nonlinear contracting system has the following properties [24, 25, 41, 46]

- convergence rates can be explicitly computed as eigenvalues of well-defined Hermitian matrices
- many combinations and aggregations of contracting systems are also contracting
- robustness to variations in dynamics can be easily quantified

These properties, and in particular the aggregation property (which goes considerably beyond standard passivity-like results) are particularly attractive for studying biological systems [41].

Historically, ideas closely related to contraction can be traced back to [17] and even to [23] (see also [4, 31], and e.g. [19, 26], for a more exhaustive list of related references). For autonomous systems and with constant metrics, the basic nonlinear contraction result reduces to Krasovskii's theorem [42] in the continous-time case, and to the contraction mapping theorem in the discrete-time case [6, 24].

In this chapter, after briefly reviewing the main concepts and results concerning contraction theory, we will introduce an algorithmic procedure to assess whether a system of interest in contracting. The key idea is to make use of non-Euclidean norms and matrix measures to prove contraction. This allows the derivation of a graphical algorithm that can be effective in determining conditions for a biological system (or network) under investigation to be contracting. We will then apply the results to three representative biological applications. The theoretical derivation will be illustrated by numerical simulations. The results reviewed in the chapter are based on those recently presented in [37–40].

Contraction Theory: An Overview

Basic Results

The basic result of nonlinear contraction analysis [24] which we shall use in this paper can be stated as follows.

Theorem 1 (Contraction). *Consider the m-dimensional deterministic system*

$$\dot{x} = f(x,t) \tag{5.1}$$

where f is a smooth nonlinear function. The system is said to be contracting if any two trajectories, starting from different initial conditions, converge exponentially

Table 5.1 Standard matrix measures

Vector norm, $\|x\|$	Induced matrix measure, $\mu(A)$
$\|x\|_1 = \sum_{j=1}^{m} \|x_j\|$	$\mu_1(A) = \max_j \left(a_{jj} + \sum_{i \neq j} \|a_{ij}\|\right)$
$\|x\|_2 = \left(\sum_{j=1}^{n} \|x_j\|^2\right)^{\frac{1}{2}}$	$\mu_2(A) = \max_i \left(\lambda_i \left\{\frac{A + A^*}{2}\right\}\right)$
$\|x\|_\infty = \max_{1 \leq j \leq m} \|x_j\|$	$\mu_\infty(A) = \max_i \left(a_{ii} + \sum_{j \neq i} \| a_{ij} \|\right)$

towards each other. A sufficient condition for a system to be contracting is that there exists a constant invertible matrix Θ *such that the so-called generalized Jacobian*

$$F(x,t) = \Theta \, \frac{\partial f}{\partial x}(x,t) \, \Theta^{-1} \tag{5.2}$$

verifies

$$\exists \lambda > 0, \ \forall x, \ \forall t \geq 0, \ \ \mu(F(x,t)) \ \leq \ -\lambda$$

where μ *is one the the standard matrix measures in Table 5.1. The scalar* λ *defines the contraction rate of the system.*

For convenience, in this paper we will also say that a *function* $f(x,t)$ is contracting if the system $\dot{x} = f(x,t)$ satisfies the sufficient condition above. Similarly, we will then say that the corresponding Jacobian *matrix* $\frac{\partial f}{\partial x}(x,t)$ is contracting.

We shall also use the following two properties of contracting systems, whose proofs can be found in [24, 41].

Hierarchies of contracting systems Assume that the Jacobian of (5.1) is in the form

$$\frac{\partial f}{\partial x}(x,t) \ = \ \begin{bmatrix} J_{11} & J_{12} \\ 0 & J_{22} \end{bmatrix} \tag{5.3}$$

corresponding to a hierarchical dynamic structure. The J_{ii} may be of different dimensions. Then, a sufficient condition for the system to be contracting is that (1) the Jacobians J_{11}, J_{22} are contracting (possibly with different Θ's and for different matrix measures), and (2) the matrix J_{12} is bounded.

Periodic inputs Consider the system

$$\dot{x} = f\,(x, r(t)) \tag{5.4}$$

where the input vector $r(t)$ is periodic, of period T. Assume that the system is contracting (i.e. that the Jacobian matrix $\frac{\partial f}{\partial x}(x, r(t))$ is contracting for any $r(t)$). Then the system state $x(t)$ tends exponentially towards a periodic state of period T.

Partial Contraction A simple yet powerful extension to nonlinear contraction theory is the concept of *partial* contraction [46].

Theorem 2 (Partial contraction). *Consider a smooth nonlinear m-dimensional system of the form $\dot{x} = f(x, x, t)$ and assume that the so-called virtual system $\dot{y} = f(y, x, t)$ is contracting with respect to y. If a particular solution of the auxiliary y-system verifies a smooth specific property, then all trajectories of the original x-system verify this property exponentially. The original system is said to be* partially contracting.

A Graphical Approach to Prove Contraction

The use of matrix measures and norms induced by non-Euclidean vector norms (such as μ_1, μ_∞, $\|\cdot\|_1$, $\|\cdot\|_\infty$) can be effectively exploited to obtain alternative conditions to check for contraction of a dynamical system of interest.

In this section we show that by means of these measures and norms, it is possible to obtain a graphical procedure for showing that a system is contracting in a constant diagonal metric, or for imposing such property (see [35, 39] for further details). The 'qualitative' nature of the approach, combined with the aggregation properties of contracting systems, favors the flexibility and conceptual robustness desirable in studying biological systems [1], as will be shown later in the section "Entrainment and Synchronization of Biological Systems".

The outcome of the procedure is to provide a set of conditions on the elements of the system Jacobian, J, (and hence on the dynamics of $f(\cdot,\cdot)$) that can be used to prove contraction. Notice that (5.1) can represent, for instance, a closed loop control system, in which the control input is a function of the system state. Thus, the procedure presented here may be used both for checking (e.g. in a system analysis context) and for improving contractivity and hence some desired behavior (e.g. in a synthetic biology context).

Outline

The first step of the procedure is to differentiate the system of interest, in order to obtain the Jacobian matrix, $J := \frac{\partial f}{\partial x}$:

$$\begin{bmatrix} J_{1,1}(x,t) & J_{1,2}(x,t) & \dots & J_{1,m}(x,t) \\ J_{2,1}(x,t) & J_{2,2}(x,t) & \dots & J_{2,m}(x,t) \\ \dots & \dots & \dots & \dots \\ J_{m,1}(x,t) & J_{m,2}(x,t) & \dots & J_{m,m}(x,t) \end{bmatrix} \tag{5.5}$$

which is, in general, state/time dependent.

The next step is to construct a directed graph from the system Jacobian. To this aim, we first derive an adjacency matrix from J, say $\mathcal{A}$, using the following rules:

1. initialize $\mathcal{A}$ so that $\mathcal{A}(i,j) = 0, \forall i, j$;
2. for all $i \neq j$, set $\mathcal{A}(i,j) = \mathcal{A}(j,i) = 1$ if either $J_{i,j}(x,t) \neq 0$, or $J_{i,j}(x,t) \neq 0$.

Such a matrix describes an undirected graph (see e.g. [15]), say $\mathcal{G}(\mathcal{A})$. The second step in the procedure is then to associate directions to the edges of $\mathcal{G}(\mathcal{A})$ to obtain a directed graph, say $\mathcal{G}_d(\mathcal{A})$. This is done by computing the quantity

$$\alpha_{i,j}(x,t) = \frac{|J_{i,j}(x,t)|}{|J_{i,i}(x,t)|}(m - n_{0i} - 1). \tag{5.6}$$

In the above expressions n_{0i} is the number of zero elements on the i-th row of $\mathcal{A}$. (Note that if $J_{i,i}(x,t) = 0$ for some i, then, before computing (5.6), the system parameters/structure must be engineered so that $J_{i,i}(x,t) \neq 0$, for all i.)

The directions of the edges of $\mathcal{G}_d(\mathcal{A})$ are then obtained using the following simple rule:

the edge between node i and node j is directed from i to j if the quantity $\alpha_{i,j}(x,t) < 1$ while it is directed from j to i if $\alpha_{i,j}(x,t) \geq 1$.

Note that, the quantities $\alpha_{i,j}(x,t)$ will be in general time-dependent, therefore the graph directions might be time-varying.

Once the directed graph $\mathcal{G}_d(\mathcal{A})$ has been constructed, contraction is then guaranteed under the following conditions:

1. uniform negativity of all the diagonal elements of the Jacobian, i.e. $J_{i,i}(x,t) < 0$ for all i;
2. for all t, the directed graph $\mathcal{G}_d(\mathcal{A})$ does not contain loops of any length and $\alpha_{ij}(x,t)\alpha_{ji}(x,t) \leq 1$ for any $i \neq j$.

Note that, when the above conditions are not satisfied, our approach can be used to impose contraction for the system of interest by:

1. using, if possible, a control input to impose the first condition of the above procedure for all the elements $J_{i,i}(x,t)$ that do not fulfill it;
2. *re-direct* (using an appropriate control input, or tuning system parameters) some edges of the graph $\mathcal{G}_d(\mathcal{A})$ in order to satisfy the *loopless* condition;
3. associate to each reverted edge (e.g. the edge between node i and node j) one of the following inequalities:
   - $\alpha_{i,j}(x,t) \geq 1$, if the edge is reverted from j to i;
   - $\alpha_{i,j}(x,t) < 1$, if the edge is reverted from i to j;
   - ensure that $\alpha_{ij}(x,t)\alpha_{ji}(x,t) \leq 1$.

We remark here that the procedure presented above is based on the use of $\mu_\infty\left(\Theta J \Theta^{-1}\right)$ for proving contraction. Other matrix measures and norms can also be used. In particular, it is easy to prove that, using $\mu_1\left(\Theta J \Theta^{-1}\right)$, yields the same procedure applied on J^T. If this is the case, the resulting procedure will follow the same logical steps as those presented above, with the only difference being the expression of $\alpha_{i,j}(x,t)$:

$$\alpha_{i,j}(x,t) := \frac{\left|J_{j,i}(x,t)\right|(m - c_{0i} - 1)}{\left|J_{i,i}(x,t)\right|} \tag{5.7}$$

where c_{0i} denotes the number of zero elements of the i-th column of J.

Entrainment and Synchronization of Biological Systems

In this Section, we illustrate two applications of the graphical approach presented above.

Entrainment of Transcriptional Modules

We start with studying a general externally-driven transcriptional module ubiquitous in both natural and synthetic biology [9]. We assume that the rate of production of a transcription factor X is proportional to the value of a time dependent input function, $u(t)$, and X is subject to degradation and/or dilution at a linear rate (more general models are analyzed in [40]). The signal $u(t)$ might be an external input, or it might represent the concentration of an enzyme or of a second messenger that activates X. In turn, X drives a downstream transcriptional module by binding to a promoter (or substrate) denoted by e, with concentration $e = e(t)$. The binding reaction is reversible and Y denotes the complex protein-promoter. We remark that, as the promoter is not subject to decay, its total concentration, i.e. $E_T = e + Y$, is conserved.

In [9] the following mathematical model was analyzed:

$$\begin{aligned} \dot{x} &= u(t) - \delta x + k_1 y - k_2 (E_T - y) x \\ \dot{y} &= -k_1 y + k_2 (E_T - y) x \end{aligned} \tag{5.8}$$

where: x denotes the concentration of X, y denotes the concentration of Y, k_1 and k_2 are the binding and dissociation rates associated to the reaction $X + e \leftrightarrows Y$.

We will show, using the procedure presented in section "Outline", that this system is contracting. This immediately implies in turn that, when forced by a periodic input $u(t)$, system (5.8) tends globally exponentially to a periodic solution of the same period as $u(t)$. That is, the system becomes entrained to any periodic input. This property is often a desirable property for biological systems: many important activities of life are, in fact, regulated by periodic, clocklike rhythms. We can think for example of the suprachiasmatic nucleus (SCN), whose activity is regulated by daily dark-light cycles (see e.g. [43]). Contraction analysis of more general transcriptional modules is presented and extensively studied in [40]. Computing the Jacobian of (5.8) yields

$$J = \begin{bmatrix} -\delta - k_2 (E_T - y) & k_1 + k_2 x \\ k_2 (E_T - y) & -k_1 - k_2 x \end{bmatrix} \tag{5.9}$$

In this case, the graph $G_d(A)$ associated to J contains only two nodes, labeled as 1 and 2. Thus, the only possible loop in such a graph has length 2. To avoid the presence of such a loop, we have to ensure that the direction determined by $\alpha(1,2)$ is the same as that determined by $\alpha(2,1)$. Computation of these two quantities yields

$$\alpha_{1,2} = \frac{k_2 (E_T - y)}{\delta + k_2 (E_T - y)} < 1$$

and

$$\alpha_{2,1} = \frac{k_1 + k_2 x}{k_1 + k_2 x} = 1$$

Following the schematic procedure of section "Outline", this in turn implies that the directions determined by $\alpha(1,2)$ and $\alpha(2,1)$ are the same and that $\alpha_{1,2}\alpha_{2,1} < 1$. In particular, the unique edge of the graph is directed from node 1 to node 2 and no loop can be present. Contraction is then proven.

We refer the interested reader to [40] for a detailed contraction analysis of the transcriptional module (5.8) using a matrix measure induced by a non-Euclidean norm. In the same paper, some generalizations are provided including: the case where X is activated by enzyme kinetics, analysis of the interconnections of models (5.8), analysis of a larger class of nonlinear systems presenting a structure similar to that of (5.8). An application to synthetic biology is also presented: specifically, it is shown how to use the proposed methodology in order to entrain a population of Repressilators.

Synchronization of Biological Systems

The problem that we address in this Section is that of tuning the parameters of synthetic biological oscillators so that, when coupled, they self synchronize (see [35, 37, 39] for further details). We show that using the graphical approach of section "Outline" one obtains a set of conditions for synchronization that naturally map onto the biochemical parameters of each of the oscillators in the network of interest. As a representative example, we consider a network of Repressilators.

The Repressilator is a synthetic biological circuit of three genes inhibiting each other in a cyclic way [10]. As shown in Fig. 5.1, gene *lacI* (associated to the

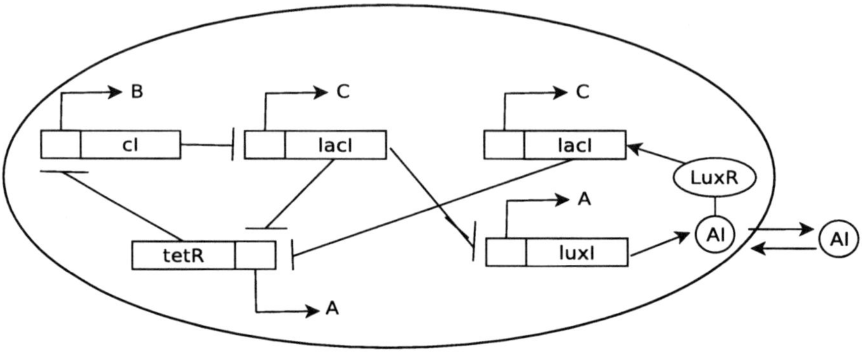

Fig. 5.1 Repressilator circuit and coupling mechanism

state-variable c_i in our model) expresses protein LacI (C_i), which inhibits transcription of gene *tetR* (a_i). This translates into protein TetR (A_i), which inhibits transcription of gene cI (b_i). Finally, the protein CI (B_i) translated from *cI* inhibits expression of *lacI*, completing the cycle. In [11], a modular addition to the *classical* Repressilator circuit is proposed with the aim of coupling different oscillators using the quorum sensing mechanism.

Quorum sensing provides a broadcast strategy for the exchange of information between bacteria (for further details see section "Generic Quorum Sensing Networks"). One could think of bacteria as nodes in a network that becomes fully connected via an all-to-all topology when quorum sensing is present.

To model the dynamics of gene expression in the cell, one must keep track of the temporal evolution of all $mRNA$ and protein concentrations. Note that, for the sake of simplicity, variations in the cell density are neglected here. The resulting mathematical model for the network is

$$\begin{aligned}
\dot{a}_i &= -a_i + \frac{\alpha}{1 + C_i^2} \\
\dot{b}_i &= -b_i + \frac{\alpha}{1 + A_i^2} \\
\dot{c}_i &= -c_i + \frac{\alpha}{1 + B_i^2} + \frac{kS_i}{1 + S_i} \\
\dot{A}_i &= \beta_A a_i - d_A A_i \\
\dot{B}_i &= \beta_B b_i - d_B B_i \\
\dot{C}_i &= \beta_C c_i - d_C C_i \\
\dot{S}_i &= -k_{s0} S_i + k_{s1} A_i - \eta (S_i - S_e) \\
\dot{S}_e &= -k_{se} S_e + \eta_{ext} \sum_{j=1}^{N} (S_j - S_e)
\end{aligned} \tag{5.10}$$

having chosen the Hill coefficient equal to 2 as in [11]. We assume that the Repressilator circuits on which the approach is applied are all identical. Referring to [11] we set the parameters $\beta_A = \beta_B = \beta_C = 2$. In (5.10) the dynamical equations corresponding to the Repressilator circuits, i.e. the intracellular species concentrations, are denoted with the subscript i, while S_e is the dynamical equation for the coupling auto-inducer.

The network of interest is an all-to-all network. Hence, the virtual system can be chosen as having the same dynamics as the individual Repressilator circuit, forced by the external coupling signal S_e (see e.g. (see [35, 36, 39]) for further details), i.e.

$$\begin{aligned}
\dot{a} &= -a + \frac{\alpha}{(1 + C^2)} \\
\dot{b} &= -b + \frac{\alpha}{(1 + A^2)} \\
\dot{c} &= -c + \frac{\alpha}{(1 + B^2)} + \frac{(kS_i)}{(1 + S_i)} \\
\dot{A} &= \beta_A a - d_A A \\
\dot{B} &= \beta_B b - d_B B \\
\dot{C} &= \beta_C c - d_C C \\
\dot{S} &= -k_{s0} S + k_{s1} A - \eta (S - S_e) \\
\dot{S}_e &= -k_{se} S_e + \eta_{ext} (S_1 + \ldots + S_N) - \eta_{ext} N S_e
\end{aligned} \tag{5.11}$$

Indeed, by direct inspection it is easy to check that, by substituting the state variables of the nodes dynamics for the virtual variables (i.e. $[a_i, b_i, c_i, A_i, B_i, C_i, S_i, S_e]$ for $[a, b, c, A, B, C, S, S_e]$), the equations of the each Repressilator circuit in the network are obtained. In this sense, the virtual system embeds the trajectories of all network oscillators. Hence, contraction of (5.11) implies synchronization of (5.10).

We can now check contraction of (5.11) using the graphical approach of section "Outline". Computing the Jacobian matrix of the virtual system yields

$$J = \begin{bmatrix}
-1 & 0 & 0 & 0 & 0 & f_1(C) & 0 & 0 \\
0 & -1 & 0 & f_1(A) & 0 & 0 & 0 & 0 \\
0 & 0 & -1 & 0 & f_1(B) & 0 & f_2(S) & 0 \\
\beta & 0 & 0 & -\beta & 0 & 0 & 0 & 0 \\
0 & \beta & 0 & 0 & -\beta & 0 & 0 & 0 \\
0 & 0 & \beta & 0 & 0 & -\beta & 0 & 0 \\
0 & 0 & 0 & k_{s1} & 0 & 0 & -k_{s0} - \eta & \eta \\
0 & 0 & 0 & 0 & 0 & 0 & 0 & -k_{se} - k_{diff}
\end{bmatrix} \tag{5.12}$$

where f_1 and f_2 denote the partial derivatives of decreasing and increasing Hill functions with respect to the state variable of interest and $k_{diff} = \eta_{ext} N$. Note that the Jacobian matrix J has the same structure as (5.3) with

$$J_{11} := \begin{bmatrix} -1 & 0 & 0 & 0 & 0 & f_1(C) & 0 \\ 0 & -1 & 0 & f_1(A) & 0 & 0 & 0 \\ 0 & 0 & -1 & 0 & f_1(B) & 0 & f_2(S) \\ \beta & 0 & 0 & -\beta & 0 & 0 & 0 \\ 0 & \beta & 0 & 0 & -\beta & 0 & 0 \\ 0 & 0 & \beta & 0 & 0 & -\beta & 0 \\ 0 & 0 & 0 & k_{s1} & 0 & 0 & -k_{s0} - \eta \end{bmatrix},$$

$J_{12} := \left[0\,0\,0\,0\,0\,0\,\eta\right]^T$, $J_{22} := -K_q$. Thus it represents a hierarchical combination of dynamical systems, [24]. Furthermore, notice that J_{22} (associated to the quorum sensing dynamics) is negative, i.e. such dynamics is contracting. This implies that the overall dynamics of the virtual system is contracting if the submatrix J_{11} is contracting (see section "Basic Results"). Thus, our approach can be applied directly onto the submatrix $\tilde{J} := J_{11}$.

The diagonal elements of $\tilde{J}$ are all negative, thus (see section "Outline"), $\mathcal{G}_d(\mathcal{A})$ has to be constructed. In so doing, matrix $\mathcal{A}$ is derived:

$$\mathcal{A} = \begin{bmatrix} 0 & 0 & 0 & 1 & 0 & 1 & 0 \\ 0 & 0 & 0 & 1 & 1 & 0 & 0 \\ 0 & 0 & 0 & 0 & 1 & 1 & 1 \\ 1 & 1 & 0 & 0 & 0 & 0 & 1 \\ 0 & 1 & 1 & 0 & 0 & 0 & 0 \\ 1 & 0 & 1 & 0 & 0 & 0 & 0 \\ 0 & 0 & 1 & 1 & 0 & 0 & 0 \end{bmatrix} \tag{5.13}$$

From (5.13), $\mathcal{G}(\mathcal{A})$ is obtained as shown in Fig. 5.2 (*left panel*).

Then, computation of coefficients $\alpha(i, j)$ (reported in Table 5.2) provides the directions of the edges of $\mathcal{G}(\mathcal{A})$. Notice that the elements of the left column of Table 5.2 are all state-dependent. This implies that the directions of the corresponding edges in $\mathcal{G}_d(\mathcal{A})$ can be time-varying as they are associated to conditions

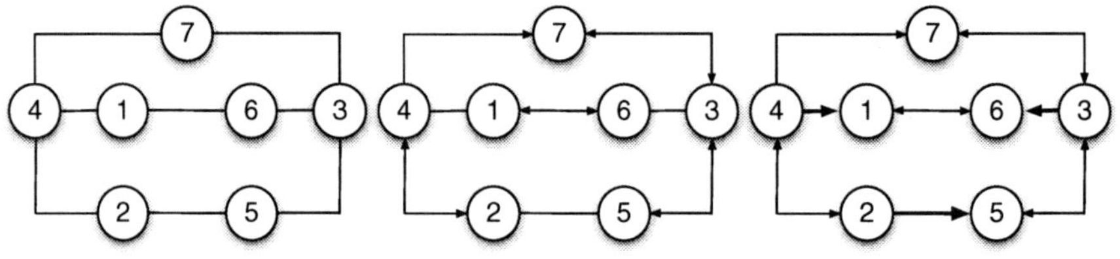

Fig. 5.2 Graphs associated to J: $\mathcal{G}_{un}(\mathcal{A})$ (*left panel*); $\mathcal{G}_d(\mathcal{A})$ with state dependent edges (*central panel*); choice for $\mathcal{G}_d(\mathcal{A})$ (*right panel*)

Table 5.2 Set of coefficients $\alpha\,(i,j)$

$\alpha\,(i,j)$	Algebraic expression	$\alpha\,(i,j)$	Algebraic expression
$\alpha\,(1,6)$	$\frac{2\alpha C}{(1+C^2)^2}$	$\alpha\,(4,1)$	$\frac{d_A}{\beta_A}$
$\alpha\,(2,4)$	$\frac{2\alpha A}{(1+A^2)^2}$	$\alpha\,(5,2)$	$\frac{d_B}{\beta_B}$
$\alpha\,(3,5)$	$\frac{4\alpha B}{(1+B^2)^2}$	$\alpha\,(6,3)$	$\frac{d_C}{\beta_C}$
$\alpha\,(3,7)$	$\frac{2K}{(1+S_i)^2}$	$\alpha\,(7,4)$	$\frac{K_{s1}}{K_{s0}+\eta}$

Table 5.3 Constraints on the biochemical parameters imposed by the graphical approach

Direction between node i and node j	Constraint
from node 4 to node 1	$\frac{d_A}{\beta_A} > 1$
from node 2 to node 5	$\frac{d_B}{\beta_B} < 1$
from node 3 to node 6	$\frac{d_C}{\beta_C} < 1$

which are functions of the state. Moreover, due to biochemical constraints [11], $\alpha\,(7,4) < 1$. However, the other coefficients in the table can be easily tuned since they depend only on biochemical parameters of the network. In Fig. 5.2 (*central panel*), a *partially directed* graph is shown, obtained by assigning directions to the edges between nodes corresponding to the first four rows and the last row of Table 5.2. Notice that the edges associated to state-dependent conditions are all denoted with a double arrow, as the directions of these links might vary in time. The design task is then to use coefficients $\alpha\,(4,1)$, $\alpha\,(5,2)$, $\alpha\,(6,3)$ to avoid the formation of loops as required by the graphical approach. A possible choice is presented in the right panel of Fig. 5.2. The inequalities associated to the new directions are reported in Table 5.3. To satisfy these constraints we can choose $d_a = 2\beta_A$, $d_B = 0.5\beta_B$, $d_C = 0.4\beta_C$. Simulation results, shown in Fig. 5.3, confirm that under these conditions synchronization is indeed achieved.

Generic Quorum Sensing Networks

In section "Entrainment and Synchronization of Biological Systems" we studied the problem of synchronizing a population of Repressilators by tuning their biochemical parameters. One of the main features of such a network was that coupling between nodes occurred by means of an *autoinducer* molecule. That is, network nodes make use of the environment where they live to communicate and hence to achieve a coordinated, synchronous, behavior.

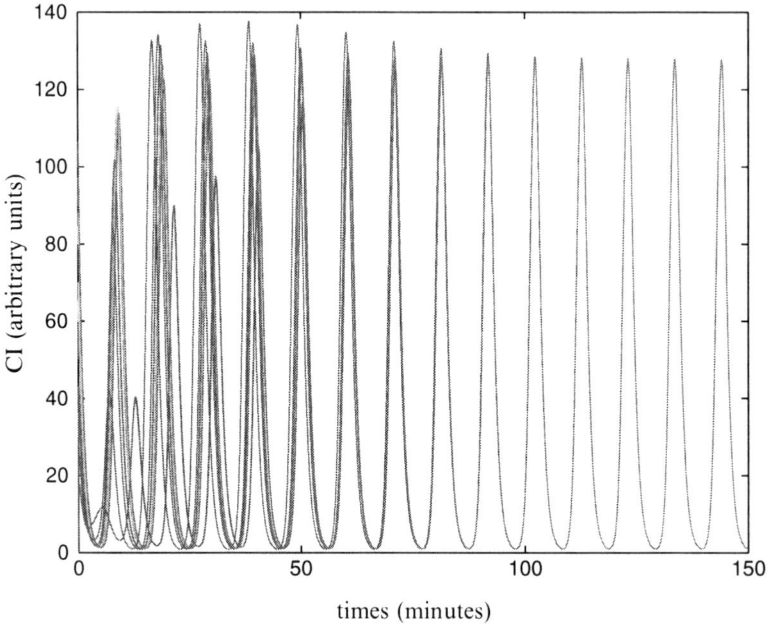

Fig. 5.3 Synchronization regime emerging for network (5.10)

This is a common feature of many natural synchronization phenomena, where communication between individual elements occurs not directly, but rather through the environment. One of these instances is bacterial quorum sensing, where bacteria release signaling molecules in the environment which in turn are sensed and used for population coordination. This mechanism [27,29,30] is believed to play a key role in bacterial infections, as well as e.g. in bioluminescence and biofilm formation [3,28]. In a neuronal context, a mechanism similar to that of quorum sensing may involve *local field potentials*, which may play an important role in the synchronization of clusters of neurons [2, 8, 32, 44], or it may occur through a different level in a cortical hierarchy [7, 13, 14, 20, 48]. Other examples of such a mechanism are the synchronization of chemical oscillations of catalyst-loaded reactants in a medium of catalyst-free solution [45], cold atoms interacting with a coherent electromagnetic field [18] and the onset of coordinated activity in a population of micro-organisms living in a shared environment [16, 34]. Besides its biological pervasiveness, quorum sensing may also be viewed as an astute *computational* tool. Specifically, the use of a shared variable significantly reduces the number of links required to achieve a given connectivity [33, 46].

In this Section, sufficient conditions for the coordination of nodes communicating through quorum-sensing-like mechanisms are presented which generalize the analysis of section "Entrainment and Synchronization of Biological Systems". Those results, based on [38], can be used both to study natural networks and to guide the design of communication mechanisms in synthetic or partially synthetic networks.

From a network dynamics viewpoint, the key characteristic of quorum sensing-like mechanisms lies in the fact that communication between nodes (e.g. bacteria) occurs by means of a shared quantity (e.g. the autoinducer concentration), typically in the environment. Furthermore, the production and degradation rates of such a quantity are affected by all the nodes of the network. Therefore, a detailed model of such a mechanism needs to keep track of the temporal evolution of the shared quantity, resulting in an additional set of ordinary differential equations:

$$\begin{aligned} \dot{x}_i &= f(x_i, z, t) \qquad i = 1, \ldots, N \\ \dot{z} &= g(z, \Psi(x_1, \ldots, x_N), t) \end{aligned} \tag{5.14}$$

Here, the set of state variables of the i-th node, i.e. the i-th element composing the network, is x_i, while the set of the state variables of the common (shared) medium dynamics is z and the number of nodes communicating over the common medium is N. That is, in terms of the Repressilator example of section "Entrainment and Synchronization of Biological Systems", x_i denotes the concentrations of the biochemical species $[a_i, b_i, c_i, A_i, B_i, C_i, S_i]$ composing the Repressilator and coupling circuit of (5.10), while z denotes the extracellular autoinducer concentration, S_e in (5.10).

Notice that the nodes dynamics and the medium dynamics can be of different dimensions (e.g. $x_i \in \mathbb{R}^n$, $z \in \mathbb{R}^d$). The dynamics of the nodes affect the dynamics of the common medium by means of some (coupling, or input) function, $\Psi : \mathbb{R}^{Nn} \to \mathbb{R}^d$. These functions may depend only on some of the components of the x_i or of z (as the example in section "Controlling Synchronization of Genetic Oscillators" illustrates). A simplified version of the above model was recently analyzed by means of the graphical approach presented above(see [35]).

We remark here that in the case of diffusive-like coupling between nodes and the common medium, system (5.14) reduces to:

$$\begin{aligned} \dot{x}_i &= f(x_i, t) + k_z(z) - k_x(x_i) \qquad i = 1, \ldots, N \\ \dot{z} &= g(z, t) + \sum_{j=1}^{N} \left[u_x(x_j) - u_z(z)\right] \end{aligned} \tag{5.15}$$

For instance, in the mathematical model of the network of Repressilators coupled by means of an autoinducer, i.e. (5.10) we have $u_x(x_j) - u_z(z) := \eta_{ext}(S_j - S_e)$ and:

$$f(x_i, t) := \begin{bmatrix} -a_i + \dfrac{\alpha}{1 + C_i^2} \\ -b_i + \dfrac{\alpha}{1 + A_i^2} \\ -c_i + \dfrac{\alpha}{1 + B_i^2} + \dfrac{kS_i}{1 + S_i} \\ \beta_A a_i - d_A A_i \\ \beta_B b_i - d_B B_i \\ \beta_C c_i - d_C C_i \\ -k_{s0} S_i + k_{s1} A_i \end{bmatrix} \quad k_z(z) - k_x(x_i) := \begin{bmatrix} 0 \\ 0 \\ 0 \\ 0 \\ 0 \\ 0 \\ -\eta(S_i - S_e) \end{bmatrix}$$

The following result is a sufficient condition for convergence of all nodes trajectories of (5.14) towards each other (see [38] for a proof).

Theorem 3. *All nodes trajectories of network (5.14) globally exponentially converge towards each other if the function* $f(x, v(t), t)$ *is contracting for any* $v(t) \in \mathbb{R}^d$.

Theorem 3 provides a sufficient condition on the network nodes' dynamics, i.e. $f(x, v, t)$, ensuring synchronization. That is, under the condition of Theorem 3, network synchronization is attained regardless of the common medium dynamics. With the next result, we show that such dynamics can be used (and becomes indeed *crucial*) to guarantee some desired property on the steady state synchronous evolution. Specifically, we consider the problem of synchronizing a network of interest onto a periodic orbit having some desired period, T. To this aim, the idea is to use some T-periodic control input, say $r(t)$, that acts on the dynamics of the media shared by network nodes. A related problem has been recently addressed in [40] in the context of entrainment of biochemical systems.

Theorem 4. *Consider the following network*

$$\begin{aligned} \dot{x}_i &= f(x_i, z) \qquad i = 1, \ldots, N \\ \dot{z} &= g(z, \Psi(x_1, \ldots, x_N)) + r(t) \end{aligned} \tag{5.16}$$

where $r(t)$ *is a* T*-periodic signal. All the nodes of the network synchronize onto a periodic orbit of period* T, *say* $x_T(t)$, *if:*

- $f(x_i, v(t))$ *is a contracting functions;*
- *the reduced order system* $(x_c(t) \in \mathbb{R}^n)$

$$\begin{aligned} \dot{x}_c &= f(x_c, z) \\ \dot{z} &= g(z, \Psi(x_c, \ldots, x_c)) + r(t) \end{aligned}$$

is contracting.

We refer the interested reader to [38] for the proofs of the above results. In the above cited paper some extensions are presented which provide sufficient conditions for (cluster) synchronization of quorum sensing networks consisting of multiple nodes and media.

Controlling Synchronization of Genetic Oscillators

As a representative application of the results presented in section "Generic Quorum Sensing Networks" we consider the problem of synchronizing a population of genetic oscillators onto a periodic orbit of desired period (see [38]).

We consider the genetic circuit presented in [22] (a variant of [21]), see Fig. 5.4, *top panel*. Such a circuit is composed of two engineered gene networks that have been experimentally implemented in *E. coli*; namely: the toggle switch [12] and

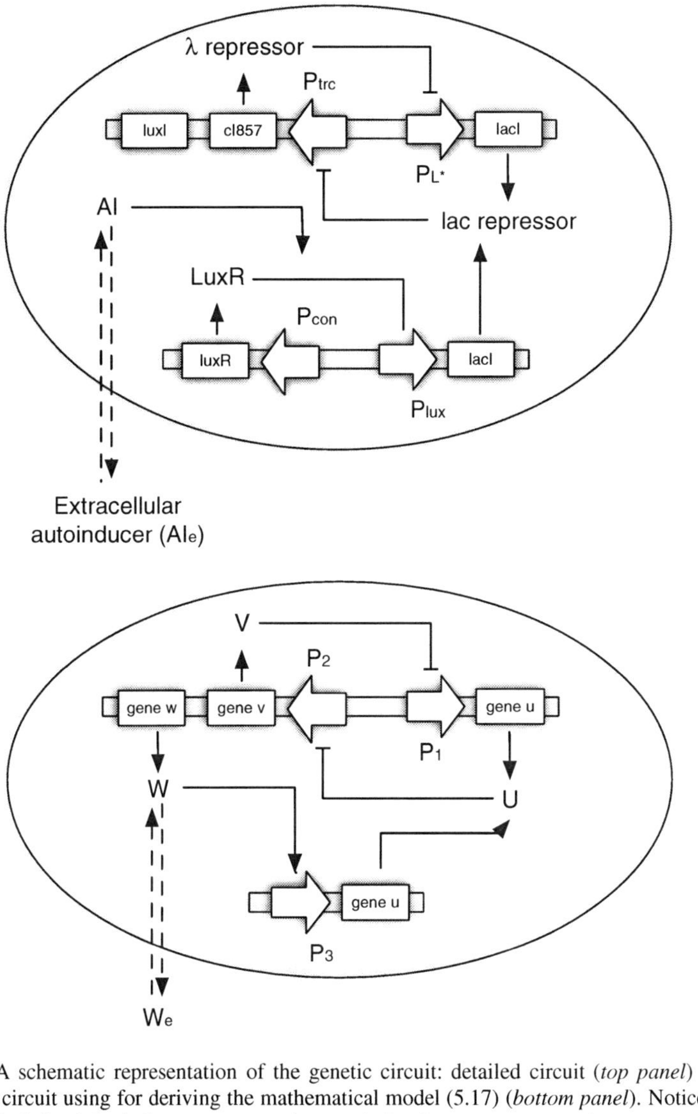

Fig. 5.4 A schematic representation of the genetic circuit: detailed circuit (*top panel*) and the simplified circuit using for deriving the mathematical model (5.17) (*bottom panel*). Notice that in the simplified circuit both the promoters and transcription factors are renamed

an intercell communication system [47]. The toggle switch is composed of two transcription factors: the *lac* repressor, encoded by gene *lacI*, and the temperature-sensitive variant of the λcI repressor, encoded by the gene *cI857*. The expressions of *cI8547* and *lacI* are controlled by the promoters P_{trc} and P_{L*} respectively (for further details see [22]). The intercell communication system makes use of components of the quorum-sensing system from *Vibro fischeri* (see e.g. [30] and references therein). Such a mechanism allows cells to sense population density through the transcription factor LuxR, which is an activator of the genes expressed by the P_{lux} promoter, when a small molecule AI binds to it. This small molecule, synthesized by the protein LuxI, is termed as autoinducer and it can diffuse across the cell membrane.

In [22], the following dimensionless simplified model is analyzed (see Fig. 5.4, *bottom panel*):

$$\dot{u}_i = \frac{\alpha_1}{1+v_i^\beta} + \frac{\alpha_3 w_i^\eta}{1+w_i^\eta} - d_1 u_i \tag{5.17a}$$

$$\dot{v}_i = \frac{\alpha_2}{1+u_i^\gamma} - d_2 v_i \tag{5.17b}$$

$$\dot{w}_i = \varepsilon\left(\frac{\alpha_4}{1+u_i^\gamma} - d_3 w_i\right) + 2d\,(w_e - w_i) \tag{5.17c}$$

$$\dot{w}_e = \frac{D_e}{N}\sum_{i=1}^{N}(w_i - w_e) - d_e w_e \tag{5.17d}$$

where u_i, v_i and w_i denotes the (dimensionless) concentrations of the *lac* repressor, λ repressor and LuxR-AI activator respectively. The state variable w_e denotes instead the (dimensionless) concentration of the extracellular autoinducer.

The control mechanism that we use here is an exogenous signal acting on the extracellular autoinducer concentration, see also [40]. That is, the idea is to modify (5.17d) as follows

$$\dot{w}_e = \frac{D_e}{N}\sum_{i=1}^{N}(w_i - w_e) - d_e w_e + r\,(t) \tag{5.18}$$

where $r\,(t)$ is some T-periodic signal. The set up that we have in mind here is illustrated in Fig. 5.5, where multiple copies of the genetic circuit of interest share the same surrounding solution, on which $r\,(t)$ acts. From the technological viewpoint, $r\,(t)$ can be implemented by controlling the temperature of the surrounding solution, and/or using e.g. the recently developed microfluidics technology (see e.g. [5] and references therein).

In what follows, we will use Theorem 3 to find a set of biochemical parameters that ensure synchronization of (5.17a)–(5.17d). This, using the results of the above Section, immediately implies that the forced network (5.17a)–(5.17c), (5.18) globally exponentially converges towards a T-periodic steady state behavior.

System (5.17) has the same structure as (5.15), with $x_i = [u_i, v_i, w_i]^T$, $z = w_e$, and:

$$f\,(x_i, t) = \begin{bmatrix} \dfrac{\alpha_1}{1+v_i^\beta} + \dfrac{\alpha_3 w_i^\eta}{1+w_i^\eta} - d_1 u_i \\ \dfrac{\alpha_2}{1+u_i^\gamma} - d_2 v_i \\ \varepsilon\left(\dfrac{\alpha_4}{1+u_i^\gamma} - d_3 w_i\right) \end{bmatrix}$$

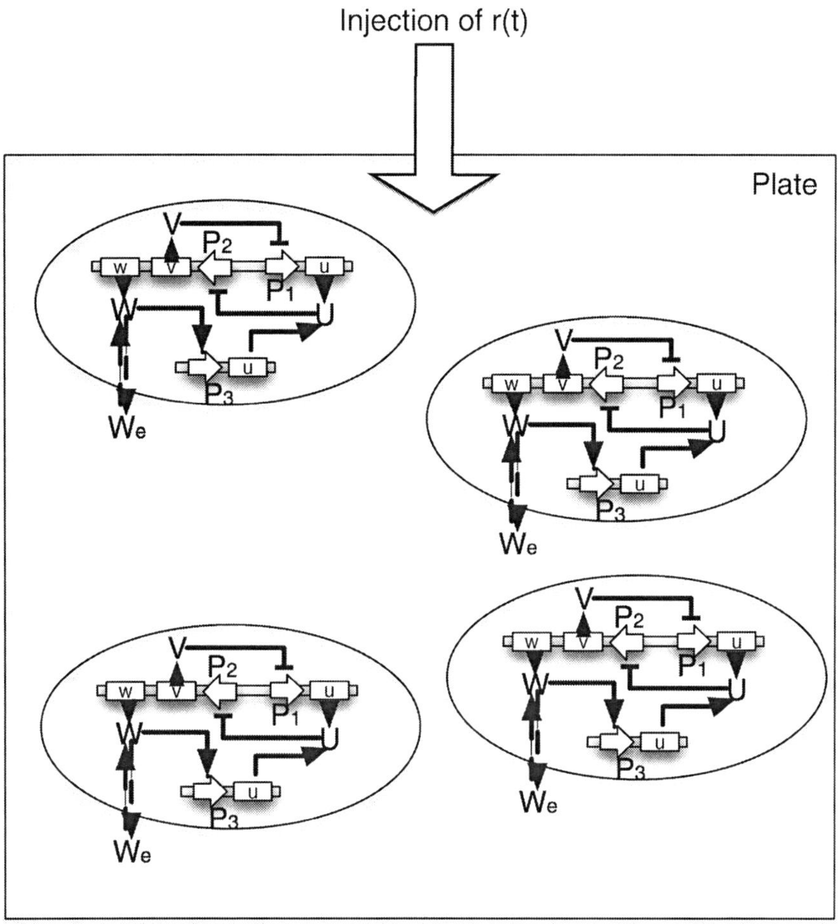

Fig. 5.5 Network control setup

$$k_z(z) - k_x(x_i) = \begin{bmatrix} 0 \\ 0 \\ 2d\,(w_e - w_i) \end{bmatrix}$$

$$g(z,t) = -d_e w_e$$

$$\sum\nolimits_{i=1}^{N} \left[u_x(x_i) - u_z(z)\right] = \frac{D_e}{N} \sum\nolimits_{i=1}^{N} (w_i - w_e)$$

The hypotheses of Theorem 3 are fulfilled if:

1. $f(x_i, t) - k_x(x_i)$ is contracting;
2. $g(z,t) - N u_z(z)$ is contracting.

That is, contraction is ensured if there exist some matrix measures, μ_* and μ_{**}, such that

$$\mu_* \left(f\left(x_i, t\right) - k_x\left(x_i\right)\right) \quad \text{and} \quad \mu_{**}\left(g\left(z, t\right) - N u_z\left(z\right)\right)$$

are uniformly negative definite. We use the above two conditions in order to obtain a set of biochemical parameters ensuring node convergence. A possible choice for the above matrix measures is $\mu_* = \mu_{**} = \mu_1$ (see [39,40]). Clearly, other choices for the matrix measures μ_* and μ_{**} can be made, leading to different algebraic conditions, and thus to (eventually) a different choice of biochemical parameters.

We assume that $\beta = \eta = \gamma = 2$, and show how to find a set of biochemical parameters satisfying the above two conditions.

Condition 1. Differentiation of $\frac{\partial f}{\partial x_i} - \frac{\partial k_x}{\partial x_i}$ yields the Jacobian matrix (where the subscripts have been omitted)

$$J_i := \begin{bmatrix} -d_1 & \dfrac{-2\alpha_1 v}{\left(1+v^2\right)^2} & \dfrac{2\alpha_3 w}{\left(1+w^2\right)^2} \\ \dfrac{-2\alpha_2 u}{\left(1+u^2\right)^2} & -d_2 & 0 \\ \dfrac{-2\varepsilon\alpha_4 u}{\left(1+u^2\right)^2} & 0 & -\varepsilon d_3 - 2d \end{bmatrix} \tag{5.19}$$

Now, by definition of μ_1, we have:

$$\mu_1\left(J_i\right) = \max\left\{-d_1 + \frac{2\alpha_2 u}{\left(1+u^2\right)^2} + \frac{2\varepsilon\alpha_4 u}{\left(1+u^2\right)^2},\right.$$
$$\left. -d_2 + \frac{2\alpha_1 v}{\left(1+v^2\right)^2}, -\varepsilon d_3 - 2d + \frac{2\alpha_3 w}{\left(1+w^2\right)^2}\right\}$$

Thus, J_i is contracting if $\mu_1\left(J_i\right)$ is uniformly negative definite. That is,

$$\begin{aligned} & -d_1 + \frac{2\alpha_2 u}{\left(1+u^2\right)^2} + \frac{2\varepsilon\alpha_4 u}{\left(1+u^2\right)^2} \\ & -d_2 + \frac{2\alpha_1 v}{\left(1+v^2\right)^2} \\ & -\varepsilon d_3 - 2d + \frac{2\alpha_3 w}{\left(1+w^2\right)^2} \end{aligned} \tag{5.20}$$

are all uniformly negative. Notice now that the maximum of the function $a\left(v\right) = \frac{\bar{a}v}{\left(1+v^2\right)^2}$ is $\hat{a} = \frac{3\sqrt{3}\bar{a}}{16}$. Thus, the set of inequalities (5.20) is fulfilled if:

$$-d_1 + \frac{6\alpha_2\sqrt{3}}{16} + \frac{6\varepsilon\alpha_4\sqrt{3}}{16}$$

$$-d_2 + \frac{6\alpha_1\sqrt{3}}{16}$$

$$-\varepsilon d_3 - 2d + \frac{6\alpha_3\sqrt{3}}{16} \tag{5.21}$$

are all uniformly negative.

Condition 2. In this case it is easy to check that the matrix $J_e := \frac{\partial g}{\partial z} - N\frac{\partial u}{\partial z}$ is contracting for any choice of the (positive) biochemical parameters D_e, d_e.

Thus, we can conclude that any choice of biochemical parameters fulfilling (5.21) ensures synchronization of the network onto a periodic orbit of period T. In [22], it was shown that a set of parameters for which synchronization is attained is: $\alpha_1 = 3, \alpha_2 = 4.5, \alpha_3 = 1, \alpha_4 = 4, \varepsilon = 0.01, d = 2, d_1 = d_2 = d_3 = 1$. We now use the guidelines provided by (5.21) to make a minimal change of the parameters values ensuring network synchronization with steady state oscillations of period T. Specifically, such conditions can be satisfied by setting $d_1 = 6$, $d_2 = 2$. Fig. 5.6 shows the behavior of the network for such a choice of the parameters.

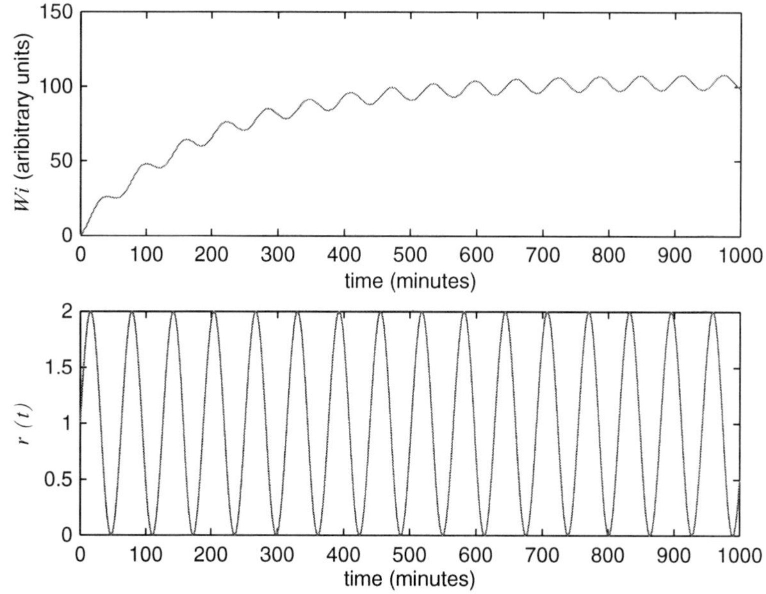

Fig. 5.6 Behavior of (5.17a)–(5.17c), (5.18), when forced by $r(t) = 1 + \sin(0.1t)$. Notice that the nodes have initial different conditions, and that they all converge onto a common asymptotic having the same period as $r(t)$

Conclusions

We discussed how contraction theory can be used as an effective tool to investigate the properties of biological systems and networks based on a number of recent results [37–40]. In particular, after introducing the basic results concerning contraction theory, we presented a graphical algorithm that can be used to assess effectively whether a given system (or network) of interest is contracting. Such a methodology was based on the use of non-Euclidean norms and matrix measures and was applied to a set of representative biological applications. Specifically, the problems were studied of assessing entrainment and synchronization of biological circuits. As a testbed example, we considered the synchronization of networks of Repressilators coupled via quorum sensing. Firstly, we showed that, using contraction, it is indeed possible to derive conditions on the biochemical parameters of each individual oscillator in order to guarantee convergence towards a synchronous evolution. Then, we focussed on quorum sensing, giving conditions on the coupling parameters that can be used not only to synchronize but also to control the biological network towards some desired evolution. As discussed in this chapter, contracting systems share a number of useful properties that can be exploited in biological applications and their applications in biology is just at the beginning.

References

1. Alon U (2007) Network motifs: theory and experimental approaches. Nature 450:450–461
2. Anastassiou C, Montgomery SM, Barahona M, Buzsaki G, Koch C (2010) The effect of spatially inhomogeneous extracellular electric fields on neurons. J Neurosci 30:1925–1936
3. Anetzberger C, Pirch T, Jung K (2009) Heterogeneity in quorum sensing-regulated bioluminescence of vibro harvey. Mol Microbiol 2:267–277
4. Angeli D (2002) A Lyapunov approach to incremental stability properties. IEEE Trans Automat Contr 47:410–321
5. Beebe DJ, Mensing G, Walker G (2002) Physics and applications of microfluidics in biology. Annu Rev Biomed Eng 4:261–286
6. Bertsekas D, Tsitsiklis J (1989) Parallel and distributed computation: numerical methods. Prentice-Hall, Upper Saddle River
7. Borenstein E, Ullman S (2008) Combined top–down/bottom–up segmentation. IEEE Trans Pattern Anal Mach Intell 30:2109–2125
8. Boustani SE, Marre O, Behuret P, Yger P, Bal T, Destexhe A, Fregnac Y (2009) Network–state modulation of power-law frequency-scaling in visual cortical neurons. PLoS Comput Biol 5:e1000519
9. Del Vecchio D, Ninfa AJ, Sontag ED (2008) Modular cell biology: retroactivity and insulation. Nat Mol Syst Biol 4:161
10. Elowitz MB, Leibler S (2000) A synthetic oscillatory network of transcriptional regulators. Nature 403:335–338
11. Garcia-Ojalvo J, Elowitz MB, Strogatz SH (2004) Modeling a synthetic multicellular clock: repressilators coupled by quorum sensing. Proc Natl Acad Sci U S A 101:10955–10960
12. Gardner T, Cantor C, Collins J (2000) Construction of a genetic toggle in Escherichia coli. Nature 403:339–342

13. George D, Hawkins J (2009) Towards a mathematical theory of cortical micro-circuits. PLoS Comput Biol 5:e1000532
14. Gigante G, Mattia M, Braun J, DelGiudice P (2009) Bistable perception modeled as competing stochastic integrations at two levels. PLoS Comput Biol 5:e1000430
15. Godsil C, Royle G (2001) Algebraic graph theory. Springer, New York
16. Gregor T, Fujimoto K, Masaki N, Sawai S (2010) The onset of collective behavior in social Amoebe. Science 328:1021–1025
17. Hartman P (1961) On stability in the large for systems of ordinary differential equations. Canadian J Math 13:480–492
18. Javaloyes J, Perrin M, Politi A (2008) Collective atomic recoil laser as a synchronization transient. Phys Rev E 78:011108
19. Jouffroy J (2005) Some ancestors of contraction analysis. In: Proceedings of the 44th IEEE Conference on Decision and Control, Seville, Spain pp 5450–5455
20. Kandel E, Schwartz J, Jessel T (2000) Principles of neural science, 4th edn. Oxford University Press, McGraw-Hill, New York, USA
21. Kobayashi H, Kaern M, Araki M, Chung K, Gardner T, Cantor C, Collins J (2004) Programmable cells: interfacing natural and engineered gene networks. Proc Natl Acad Sci USA 101:8414–8419
22. Kuznetsov A, Kaern M, Kopell N (2004) Synchrony in a population of hysteresis-based genetic oscillators. SIAM J Appl Math 65:392–425
23. Lewis DC (1949) Metric properties of differential equations. Am J Math 71:294–312
24. Lohmiller W, Slotine JJE (1998) On contraction analysis for non-linear systems. Automatica 34:683–696
25. Lohmiller W, Slotine JJE (2000) Nonlinear process control using contraction theory. AIChE J 46:588–596
26. Lohmiller W, Slotine JJ (2005) Contraction analysis of non-linear distributed systems. Int J Control 78:678–688
27. Miller M, Bassler B (2001) Quorum sensing in bacteria. Annu Rev Microbiol 55:165–199
28. Nadell CD, Xavier J, Levin SA, Foster KR (2008) The evolution of quorum sensing in bacteria biofilms. PLoS Comput Biol 6:e14
29. Nardelli C, Bassler B, Levin S (2008) Observing bacteria through the lens of social evolution. J Biol 7:27
30. Ng W, Bassler B (2009) Bacterial quorum-sensing network architectures. Ann Rev Genet 43:197–222
31. Pavlov A, Pogromvsky A, van de Wouv N, Nijmeijer H (2004) Convergent dynamics, a tribute to Boris Pavlovich Demidovich. Syst Control Lett 52:257–261
32. Pesaran B, Pezaris J, Sahani M, Mitra P, Andersen R (2002) Temporal structure in neuronal activity during working memory in macaque parietal cortex. Nature 5:805–811
33. Pham QC, Tabareau N, Slotine JJE (2009) A contraction theory approach to stochastic incremental stability. IEEE Trans Automat Contr 54:816–820
34. Prindle A, Hasty J (2010) Stochastic emergence of groupthink. Science 328:987–988
35. Russo G, di Bernardo M (2009) An algorithm for the construction of synthetic self synchronizing biological circuits. In: International symposium on circuits and systems. Taipei, Taiwan pp 305–308
36. Russo G, di Bernardo M (2009) Contraction theory and the master stability function: linking two approaches to study synchronization in complex networks. IEEE Trans Circuit Syst II 56:177–181
37. Russo G, di Bernardo M (2009) How to synchronize biological clocks. J Comput Biol 16:379–393
38. Russo G, Slotine J (2010) Global convergence of quorum-sensing networks, to be published in Phys Rev E 82:041919
39. Russo G, di Bernardo M, Slotine J (2011) A graphical algorithm to prove contraction of nonlinear circuits and systems. IEEE Trans Circuit Syst I 58:336–348
40. Russo G, di Bernardo M, Sontag E (2010) Global entrainment of transcriptional systems to periodic inputs. PLoS Comput Biol 6:e1000739

41. Slotine J (2003) Modular stability tools for distributed computation and control. Int J Adapt Control Signal Process 17:397–416
42. Slotine JJE, Li W (1990) Applied nonlinear control. Prentice-Hall, Englewood Cliffs
43. Stokkan KA, Yamazaki S, Tei H, Sakaki Y, Menaker M (2001) Entrainment of the circadian clock in the liver by feeding. Science 19:490–493
44. Tabareau N, Slotine J, Pham Q (2010) How synchronization protects from noise. PLoS Comput Biol 6:e1000637
45. Toth R, Taylor AF, Tinsley MR (2006) Collective behavior of a population of chemically coupled oscillators. J Phys Chem 110:10170–10176
46. Wang W, Slotine JJE (2005) On partial contraction analysis for coupled nonlinear oscillators. Biol Cyber 92:38–53
47. You L 3rd, Cox RS, Weiss R, Arnold FH (2004) Programmed population control by cell-cell communication and regulated killing. Nature 428:868–871
48. Yu G, Slotine JJE (2009) Visual grouping by oscillator networks. IEEE Trans Neural Netw 20:1871–1884

Part II
Modularity and Abstraction

Chapter 6
Toward Modularity in Synthetic Biology: Design Patterns and Fan-out

Kyung Hyuk Kim, Deepak Chandran, and Herbert M. Sauro

Abstract Modularity is a concept that is widely used in biological science with various interpretations. In this chapter we will first give a general overview of modularity in biology, and later focus on modularity in synthetic biology. In engineering, a module is a component whose intrinsic functionality is independent of its surrounding milieu. In biology, however, modularity is less clear-cut; for example, modules can be classified by network interactions or by functional distinctiveness such as the reuse of protein domains. In synthetic biology the question of modularity is more closely related to engineering where functional independence is important. One way of defining synthetic modules is by specifying a generic pattern of regulations that results in desired functionalities, which we term a design pattern. In this perspective, connections between modules are described by the regulatory links, which are represented by molecular reactions. Under these reactions, the output of an upstream module – the concentration of regulating molecules – is sequestered by the input of the downstream module. This sequestration can cause changes in the upstream module function. We quantify the maximally tolerable load from the downstream input, which we term gene circuit fan-out. We provide an efficient and practical way of estimating the fan-out by experiment.

Keywords Module · Design pattern · Fan-out · Retroactivity · Electrical circuit · Gene expression noise · Stochasticity

Introduction

Why do we need to define the notion of 'modularity' in biology? There are two answers. The first answer is that recognizing modules in a biological system may allow us to reduce the system complexity by decomposing it into smaller

K.H. Kim (✉)
Department of Bioengineering, University of Washington, William H. Foege Building, Box 355061, Seattle, WA 98195-5061, USA
e-mail: kkim@uw.edu

H. Koeppl et al. (eds.), *Design and Analysis of Biomolecular Circuits: Engineering Approaches to Systems and Synthetic Biology*, DOI 10.1007/978-1-4419-6766-4_6,

more-manageable parts. The second is that understanding how to construct modules will allow synthetic biologists to engineer a biological system more efficiently by reusing existing modules. These two answers are not entirely distinct. Understanding modularity in natural systems will inevitably help synthetic biologists design artificial modules. Similarly, engineering synthetic modules enables us to gain better understanding of the requirements for modules, allowing us to recognize modules in natural systems.

The term modularity has different meanings for different biological systems of interest. Modules in natural systems can be clusters of interacting proteins or members of a complex pathway. Modules may also be patterns of biological interaction that are repeated in different context and provide similar functions. In synthetic biology, a module is a biological component that maintains its defined function. In order for a module to retain its function in different surrounding milieu, it must have functional independence. The majority of this chapter will discuss what factors can disrupt a module's function and how to maintain functional independence.

Modules in Natural Systems

The definition of a module and its interpretation are context-dependent in biology. In this section, we will discuss four different definitions and interpretations. The first two are concepts derived from classical graph theory, and the rest from computer and electrical engineering.

Modules as Physically Interacting Molecules

With the coming of high-throughput methods, networks of interacting proteins have been constructed for various organisms. These networks do not indicate the cause or effect of an interaction; they state the fact that two molecules are capable of interaction due to their physical structure. The networks formed by these interactions have some resemblances of other evolved networks, such as social networks or the Internet [1]. Such networks are often described by key words such as 'scale-free' and 'small-world'. These terms follow from the fact that the networks have fractal-like properties and that the path to reach any molecule from any other molecule is usually very short (hence it's a 'small world'). The distribution of highly interacting and weakly interacting molecules generally follow a power law distribution, and the nature of such networks have been thoroughly studied in graph theory [3, 12]. At the same time, they also have distinct properties, such as the distribution of network 'motifs', which are significantly recurring network subgraphs [2].

Modules in graph theory are based on clustering properties of nodes in a network [33]. For example, in social networks, it is common to see clusters of individuals who are all socially acquainted with each other. Further, if additional information such as ethnicity or gender are available, different clusters often show enrichment of different features. Such clustering and enrichment are seen in interacting proteins; clusters of interacting proteins often belong to the same biological process [44], such as DNA replication or stress-response. These clusters that are enriched in a biological process can be defined as biological modules. In this case, the concept of a module is defined by physical interactions of related biological molecules. Since the physical interactions do not imply any particular function describing an input-output relationship, this definition of modularity is less useful for the purpose of engineering.

Modules as Temporally Interacting Genes

Another definition of a module is a set of molecules that are temporally correlated [39]. In other words, if one molecule of a set is present at a particular point in time, then it is likely that the other molecules are also present. Similarly, if one is absent, then it is likely that the others are absent as well. These correlations are normally found from microarray data measuring expression of thousands of genes across multiple experimental conditions [26]. When we see such correlations, the most obvious hypothesis is that these correlated molecules are involved in common cellular processes. Interestingly, many of these correlations are conserved across multiple species [46], adding to the hypothesis that the common processes are used as a module. Combining physical interactions and temporal correlations can often result in more compact and meaningful modules [26]. Such definitions can be useful in identifying components of natural systems that are related to a particular pathway of interest. For example, suppose a metabolic engineer is interested in optimizing production of a metabolite. For such an engineering goal, understanding the biological players in the pathway will be crucial. Physical and temporal correlations from high-throughput data will be useful for such purpose. Nonetheless, this definition of modularity does not provide information on the functional input-output relationship of the module, and therefore the definition is not entirely useful to synthetic biology.

Modules as an Input-Output System

A module can be defined as a functional component displaying a certain input-output characteristics [36]. For example, the protein receptors and catalytic enzymes involved in quorum sensing can be considered a potential module. This is because the entire quorum sensing system has a specific ‘core’, or module, that can be reused across different species to provide the same functionality in different context. Weiss et al. [6] demonstrated that the quorum sensing system from *A. thaliana* can be placed inside yeast cells to provide the cells with quorum sensing capability. This

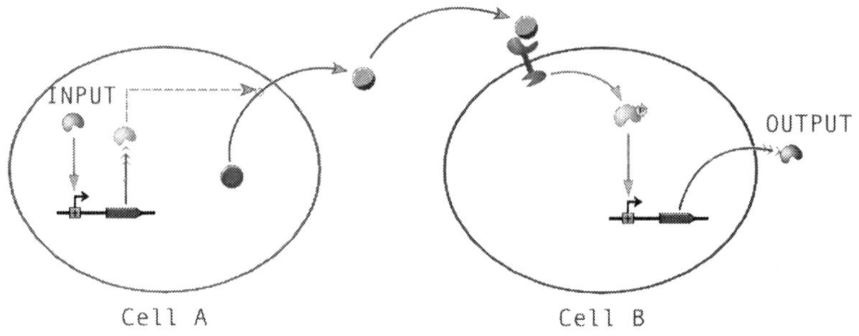

Fig. 6.1 Quorum sensing as a module. The quorum sensing system spans two or more cells. The input for the system is the transcription factor and the promoter that it controls. The output is defined as the downstream gene or the protein that the gene encodes. The input of the system drives the production of an enzyme that produces the quorum sensing molecule. The receiving cell contains a receptor that binds the specific signal molecule and triggers a transcriptional response, which is the output of the module

demonstrates that the function of the quorum sensing system is fairly independent of the host system, i.e. plant or yeast, although the system details, such as diffusion rates and binding affinities, will probably differ between yeast and plants. The quorum sensing system can be defined as a module that allows an input transcription factor to control the production of an output protein in another cell (Fig. 6.1). The definition of the module may include additional details such as time delay between the input and output.

Another good example of an input-output module are the numerous two-component systems in bacteria. The two-component systems consist of a membrane receptor that receives a specific extra-cellular signal and phosphorylates a protein inside the cell. The phosphorylated protein can be a transcription factor that upregulates one or more target genes. The two-component system is a module because each of those systems can be 'rewired' to control the expression of different downstream genes [50].

Modules as Design Patterns

The concept of 'design pattern' is frequently used in computer science [13], where it refers to generalized solutions to specific types of computer programming problems. The term has been used in biology [22] to capture the same concept. In this chapter, we define design patterns in the context of biology as general solutions to achieve biological objectives. Examples of the objectives can be specific types of signal-response curves, dynamic behaviors such as oscillations, or population level distribution of phenotypes.

Natural biological systems often contain design patterns that are repeatedly present in different contexts, and each pattern usually has specific characteristic

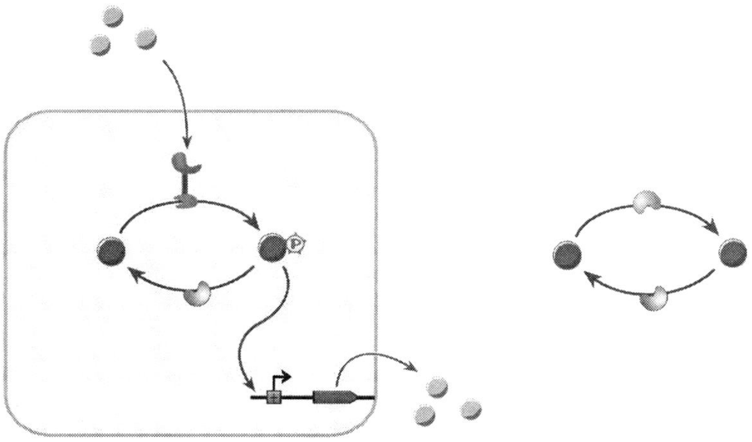

Fig. 6.2 A two-component system module and its design pattern. The module (*left*) convert extracellular signal to gene transcription and produce a sigmoidal response by using a design pattern: activation-deactivation loop (*right*). The boxed region is the module, and the molecules outside the boxed region are the inputs and outputs. The design pattern (*right*) can show a protein switching states; the state transition is catalyzed by enzymes. The sigmoidal response – design objective – is achieved when one of the enzymes is used as the 'input' and one of the states of protein is used as the 'output'. The two-component module (*left*) is a special case of the design pattern shown to the right

features (objectives). For example, the entire class of two-component systems follow a generic design pattern of an activation-deactivation loop (Fig. 6.2), which results in a sigmoidal input-output response [35]. While each two-component system has a specific molecule that acts as the input, the design pattern itself does not specify the type of molecule that acts as the input signal or the downstream protein.

Design patterns can include network features such as feedback regulation, which is commonly observed in natural systems as well as in synthetic systems. Negative feedback can also be used to maintain oscillations within a cell, and all natural oscillations that have been investigated has some form of negative feedback [48]. In contrast, positive feedback can result in bistability [48]. An interesting design pattern that combines oscillations and positive feedback is the one that causes segmentation in insects and mammal bodies [7]. In this design, the oscillations are interrupted by a morphogen gradient that recedes during the growth of the organism. As the morphogen leaves the system, the oscillations stop. However, the stopping of the oscillation triggers a switch, which causes the cell to differentiate into different types depending on the phase of the oscillation. The net effect is that an oscillation in time is converted to an oscillation in space. The design pattern, oscillator and positive feedback, is used for the objective, conversion of a temporal oscillation to a spatial one.

The definition of design patterns can overlap with that of network motifs [43], although design patterns can be more abstract than network motifs. Network motifs are frequently occurring network architectures. A design pattern need not have a specific network architecture. This can be demonstrated by the class of biological networks that have a non-monotonic input-output response. In these networks, the input acts as an activator until a specific threshold, after which it acts like a repressor. There are many network architectures that can give this response [11], but the general design pattern is the same: the input needs to upregulate and downregulate the output in some way such that the upregulation dominates initially and the downregulation dominates after the specific threshold, causing the non-monotonic behavior. The incoherent feed-forward motif [43] is a specific architecture, or network motif, that follows this design pattern, and it has been found to exhibit non-monotonic input-output behavior [11]. The design pattern can be the same as this motif, but it can also be something much more general. For example, a network where the downregulation and upregulation is achieved through protein phosphorylation would not be the same network motif as the one without the phosphorylation, but it would be the same design pattern.

Identifying design patterns in nature can greatly benefit biological engineering. There are many themes that are often repeated in nature. For example, more than half the genes in *E. coli* genetic regulatory networks have some form of autoregulation [43,47]. When we understand the reason for these many auto-regulations, we might identify certain design patterns that nature uses in constructing gene regulatory networks. There may be similar repeated themes in protein networks as well. Identifying design patterns that satisfy desired objectives will allow engineers to decompose a challenging problem into manageable subproblems.

Modules in Synthetic Systems

The last two definitions of modularity, namely input-output systems and design patterns, overlap with the concept of modularity in synthetic biology. Synthetic biologists often think of modules as input-output systems, which is exemplified by the several Boolean logic abstractions that can be observed in synthetic genetic networks [24]. Similarly, design patterns such as feedback have been repeatedly used in synthetic biology to achieve specific design objectives such as bistability [15, 18] and oscillations [14, 45, 49]. Additionally, concepts from classical control theory have been used in conjunction with knowledge of biological systems to construct simple devices such as linear amplifiers [27].

When modules are connected, the module interface can be described by reaction processes. It is important to realize that the reactions between molecules cause the involved reactants to become converted or sequestered during the process, e.g., through enzymatic reactions or binding-unbinding reactions. As a result, when we consider a functional module, the output of one module can be affected by downstream modules, hence disrupting the function of the upstream. The remainder of this chapter will discuss when such disruptions can occur and how they can be prevented.

Module Interface Condition: Fan-Out

In electrical engineering there exist guidelines and published constraints on how many electronic parts can be connected and driven from a source. For example, in analog circuits the impedance at the input is designed to be matched roughly at ten times the impedance at the driving circuit. In digital circuits, such as TTL (transistor-transistor logic) circuits (Lancaster DE: TTL Cookbook Indianapolis: Sams HW; 1974), the fan-out and fan-in – the maximum numbers of downstream and upstream logic gates that can be connected to – are specified for a given electronic module. Satisfying such constraints is crucial for the expected circuit functionality.

Similar criteria for connecting two synthetic biological modules has been proposed recently [8, 17, 20, 21]. The fan-out of a genetic circuit has been defined as the maximum number of downstream promoters that can be driven from an upstream circuit signal without significant time-delay or signal attenuation [20]. The fan-out was shown to be closely related to retroactivity proposed by Del Vecchio et al. [8]. Here we will show how the fan-out is quantified and estimated.

Module Interface Process

Transcription factors (TFs) play a role of connecting two synthetic gene circuits. The connection can be described by a set of reaction processes: transcription, translation, degradation, and downstream-module promoter regulation. This set of reactions will be called module interface processes (MIPs) (see Fig. 6.3).

Mapping Between a MIP and an RC-Circuit

We will show how a MIP can be mapped to an electric circuit composed of a resistor and a capacitor connected in series – RC circuit, and use this mapping for the interpretation of retroactivity [8].

Isolated Case When an upstream output is isolated, the interface process can be considered as a simple TF translation-degradation process (Fig. 6.4a, b). The concentration X of the TF follows

$$\frac{dX}{dt} = \alpha(t) - \gamma X, \tag{6.1}$$

with $\alpha(t)$ the translation rate and γ the degradation rate constant. Consider a RC circuit shown in Fig. 6.4c. The voltage V_{in} is applied to both the resistor and capacitor:

$$V_{in} = RI + V_{out}, \tag{6.2}$$

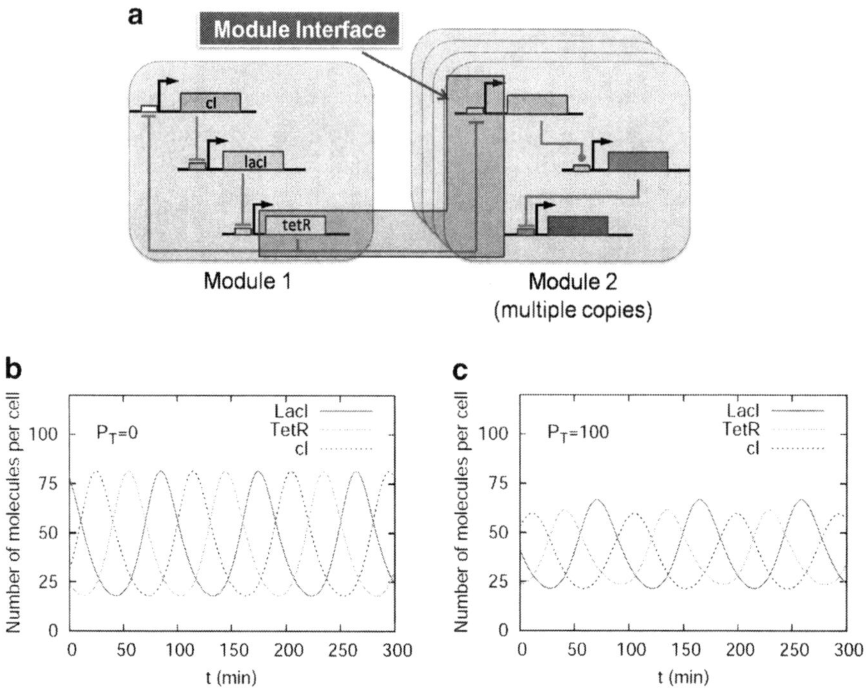

Fig. 6.3 Module interface process for gene circuits: (**a**) TetR repressors of the repressilator [10] drives a multiple copy of a downstream module. The downstream module can be considered to be placed on a plasmid. (**b** and **c**) The amplitude and period of signals in the upstream module can be changed as the load from the downstream increases. The BioModel BIOMD0000000012 [23] was used for the repressilator (refer to the supplementary information in [20])

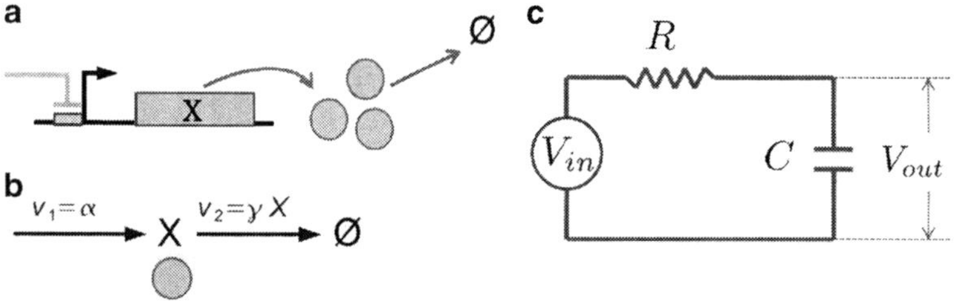

Fig. 6.4 Isolated module output: A translation-degradation process for X as shown in (**a**) can be simplified by a reaction process shown in (**b**). The process can be mapped to an RC-circuit as shown in (**c**) by $V_{out} = X$, $V_{in} = \alpha/\gamma$, and $RC = 1/\gamma$

with I the current and V_{out} the voltage drop across the capacitor. The current is equal to the rate of charge accumulation (Q) in the capacitor: $I = dQ/dt$, and an increment in the charge dQ increases the voltage drop across the capacitor: $dQ = CdV_{out}$, with C capacitance [28]. Thus, I can be expressed as

$$I = C\frac{dV_{out}}{dt}.$$

By substituting this into Eq. 6.2, we obtain

$$\frac{dV_{out}}{dt} = \frac{V_{in}}{RC} - \frac{V_{out}}{RC}, \tag{6.3}$$

where RC is known as the response time of the circuit [28]. From Eqs. 6.1 and 6.3, we obtain the following correspondence: $X = V_{out}$, $\alpha = V_{in}/RC$, and $\gamma = 1/RC$, and the response time τ_0 is expressed as

$$\tau_0 = RC = \frac{1}{\gamma}. \tag{6.4}$$

Thus, the MIP in the isolated case can be directly mapped to the RC-circuit.

Connected Case When two modules are connected (Fig. 6.5a), the upstream module output TFs are sequestered by the downstream promoters. This was shown to slow down the interface dynamics and this effect was quantified by retroactivity [8]. In [8], they assumed that the binding-unbinding process of the TF is fast enough that the process reaches the quasi-steady state and also assumed that the lifetime of the bound TF is much longer than that of the unbound TFs.

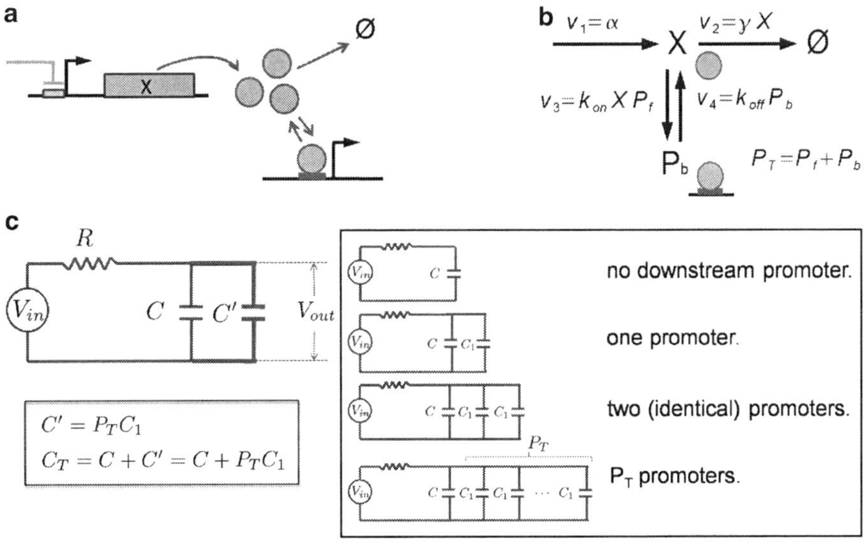

Fig. 6.5 Connected Module Output: The upstream TFs (X) regulate the downstream promoter (**a**). The corresponding MIP can be modeled as the reaction process shown in (**b**), where P_f, P_b, and P_T denote the numbers of free, bound, and total promoters, respectively. The reaction process is mapped to an RC-circuit with an increased capacitance by C' as shown in (**c**). It is shown that C' increases linearly with P_T (refer to [20] for the proof)

Specifically, they showed that the free TF concentration X changes in time by the following equation

$$\frac{dX}{dt} = (1 - \mathcal{R}(X))(\alpha - \gamma X), \tag{6.5}$$

where $\mathcal{R}(X)$ is the retroactivity that is always < 1 and non-negative. The extra factor $1 - \mathcal{R}$, that is also < 1 and non-negative, appears when compared with the isolated case. This is mathematically why the dynamics of X slows down. The slow-down is related to the increase in the apparent life time of X:

$$\tau_a \equiv \frac{1}{(1 - \mathcal{R})\gamma}. \tag{6.6}$$

We will consider the MIP shown in Fig. 6.5a, b, and map it to a circuit as shown in Fig. 6.5c. In the circuit, the total capacitance becomes the sum of the two capacitances:

$$C_T = C + C'.$$

and the response time becomes RC_T:

$$\tau = RC_T.$$

Thus, the output voltage is governed by the following equation:

$$\frac{dV_{out}}{dt} = \frac{V_{in}}{RC_T} - \frac{V_{out}}{RC_T} = \left[1 - \frac{C'}{C + C'}\right]\left[\frac{V_{in}}{RC} - \frac{V_{out}}{RC}\right]. \tag{6.7}$$

From Eqs. 6.5 and 6.7, we obtain

$$\mathcal{R} = \frac{C'}{C + C'}, \tag{6.8}$$

and find that the response time τ corresponds to τ_a (Eq. 6.6):

$$\tau = RC_T = \frac{1}{(1 - \mathcal{R})\gamma}\tau_a. \tag{6.9}$$

We have shown that connecting downstream promoters is equivalent to connecting extra capacitors in parallel. These extra capacitors increase the total capacitance of the circuit, which means it takes a longer time to fully charge, resulting in the longer response time. Biologically, the bound promoters sequester free TFs into a nearly non-degradable state (one of the assumptions taken in [8]). This causes the apparent lifetime to increase and the interface dynamics to slow down.

How is the response time τ related with the load from the downstream, i.e., the number of promoters P_T? The response time was shown to increase with P_T [20] as

$$\tau_{P_T} = R(C + P_T C_1), \tag{6.10}$$

with C_1 a proportionality constant. By comparing Eq. 6.10 with Eq. 6.9 the total capacitance can be obtained as

$$C_T = C + P_T C_1.$$

This indicates that a unit load of a single downstream promoter is C_1. This linearity appears since each downstream promoter acts as an independent sequestrator of the upstream output TFs. Note that the linearity does not come from any linearization approximation.

Gene Circuit Fan-Out

A gene circuit fan-out is defined by the maximum number of promoters in a downstream module that the upstream output can regulate without significant signal distortion. For example, we consider a MIP described in Fig. 6.3. When the upstream output TetR regulates downstream *tetR*-promoters, the oscillation amplitude of the TetR concentration can be significantly changed. In Fig. 6.3b, c, it was decreased by 40% when P_T was increased from 0 to 100. Our interest is to quantify the maximum value of P_T that the upstream module can tolerate.

Consider again the RC-circuit shown in Fig. 6.5c as a map of the simple MIP shown in Fig. 6.5a, and analyze the circuit frequency response between the input and output voltages. The capacitor of the circuit acts as a low pass filter: The high frequency components of the input signal are suppressed in the output due to the capacitor response time (charging time). When the capacitance no longer responds as fast as the input signal changes (Fig. 6.6b), the corresponding frequency is called the *cut-off frequency* (ω_c) (Fig. 6.6b) and this corresponds to the inverse of the response time [28]:

$$\omega_c = \frac{1}{RC_T}.$$

Let us assume that the upstream module is the repressilator. In [10], the oscillator could generate a frequency $1/150\,\text{min}^{-1}$. The time delay in genetic regulation may apply a certain upper limit ω_{max} in the frequency. If ω_{max} is smaller than the cut-off frequency ω_c, the repressilator output will operate without any significant signal loss. As the number of the downstream promoters increases, the cut-off frequency ($\omega_c = 1/RC_T$) decreases, and when ω_c becomes smaller than ω_{max}, the signal output will start to be suppressed. Thus, it is desirable that the total number of promoters must be smaller than a certain value. This will be called the *fan-out*. The fan-out denoted by $F_{\omega_{max}}$ is obtained where ω_c equals ω_{max} by solving for P_T:

$$F_{\omega_{max}} = \frac{C}{C_1}\left[\frac{1/\tau_0}{\omega_{max}} - 1\right]. \tag{6.11}$$

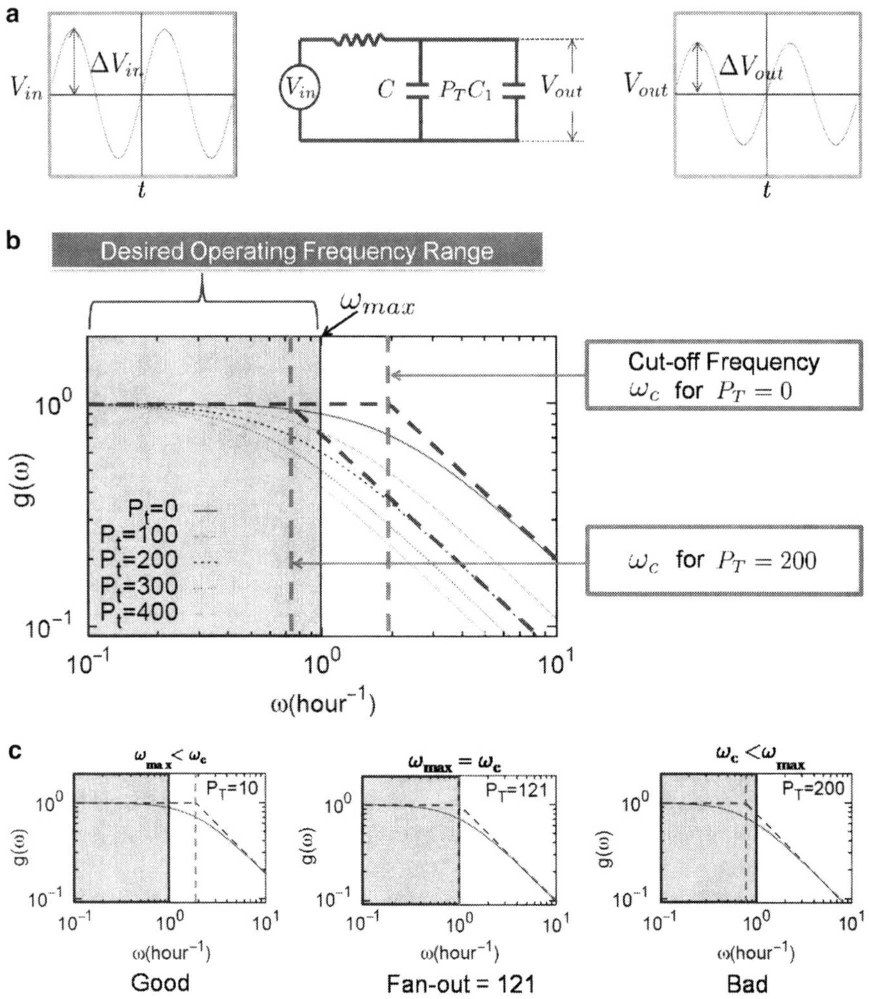

Fig. 6.6 Frequency response of the RC circuit shown in Fig. 6.5c: (**a**) The signal gain $g(\omega)$ is defined by the ratio of the oscillation amplitude of the output signal (V_{out}) to that of the input (V_{in}): $g(\omega) = \dfrac{\Delta V_{out}(\omega)}{\Delta V_{in}(\omega)}$, and is described by $g(\omega) = \sqrt{1 + \omega^2/(RC_T)^2}^{-1}$ [28]. (**b**) The cut-off frequency ($\omega_c = 1/RC_T$) decreases as P_T increases. We assume that the output signal is desired to be operated with the frequency less than a maximum operating frequency ω_{max} equal to 1 h^{-1}. (**c**) Fan-out is defined when $\omega_c = \omega_{max}$. Parameters of the model: dissociation constant of the TF, $K_d = k_{off}/k_{on} = 1$ nM [$k_{on} = 10$(1/nM/h), $k_{off} = 10$ (1/h)], $\gamma = 2$(1/h), $\alpha = 20$(nM/h)

In the fan-out equation (6.11), there are two unknown parameters: C/C_1, and τ_0. These can be experimentally estimated by performing two independent experiments with and without any downstream module. In each experiment we measure the corresponding response time: τ_0 and τ_{P_T} (the operational method for measuring the response time will be presented later in this chapter). τ_0 can be measured, and now how can we estimate the other unknown C/C_1 from τ_{P_T}? If we know a priori the copy number of the promoters P_T, we can obtain the value of C/C_1 from Eq. 6.10. If the promoters are placed on plasmids, the copy number of the plasmids can be estimated depending on what type of origin of replication is used, and thus the copy number of the promoters P_T can be known. By calculating τ_0/RC_1, we can estimate the other unknown, C/C_1. The proposed estimation method for the fan-out is very efficient in that a series of experiments for different values of P_T do not need to be performed. What we need is just two experiments.

Gene Circuit Fan-Out in more General Interfaces

We have hitherto considered a simple MIP without feedback and where the degradation rate is assumed to be first-order. It was shown that the same or similar fan-out function as Eq. 6.11 can be used in more general conditions [20] which include

- TFs are oligomer and under enzyme-mediated degradation and self-regulation (Fig. 6.7a). The same fan-out expression as equation (6.11) holds.
- TFs regulate multiple promoters having different affinities. When there are two kinds of promoters, MIP can be mapped to an RC-circuit having two different capacitances connected in parallel to C as shown in Fig. 6.7b. The fan-out of each promoter was shown to satisfy the following functional relationship between F_1 and F_2 [20]:

$$1/\omega_{max} = \tau_0 \left(1 + F_1 \frac{C_1}{C} + F_2 \frac{C_2}{C}\right). \tag{6.12}$$

  We note that the fan-out is not a single number but is given by a functional relationship between F_i's: We need to balance the number of plasmids of different kinds depending on its unit load on the retroactivity, i.e., C_i/C.
- TFs regulate multiple operators (Fig. 6.7c). Regardless the number of the operators, the same fan-out function as Eq. 6.11 is obtained.
- Multiple output signals are used. When two output TFs (X and Z) regulate a downstream promoter independently (Fig. 6.7d), i.e., if there is no overlap between the operator regions and somehow X does not interfere with the operator region of Z and vice versa, the fan-out corresponding to each output TF can be obtained.

Fan-Out is Enhanced with Inhibitory Auto-Regulation

To increase the fan-out, there are two ways based on the fan-out equations (6.11): increasing C/C_1 or $1/\tau_0$. To increase $1/\tau_0$, we can apply a negative feedback on the translation of X or a positive feed-forward on the degradation rate. With either

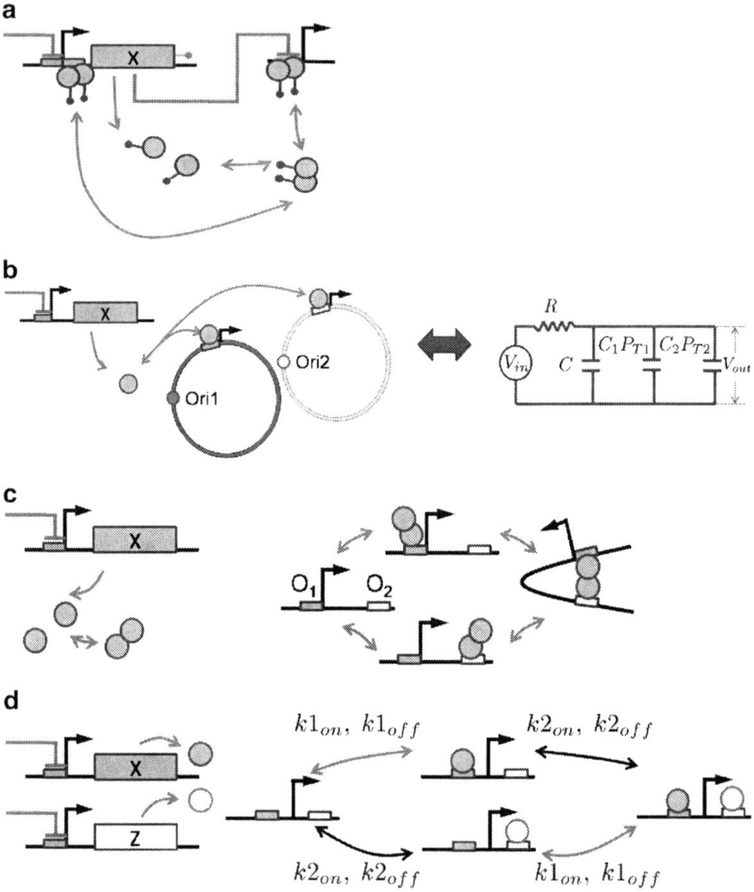

Fig. 6.7 Module interface processes that the fan-out function Eqs. 6.11 and 6.12 can be applied to: (**a**) An oligomer TF is degraded by proteases. (**b**) A TF can bind two different promoter plasmids having different binding affinities and different origins of replication. This can be mapped to an RC-circuit with two different capacitances connected in parallel. (**c**) An Oligomer TF can bind multiple operators. (**d**) Each different TF binds to its specific operator without affecting the binding affinity of the other

of these two feedback loops applied, the concentration level of X will be decreased. To prevent this decrease it is desirable to amplify the translation rate. This is exactly the same mechanism proposed by Del Vecchio et al [8] to reduce retroactivity; when the retroactivity is reduced, the additional load from the downstream can be applied to achieve the same signal output attenuation.

This mechanism of inhibitory auto-regulation is frequently found in *Escherichia coli* transcription factors regulating a set of operons, e.g., for amino-acid biosynthesis where a single TF may control multiple targets, likewise for flagella formation [43]. Such motifs are called single-input-module motifs [43].

The concept of fan-out is not limited to gene regulatory circuits. In principle, as long as the same class of interface processes are found regardless of the type of biological systems, the fan-out and retroactivity concepts can be applied [8, 36]. For example, in the eukaryotic MAPK pathway, doubly phosphorylated MAPK can activate a number of downstream proteins and transcription factors in the nucleus. This MAPK regulation can be described by the module interface process similar to the one shown in Fig. 6.7b (in this case, many promoter plasmids instead of the two). In the MAPK pathway, there is a negative feedback from MAPK to the phosphorylation of MAPKKK [37, 38]. The negative feedback increases the fan-out of the MAPK module thereby permitting MAPK to effectively regulate multiple targets and multiple homologous binding sites.

Measuring the Time Constant τ from Gene Expression Noise

In the previous section, we introduced the concept of fan-out and quantified it for various gene circuit module interfaces. To quantify the fan-out, we need to estimate the response time τ for the case when the downstream is connected and disconnected. In this section, we focus on an operational method for measuring the response time by using the stochastic nature of gene circuits. Gene expression is known to show significant stochastic fluctuations (for review, [19, 31, 32, 42]), which often contains useful information [9, 25]. Because the noise can be considered an outcome of continuous perturbations (generated from both intrinsic and extrinsic sources), it can be used to obtain the systems dynamical response to the perturbations.

The response time of the circuit output was shown to be closely related to its correlation time. The fact that the response time can increase significantly with the downstream load indicates that the output noise can show much longer correlation time, i.e., fluctuate much more slowly. Thus it was proposed that the response time can be estimated from the measurable changes in the noise correlation time [21]. The proposed method does not require any externally manipulated signals or pulses, but rather uses the noise present inherently in the system, yielding a practical estimation approach.

Noise Correlation Time

In this section, we consider the MIP described in Fig. 6.8b in the stochastic regime. The concentration level of TFs that are bound to their specific promoters, P_b, fluctuates stochastically. The fluctuations are composed of two types of fluctuations, fast and slow. The fast one comes from the rapid binding-unbinding reactions and the other from the slow translation-degradation processes. We are interested in the time-scale of the slow process and assume quasi-equilibrium in the

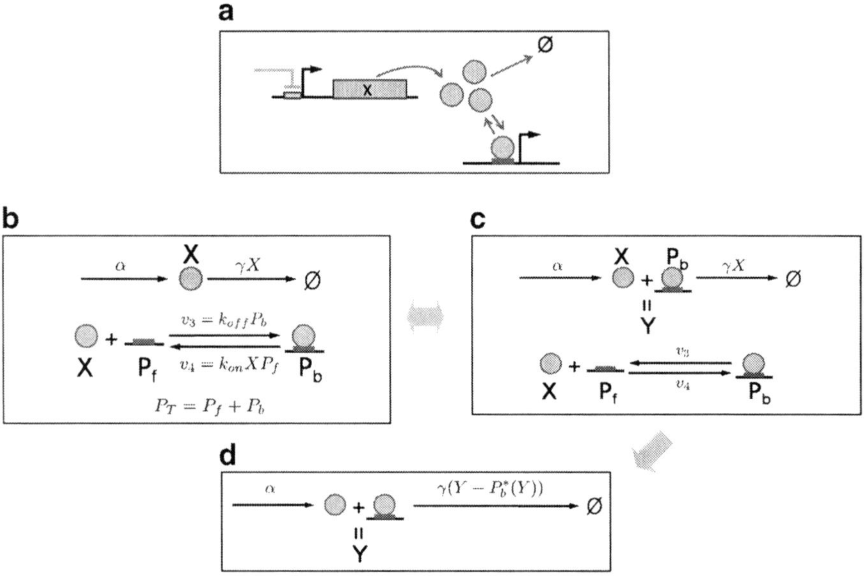

Fig. 6.8 Reaction models for a MIP: (**a**) Monomer transcription factors regulate promoters located in a downstream module and this process can be modeled as shown in (**b**). This reaction model can be equivalently described by (**c**). Y denotes the total copy number of the TFs. (**d**) The reaction model (**b**) and (**c**) can be simplified under the quasi-equilibrium assumption for P_b

binding-unbinding processes (the quasi-equilibrium approximation is introduced for formulating the concept of stochastic retroactivity, but not for simulations). Under this assumption, we replace P_b with the mean value of P_b over the fast fluctuations (Fig. 6.8d).

Isolated Case Consider first the isolated case. In the stochastic description, stochastic fluctuations in X, that deviate from the stationary state mean value, will spend a time $1/\gamma$ typically in reaching the mean value (Fig. 6.10), so that the autocorrelation function $G_X(\tau)$ becomes significant up to the time interval $1/\gamma$ (called the correlation time; see Fig. 6.10): mathematically, $G_X(\tau) = G_X(0)e^{-\tau/T_i}$ with $T_i \equiv 1/\gamma$ [4]. For the isolated case, the correlation time in the stochastic framework equals the response time in the deterministic framework [4].

Connected Case The degradation rate of Y $(=\gamma[Y - P_b^*(Y)]$ as shown in Fig. 6.8d) can become highly nonlinear: When the number of TFs is less than the number of their specific promoters, most TFs are bound and less likely to degrade. When the number of TFs are much larger than the number of the promoters, most TFs are unbound and can degrade. This is why the degradation rate can become highly nonlinear when the binding affinity of the TFs are strong (see Fig. 6.9b). For example, consider the case that the total copy number, Y, fluctuates for most of the time between 99 and 102 as shown in Fig. 6.9c. When the value of Y is between 100 and 102, the corresponding degradation rate has a approximate slope

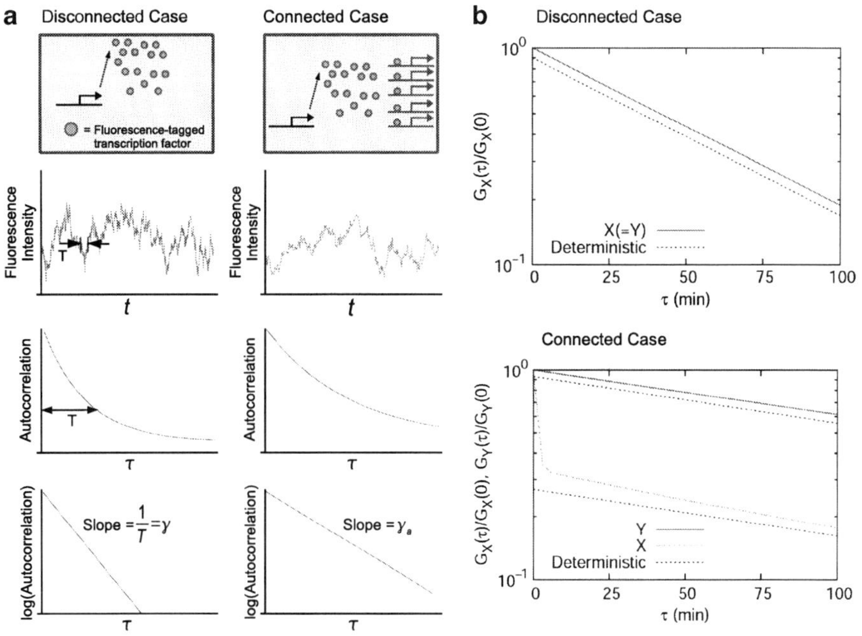

Fig. 6.9 Stochastic fluctuations in a fluorescence intensity from fluorescence-tagged TFs and its autocorrelation function: (**a**) The intensity can be represented by the total number of the TFs, free and bound. If the autocorrelation $G_Y(\tau)$ follows an exponential function $G_Y(0)e^{-\frac{\tau}{T}}$, the correlation time corresponds to T [4]. The autocorrelation can show longer correlations when the upstream module regulates the downstream. (**b**) For the connected case (Fig. 6.8b; here we do not assume the quasi-equilibrium of P_b), the autocorrelation function of Y approximates an exponential function and its correlation time also approximates the response time measured in the deterministic case. However, the approximation does not hold for X. The lines labeled 'Deterministic' are drawn for comparison purposes. We used the Gillespie stochastic simulation algorithm [16]. Parameters: (**b**) $K_d = 0.1$ nM with $k_{on} = 10\,\text{nM}^{-1}\text{h}^{-1}$, $\langle X \rangle = \alpha/\gamma = 2$ nM ($\alpha = 2$ nM h^{-1}, $\gamma = 1\,\text{h}^{-1}$), and $P_T = 0$ nM (Disconnected Case) and 100 nM (Connected Case)

of γ, indicating that the corresponding noise correlation time is approximately equal to $\simeq 1/\gamma$. However, when the value of Y is between 99 and 100, the slope drops significantly, indicating that the correlation time is much larger than $1/\gamma$. Thus, as the net effect of the two, the apparent correlation time increases. This result reflects the increase in the response time in the deterministic framework, i.e., retroactivity.

We define the correlation time T as the slope of the autocorrelation of signal Y (note for the isolated case that $Y = X$):

$$1/T \equiv -\frac{d \log G_Y(\tau)}{d\tau}. \tag{6.13}$$

In the estimation of correlation times, we recommend to use the signal Y rather than X. There are two reasons for this. The first is that Y can be observed experimentally when the output TF is tagged for fluorescence and the bound TFs have the same fluorescence intensity as the free TFs. The second is that Y is the variable

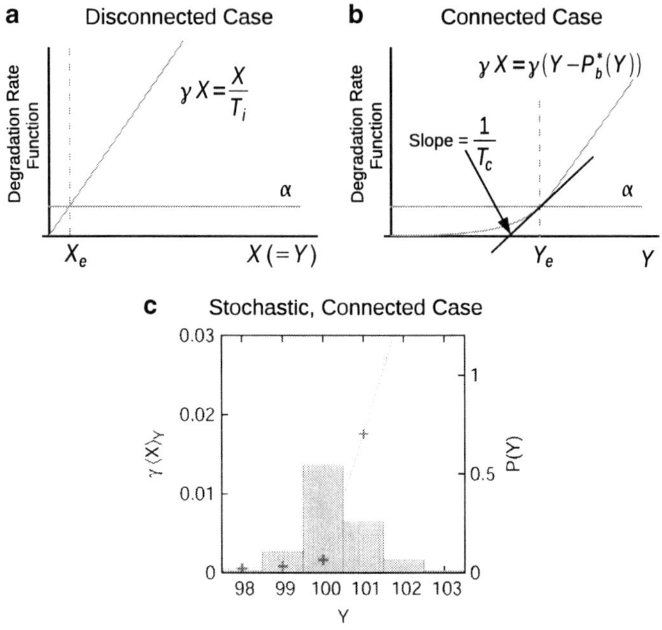

Fig. 6.10 Degradation rate functions (γX) for the MIP shown in Fig. 6.8d. (**a** and **b**) The rate functions are compared for the deterministic isolated and connected cases. X_e and Y_e represent the equilibrium values of X and Y. (**c**) In the stochastic framework, the degradation rate function $\gamma\langle X\rangle_Y$ can become highly nonlinear for the probable region of Y. The average copy number of the unbound TFs $\langle X\rangle_Y$ is computed for the different values of the total copy number (Y), and the probability distribution function of Y, $P(Y)$, is numerically computed for the process shown in Fig. 6.8d based on the Gillespie stochastic simulation algorithm [16]. Parameters: $K_d = 1$ pM, and $P_T = 100$ nM [$\alpha = 0.5\,\text{nM}\,\text{h}^{-1}$, $\gamma = 1\,\text{h}^{-1}$, $k_{on} = 10\,\text{nM}^{-1}\,\text{h}^{-1}$, $k_{off} = 0.01\,\text{h}^{-1}$]. We set the volume of the host cell (e.g. *E. coli*) roughly equal to 1 μm^3, and a copy number of one corresponds to 1 nM. As a result we interchange the unit of nM with that of copy number

relevant at the time scale of our interest (of the order of cell-doubling time or less). In the time scale of our interest, the fast binding-unbinding reactions occur many times, resulting in rapid fluctuations in X and the fluctuations can be considered averaged out in this slow time scale. Thus it is natural to consider a variable that does not fluctuate due to the binding-unbinding process. The total number Y satisfies this property. Thus, the variable Y was considered a pure slow mode [8, 30]. Using the signal Y results in more accurate estimates of correlation times when compared with the case of using X (e.g., see Fig. 6.10b). For the case of X, the autocorrelation of X is strongly affected by the fast binding-unbinding reactions and this is why we have used the autocorrelation of Y.

Simulation Results We have numerically estimated the correlation time by performing stochastic simulations. We have used parameter values appropriate for degradation tagged TFs in *E. coli* host cells: The average copy number of the TF is set equal to 2 and the dissociation constant K_d of the TF specific promoters between 0.001 and 100 nM [29,34,40,41] and the average copy number of plasmids containing the specific promoters to 1 and 100. We have fitted the autocorrelation of signals X and Y to exponential functions. The fit turns out much better for the signal Y (one specific example is shown in Fig. 6.10). The noise correlation time matches well with the response time estimated in the deterministic framework. For more detailed discussion on the simulation results, we refer to [21].

Consideration of Extrinsic Noise We have not hitherto considered any extrinsic noise, which appears due to cell replication and environmental fluctuations. Such extrinsic noise has been shown to affect the autocorrelation functions [5, 9, 34, 51] and thus needs to be taken into account for estimating the response time.

If a transcription factor with a fluorescence marker is tagged for degradation, the lifetime of the TF can be comparable to the cell doubling time. Then, the autocorrelation function of the fluorescence emitted from the TF can be fitted to the multi-exponential function [21]:

$$G_Y(\tau) = Ae^{-\gamma_E \tau} + Be^{-(\gamma_E + 1/T)\tau}, \tag{6.14}$$

with $\gamma_E = \log(2)/T_d$ (T_d is a cell doubling time and can be independently measured by experiment) and T the correlation time. The above form of the autocorrelation has been investigated in its Fourier transform (power spectral density) by Austin et al [5] for a half-life reduced GFP variant. By fitting the above nonlinear function Eq. 6.14 to experimentally estimated autocorrelations, the correlation time T can be obtained.

Fan-Out/Retroactivity Estimation

Let us consider the simple MIP shown in Fig. 6.5a as a model for TFs in *E. coli* (without considering any extrinsic noise). We performed the stochastic simulations with and without any downstream-module promoter ($P_T = 100$ and 0) for experimentally reasonable parameter values ($\alpha = 20\,\text{nM}\,\text{h}^{-1}$, $\gamma = 2\,\text{h}^{-1}$, $k_{on} = 10\,\text{nM}^{-1}\,\text{h}^{-1}$, and $k_{off} = 10\,\text{h}^{-1}$). For the simulation, we used the standard Gillespie method [16]. The concentration levels of the total TF was recorded 100 times over 2 h. We observed the autocorrelations of the output signals and fitted them to exponential functions: $G(\Delta t) = A\exp(-\Delta t/\tau)$ with τ a correlation time (we conducted a linear fit in the log-scale in the y-axis and the normal scale in the x-axis and obtained the correlation time τ from the fitted slope). We obtained the error bar of the time constant from ten independent replicates of the autocorrelation [20].

When the translation rate α was set to $20\,\text{nM}\,\text{h}^{-1}$, we obtained $\tau_0 = 0.52 \pm 0.06$ h and $\tau_{100} = 0.9 \pm 0.1$ h. We obtained $C/C_1 = 140 \pm 20$, by using

$$\frac{C}{C_1} = P_T \frac{RC}{RC_T - RC}\bigg|_{P_T=100} = P_T \frac{\tau_0}{\tau_{P_T} - \tau_0}\bigg|_{P_T=100},$$

where we used $C_T - C = P_T C_1$. From Eq. 6.11, we obtained the fan-out function for this MIP:

$$F_{\omega_{max}} = 140\,[\pm 20]\left(\frac{1/0.9\,[\pm 0.1]}{\omega_{max}} - 1\right).$$

If the maximum operating frequency ω_{max} of the upstream module is $1\,\mathrm{h}^{-1}$, the fan-out estimate is $F = 130 \pm 20$. This means that we can use promoter plasmids with low, medium, and high copy numbers without affecting the upstream module, if a single TF-specific operator site resides on a plasmid. We can also estimate the retroactivity by using $\mathcal{R} = (C_T - C)/C_T = (\tau_T - \tau_0)/\tau_T$: $\mathcal{R} = 0.4 \pm 0.1$.

If we reduce the translation rate by half (now, $\alpha = 10\,\mathrm{nM\,h}^{-1}$) the free TF concentration decreases by half. As the concentration decreases, the fraction of the TF that are bound increases, resulting in higher retroactivity [8] and lower fan-out. We estimated $\tau_0 = 0.52 \pm 0.07\,\mathrm{h}$ and $\tau_{100} = 1.75 \pm 0.04\,\mathrm{h}$. For the same $\omega_{max} = 1\,\mathrm{h}^{-1}$ we obtained the fan-out: $F = 40 \pm 1$. This would mean that we could safely use only low copy number plasmids. The retroactivity is estimated to be 0.70 ± 0.05.

Summary

In this chapter we first gave a general overview of the concept of modularity in different biological contexts and introduced a modularity concept called design pattern that describes a generalized network architecture for achieving certain types of design objectives. The modules in synthetic biology conform to the notion of design pattern because synthetic biology is concerned with design of novel networks for achieving specific goals. Later, we considered synthetic circuits in terms of a functional module and investigated what conditions the modules require for minimizing interference between modules. We have introduced the concept of fan-out, which quantify the maximum load from a downstream module that can be tolerated by the upstream module. We have proposed an efficient operational method to estimate the fan-out experimentally by minimizing the number of experiments significantly and by utilizing gene expression noise that is present inherently. We have shown that the fan-out can be enhanced by self-inhibitory regulation on the output. In the estimation process of the fan-out, the retroactivity can also be estimated. This study provides a way for quantifying the level of modularity in gene regulatory circuits and helps characterize and design module interfaces and therefore the modular construction of gene circuits.[1]

References

1. Albert R, Barabási AL (2002) Statistical mechanics of complex networks. Rev Mod Phys 74(1):47
2. Alon U (2003) Biological networks: the tinkerer as an engineer. Science 301(5641):1866–1867

[1] Some of the content of this chapter can be found in [20, 21].

3. Amaral LAN, Scala A, Barthélémy M, Stanley HE (2000) Classes of small-world networks. Proc Natl Acad Sci U S A 97(21):11149
4. Anishchenko VS, Astakhov V, Neiman A, Vadivasova T, Schimansky-Geier L (2002) Nonlinear dynamics of chaotic and stochastic systems. Springer-Verlag, Berlin, New York
5. Austin DW, Allen MS, McCollum JM, Dar RD, Wilgus JR, Sayler GS, Samatova NF, Cox CD, Simpson ML (2006) Gene network shaping of inherent noise spectra. Nature 439:608–611
6. Chen MT, Weiss R (2005) Artificial cell-cell communication in yeast Saccharomyces cerevisiae using signaling elements from Arabidopsis thaliana. Nat Biotechnol 23(12):1551–1555
7. Cooke J, Zeeman EC (1976) A clock and wavefront model for control of the number of repeated structures during animal morphogenesis. J Theor Biol 58(2):455–476
8. Del Vecchio D, Ninfa AJ, Sontag ED (2008) Modular cell biology: retroactivity and insulation. Mol Syst Biol 4:161
9. Dunlop MJ, Cox III RS, Levine JH, Murray RM, Elowitz MB (2008) Regulatory activity revealed by dynamic correlations in gene expression noise. Nat Genet 40(12):1493–1498
10. Elowitz MB, Leibler S (2000) A synthetic oscillatory network of transcriptional regulators. Nature 403(6767):335–338
11. Entus R, Aufderheide B, Sauro HM (2007) Design and implementation of three incoherent feed-forward motif based biological concentration sensors. Syst Synth Biol 1(3):119–128
12. Faloutsos M, Faloutsos P, Faloutsos C (1999) On power-law relationships of the internet topology. In: Proceedings of the Conference on Applications, Technologies, Architectures, and Protocols for Computer Communication. ACM, Cambridge, MA, USA, pp 251–262
13. Gamma E, Helm R, Johnson R, Vlissides J (2002) Design patterns. Addison-Wesley, Reading
14. Garcia-Ojalvo J, Elowitz MB, Strogatz SH (2004) Modeling a synthetic multicellular clock: repressilators coupled by quorum sensing. Proc Natl Acad Sci U S A 101(30):10955–10960
15. Gardner TS, Cantor CR, Collins JJ (2000) Construction of a genetic toggle switch in Escherichia coli. Nature 403(6767):339–342
16. Gillespie DT (1977) Exact stochastic simulation of coupled chemical reactions. J Phys Chem 81:2340–2361
17. Jayanthi S, Del Vecchio D (2009) On the compromise between retroactivity attenuation and noise amplification in gene regulatory networks. In: Proceedings of the 48th IEEE conference on decision and control, 15–18 December. Shanghai, China 2010, pp 4565–4571
18. Judd EM, Laub MT, McAdams HH (2000) Toggles and oscillators: new genetic circuit designs. Bioessays 22(6):507–509
19. Kærn M, Elston TC, Blake WJ, Collins JJ (2005) Stochasticity in gene expression: from theories to phenotypes. Nat Rev Genet 6(6):451–464
20. Kim KH, Sauro HM (2010) Fan-out in Gene Regulatory Networks. J Biol Eng 4:16
21. Kim KH, Sauro HM (2011) Measuring Retroactivity from Noise in Gene Regulatory Networks. Biophys J 100:1167–1177
22. Konopka AK (2007) Systems biology: principles, methods, and concepts. CRC Boca Raton : CRC Press/Taylor & Francis
23. Le Novère N, Bornstein B, Broicher A, Courtot M, Donizelli M, Dharuri H, Li L, Sauro H, Schilstra M, Shapiro B, Snoep JL, Hucka M (2006) BioModels database: a free, centralized database of curated, published, quantitative kinetic models of biochemical and cellular systems. Nucleic Acids Res 34(Database Issue):D689–D691
24. Lou C, Liu X, Ni M, Huang Y, Huang Q, Huang L, Jiang L, Lu D, Wang M, Liu C et al (2010) Synthesizing a novel genetic sequential logic circuit: a push-on push-off switch. Mol Syst Biol 6(1):350
25. Munsky B, Trinh B, Khammash M (2009) Listening to the noise: random fluctuations reveal gene network parameters. Mol Syst Biol 5:318
26. Murali T, Rivera C (2007) Network legos: building blocks of cellular wiring diagrams. In: Research in computational molecular biology 11th Annual International Conference, RECOMB 2007, April 20–25. Oakland, CA, USA, pp 47–61
27. Nevozhay D, Adams RM, Murphy KF, Josić K, Balázsi G (2009) Negative autoregulation linearizes the dose–response and suppresses the heterogeneity of gene expression. Proc Natl Acad Sci U S A 106(13):5123–5128

28. Nilsson JW, Riedel SA (2008) Electric circuits. Pearson Prentice Hall, Upper Saddle River, NJ
29. Pompeani AJ, Irgon JJ, Berger MF, Bulyk ML, Wingreen NS, Bassler BL (2008) The *Vibrio harveyi* master quorum-sensing regulator, LuxR, a TetR-type protein is both an activator and a repressor: DNA recognition and binding specificity at target promoters. Mol Microbiol 70(1):76–88
30. Rao CV, Arkin AP (2003) Stochastic chemical kinetics and the quasi-steady-state assumption: application to the Gillespie algorithm. J Chem Phys 118:4999–5010
31. Rao CV, Wolf DM, Arkin AP (2002) Control, exploitation and tolerance of intracellular noise. Nature 420(6912):231–237
32. Raser JM, O'Shea EK (2004) Control of stochasticity in eukaryotic gene expression. Science 304(5678):1811–1814
33. Reichardt J, Bornholdt S (2006) Statistical mechanics of community detection. Phys Rev E 74(1):16110
34. Rosenfeld N, Young JW, Alon U, Swain PS, Elowitz MB (2005) Gene regulation at the single-cell level. Science 307(5717):1962–1965
35. Russo FD, Silhavy TJ (1993) The essential tension: opposed reactions in bacterial two-component regulatory systems. Trends Microbiol 1(8):306–310
36. Sauro HM (2008) Modularity defined. Mol Syst Biol 4(1):166
37. Sauro HM, Ingalls B (2007) MAPK cascades as feedback amplifiers. arXiv:0710.5195[q-bio.MN]
38. Sauro HM, Kholodenko BN (2004) Quantitative analysis of signaling networks. Prog Biophys Mol Biol 86(1):5–43
39. Schmitt WA, Raab RM, Stephanopoulos G (2004) Elucidation of gene interaction networks through time-lagged correlation analysis of transcriptional data. Genome Res 14(8):1654
40. Scholz O, Schubert P, Kintrup M, Hillen W (2000) Tet repressor induction without Mg^{2+}. Biochemistry 39(35):10914–10920
41. Setty Y, Mayo AE, Surette MG, Alon U (2003) Detailed map of a cis-regulatory input function. Proc Natl Acad Sci U S A 100(13):7702–7707
42. Shahrezaei V, Ollivier JF, Swain PS (2008) Colored extrinsic fluctuations and stochastic gene expression. Mol Syst Biol 4:196
43. Shen-Orr SS, Milo R, Mangan S, Alon U (2002) Network motifs in the transcriptional regulation network of Escherichia coli. Nat Genet 31(1):64–68
44. Spirin V, Mirny LA (2003) Protein complexes and functional modules in molecular networks. Proc Natl Acad Sci U S A 100(21):12123–12128
45. Stricker J, Cookson S, Bennett MR, Mather WH, Tsimring LS, Hasty J (2008) A fast, robust and tunable synthetic gene oscillator. Nature 456(7221):516–519
46. Stuart JM, Segal E, Koller D, Kim SK (2003) A gene-coexpression network for global discovery of conserved genetic modules. Science 302(5643):249–255
47. Thieffry D, Huerta AM, Pérez-Rueda E, Collado-Vides J (1998) From specific gene regulation to genomic networks: a global analysis of transcriptional regulation in Escherichia coli. Bioessays 20(5):433–440
48. Thomas R, D'Ari R (1990) Biological feedback. CRC Press, Boca Raton, Florida, USA, p 316
49. Tigges M, Marquez-Lago TT, Stelling J, Fussenegger M (2009) A tunable synthetic mammalian oscillator. Nature 457(7227):309–312
50. Voigt CA (2006) Genetic parts to program bacteria. Curr Opin Biotech 17(5):548–557
51. Weinberger LS, Dar RD, Simpson ML (2008) Transient-mediated fate determination in a transcriptional circuit of HIV. Nat Genet 40(4):466–470

Chapter 7
Retroactivity as a Criterion to Define Modules in Signaling Networks

Julio Saez-Rodriguez, Holger Conzelmann, Michael Ederer, and Ernst Dieter Gilles

Abstract The concept of modularity has been widely studied in the context of molecular biology. Since engineering sciences are used to work in a modular manner, it is tempting to approach the definition of biological modules from an engineering perspective. From a system-theoretical point of view an interesting criterion might be the definition of modules where both the input signals and the output signals are unidirectional, that is, there is no retroactivity. In this chapter, we review the applicability of this concept to biological networks. We start describing which biochemical situations can lead to absence of retroactivity. Then, we show how this concept can be automatized into an algorithm to decompose biochemical networks into modules so that the retroactivity among the modules is minimized. This decomposition facilitates the analysis of complex models because the modules can, to some degree, be studied separately. We complement this analysis with a consideration of retroactivity in signal transduction processes using a domain-oriented description. Finally, we explore the interplay between retroactivity and thermodynamics in the domain-oriented description, and show how the binding site phosphorylation is a mechanism that is able to realize unidirectional signal transduction.

Keywords Retroactivity · Modularity · Wegscheider condition · Domain-oriented modeling · Signaling · Thermodynamics · Systems-theory · Network theory · Unidirectionality · Futily cycles · Phosphorylation · MAPK · Michaelis-Menten

Introduction

The definition of functional units, i.e. entities whose function is separable from those of other units, has been proposed as a promising rationale for the analysis of large biochemical networks [1, 12, 25]. This modular approach follows a simple rationale:

J. Saez-Rodriguez (✉)
European Bioinformatics Institute (EMBL-EBI) and EMBL-Heidelberg, Genome Biology Unit, Wellcome Trust Genome Campus, Cambridge CB10 1SD, UK
e-mail: saezrodriguez@ebi.ac.uk

H. Koeppl et al. (eds.), *Design and Analysis of Biomolecular Circuits: Engineering Approaches to Systems and Synthetic Biology*, DOI 10.1007/978-1-4419-6766-4_7,

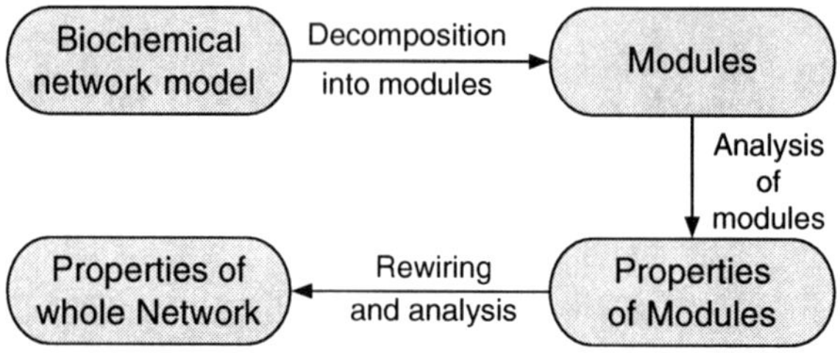

Fig. 7.1 General procedure of a modular analysis approach. The approach starts with the decomposition of the model of a (typically large) network in a suitable manner into smaller, easier to analyze subunits. Upon a thorough analysis of the resulting modules, they can be rewired together (either in their original form or in a reduced one [7]), and new insights into the network as a whole can be obtained [22]

divide and win. As depicted in Fig. 7.1, by decomposing a system into subunits, one obtains modules which are significantly easier to handle. Once these *relatively* simple units are well understood, they can be re-assembled in order to analyze the emergent properties of the resulting systems [16]. Furthermore, one could set up a kit of reusable elements, simplifying the setup of models, since many parts of biological networks are found in several signal transduction pathways.

From this perspective of facilitating the analysis, what would be a good criterion to define these modules? Since engineering sciences are used to work in a modular manner, it is tempting to look at how modules are defined and constructed in a technical context. From a system-theoretical point of view an interesting criterion might be the definition of elements where both the input and the output are unidirectional, that is, it does not exist retroactivity[1] [22, 33]. This is actually the form in which most technical systems are devised, facilitating their analysis and design: for example, a thermometer is constructed in such a way that it receives information about the temperature of a certain object, but it does not affect significantly the energy (and thus temperature) of the object itself.

These effects of elements downstream back to elements upstream have been extensively studied [26, 33], and the effects of retroactivity on the behavior of biochemical networks will be discussed in the chapter by Del Vecchio. Its impact on the identification of biochemical systems is discussed in the chapter by Sontag, and the connections between modularity and synthetic biology in the chapter of Chandran et al. Here, we provide a complementary angle on retroactivity reviewing the different biochemical situations that can lead to absence of retroactivity [23], and how to employ them to define modules within network [24]. We will approach this point from different angles, from a bipartite representation of biochemical networks based on the network theory of [11] to thermodynamic considerations applied to a domain-oriented representation of biochemical systems [8].

[1] We coined the term retroactivity in [22, 23] as a translation of the German word Rückwirkung, that can be more accurately translated by the longer expression 'backwards effect'.

The Absence of Retroactivity as a Criterion to Demarcate Modules

In this section we will introduce the definition of retroactivity, its description in terms of network theory, and its implementation into an automatic algorithm.[2]

Consider a biochemical network as a general non-linear dynamical system described by a set of ordinary differential equations (ODEs) of the form

$$\dot{\vec{c}} = \frac{d\vec{c}}{dt} = f(\vec{c}, \vec{u}, \vec{p}), \tag{7.1}$$

where $\frac{d\vec{c}}{dt} \in I\!R^n$ is the vector of the time derivatives of the concentrations c_i, $\vec{u}$ the vector of inputs and $\vec{p}$ the vector of parameters. A vector of outputs $\vec{y} = g(\vec{c})$ may also be defined. If we would like to decompose this system into two subsystems $\vec{c_1}$ and $\vec{c_2}$ that are *decoupled* we would need to find two subsets of states c_i so that

$$\begin{aligned} \dot{\vec{c_1}} &= f_1(\vec{c_1}, \vec{u}, \vec{p}) \\ \dot{\vec{c_2}} &= f_2(\vec{c_1}, \vec{c_2}, \vec{u}, \vec{p}). \end{aligned} \tag{7.2}$$

The decomposition into decoupled systems of the form of Eq. 7.2 is a well studied problem in the field of systems theory [32]. However, biochemical – in particular signaling – networks are characterized by a high degree of coupling, so that a clean decomposition in the form of Eq. 7.2 is in most cases not possible.

We therefore introduce a less strict requirement, which we shall call the absence of retroactivity [22–24]: two modules $\vec{c_1}$ and $\vec{c_2}$ are connected without retroactivity if there is no pair of species (states), one in each module, which influence each other (see Fig. 7.2), i.e.

$$\nexists(i, j) : \dot{c}_{1i} = f_{1i}(c_{2j}, \ldots) \wedge \dot{c}_{2j} = f_{2j}(c_{1i}, \ldots) \tag{7.3}$$

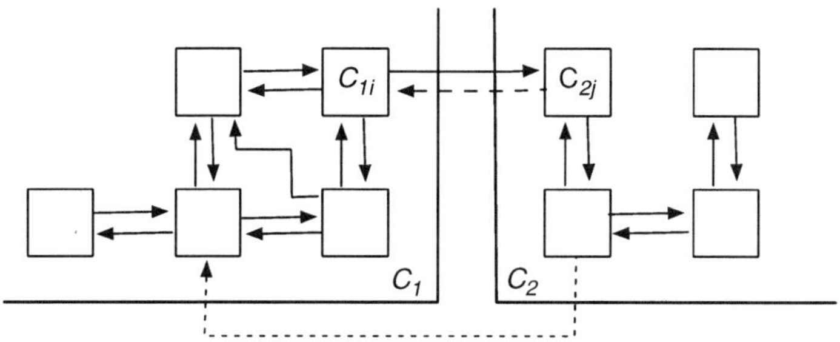

Fig. 7.2 Schematic representation of the concept of retroactivity. If the state c_{1i} influences the submodule state c_{2j} (*solid line*), but the state c_{2j} does not directly influence c_{1i} (*dotted line*), the connection between $\vec{c_1}$ and $\vec{c_2}$ is free of retroactivity, even if a *unidirectional* feedback from another element $\vec{c_2}$ to $\vec{c_1}$ (*dashed-dotted line*) is present [22]

[2] This section summarizes the work described in [23, 24]. Some portions of the text and figures are reproduced, with permission from Elsevier and Oxford University Press, respectively.

where c_{1i} is the ith element of $\vec{c}_1$ and c_{2j} is the jth element of $\vec{c}_2$. Thus, instead of the *global* decoupling between the modules as a whole imposed in Eq. 7.2, we just require a *local* unidirectionality between all elements of one module and all elements of the other module. A *feedback* effect between different components of the modules does not imply a retroactive effect [22].

Network Theory and Retroactivity

Biochemical systems are often described as two sets of elements, species (substances) and reactions. This leads Eq. 7.1 to take the form

$$\frac{d\vec{c}}{dt} = \dot{\vec{c}} = N\,\vec{v}, \tag{7.4}$$

where $\vec{v}(\vec{c}, \vec{u}, \vec{p}) \in I\!R^m$ is the vector of the m reactions, and $N \in I\!R^{n\times m}$ the stoichiometric matrix [13]. A suitable frame to describe such bipartite systems is provided by the network theory introduced by [11], that consists of a combination of two types of elementary units: *components*, which store physical quantities and storage-free *coupling elements*, which describe the interactions between the components. These elements can be aggregated into higher-order units in a hierarchical fashion. Components and coupling elements are connected by two types of signal vectors: potential vectors (from components to coupling elements) and current vectors (from coupling elements to components). In biochemical networks the compounds (species) are the components, the reactions the coupling elements, potential vectors carry information about the concentrations from the compounds to the reactions and current vectors would bring information about the rates back to the compounds (see Fig. 7.3), leading to a system of differential equations of the form of Eq. 7.4. This bipartite description is useful to cleanly characterize, from a biochemical point of

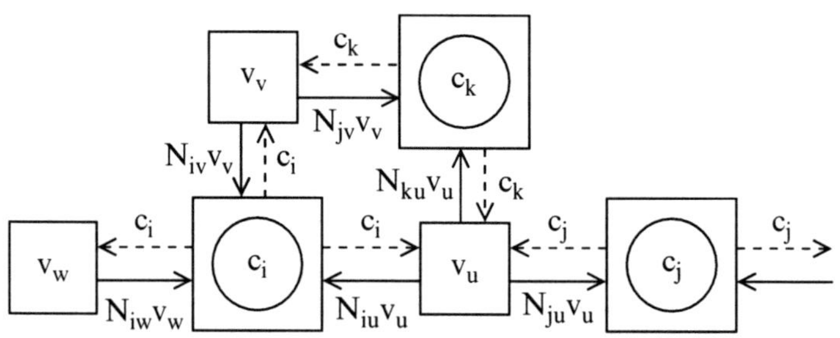

Fig. 7.3 Representation of a certain set of biochemical reactions according to the network theory, from [24]. *Dashed lines* represent potential (concentration) vectors *solid thin lines* current (rates) vectors

view, the coupling among modules. The different cases that lead to unidirectional effects can be separated between those where a potential vector and a current vector can be neglected.

Neglect of a Current The balance of the concentration of a storage c_i is a function of the reaction rates (Eq. 7.4),

$$\frac{dc_i}{dt} = \dot{c}_i = N_{i1}v_1 + N_{i2}v_2 + \cdots + N_{ik}v_k + \cdots + N_{im}v_m.$$

Therefore, a reaction v_u does not influence *significantly* a storage c_i (see Fig. 7.3) if its contribution to the balance is negligible, that is to say, if

$$N_{iu}v_u \ll \sum_{k=1}^{m} N_{ik}v_k \Rightarrow \frac{N_{iu}v_u}{\sum_{k=1}^{m} N_{ik}v_k} = \widehat{g}_{iu}^{c} \ll 1, \tag{7.5}$$

that corresponds to neglecting the arrow from v_u to c_i. This is the case if the amount of a compound that is consumed or produced in a reaction is negligible compared to the total amount. This definition might lead to numerical problems since in steady state $\frac{dc_i}{dt} = \sum_{k=1}^{m} N_{ik}v_k = 0$. Using the absolute values would circumvent this problem; we use thus

$$\frac{|N_{iu}v_u|}{\sum_{k=1}^{m} |N_{ik}v_k|} = g_{iu}^{c} \ll 1, \tag{7.6}$$

which is a more strict condition than Eq. 7.5. Therefore, the time-dependent function g_{iu}^{c} defines the effect of the reaction v_u on a storage c_i.

A very common biological process that deserves a more detailed description is an enzymatic reaction. If we consider the general case where a compound S is transformed into P, by reaction with another compound E, being E regenerated in an additional step (Fig. 7.4(a)) defined by the equations

$$E' \rightleftharpoons E \tag{7.7}$$

and

$$S + E \rightleftharpoons SE \rightleftharpoons P + E', \tag{7.8}$$

we obtain a highly interconnected systems, without unidirectional connections. If the second step of the second reaction (Eq. 7.8) is considered irreversible we obtain

$$S + E \rightleftharpoons SE \rightharpoonup P + E' \tag{7.9}$$

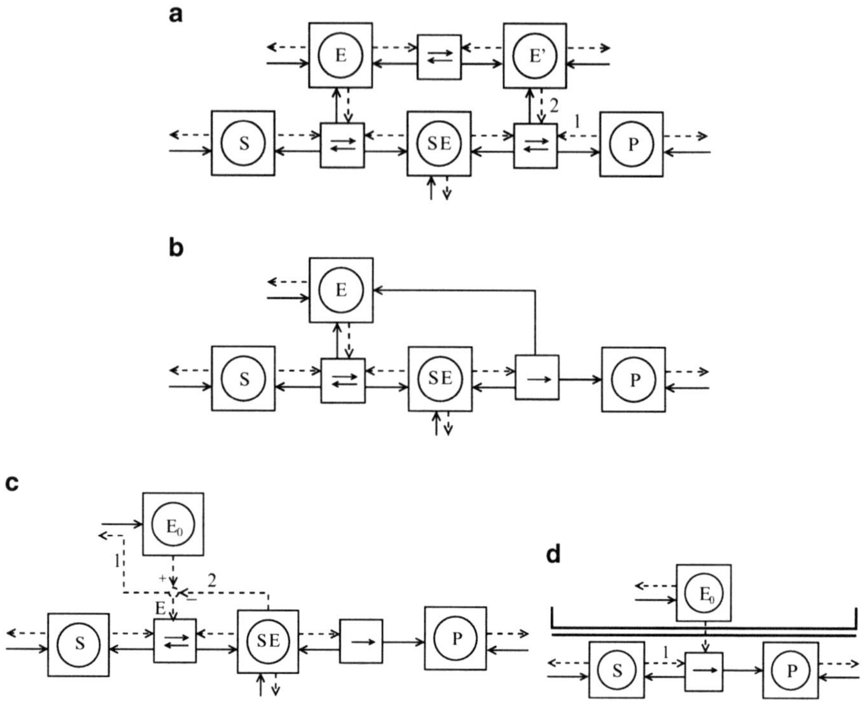

Fig. 7.4 Representation of different schemes of enzymatic reactions according to the network theory, from [23], where more cases as discussed. *Dashed lines* represent potential (concentration) vectors, *solid thin lines* current (rates) vectors, and *solid thick lines* the borders of the modules; (**a**) enzymatic reaction system defined by Eqs. 7.7 and 7.8; (**b**) system defined by Eq. 7.10; (**c**) same system as in (**b**) but with a change of variable $E_0 = E + SE$; (**d**) system defined by Eq. 7.11

instead of Eq. 7.8. The representation of the new system is obtained by deleting the vectors 1 and 2 in Fig. 7.4(a). In this system, there is a unidirectional connection defined by the irreversible step, but the connection between E and the reaction $S \rightleftharpoons P$ has still retroactivity (see Fig. 7.4(a)). If, additionally, $E = E'$, the system

$$S + E \underset{k_{-1}}{\overset{k_1}{\rightleftharpoons}} SE \overset{k_2}{\rightharpoonup} P + E \tag{7.10}$$

is obtained, which is shown in Fig. 7.4(b) and represents the irreversible conversion of S into P catalyzed by an enzyme E. Defining a new variable $E_0 = E + SE$ we obtain an alternative representation (Fig. 7.4(c)). Analyzing this schema we see that a connection free of retroactivity from the enzyme to the reaction can be achieved if:

1. The external reactions that influence E_0 are not influenced by E (e.g. they are irreversibles), that is, we can neglect vector 1 in Fig. 7.4(c).
2. The dynamics of the compound SE can be neglected (i.e. $dc_{SE}/dt \approx 0$, which means that the vector 2 in Fig. 7.4(c) is negligible). This approximation is known

as the quasi-steady-state assumption, and leads to the reduced system (see for example [30])

$$S \rightarrow P, \tag{7.11}$$

following the reaction rate r the classical Michaelis Menten equation [29]

$$v = \frac{V_{max} \cdot S}{K_m + S} = \frac{k_2 \cdot E_0 \cdot S}{K_m + S}, \tag{7.12}$$

where $K_m = (k_{-1} + k_2)/k_1$. We obtain thus a connection free of retroactivity by absence of a current vector, as represented in Fig. 7.4(d). If, additionally, the enzyme is saturated by the substrate ($K_m \ll S$), then the reaction rate r becomes

$$r = k_2 E_0 \tag{7.13}$$

and the system can be represented as in Fig. 7.4(d) deleting the vector 1, leading to an additional connection free of retroactivity between the reaction r and the substrate S.

The assumption $dc_{SE}/dt \approx 0$ is correct for the system defined in Eq. 7.10 if $\varepsilon \ll 1$, where $\varepsilon = E_0/(K_m + S_0)$, being E_0 and S_0 the total concentration of E and S, respectively [30]. This condition is fulfilled if $E_0 \ll S_0$ and if $E_0 \ll K_m$. $E_0 \ll S_0$ (much less enzyme than substrate, a usual situation in many *in vitro* experiments) is the usual assumption for the application of Michaelis Menten equation.

The condition $E_0 \ll K_m$ can be rewritten as $E_0 k_1 \ll k_2 + k_{-1}$. Since k_1 is the kinetic constant for the formation of the complex SE, and k_{-1} and k_2 the kinetic constants for the dissociation of the complex SE (Eq. 7.10), this condition can be interpreted as the decomposition of SE being much faster than the formation of SE.

The Michaelis Menten expression (Eq. 7.12) is widely used for enzymatic reactions without considering whether the assumptions described above are fulfilled or not. This implies that the substrate has a 'high impedance', and it therefore measures the activity of the enzyme, without affecting it. However, the retroactive effects due to the sequestration of enzymes can affect significantly the behavior of the system [2].

Neglect of a Potential Since in the general case the reaction rates $v_u(\vec{c}, \vec{p}, \vec{u})$ are *not* a linear function of the concentrations, the effect of a storage c_j on a reaction v_u can be estimated by the derivative

$$\varepsilon^*_{uj} = \frac{\partial v_u}{\partial c_j} \tag{7.14}$$

This derivative is known as (unscaled) elasticity in the field of Metabolic Control Analysis [13], and we adopt this nomenclature here. For example, in the case of a reaction v_1

$$c_1 \underset{k_2}{\overset{k_1}{\rightleftharpoons}} c_2 \tag{7.15}$$

following simple mass action law kinetics, $\varepsilon^*_{11} = k_1$ and $\varepsilon^*_{12} = -k_2$. More appropriate is the use of the scaled elasticity, defined by $\varepsilon^*_{uj} c_j$. Accordingly, a storage c_j does not affect significantly a reaction v_u (see Fig. 7.3) if its effect is negligible when compared to that of the rest of storages, i.e.

$$|\varepsilon^*_{uj}|c_j \ll \sum_{k=1}^{n} |\varepsilon^*_{uk}|c_k \Rightarrow \frac{|\varepsilon^*_{uj}|c_j}{\sum_{k=1}^{m} |\varepsilon^*_{uk}|c_k} = g^p_{uj} \ll 1. \tag{7.16}$$

The time-dependent function g^p_{uj} defines thus the effect of a storage c_j on a reaction v_u. Note that, as in the case of the neglect of a current g^c, we use absolute values. If we divide numerator and denominator by v_u we obtain

$$g^p_{uj}(t) = \frac{\frac{|\varepsilon^*_{uj}|c_j}{|v_u|}}{\sum_{k=1}^{m} \frac{|\varepsilon^*_{uk}|c_j}{|v_u|}} = \frac{|\varepsilon_{uj}|}{\sum_{k=1}^{m} |\varepsilon_{uk}|}, \tag{7.17}$$

where $\varepsilon_{uj}(t)$ is the scaled elasticity, $\varepsilon_{uj} = \frac{\partial v_u}{\partial c_j}\frac{c_j}{v_u} = \frac{\partial v_u / v_u}{\partial c_j / c_j}$ [13]. A potential can be neglected if the concentration of one compound does not affect the rate of a reaction where it is consumed or produced (e.g. vector 1 in Fig. 7.4(a)). An example is an irreversible reaction, where the product does not affect the reaction rate.

An Algorithm to Identify Modules Minimizing Retroactivity

In the previous section we saw that a system shows a junction free of retroactive effects if a potential or current vectors can be neglected. We further analyzed typical biochemical cases. We also discussed how to formalize within the framework of network theory the concept of retroactivity, using the time-dependent matrices $g^p : I\!R^+ \rightarrow I\!R^{n \times m}$ (Eq. 7.6) and $g^c : I\!R^+ \rightarrow I\!R^{m \times n}$ (Eq. 7.17). In this section, we will see how this can be implemented in an algorithm to detect modules based on the retroactivity.

The values of g^c_{ij} or g^c_{uj} are strictly zero if there is a *structural* absence of retroactivity. For example, a strictly irreversible reaction will lead to $\varepsilon_{uj} = 0$ between the reaction u and the product j, and an enzymatic reaction modeled with Michaelis Menten kinetics will result in $N_{iu} = 0 \Rightarrow g^c_{iu}(t) = 0$ between the enzyme i and the reaction u.

However, in many other cases, g^c or g^p might not be strictly zero but have very low values. For those cases, a criterion determining what is low enough is required. In some cases a simple criterion such as the maximal or the average value be lower than a certain threshold at a characteristic trajectory could be a reasonable criterion. Alternatively, one could perform more exhaustive analyses using e.g. Monte Carlo methods to explore the parameter space. To determine the stricter, structural absence of retroactivity only the structure of the network is needed, while for the approximate, kinetic-dependent absence of retroactivity a parameterized model and a particular operating condition have to be defined.

Definition of the Retroactivity Matrix Once a certain criterion has been applied, the matrices $R^p \in \{0,1\}^{n\times m}$ and $R^c \in \{0,1\}^{m\times n}$ can be obtained which define which storages affect which reactions and vice versa. Now, at the connection between a storage c_i and a reaction v_j following cases are possible:

$$\begin{array}{ll} \hline R^p_{ij}=1,\ R^c_{ji}=1 & \text{retroactive connection} \\ R^p_{ij}=1,\ R^c_{ji}=0 & \text{unidirectional connection (neglect of current)} \\ R^p_{ij}=0,\ R^c_{ji}=1 & \text{unidirectional connection (neglect of potential)} \\ R^p_{ij}=0,\ R^c_{ji}=0 & \text{absence of connection} \\ \hline \end{array} \tag{7.18}$$

Importantly, if one considers only the *structural* retroactivity, since $N_{ij}=0 \Rightarrow R^c_{ij}=0$ and $\varepsilon_{ij}=0 \Rightarrow R^p=0$, R^c and R^p correspond to the matrices expressing the occupancy of N and ε (what we shall call the indicator matrices N^I and ε^I), respectively:

$$R^c_{ij}=N^I_{ij}=\begin{cases}0 & \text{if } N_{ij}=0\\ 1 & \text{else}\end{cases} \tag{7.19}$$

and, analogously,

$$R^p_{ij}=\varepsilon^I_{ij}=\begin{cases}0 & \text{if } \varepsilon_{ij}=0\\ 1 & \text{else}\end{cases} \tag{7.20}$$

From now on we shall concentrate on the structural case, but the same arguments can be applied to the general case.

Now, defining a matrix $R \in \{0,1,2,3\}^{n\times m}$, which we shall call the retroactivity matrix, so that $R_{ij}=R^p_{ij}+2R^c_{ji}$, we would obtain the information about the retroactivity in a compact manner:

$$\begin{array}{ll} \hline R_{ij}=0 & \text{absence of connection} \\ R_{ij}=1 & \text{unidirectional connection (potential)} \\ R_{ij}=2 & \text{unidirectional connection (current)} \\ R_{ij}=3 & \text{retroactive connection} \\ \hline \end{array} \tag{7.21}$$

Posing the Minimization of Retroactivity as a Community Detection Problem The next step should be to try to demarcate the modules in such a way that the number of retroactive connections (where $R_{ij}=3$) among modules is minimized (ideally zero), and maximized inside the modules. Considering the matrix $R^r \in \{0,1\}^{n\times m}$ where

$$R^r_{ij}=\begin{cases}1 & \text{if } R_{ij}=3\\ 0 & \text{else,}\end{cases} \tag{7.22}$$

clustering techniques can be applied. The methods of Newman and colleagues, relying on the maximization of a mathematical value of the modularity [19], seem particularly suited for this task, since we are trying to minimize the number of retroactive connections. However, these approaches consider mostly (undirected) interaction graphs of networks, but here we are dealing with a bipartite graph (since there are two kinds of nodes: storages and reactions). Unfortunately, clustering algorithms for bipartite graphs are much less developed. A detour to circumvent this problem would be to convert the information about the relationships between the storages and reactions coded in R into a quadratic matrix defining the connections among one type of elements (storages). This information defines a monopartite graph.

A natural monopartite graph would be one reflecting the reciprocal influence among the compounds. A compound c_i does not influence directly a compound c_j if there is no connection from c_i to c_j through any reaction. The influence of c_i on c_j via the reaction v_v is determined by $R^c_{jv} \cdot R^p_{vi}$: the influence of c_i on v_v (R^p_{vi}) multiplied by the influence of v_v on c_j (R^c_{jv}). Thus, extending this argument to all reactions, one gets that the influence of c_i on c_j reads

$$R^c_{j1} \cdot R^p_{1i} + R^c_{j2} \cdot R^p_{2i} + \cdots + R^c_{jm} \cdot R^p_{mi} = \sum_{k=1}^{m} R^c_{jk} \cdot R^p_{ki}, \tag{7.23}$$

and thus, if this expression is equal to zero, there is no influence of c_i on c_j. And for the structural case, from Eqs. 7.19 and 7.20,

$$\sum_{k=1}^{m} N^I_{jk} \cdot \varepsilon^I_{ki} = 0 \tag{7.24}$$

There is a close relationship between this expression and the Jacobian $J \in \mathbb{R}^{n\times n}$ with

$$J_{ij} = \frac{\partial f_i}{\partial c_j} \tag{7.25}$$

of a system of differential equations in the form of Eq. 7.1. The sign of J_{ij} informs whether c_j has a direct positive or negative influence on c_i, and a matrix considering the sign of these entries can be seen as the adjacency matrix of the underlying interaction graph [14]. The relation to Eq. 7.24 is simply obtained by deriving f_i with respect to c_j. From Eq. 7.4 it results

$$\frac{\partial f_i}{\partial c_j} = N_{i1}\frac{\partial v_1}{\partial c_j} + N_{i2}\frac{\partial v_2}{\partial c_j} + \cdots + N_{ik}\frac{\partial v_k}{\partial c_j} + \cdots + N_{im}\frac{\partial v_m}{\partial c_j}, \tag{7.26}$$

that is

$$J_{ij} = N_{i1}\varepsilon^*_{1j} + N_{i2}\varepsilon^*_{2j} + \cdots + N_{ik}\varepsilon^*_{ij} + \cdots + N_{im}\varepsilon^*_{mj}, \tag{7.27}$$

or in compact manner

$$J = N\varepsilon^*. \tag{7.28}$$

Therefore, for the structural case that we are considering, the indicator matrix of the Jacobian $J^I \in \{0,1\}^{n \times n}$,

$$J^I_{ij} = \begin{cases} 1 & \text{if } J_{ij} = \sum_{k=1}^{m} N_{ik} \cdot \varepsilon_{kj} \neq 0 \\ 0 & \text{else} \end{cases} \tag{7.29}$$

would provide a starting point for these algorithms.

To identify modules based on retroactivity the presence of a unidirectional connection is equivalent to no connection at all. The connection between two compounds c_i and c_j is retroactive if $J^I_{ij} = J^I_{ji} = 1$. If $J^I_{ij} = J^I_{ji} = 0$, there is no connection between the elements, and if $J^I_{ij} = 1$ and $J^I_{ji} = 0$ there is an unidirectional connection from c_i to c_j (involving either the neglect of a current or of a potential). The retroactivity can thus be captured via a symmetric matrix $J^{IR} \in \{0,1\}^{n \times n}$, so that

$$J^{IR}_{ij} = J^{IR}_{ji} = \begin{cases} 1 & \text{if } J^I_{ij} = J^I_{ji} = 1 \\ 0 & \text{else} \end{cases} \tag{7.30}$$

This has also an advantage for applying Newman's algorithms, since they are devised for symmetric matrices. Newman defines modularity as the number of edges within modules with respect to the number of edges within modules expected for a random network [19], which reads for a network decomposed in two modules

$$Q = \frac{1}{4m} \sum_{ij}^{n} \left(A_{ij} - \frac{k_i k_j}{2m} \right) (s_i s_j + 1) \tag{7.31}$$

with $s_i = 1$ if i belongs to module 1 and $s_i = -1$ if it belongs to module 2. A is the adjacency matrix, and k_i the number of edges connected to a node i. We will use the Newman definition for Modularity (Eq. 7.31) applied to J^{IR}. A number of algorithms can be used to optimize the modularity of Eq. 7.31, and can be applied in our context [24].

Characterization of the Connections Among the Modules After applying a modularity analysis, one would like to know what kind of connections couple the different modules. This information is, however, not present in the matrix J^I alone. One can obtain this information from the matrices $R^p(\varepsilon^*)$ and $R^c(N)$. Alternatively, one can deduce the nature of the connections from J^I and N. The latter has the advantage that it is not required to compute ε^* if it is not available (J^I is available from the previous steps and N is easily obtainable). Additionally, one remains in the monopartite (concentrations) description and does not need to go back to the bipartite (concentrations+reactions) formalism. The matrix $N^C = N(-N)^T$ is a symmetric matrix so that $N^C_{ij} = N^C_{ji} \neq 0$ if there is a mass flux between the species i and j, and is 0 otherwise [9]. Considering conjunctly its indicator matrices $N^{CI} \in \{0,1\}^{n \times n}$ and J^I the different cases can be retrieved (see Table 7.1), allowing us to characterize the connection among the modules.

Table 7.1 Types of connections between two species as a function of J^{IR} and N^{CI} from [24]. All possible dependencies between two storages i and j can be unambiguously determined by the values of J_{ij}^{IR}, J_{ji}^{IR}, N_{ij}^{CI} and N_{ji}^{CI}

J^{I}_{ij}	0	0	1 0	1 0	1	1
J^{I}_{ji}	0	0	0 1	0 1	1	1
N_{ij}^{CI}	0	1	0 0	1 1	0	1
N_{ji}^{CI}	0	1	0 0	1 1	0	1
	No connection	Products of common irrev. reaction	Control by potential (enzyme)	Control by current (irrev. reaction)	Reciprocal control by potential (enzyme)	Coupled storages (rev. reaction)

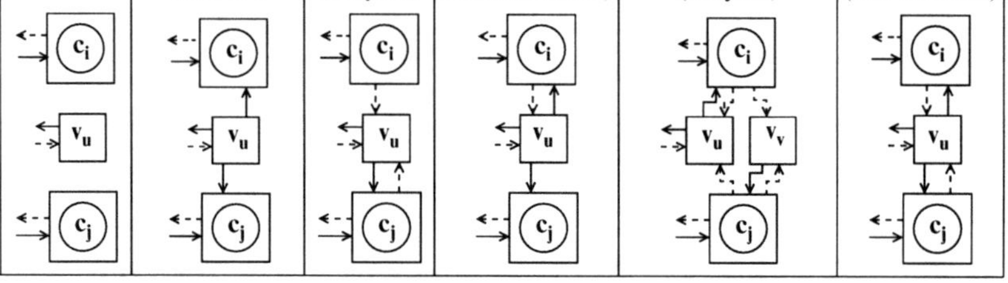

Integration of the Algorithm into ProMoT The identification of the modules underlying a given model provides useful insights into its structure. Furthermore, it allows one then to subsequently analyze the system in a modular manner. For example, one would like to consider only one module in isolation, or the combination of several of them, eventually testing different variants considering different connections between modules, a reduced version of one of the modules, etc [22]. Rewriting by hand the model to consider all these possibilities is a cumbersome and error prone task. Therefore, it would be convenient to have at one's disposal a framework where these tasks can be performed in an automatic manner. The modeling tool ProMoT (Process Modeling Tool) [18] provides a natural environment for such a modular modeling: ProMoT is based on an object-oriented modeling concept, facilitating the reuse and combination of modules. For this purpose, different libraries of models and modules can be implemented, which can be easily combined via a user interface. Thus, ProMoT allows to intuitively implement models in a modular and hierarchical manner. Therefore, an extension for ProMoT was developed, which allows to import models modularly decomposed according to the procedure described above. Currently, the modularity analysis is performed in Matlab [17] and the results are imported into ProMoT. A full integration of those methods into ProMoT is to be performed in the close future. A similar implementation should be feasible in another modular modeling language such as antimony [31].

In the following, the applicability of the criteria to real cases will be exemplified using the Epidermal Growth Factor signaling as an example.

Case Studies The algorithm was benchmarked against different models, describing signal transduction systems of different complexity, and formulated with either mass-action law kinetics or the Michaelis-Menten simplification [24]. The systems modeled ranged from a simple Mitogen-activated Protein Kinase (MAPK) cascade, to a complex signaling network downstream of the Epidermal Growth Factor Receptor (EGFR).

For the later, we considered in particular two models of complementary complexity: (1) a very comprehensive network [20], one of the largest models of a signaling network, within a kinetic (stoichiometric) framework, and (2) a highly entangled model of the EGFR induced MAPK cascade, including receptor internalization process [27]. Because the model of Oda et al only contains a description of the network elements and the reactions among them, we performed an analysis based on the structural retroactivity. The decomposition as outlined above was able to separate the model (comprising over 200 reactions and 300 species) into 55 modules [24].

In the case of the model of Schoeberl et al (comprising 94 species, highly entangled via mass-action reactions), an analysis of the dynamic retroactivity was also possible, since it is a fully parameterized model. The modules obtained corresponded well with those defined manually before using the retroactivity criterion [23]. Furthermore, a simple analysis of the input/output behavior of these modules was performed [22], and a model reduction was performed based on these modules: if one can identify simple systems whose dynamic behavior is similar to that of certain modules, they can replace them within the large model, leading to a reduction of the model size [7].

These results showed the applicability of our algorithm to networks of realistic size. We refer the reader to [24] for a more detailed description of these analysis.

Domain Oriented Modeling and Retroactivity

The previous section covered cases where the absence of retroactivity was directly apparent from the kinetic rate equations. Two cases were distinguished: (1) The influence of a rate on a concentration can be neglected if the rate is small compared to the other rates that appear in the balance equation. (2) The influence of a concentration on a rate can be neglected if the elasticity of the rate with respect to the concentration is small compared to the elasticities of the rate with respect to the other concentrations. These criteria are dependent on the choice of variables made during the modeling process, i.e. the choice of species and reactions. Transformations of concentrations and fluxes may reveal an absence of retroactivity in systems where the original kinetic rate equations seem to be strongly coupled. A simple example for this case was already discussed in Fig. 7.4(d). The unidirectional connection between the enzyme and the reaction becomes only apparent when the system is formulated in terms of the total enzyme concentration $E_0 = E + ES$ and is not visible in the concentrations of the enzyme species E and ES. In this section, we show that such cases regularly occur in signal transduction networks that are based on protein-protein interactions.

Here, we adapt the point of view that domains instead of molecular species are the fundamental elements in signal transduction, which has been introduced

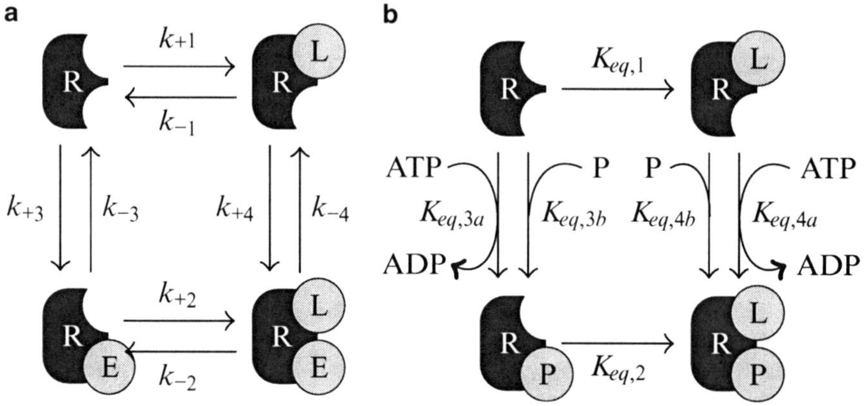

Fig. 7.5 Example systems from section "Domain Oriented Modeling and Retroactivity". (**a**) A receptor R with two binding sites for L and E, respectively. The *arrows* indicate forward and backward reactions. (**b**) A receptor with a binding site for L and a phosphorylation site. The reactions are reversible and the *arrows* indicate the positive direction of the reactions. The numbers on the equilibrium constants K_{eq} correspond to the numbers of reactions in the text (see Eq. 7.37)

by [21]. These domains can be either occupied by other proteins or can undergo post-translational modifications like phosphorylation. We define a binding process as the sum of all reactions that change the level of occupancy of the considered domain. Analogously, we define a modification process as the sum of all reactions changing the degree of modification of a domain. Two arbitrary processes, no matter if binding or modification processes, may be either completely independent, interact unidirectionally or mutually. These different types of interactions shall be exemplified considering a simple example which is taken from [6]. In this example one considers a receptor R, which provides two binding domains for two ligands L and E (see Fig. 7.5a). Hence, the system comprises two binding processes, namely L-binding and E-binding. In this case the reaction system consists of four reversible reactions (two describing L-binding to R_{00} and R_{0E}, and two describing E binding to R_{00} and R_{L0}), for which the following reaction rates can be formulated using mass action kinetics:

$$r_1 = k_{+1} \cdot R_{00} \cdot L - k_{-1} \cdot R_{L0}, \qquad r_2 = k_{+2} \cdot R_{0E} \cdot L - k_{-2} \cdot R_{LE},$$
$$r_3 = k_{+3} \cdot R_{00} \cdot E - k_{-3} \cdot R_{0E}, \qquad r_4 = k_{+4} \cdot R_{L0} \cdot E - k_{-4} \cdot R_{LE}. \qquad (7.32)$$

We assume that the concentrations in this and the following kinetic equations are dimensionless. Thus, the rate constants $k_{\pm i}$ have all the same dimensions. For the sake of simplicity, we assume that the concentrations of free L and E are constant. The results can be generalized to the case where the ligand concentrations vary or where more than two ligands are involved. According to [6] the following process interaction types can be distinguished:

- **non-interacting processes:** Complete independence implies that the kinetic association and dissociation constants of one domain does not change upon ligand binding on the other domain. Hence, it follows for the parameters $k_{+2} = k_{+1}, k_{-2} = k_{-1}, k_{+4} = k_{+3}$ and $k_{-4} = k_{-3}$.
- **unidirectionally interacting processes:** The binding of one ligand, say ligand L, is not influenced by the binding of the other one, but binding of L does change the kinetic properties of the other domain. In this case only the conditions $k_{+2} = k_{+1}$ and $k_{-2} = k_{-1}$ have to be fulfilled.
- **mutually interacting processes:** In this general case, binding of a ligand has an influence on binding of the other ligand and vice versa. Therefore, all parameters can have different values.

Under certain reasonable conditions the absence of interactions or the occurrence of unidirectionality allow model reduction and modularization [3, 4, 8, 10, 15].

The method of [8] assumes unidirectionally interacting processes and is based on a hierarchical transformation of the system. This transformation leads to new variables that correspond to the total concentrations of proteins and the degrees of occupancy of the binding sites. For example, the system in Eq. 7.32 is transformed into the new coordinates $z_0 = R_{00} + R_{L0} + R_{0E} + R_{LE}$ (total receptor concentration), $z_L = R_{L0} + R_{LE}$ (concentration of R-L complexes), $z_E = R_{0E} + R_{LE}$(concentration of R-E complexes) and $z_{LE} = R_{LE}$ (concentration of R-L-E complexes). The corresponding balance equations are

$$\begin{aligned} \dot{z}_0 &= 0, \\ \dot{z}_L &= r_1 + r_2 = k_{+1} \cdot (z_0 - z_L) \cdot L - k_{-1} \cdot z_L = f_L(z_0, z_L), \\ \dot{z}_E &= r_3 + r_4 = f_E(z_0, z_L, z_E, z_{LE}) \\ \dot{z}_{LE} &= r_2 + r_4 = f_{LE}(z_0, z_L, z_E, z_{LE}), \end{aligned} \tag{7.33}$$

where f_L, f_E and f_{LE} are functions of the indicated variables. In this representation the unidirectional connection from the L binding module $\{z_0, z_L\}$ to the E-binding module $\{z_E, z_{EL}\}$ becomes clear (cf. Eq. 7.2). These considerations are generalizable to more complex systems and can be exploited for constructing reduced models of large systems [8]. Therefore, unidirectionality, i. e. the absence of retroactivity as defined by Eq. 7.2, may occur ubiquitously in signal transducing systems but a transformation is needed to reveal the unidirectionality. The resulting modules are not defined as a set of species and reactions, but rather a set of binding domains and processes. In the following sections, we discuss under which conditions such unidirectional interactions can occur.

Thermodynamic Constraints

The second law of thermodynamics in conjunction with the principle of microscopic reversibility demands that, for a closed reaction system, a state of thermodynamic equilibrium exists that is a stable steady state where all thermodynamic forces and fluxes vanish. In thermodynamic equilibrium the forward rate through any reaction is equal to the corresponding backward rate such that the overall rate vanishes. Although biochemical networks are usually not closed and do not reach a thermodynamic equilibrium, the laws of thermodynamics restrict their behavior. The thermodynamic restrictions result from the fact that all thermodynamic forces and fluxes are directed towards a state of thermodynamic equilibrium: if the systems would be isolated from the environment, it will approach thermodynamic equilibrium. This results in constraints on the systems dynamics even if thermodynamic equilibrium is never reached in an open system with energy and mass exchange with the environment.

If a network can be described by generalized mass-action kinetics, these relations lead to the generalized Wegscheider conditions [13, 28]

$$B^T \cdot \ln(\vec{K}_{eq}) = 0, \tag{7.34}$$

where the logarithm of $\vec{K}_{eq}$ has to be taken element-wise. The vector of equilibrium constants $\vec{K}_{eq}$ is defined by $K_{eq,i} = k_{+i}/k_{-i}$, and if we use dimensionless equations as in Eq. 7.32, the equilibrium constants $\vec{K}_{eq}$ are dimensionless. B is a kernel matrix of the stoichiometric matrix N with $N \cdot B = 0$, so that the columns of the matrix B correspond to linearly independent cycles in the network. Therefore, Eq. 7.34 states that the product of the equilibrium constants along any stoichiometric cycle is unity.

Thermodynamic Restrictions on Process Interactions

For the example network (Fig. 7.5a), the Wegscheider conditions read

$$\frac{K_{eq,1} \cdot K_{eq,4}}{K_{eq,2} \cdot K_{eq,3}} = 1. \tag{7.35}$$

The Wegscheider conditions can be rewritten as $a_{LE} = K_{eq,4}/K_{eq,3} = K_{eq,2}/K_{eq,1}$ where a_{LE} is a constant. The constant $K_{eq,1}$ describes the equilibrium of L binding to the empty receptor R_{00} and $K_{eq,3}$ describes the equilibrium of E binding to the empty receptor R_{00}. The equilibrium constants of L-binding to R_{0E} and E-binding to R_{L0} can now be described by

$$K_{eq,2} = K_{eq,1} \cdot a_{LE} \quad \text{and} \quad K_{eq,4} = K_{eq,3} \cdot a_{LE}. \tag{7.36}$$

The factor a_{LE} describes the interaction of L and E at the receptor, that is,the change of the equilibrium of E and L binding if L or E are already bound, respectively. For non-interacting binding processes $a_{LE} = 1$. Note that the effect of bound L on bound

E is equal to the effect of bound E to bound L. This imposes strong restrictions on unidirectional signal transduction. A unidirectional interaction $L \rightarrow E$ can be realized if $k_{+2} = k_{+1}, k_{-2} = k_{-1}$ (i.e. $K_{eq,2} = K_{eq,1}$) and $k_{+4} \neq k_{+3}, k_{-4} \neq k_{-3}$ (which in general implies $K_{eq,4} \neq K_{eq,3}$). However, these requirements contradict the Wegscheider condition, since from $K_{eq,2} = K_{eq,1}$ it follows $a_{LE} = 1$, while $K_{eq,4} \neq K_{eq,3}$ would require $a_{LE} \neq 1$. Thus, thermodynamics does not allow unidirectional interactions between binding processes at the level of equilibrium constants. One special case of unidirectional interaction is allowed, though, if $k_{+4} = k_{+3}\,b \neq k_{+3}$ and $k_{-4} = k_{-3}\,b \neq k_{-3}$, where $b > 0$. In this case, the equilibrium constants $K_{eq,4}$ and $K_{eq,3}$ are equal but the two reactions proceed with different velocities. In the example system, this would yield a transient influence of the amount of bound L on the binding velocity of E. In equilibrium both processes are independent, i. e. the degree of E-binding does not depend on the degree of L-binding.

These restrictions also hold for scaffold proteins with higher numbers of binding domains [5]. Unidirectionality is highly restricted and only feasible in terms of reaction velocities but not of equilibrium constants. In the following section, we will show how unidirectional signal transduction can be achieved by making use of futile cycles that keep the system away from equilibrium.

Unidirectionality and Futile Cycles

The main idea of this chapter is that unidirectionality is an important feature of signal transducing networks of all kinds because it defines clearly distinguishable modules, but we have seen above that unidirectionality on the level of equilibrium constant contradicts the laws of thermodynamics. To achieve a sustained unidirectional signal transduction the system needs to be kept away from the thermodynamic equilibrium by an external force. We will see how the ATP/ADP gradient can act as this external force, but it may be any other subsystem that is able to constantly deliver energy.

Consider the reaction network describing a scaffold protein R with two binding domains (see Fig. 7.5b). One of the two domains can bind a ligand L and the other domain can be phosphorylated. This is a fairly simple case, when compared to realist systems such as receptor tyrosine kinases, but it is sufficient to demonstrate the general principle.

The scaffold possesses two binding domains, as the one described in Eq. 7.32, but the reaction network is different. The complete mechanistic model comprises four phosphorylation reactions, two reactions in which ATP is converted to ADP (r_3 and r_4), and two in which free phosphates bind to and dissociate from the domain (r_1 and r_2; see Fig. 7.5b). The equilibrium of the reactions $3a$ and $4a$ typically lies very much on the product side (more R0P than R00) because the ATP/ADP gradient drives these reactions. The equilibrium of $3b$ and $4b$ typically lies very much on the educt side (more R00 than R0P) because the high binding energy of the phospho group. This reaction network consists of two independent true reaction cycles which means that one has two Wegscheider conditions, namely $K_{eq,1} \cdot K_{eq,4a} \cdot K_{eq,2}^{-1} \cdot K_{eq,3a}^{-1} = 1$ and $K_{eq,1} \cdot K_{eq,4b} \cdot K_{eq,2}^{-1} \cdot K_{eq,3b}^{-1} = 1$. Assume that ligand binding is not influenced by the degree of phosphorylation, i. e. $k_{+1} = k_{+2}$, $k_{-1} = k_{-2}$ and $K_{eq,1} = K_{eq,2}$. As a result the thermodynamic constraints imply that both phosphorylation reactions in which ATP is converted to ADP must have the same equilibrium constants $K_{eq,3a} = K_{eq,4a}$ and both reactions in which free phosphates bind or dissociate also must have the same equilibrium constants $K_{eq,3b} = K_{eq,4b}$. Let's assume that these reactions proceed with different velocities depending on whether the scaffold has bound L or not, e.g. because L recruits or activates a kinase or L stabilizes R in a form that is able to recruit or activate a kinase there is no violation of the Wegscheider conditions since these are two independent cycles. Hence, the six reactions of the regarded system can be written as

$$\begin{aligned}
r_1 &= k_{+1} \cdot L \cdot R_{00} - k_{-1} \cdot R_{L0} \\
r_2 &= k_{+1} \cdot L \cdot R_{0P} - k_{-1} \cdot R_{LP} \\
r_{3a} &= k_{+3a} \cdot ATP \cdot R_{00} - k_{-3a} \cdot ADP \cdot R_{0P} \\
r_{4a} &= x \cdot (k_{+3a} \cdot ATP \cdot R_{L0} - k_{-3a} \cdot ADP \cdot R_{LP}) \\
r_{3b} &= k_{+3b} \cdot P \cdot R_{00} - k_{-3b} \cdot R_{0P} \\
r_{4b} &= y \cdot (k_{+3b} \cdot P \cdot R_{L0} - k_{-3b} \cdot R_{LP}) ,
\end{aligned} \tag{7.37}$$

in which x and y are real positive numbers describing the change of velocity for both forward and backward reactions that is caused by the ligand L. If we take the reasonable assumption that the concentrations of ATP, ADP and P in the cell are approximately constant, we can combine the reactions r_{3a} and r_{3b} as well as r_{4a} and r_{4b} to two 'virtual' phosphorylation reactions

$$\begin{aligned}
r_3^* &= \underbrace{(k_{+3a} \cdot ATP + k_{+3b} \cdot P)}_{=k_{+3}^*} \cdot R_{00} - \underbrace{(k_{-3a} \cdot ADP + k_{-3b})}_{=k_{-3}^*} \cdot R_{0P} \\
r_4^* &= \underbrace{(x \cdot k_{+3a} \cdot ATP + y \cdot k_{+3b} \cdot P)}_{=k_{+4}^*} \cdot R_{L0} \\
&\quad - \underbrace{(x \cdot k_{-3a} \cdot ADP + y \cdot k_{-3b})}_{=k_{-4}^*} \cdot R_{LP}.
\end{aligned} \tag{7.38}$$

The asterisk indicates that the described reaction is a virtual reaction. The resulting reduced network structure now only consists of one single cycle as in Fig. 7.5a. The virtual equilibrium constants of the two reactions r_3^* and r_4^* which we define as the quotient of k_{+i}^* and k_{-i}^*

$$K_{eq,3}^* = \frac{k_{+3a} \cdot ATP + k_{+3b} \cdot P}{k_{-3a} \cdot ADP + k_{-3b}}, \qquad K_{eq,4}^* = \frac{x \cdot k_{+3a} \cdot ATP + y \cdot k_{+3b} \cdot P}{x \cdot k_{-3a} \cdot ADP + y \cdot k_{-3b}} \tag{7.39}$$

are not identical. These constants describe the steady state of the futile cycles and not a thermodynamic equilibrium of the system. Importantly, although the reduced network has the same structure as Fig. 7.5a, the corresponding Wegscheider condition does not need to be fulfilled, and the system can realize an unidirectional process interaction from the binding of L to the phosphorylation of R. The constant energy input from the ATP/ADP gradient keeps the two stoichiometric cycles away from equilibrium and thus drives a futile cycle. This mechanism circumvents the thermodynamic restrictions on unidirectional signal transduction that hold for systems where the cycles are not driven by an external gradient (cf. Fig. 7.5a).

The considerations above show that unidirectional signal transduction at scaffold proteins is possible and likely under reasonable assumptions. Requirements are that the modification of a scaffold protein is realized by at least two different reactions (which corresponds to the existence of a so-called futile cycle), and the corresponding substrates required for the modification (like ATP, ADP and P) are approximately constant. This may be one of the reasons for the ubiquitous occurrence of binding site phosphorylations in signal transducing networks.

Conclusion

The absence of retroactivity, that is, that the signal transmission is not influenced by the state of the receiver but only by the emitter, is an important feature of technical signal transmission systems. The question whether biological signal transduction networks can and do utilize unidirectional interactions is of special interest, since it would allow to deploy a battery of analysis methods developed in control theory.

In this chapter we have reviewed different biochemical cases that can lead to this unidirectionality or absence of retroactivity [23], using the network theory as a bipartite framework that facilitates the characterization of the different cases. We have then seen how to define modules so that the number of retroactive connections are maximized inside the modules and minimized among the modules, and its implementation in an algorithm to automatically find modules from any biochemical network, if at least its structure is known [24].

We have then focused on signal transduction networks described in a domain-oriented manner, and have seen how unidirectionality of process interactions allows to modularize them [8]. Even though there are strong thermodynamic restrictions for mutual effects among protein domains that limit the possibility of unidirectional interactions, we have seen that this restriction can be overcome by ATP driven phosphorylation of binding sites. It is therefore tempting to speculate that the ubiquitous occurrence of phosphorylations and futile cycles in signal transduction networks is evolutionary favourable because this mechanism allows to dissect signal transduction networks into clearly distinct modules that are unidirectionally coupled.

Acknowledgments We thank Eduardo Sontag for useful discussions. The work described here was supported by DFG (FOR521) and the German Ministry of Research and Education BMBF (SysTec Initative, HepatoSys, DYNAMO Consortium).

References

1. Alon U (2007) An introduction to systems biology: design principles of biological circuits. Chapman&Hall/CRC, London, UK
2. Blüthgen N, Bruggeman FJ, Legewie S, Herzel H, Westerhoff HV, Kholodenko BN (2006) Effects of sequestration on signal transduction cascades. FEBS J 273(5):895–906
3. Borisov NM, Markevich NI, Hoek JB, Kholodenko BN (2005) Signaling through receptors and scaffolds: independent interactions reduce combinatorial complexity. Biophys J 89(2):951–966. doi:10.1529/biophysj.105.060533, http://dx.doi.org/10.1529/biophysj.105.060533
4. Borisov NM, Markevich NI, Hoek JB, Kholodenko BN (2006) Trading the micro-world of combinatorial complexity for the macro-world of protein interaction domains. Biosystems 83(2–3):152–166. doi:10.1016/j.biosystems.2005.03.006, http://dx.doi.org/10.1016/j.biosystems.2005.03.006
5. Conzelmann H (2009) Mathematical modeling of biochemical signal transduction pathways in mammalian cells: a domain-oriented approach to reduce combinatorial complexity. PhD thesis, University of Stuttgart, Germany
6. Conzelmann H, Gilles ED (2008) Functional proteomics: methods and protocols. In: Thompson JD et al. (ed) Dynamic pathway modeling of signal transduction networks – A domain-oriented approach. Humana Press, NYC, US, pp 557–576
7. Conzelmann H, Saez-Rodriguez J, Sauter T, Bullinger E, Allgoewer F, Gilles ED (2004) Reduction of mathematical models of signal transduction networks: simulation-based approach applied to egf receptor signaling. Syst Biol 1(1):159–169. doi:10.1049/sb:20045011
8. Conzelmann H, Saez-Rodriguez J, Sauter T, Kholodenko BN, Gilles ED (2006) A domain-oriented approach to the reduction of combinatorial complexity in signal transduction networks. BMC Bioinformatics 7:34. doi:10.1186/1471-2105-7-34, http://dx.doi.org/10.1186/1471-2105-7-34
9. Famili I, Palsson BO (2003) The convex basis of the left null space of the stoichiometric matrix leads to the definition of metabolically meaningful pools. Biophys J 85:16–26
10. Feret J, Danos V, Krivine J, Harmer R, Fontana W (2009) Internal coarse-graining of molecular systems. Proc Natl Acad Sci U S A 106(16):6453–6458. doi:10.1073/pnas.0809908106
11. Gilles ED (1998) Network theory for chemical processes. Chem Eng Technol 21(2):121–132
12. Hartwell L, Hopfield J, Leibler S, Murray A (1999) From molecular to modular cell biology. Nature 402(6761-supp):C47–C52
13. Heinrich R, Schuster S (1996) The regulation of cellular systems. Chapman & Hall, NYC, USA
14. Klamt S, Saez-Rodriguez J, Lindquist J, Simeoni L, Gilles ED (2006) A methodology for the structural and functional analysis of signaling and regulatory networks. BMC Bioinformatics 7:56. doi:10.1186/1471-2105-7-34
15. Koschorreck M, Conzelmann H, Ebert S, Ederer M, Gilles ED (2007) Reduced modeling of signal transduction – A modular approach. BMC Bioinformatics 8(1):336. doi:10.1186/1471-2105-8-336, http://dx.doi.org/10.1186/1471-2105-8-336
16. Lauffenburger DA (2000) Cell signaling pathways as control modules: complexity for simplicity? Proc Natl Acad Sci U S A 97(10):5031–5033
17. Mathworks (2006) Matlab. http://www.mathworks.com/
18. Mirschel S, Steinmetz K, Rempel M, Ginkel M, Gilles ED (2009) Promot: modular modeling for systems biology. Bioinformatics 25(5):687–689. doi:10.1093/bioinformatics/btp029

19. Newman MEJ (2006) Modularity and community structure in networks. Proc Natl Acad Sci U S A 103(23):8577–8582. doi:10.1073/pnas.0601602103, http://dx.doi.org/10.1073/pnas.0601602103
20. Oda K, Matsuoka Y, Funahashi A, Kitano H (2005) A comprehensive pathway map of epidermal growth factor receptor signaling. Mol Syst Biol 1:2005.0010
21. Pawson T, Nash P (2003) Assembly of cell regulatory systems through protein interaction domains. Science 300(5618):445–452. doi:10.1126/science.1083653, http://dx.doi.org/10.1126/science.1083653
22. Saez-Rodriguez J, Kremling A, Conzelmann H, Bettenbrock K, Gilles ED (2004) Modular analysis of signal transduction networks. IEEE Contr Syst Mag 24(4):35–52. doi:10.1109/MCS.2004.1316652
23. Saez-Rodriguez J, Kremling A, Gilles ED (2005) Dissecting the puzzle of life: modularization of signal transduction networks. Comput Chem Eng 29(3):619–629. doi:10.1016/j.compchemeng.2004.08.035
24. Saez-Rodriguez J, Gayer S, Ginkel M, Gilles ED (2008) Automatic decomposition of kinetic models of signaling networks minimizing the retroactivity among modules. Bioinformatics 24(16):i213–i219. doi:10.1093/bioinformatics/btn289, http://bioinformatics.oxfordjournals.org/cgi/content/abstract/24/16/i213
25. Sauro HM (2008) Modularity defined. Mol Syst Biol 4:166. doi:10.1038/msb.2008.3
26. Sauro HM, Kholodenko BN (2004) Quantitative analysis of signaling networks. Prog Biophys Mol Biol 86(1):5–43, doi:10.1016/j.pbiomolbio.2004.03.002
27. Schoeberl B, Eichler-Jonsson C, Gilles E, Müller G (2002) Computational modeling of the dynamics of the MAP kinase cascade activated by surface and internalized EGF receptors. Nat Biotechnol 20(4):370–375
28. Schuster S, Schuster R (1989) A generalization of Wegscheider's condition. Implications for properties of steady states and for quasi-steady-state approximation. J Math Chem 3(1):25–42. doi:10.1007/BF01171883, http://dx.doi.org/10.1007/BF01171883
29. Segel IH (1993) Enzyme kinetics. Behavior and analysis of rapid equilibrium and steady-State enzyme systems. Wiley, Berlin, Germany
30. Segel LA (1988) On the validity of the steady state assumption of enzyme kinetics. Bull Math Biol 50(6):579–593
31. Smith LP, Bergmann FT, Chandran D, Sauro HM (2009) Antimony: a modular model definition language. Bioinformatics 25(18):2452–2454. doi:10.1093/bioinformatics/btp401
32. Sontag ED (1998) Mathematical control theory, 2nd edn. Springer, Berlin and Heidelberg, Germany
33. Vecchio DD, Ninfa AJ, Sontag ED (2008) Modular cell biology: retroactivity and insulation. Mol Syst Biol 4:161. doi:10.1038/msb4100204

Chapter 8
The Impact of Retroactivity on the Behavior of Biomolecular Systems

A Review of Recent Results

Domitilla Del Vecchio

Abstract Modularity is a powerful property for analyzing the behavior of a system on the basis of the behavior of its components. According to this property, any two components maintain their behavior unchanged upon interconnection. Is modularity a natural property of biomolecular networks? In this review, we summarize recent theoretical and experimental results that demonstrate that the answer to this question is negative. Just as in many electrical, mechanical, and hydraulic systems, impedance-like effects, called retroactivity, arise at the interconnection of biomolecular systems and alter the behavior of connected components. Here, we illustrate the effects of retroactivity on the static characteristics and on the dynamic input/output response of biomolecular systems by employing a mixture of control theoretic tools, mathematical biology, and experimental techniques on reconstituted systems.

Keywords Modularity · Retroactivity · Insulation · Transcriptional networks · Signaling cascades.

Introduction

A common approach to either designing or analyzing a complex system is to decompose it into smaller components, or modules, whose functions are well isolated by those of the neighboring components. This approach has been employed for long time in engineering disciplines, such as electrical engineering and computer science and, more recently, it has been proposed also for the analysis of biomolecular systems [1,16,24]. Specifically, scientists have been advocating for the recognition of functional modules, which include signaling systems such as MAPK cascades and covalent modification cycles, machinery for protein synthesis, and DNA replication [3,29]. However, whether modular organization is a general property of biomolecular systems is still subject of debate. The need for understanding

D. Del Vecchio (✉)
Department of Mechanical Engineering, Massachusetts Institute of Technology,
77 Massachusetts Ave. Cambridge, 02139
e-mail: ddv@mit.edu

H. Koeppl et al. (eds.), *Design and Analysis of Biomolecular Circuits: Engineering Approaches to Systems and Synthetic Biology*, DOI 10.1007/978-1-4419-6766-4_8,

the extent of modularity in biomolecular systems has become particularly pressing when designing synthetic circuits. In synthetic biology, in fact, a number of simple functional modules, such as oscillators, toggles, and inverters, are available, but connecting these 'modules' together to engineer complex functionalities is still out of reach [2, 4, 9, 12].

The fundamental assumption made when analyzing or designing a system modularly is that the behavior of each component does not change upon interconnection. However, as it occurs in several engineering systems such as electrical, mechanical, and hydraulic systems, this assumption does not generally hold in biological systems. Upon interconnection, the behavior of an 'upstream' component (the one that sends the signal) is affected by the presence of the 'downstream' component (the one that receives the signal). Consider for example the oscillator of [4] as a source generator to be employed to synchronize a number of downstream transcriptional processes (Fig. 8.1). The oscillator is connected to these downstream processes by having one of the proteins of the oscillator, say the activator A, serve as a transcription factor for the downstream systems by binding to promoter sites in amounts p_{TOT}. These downstream processes in turn act as a load on the oscillator by using up its output protein and by thus affecting its dynamics (*right-side plot* of Fig. 8.1). We broadly call *retroactivity* the phenomenon by which the behavior of an upstream component changes upon connection to a downstream client.

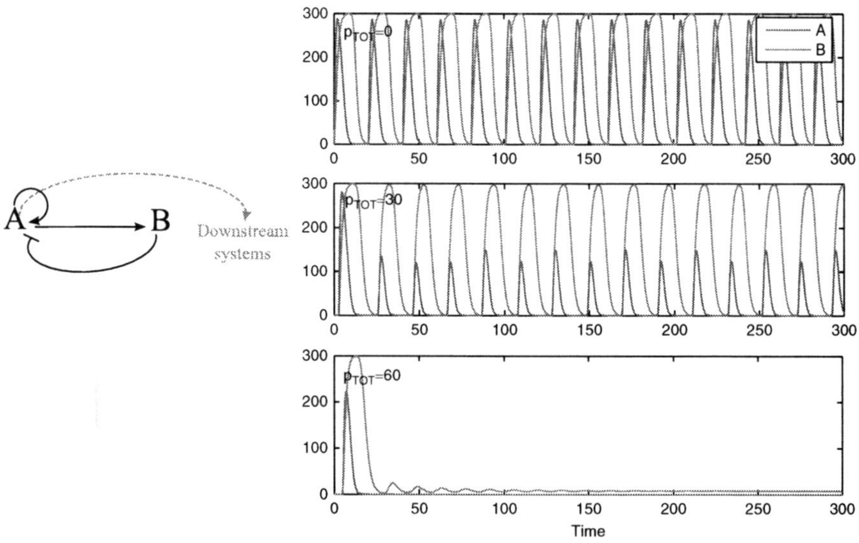

Fig. 8.1 (*Left*) Diagram representing the activator-repressor clock of [4]. This clock is composed of two proteins, A and B, in which A activates its own production and the production of B through transcriptional activation, while B represses the production of A through transcriptional repression. The downstream systems represent transcriptional components that take protein A as an input. (*Right*) In the case in which A is taken as an input to downstream systems through the binding with DNA promoter sites in total amount p_{TOT}, the behavior of the clock changes and is disrupted for high enough load

These considerations strongly motivate the need for a novel theoretical framework to formally define and quantify retroactivity effects. In this chapter, we review a recently proposed framework for studying systems with retroactivity along with theoretical and experimental findings on the effects of retroactivity on biomolecular systems [6–8, 18, 51, 52]. We illustrate this framework through a simple transcriptional system example and we then review theoretical and experimental results on the effects of retroactivity on the steady state and dynamic response of a signaling system.

This chapter is organized as follows. In section "Modeling Retroactivity", we discuss the concept of retroactivity and its general modeling. In section "Example: A Transcriptional System", we illustrate the modeling and quantification of retroactivity on a transcriptional system example. In section "Retroactivity Effects in Signaling Systems", we describe in detail the static and dynamic effects of retroactivity in signaling systems along with experimental validation on a reconstituted system. Section 8.10 concludes the chapter with a short discussion.

Modeling Retroactivity

The principle of studying complex systems through decomposition and interconnection techniques is central in control theory. Approaches based on this general principle range from passivity and more generally dissipativity-based analysis [32, 49, 53, 55, 56], to the derivation of stability properties of large interconnected systems from the graph-theoretic properties of interconnections and stability of individual systems [31, 54], to the use of backstepping feedback approaches [28, 43] based on input to state stability [46]. The work we describe here complements, but differs from, problems of optimally partitioning large networks into 'modules' for which retroactivity-like effects are minimized, which typically employ graph theoretic and statistical approaches [2, 27, 30, 35, 41, 44] . The contribution by Saez-Rodriguez et al. in this book focuses on these problems. In contrast, and similar to the work in [40], we are not concerned with network topology but with the understanding of dynamical behavior. Our ultimate goal is not top–down partitioning or to necessarily minimize retroactivity, but to formally define and characterize these effects especially in view of enabling modular assembly of synthetic biomolecular networks.

The standard model, used in any control and systems theory mathematical and engineering textbook since the 1950s, e.g. [45], is based on the view of devices described solely in terms of input channels, output channels, and state (internal, non-shared) variables. A notable exception to this standard model is found in the work of Willems [37]. Willems has emphasized the fact that, for many physical situations, directionality of signals is an artificial, and technically wrong, assumption. While agreeing with this general point of view, we argue that, in certain circumstances such as those illustrated in this work, it is appropriate to distinguish between input and output channels. Thus, instead of blurring the distinction between inputs, states,

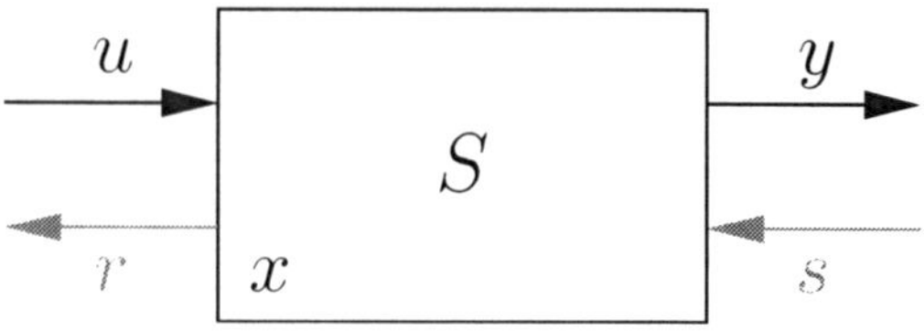

Fig. 8.2 A system model S with retroactivity. The s and r signals originate from retroactivity upon interconnection

and outputs, we keep these three distinct entities but augment the model with two additional signals, namely the retroactivities to inputs and to outputs, respectively (Fig. 8.2).

Specifically, we add an additional input, called s to the system to model any change in its dynamics that may occur upon interconnection with a downstream system. Similarly, we add to a system a signal r as another output to model the fact that when such a system is connected downstream of another system, it sends upstream a signal that alters the dynamics of the upstream system. More generally, we define a system S to have internal state x, two types of inputs (I), and two types of outputs (O): an input 'u' (I), an output 'y' (O), a *retroactivity to the input* 'r' (O), and a *retroactivity to the output* 's' (I) (Fig. 8.2). We thus represent a system S by the equations

$$\dot{x} = f(x, u, s), \; y = Y(x, u), \; r = R(x, u), \tag{8.1}$$

in which f, Y, R are arbitrary functions and the signals x, u, s, r, y may be scalars or vectors. In such a formalism, we define the input/output model of the isolated system as the one in Eqs. 8.1 without r in which we have also set $s = 0$. In practice, it is simpler to model the isolated system first, and only later model the interconnection mechanism to obtain model (8.1). Let S_i be a system with inputs u_i and s_i and with outputs y_i and r_i. Let S_1 and S_2 be two systems with disjoint sets of internal states. We define the interconnection of an upstream system S_1 with a downstream system S_2 by simply setting $y_1 = u_2$ and $s_1 = r_2$. For interconnecting two systems, we require that the two systems do not have internal states in common. For example, in the case of transcriptional components, this would mean that the two transcriptional components express different protein species; in the case of electrical circuits, this would mean that the two circuits do not share common electrical parts except for the ones that establish the interconnection mechanism.

Example: A Transcriptional System

Transcriptional networks are usually viewed as the input/output interconnection of transcriptional components, which take transcription factors as inputs and produce transcription factors as outputs [1]. We showed in [7] that the behavior of a transcriptional component in isolation differs from that of the same component when connected in the network. Specifically, consider a transcriptional component whose

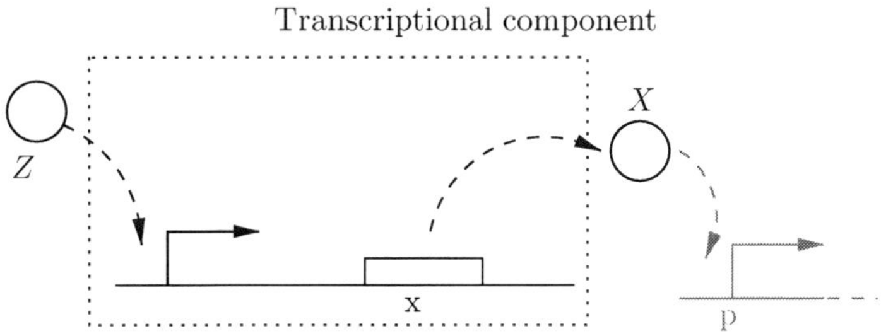

Fig. 8.3 A transcriptional component takes as input u protein concentration Z and gives as output y protein concentration X

output is connected to downstream processes (Fig. 8.3). The activity of the promoter controlling gene x depends on the amount of Z bound to the promoter. For any species X, we denote by X (*italics*) its concentration. If $Z = Z(t)$, such an activity changes with time. We denote it by $k(t)$. By neglecting the mRNA dynamics, which are not relevant to the current discussion, we can write the dynamics of X as

$$\frac{dX}{dt} = k(t) - \delta X, \tag{8.2}$$

in which δ is the decay rate of the protein. Equation 8.2 models the *isolated system* dynamics. Now, assume that X drives a downstream transcriptional system by binding to a promoter p with concentration p (8.3). The reversible binding reaction of X with p is given by X+p $\underset{k_{off}}{\overset{k_{on}}{\rightleftharpoons}}$ C, in which C is the complex protein-promoter and k_{on} and k_{off} are the binding and dissociation rates of the protein X to promoter site p. Since the promoter is not subject to decay, its total concentration p_{TOT} is conserved so that we can write $p + C = p_{TOT}$. Therefore, the new dynamics of X are governed by the equations

$$\begin{aligned} \frac{dX}{dt} &= k(t) - \delta X + \boxed{k_{off} C - k_{on}(p_{TOT} - C)X} \\ \frac{dC}{dt} &= -k_{off} C + k_{on}(p_{TOT} - C)X, \end{aligned} \tag{8.3}$$

in which $s = k_{off} C - k_{on}(p_{TOT} - C)X$ is the *retroactivity to the output*. Here, we can interpret s as being a 'flow' between the upstream and the downstream system. Equations 8.3 model the *connected system* dynamics. When $s = 0$, the first of Eqs. 8.3 reduces to the dynamics of the isolated system given in Eq. 8.2.

The effect of the retroactivity s on the behavior of X can be very large (Fig. 8.4). This is undesirable in a number of situations in which we would like an upstream system to 'drive' a downstream one as is the case, for example, when a biological oscillator has to time a number of downstream processes. We next focus on quantifying the retroactivity to the output s as function of measurable parameters (the quantification of r is similar).

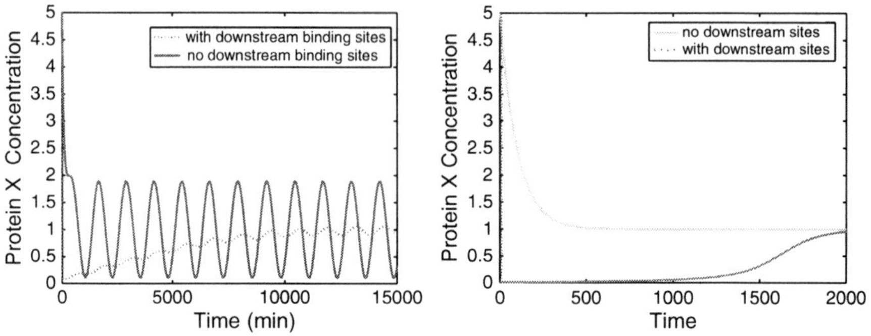

Fig. 8.4 The dramatic effect of an interconnection. Simulation results for the system in Eqs. 8.3. The solid line represents $X(t)$ originating by Eq. 8.2, while the dotted line represents $X(t)$ obtained by Eqs. 8.3. Both transient and permanent behaviors are different. Here, $k(t) = 0.01(1 + sin(\omega t))$ with $\omega = 0.005$ in the *left side plots* and $\omega = 0$ in the *right side plots*, $k_{on} = 10$, $k_{off} = 10$, $\delta = 0.01$, $p_{TOT} = 100$, $X(0) = 5$. The choice of protein decay rate (in min^{-1}) corresponds to a half life of about one hour. The frequency of oscillations is chosen to have a period of about 12 times the protein half life in accordance to what is experimentally observed in the synthetic clock of [4]

Quantification of the Retroactivity to the Output

We quantify the difference between the dynamics of X in the isolated system (8.2) and the dynamics of X in the connected system (8.3) by establishing conditions on the biological parameters that make the two dynamics close to each other. This is achieved by exploiting the difference of time scales between the protein production and decay processes and the binding/unbinding process to promoter p [1]. By virtue of this separation of time scales, we can approximate system (8.3) by a one dimensional system describing the evolution of X on the slow manifold [26]. This reduced system takes the form $\frac{d\bar{X}}{dt} = k(t) - \delta\bar{X} + \bar{s}$, where $\bar{X}$ is an approximation of X and $\bar{s}$ is an approximation of s, which can be written as $\bar{s} = -\mathcal{R}(\bar{X})(k(t) - \delta\bar{X})$ with (see [7, 8] for details)

$$\mathcal{R}(\bar{X}) = \frac{1}{1 + \frac{(1+\bar{X}/k_D)^2}{p_{TOT}/k_D}}, \tag{8.4}$$

in which $k_D = k_{off}/k_{on}$ is the dissociation constant. The expression $\mathcal{R}(\bar{X})$ quantifies the retroactivity to the output after a fast transient when $X(t) \approx \bar{X}(t)$. Retroactivity is thus low if the affinity of the binding sites p is small (k_D large) or if the signal $\bar{X}(t)$ is large enough compared to p_{TOT}. Thus, the expression of $\mathcal{R}(\bar{X})$ provides an operative quantification of retroactivity as a function of the concentration of the binding sites p_{TOT}, the dissociation constant k_D, and the range of $\bar{X}(t)$, which are all directly measurable.

Retroactivity and Noise

It is well known that biological processes are intrinsically stochastic [34, 38, 48]. Since retroactivity alters the dynamics of a biomolecular system, it may also alter its noise properties. Here, we summarize some results that appeared in [18] about the interplay between retroactivity and biological noise. One of the traditional metrics used to assess noise in many electrical engineering applications is the signal-to-noise ratio. This quantity is usually defined by taking the ratio between the power of the signal and the power of the noise. Specifically, consider periodic input signals of the form $k(t) = \bar{k} + \tilde{k}(t)$, in which $\bar{k}$ is a constant bias and $\tilde{k}(t) = A_0 sin(\omega t)$ is a periodic signal with amplitude $A_0 < \bar{k}$ and frequency ω. We assume that all the information transmitted is contained in the signal $\tilde{k}(t)$. To obtain a signal-to-noise figure of merit, the power of a signal is taken to be the square of its amplitude. The power of the noise is quantified by the steady-state variance calculated when the input is a constant and equals the bias, that is, $k(t) = \bar{k}$. Denoting A the amplitude of a signal and $\bar{\sigma}^2$ the steady-state variance, the signal-to-noise ratio is given by

$$SNR := \frac{A^2}{\bar{\sigma}^2}. \tag{8.5}$$

To calculate the value of $\bar{\sigma}^2$, we set $k(t) = \bar{k}$ and calculate the first and second order moments from the master equation by employing the linear noise approximation [11,50]. For calculating the amplitude A, we use the small signal approximation and calculate the frequency response (see [18] for details of the derivations). This leads to the signal-to-noise ratio for X given by

$$SNR(\omega) = \frac{\Omega}{k\delta} \frac{1}{1 + \frac{\omega^2}{\delta^2}(1 + R_l)^2} A_0^2, \tag{8.6}$$

in which $R_l = \frac{k_D p_{TOT}}{(\bar{k}/\delta + k_D)^2}$ and Ω is the volume. Expression (8.6) shows that for a signal with non-zero frequency retroactivity leads to a lower value of *SNR*. This is mainly due to the fact that while the amplitude of response decreases in the presence of retroactivity, the steady state variance does not depend on retroactivity. Notice that the higher the frequency, the more sensitive *SNR* is to retroactivity.

Retroactivity Effects in Signaling Systems

Cellular signaling systems cover a central role in a cell ability to respond to both internal and external input stimuli. These stimuli (often time-varying) include the transient presence of nutrients, hormonal and morphogenic signals, and the periodic excitation of cellular clocks. Numerous signaling systems consist of cycles of protein covalent modification, such as phosphorylation, and in several cases multiple cycles of covalent modification are linked to form cascade systems [39,42].

The importance of these signaling systems has long been realized, and a wealth of theoretical work has established the potential behaviors of such systems and the mechanisms by which parameters and circuitry affect system behavior [5,13,14,47]. These milestone works described how covalent modification cycles would behave in the absence of any loading caused by interconnection with downstream systems, that is, how the cycle would behave as an *isolated* signaling module. But, of course signaling systems are usually connected to the downstream targets they regulate. It is thus important to determine the effect of retroactivity by these targets on the static and dynamic response of the upstream system.

Here, we summarize the results of [51,52], which explicitly quantify the effect of retroactivity on the shape of the input-output static response of a covalent modification cycle and on the frequency response.

Model

Covalent modification cycles can be depicted according to the general scheme of Fig. 8.5, in which a signaling protein is converted from its inactive form W to its active form W* by enzyme E_1 and back to its inactive form by enzyme E_2. The converting enzymes activities can be in turn controlled by an effector through allosteric modification [10]. Here, we have denoted the effector by u and have left unspecified in the diagram whether it is an activator or a repressor of enzyme activity. The results obtained here are independent of the details of enzyme modification and we will consider different cases to ease presentation. Usually, the active protein W* transmits the signal to downstream systems (for example, other signaling targets or DNA binding sites) by binding with appropriate targets [1,22,23]. However, some signal transduction systems display downstream targets both for the active and inactive protein [21,33,36]. Hence, we analyze both cases and consider downstream targets L for the inactive protein and downstream targets N for the active protein.

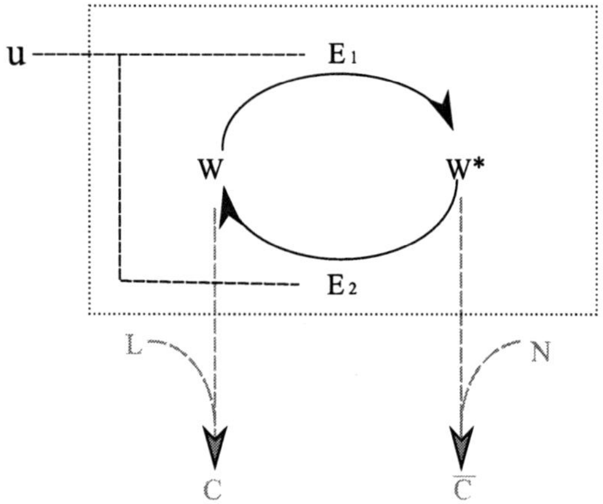

Fig. 8.5 Covalent modification cycle subject to loading due to downstream target sites N and L for the active and inactive protein species, respectively

Let C_1 denote the complex of E_1 with W and C_2 be the complex of E_2 with W^*. The standard two-step reaction model for the enzymatic reactions is given by

$$W + E_1 \underset{d_1}{\overset{a_1}{\rightleftharpoons}} C_1 \overset{k_1}{\longrightarrow} W^* + E_1 \text{ and } W^* + E_2 \underset{d_2}{\overset{a_2}{\rightleftharpoons}} C_2 \overset{k_2}{\longrightarrow} W + E_2$$

to which we add the binding reaction of W with its downstream targets L in total amount L_T and the binding of W^* with downstream targets N in total amounts N_T:

$$W + L \underset{k_{off}}{\overset{k_{on}}{\rightleftharpoons}} C \text{ and } W^* + N \underset{\bar{k}_{off}}{\overset{\bar{k}_{on}}{\rightleftharpoons}} \bar{C}.$$

The kinetic equations governing the system are given by

$$\begin{aligned}
\frac{dW}{dt} &= -a_1 W E_1 + d_1 C_1 + k_2 C_2 - k_{on} N W + k_{off} C \\
\frac{dC_1}{dt} &= a_1 W E_1 - (d_1 + k_1) C_1 \\
\frac{dW^*}{dt} &= -a_2 W^* E_2 + d_2 C_2 + k_1 C_1 - \bar{k}_{on} N W^* + \bar{k}_{off} \bar{C} \\
\frac{dC_2}{dt} &= a_2 W^* E_2 - (d_2 + k_2) C_2 \\
\frac{dC}{dt} &= k_{on} L W - k_{off} C \\
\frac{d\bar{C}}{dt} &= \bar{k}_{on} N W^* - \bar{k}_{off} \bar{C}.
\end{aligned} \tag{8.7}$$

To this differential equations, we add the algebraic equations expressing the conservation laws for the protein and the enzymes: $W_T = W + W^* + C_1 + C_2 + C + \bar{C}$, $E^*_{1T} = E_1 + C_1$, $E^*_{2T} = E_2 + C_2$, $N_T = N + \bar{C}$, $L_T = L + C$, in which we have denoted by E^*_{1T} and E^*_{2T} the total active enzyme amounts. If we assume that the allosteric effector u acts, for example, as an absolute activator for E_2 and a non-competitive inhibitor for E_1, we have that $E^*_{1T} = \frac{E_{1T}}{1+u/k'_D}$ and $E^*_{2T} = \frac{E_{2T} u}{u+\bar{k}'_D}$, in which k'_D and $\bar{k}'_D$ are the dissociation constants for the binding of u with E_1 and E_2, respectively, and E_{1T} and E_{2T} are the total amounts of enzymes [25]. This specific choice of allosteric modification does not alter the results that follow and well represents the experimental system used to test these predictions.

Steady State Effects

In order to quantify the effect of retroactivity on the static input/output characteristics of the system, we solve system (8.7) for the steady state and determine the values of W^* and W as functions of the input u, and the amount of loads L_T and N_T.

Letting $k_D := k_{off}/k_{on}$ and $\bar{k}_D := \bar{k}_{off}/\bar{k}_{on}$ and assuming that $k_D \gg W$ and that $\bar{k}_D \gg W^*$, the steady state value of C and $\bar{C}$ satisfy

$$C = \lambda W \text{ and } \bar{C} = \alpha W^*, \text{ with } \lambda = \frac{L_T}{k_D} \text{ and } \alpha = \frac{N_T}{\bar{k}_D}.$$

Note that in the case in which $\alpha = 0$, we have that $\bar{C} = 0$ and we obtain as a special case of our derivations the situation in which the load is applied only on W.

From the conservation law for W in which we have neglected the complexes C_1 and C_2 (in analogy to what is performed in [13]), we obtain that

$$W_T = W(1+\lambda) + W^*(1+\alpha). \tag{8.8}$$

Further, from setting $\frac{dC_1}{dt} = 0$ and $\frac{dC_2}{dt} = 0$, we obtain

$$C_1 = \frac{E_{1T}^* w}{K_1 + w} \text{ and } C_2 = \frac{E_{2T}^* w^*}{K_2 + w^*},$$

in which we have employed the normalized quantities

$$w^* := \frac{W^*}{W_T}, \quad w := \frac{W}{W_T}, \quad K_1 := \frac{d_1 + k_1}{a_1 W_T}, \quad K_2 := \frac{d_2 + k_2}{d_2 W_T}.$$

From the equilibrium equation $k_1 C_1 = k_2 C_2$ and the conservation law $1 = w(1+\lambda)$ $+ w^*(1+\alpha)$ with

$$S := \frac{E_{2T}^* k_2}{E_{1T}^* k_1} \text{ and } \bar{w}^* = w^*(1+\alpha)$$

we obtain that $\bar{w}^*$ satisfies the equation

$$S = \frac{(1-\bar{w}^*)(K_2(1+\alpha) + \bar{w}^*)}{\bar{w}^*(K_1(1+\lambda) + 1 - \bar{w}^*)}, \tag{8.9}$$

in which, we have that S is monotonically increasing with the input u. We have chosen to study the effects of retroactivity on the steady state value of $\bar{w}^*$ as opposed to consider w^* because in the experimental system we will illustrate, only the total modified protein $W^* + \bar{C}$ can be measured.

From expression (8.9), it is apparent that the net effect of a load on the steady state response is to increase the 'effective' normalized Michaelis-Menten constants K_1 and K_2 by factors of $(1+\lambda)$ and $(1+\alpha)$, respectively. It is well known, in turn, that the values of these constants establish the steepness of the steady state response of the cycle to the input stimulus S and that their relative values establish the point of half maximal induction [13]. We next mathematically quantify the steepness, through the response coefficient, and the point of half maximal induction, called S_{50}.

Effect of Retroactivity on Response Coefficient and S_{50} The steepness of the characteristics and the point of half maximal induction are physiologically relevant quantities in signaling systems as they determine how linear versus ultrasensitive, i.e., switch-like, the response to input stimuli is [13, 14]. We thus mathematically define the steepness and the point of half maximal induction and analytically determine how they are affected by retroactivity.

Since $\bar{w}^*$ is a decreasing function of S, the response coefficient is defined as the ratio between the value of S corresponding to 10% of the maximal value of $\bar{w}^*$, denoted S_{10}, and the value of S corresponding to 90% of the maximal value of $\bar{w}^*$, denoted S_{90}, that is,

$$R := \frac{S_{10}}{S_{90}}.$$

For a Hill equation with Hill coefficient n_H, we have that

$$R = (81)^{1/n_H},$$

that is, R decreases as the Hill coefficient n_H increases. Therefore, we can also take R as a measure of the effective Hill coefficient of a steady state response.

The maximal value of $\bar{w}^*$ corresponds to when $w = 0$ and is obtained from $1 = w(1+\lambda) + w^*(1+\alpha)$ as $\bar{w}_{max} = 1$. As a consequence, we have that

$$R = \frac{S_{10}}{S_{90}} = \frac{81(K_2(1+\alpha)+0.1)(K_1(1+\lambda)+0.1)}{(K_1(1+\lambda)+0.9)(K_2(1+\alpha)+0.9)},$$

which is a monotonically increasing function of α and λ. As a consequence, independently of where the load is applied, the steepness of the response decreases. For the case of no load, i.e., $\alpha = \lambda = 0$, the expression of R reduces to the same expression obtained by [13], while when both α and λ tend to infinity we have that $R = 81$, corresponding to Hill coefficient $n_H = 1$. That is, the response becomes hyperbolic (Michaelis-Menten type of response). In the case in which the load is applied only on W, that is, $\alpha = 0$, we obtain the same behavior for R. However, while with load applied on both W and W^* we have that R tends to 81 for large α and λ *independently* of the parameters K_1 and K_2, when the load is applied to W only, we have that $R = 81\frac{K_2+0.1}{K_2+0.9}$ for $\lambda \to \infty$, which depends on K_2 and tends to 81 only when K_2 is sufficiently large.

The expression of the half maximal induction point S_{50} is given by

$$S_{50} = \frac{(K_2(1+\alpha)+0.5)}{(K_1(1+\lambda)+0.5)},$$

which is an increasing function of α and a decreasing function of λ. In the case in which the load is applied only on W, that is, $\alpha = 0$, we obtain that S_{50} is a monotonically decreasing function of the load.

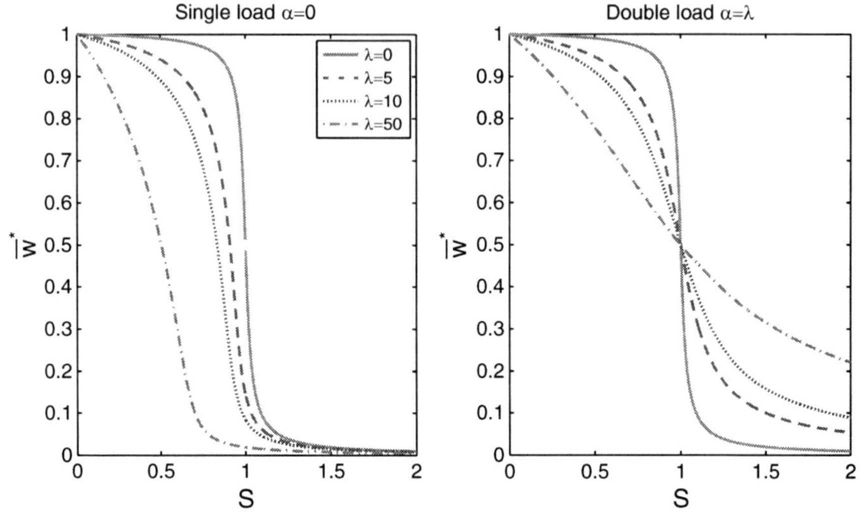

Fig. 8.6 (*Left*) Effect of the load on the steady state response of $\bar{w}^*$ to S when the load is applied only to W, that is, $\alpha = 0$. (*Right*) Effect of the load on the steady state response of $\bar{w}^*$ to S when the load is applied to both W and W*

These results are summarized in Fig. 8.6. With the load applied to W only, the effect of the load is mostly to shift the point of half maximal induction to the left. When the load is applied to both W and W* in comparable amounts, the effect of the load is mostly on reducing the steepness of the response. Hence, retroactivity from large enough loads transforms an ultrasensitive response into a more graded Michaelis-Menten type response.

Finally, we directly study the behavior of the steady state value of $\bar{w}^*$ when α and λ are varied. We thus solve Eq. 8.9 for $\bar{w}^*$, obtaining as the only root between 0 and 1 the expression

$$\bar{w}^* = \frac{(1 - \bar{K}_2 - S(\bar{K}_1 + 1)) + \sqrt{(1 - \bar{K}_2 - S(\bar{K}_1 + 1))^2 + 4(1 - S)\bar{K}_2}}{2(1 - S)}, \tag{8.10}$$

in which we have denoted $\bar{K}_2 := K_2(1 + \alpha)$ and $\bar{K}_1 := K_1(1 + \lambda)$. By computing the derivative of this expression with respect to α and λ, we have that when the load is applied to W only the steady state always decreases with the load for all values of S. By contrast, when the load is applied on both active and inactive species, the effect of the load depends on the input stimulation. Specifically, the steady state increases for large input stimulations, while it decreases for small input stimulations. This is depicted in the left plot of Fig. 8.6.

Dynamic Effects

To study the effects of retroactivity on the dynamics of the signaling system of Fig. 8.5, we consider a one-step model for the enzymatic reactions as found, for example, in [17]. Also, we assume that u is an absolute activator for E_1, while it does not regulate the activity of E_2. This substantially simplifies the analysis without affecting the end result. In this model, we neglect the complexes formed between W and E_1 and between W^* and E_2:

$$W + E_1 \xrightarrow{k_1} W^* + E_1 \text{ and } W^* + E_2 \xrightarrow{k_2} W + E_2.$$

Therefore, the new ODE model describing the covalent modification cycle is given by

$$\frac{dW^*}{dt} = k_1 \frac{E_{1T} u(t)}{k'_D + u(t)} (W_T - W^*) - k_2 E_{2T} W^*, \tag{8.11}$$

in which now $u(t)$ is a time-varying input for our study. We will refer to the ODE system model (8.11) as the *isolated system*. For shortening notation, we denote $V_1(t) := k_1 \frac{E_{1T} u(t)}{k'_D + u(t)}$ and $V_2 := k_2 E_{2T}$.

When the covalent modification cycle transmits its signal through W^* to the downstream system, we add to the isolated system model the reversible binding reaction of W^* with downstream target sites denoted p. These sites can either belong to a substrate that is modified by X_B through another covalent modification cycle as it occurs in the MAPK cascades [39, 42], or they can belong to promoter regions on the DNA if W^* is an active transcription factor [1]. We model this additional binding reaction as $W^* + p \underset{k_{off}}{\overset{k_{on}}{\rightleftharpoons}} C$, with $p + C = p_{TOT}$, in which C denotes the complex of W^* with p. The conservation law for W thus modifies to $W + W^* + C = W_T$. The new ODE model describing the covalent modification system with its downstream system is thus given by

$$\begin{aligned}\frac{dW^*}{dt} &= k_1 \frac{E_{1T} u(t)}{k'_D + u(t)} \left(W_T - W^* - \boxed{C}\right) - k_2 E_{2T} W^* \\ &\quad \boxed{-k_{on} W^* (p_{TOT} - C) + k_{off} C} \\ \frac{dC}{dt} &= k_{on} W^* (p_{TOT} - C) - k_{off} C,\end{aligned} \tag{8.12}$$

which we refer to as the *connected system*. Retroactivity enters the dynamics of the covalent modification cycle in two places indicated by the boxes. Specifically, the term in the small box causes an effect on the steady state response of the system, which we have analyzed in detail in the previous section, while the term in the large box does not have any effect on the steady state and it affects the dynamics only.

In order to precisely quantify how the dynamic response of the system is affected by retroactivity, we linearize the system about its steady state and compute the transfer function for both the isolated and connected systems. Linearization is a good approximation of the system dynamics for sufficiently small amplitudes of the input stimulus. A study on how large the amplitude of the input can be for maintaining a good approximation can be found in [15].

Isolated System For the isolated system, let $(\bar{u}, \bar{W}^*)$ be the equilibrium point and let $\tilde{u}(t) = u(t) - \bar{u}$ and $\tilde{W}^*(t) = W^*(t) - \bar{W}^*$ denote the variations about the equilibrium value. The linearized dynamics are given by

$$\dot{\tilde{W}}^* = \beta\tilde{u} - \alpha\tilde{W}^*, \tag{8.13}$$

in which we have defined

$$\beta := k_1(W_T - \bar{W}^*)\frac{E_{1T}k'_D}{(k'_D + \bar{u})^2}, \quad \alpha := \left(k_1\frac{E_{1T}\bar{u}}{k'_D + \bar{u}} + k_2E_{2T}\right). \tag{8.14}$$

Direct integration of system (8.13) starting from zero initial condition and with input $\tilde{u}(t) = 1$ leads to the time response to constant input stimuli as

$$\tilde{W}^*(t) = \frac{\beta}{\alpha}(1 - e^{-\alpha t}). \tag{8.15}$$

The *response time*, that is, the time the signal takes to rise from 10% of its final value to 90% of its final value is equal to $t_{response} = 2/\alpha$. The transfer function from $\tilde{u}$ to $\tilde{W}^*$ is given by $T(s) = \frac{\beta}{s+\alpha}$, in which $T(s) := \tilde{W}^*(s)/\tilde{u}(s)$, so that amplitude and phase lag are given by

$$A(\omega) = \sqrt{T(j\omega)T(-j\omega)} = \frac{\beta}{\sqrt{\omega^2 + \alpha^2}}$$

$$\phi(\omega) = \arctan\left(\frac{\text{Im}(T(j\omega))}{\text{Re}(T(j\omega))}\right) = \arctan(-\omega/\alpha). \tag{8.16}$$

The frequency *bandwidth*, corresponding to the value of ω such that $A(\omega)=\frac{1}{\sqrt{2}}A(0)$, is given by $\omega_{bandwidth} = \alpha$.

Connected System For the connected system, let the equilibrium point be given by $(\bar{u}, \bar{W}_c^*, \bar{C})$ and the variations about this equilibrium be denoted by $\tilde{u}(t) = u(t) - \bar{u}$, $\tilde{W}^*(t) = W^* - \bar{W}_c^*$, and $\tilde{C}(t) = C(t) - \bar{C}$. The linearized system is thus given by

$$\begin{aligned} \dot{\tilde{W}}^* &= \bar{\beta}\tilde{u} - (\alpha + \gamma)\tilde{W}^* - (\sigma + \eta)\,\tilde{C} \\ \dot{\tilde{C}} &= \gamma\tilde{W}^* - \eta\tilde{C}, \end{aligned} \tag{8.17}$$

in which we have denoted

$$\bar{\beta} := k_1(W_T - \bar{W}_c^* - \bar{C})\frac{E_{1T}k'_D}{(k'_D + \bar{u})^2}, \ \sigma := k_1\frac{E_{1T}\bar{u}}{k'_D + \bar{u}},$$
$$\gamma := k_{on}(p_{TOT} - \bar{C}), \ \ \eta := k_{on}\bar{W}_c^* + k_{off}.$$

The transfer function $T_c(s) := \tilde{W}^*(s)/\tilde{u}(s)$ is given by

$$T_c(s) = \frac{\bar{\beta}(s+\eta)}{s^2 + s(\eta+\alpha+\gamma) + \eta\alpha + \sigma\gamma}.$$

Exploiting the fact that the binding and unbinding process of a protein to binding sites is usually much faster than covalent modification reactions [10], we set $\eta = \bar{\eta}/\epsilon$ and $\gamma = \bar{\gamma}/\epsilon$, in which $\epsilon \ll 1$ and $\bar{\gamma}$ and $\bar{\eta}$ are of the same order as k_1 and k_2. By using the expressions of $\bar{\eta}$ and $\bar{\gamma}$ and setting $\epsilon = 0$, we obtain the reduced transfer function for the connected system as

$$T_c(s) = \frac{\bar{\beta}}{s(1+\mu) + \alpha + \sigma\mu}, \ \text{with } \mu = \frac{p_{TOT}k_D}{(\bar{W}_c^* + k_D)^2}.$$

Therefore, the response of $\tilde{W}^*$ to a constant input stimulus $\tilde{u}(t) = 1$ is given by (computing the inverse Laplace transform of $T_c(s)\frac{1}{s}$)

$$\tilde{W}^*(t) = \frac{\bar{\beta}}{\alpha + \sigma\mu}\left(1 - e^{-(\alpha+\sigma\mu)/(1+\mu)t}\right). \tag{8.18}$$

The response time is thus given by $t_{response,c} = \frac{2}{\alpha}\left(\frac{1+\mu}{1+\mu(\sigma/\alpha)}\right)$, which is larger than $t_{response}$ for the isolated system as $\sigma < \alpha$. Also, it is monotonically increasing with μ: for $\mu = 0$ it is equal to the response time of the isolated system while for $\mu \to \infty$ it tends to $2/\sigma$. In turn, μ monotonically increases with p_{TOT} and (for k_D sufficiently large) it also increases with $1/k_D$ (the affinity of W* to sites p). For values of k_D close to zero, the value of μ is not informative as the linear approximation does not hold. Furthermore, since $\bar{\beta} < \beta$ the amplitude of the response is also reduced for the connected system. The difference $\bar{\beta} - \beta$ is proportional to $\bar{C}$, so that the difference between the amplitude of the responses increases as p_{TOT} increases and/or k_D decreases. The amplitude and phase lag corresponding to $T_c(s)$ are given by

$$A_c(\omega) = \sqrt{T_c(j\omega)T_c(-j\omega)} = \frac{\bar{\beta}}{\sqrt{\omega^2(1+\mu)^2 + (\alpha+\sigma\mu)^2}}$$
$$\phi_c(\omega) = \arctan\left(\frac{\text{Im}(T_c(j\omega))}{\text{Re}(T_c(j\omega))}\right) = \arctan\left(\frac{-\omega(1+\mu)}{\alpha+\sigma\mu}\right), \tag{8.19}$$

so that the bandwidth of the connected system is given by $\omega_{bandwidth,c} = \alpha\frac{1+\mu(\sigma/\alpha)}{1+\mu}$. Therefore, $\omega_{bandwidth,c} < \omega_{bandwidth}$, that is, the bandwidth of the connected system is

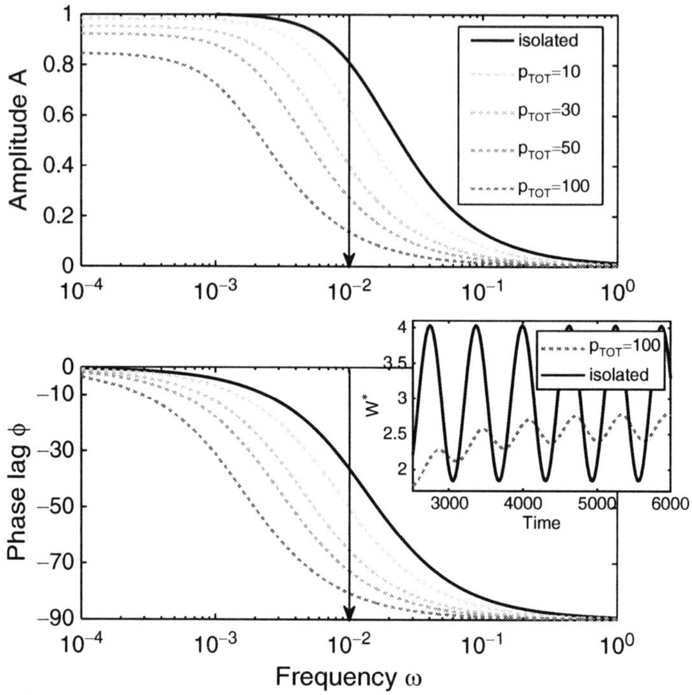

Fig. 8.7 Effect of increasing the amount of p_{TOT} on the frequency response of the system. The parameters are $k_1 = k_2 = 0.01$, $E_{1T} = 0.075$, $W_T = 600$, $E_{2T} = 1.36$, $k'_D = 100$, $k_{on} = 50$, and $k_{off} = 50$. The small panel shows simulation results for the input frequency as indicated by the arrow in the *left plots* for the value $p_{TOT} = 100$

strictly smaller than the bandwidth of the isolated system and the connected system displays a phase lag with respect to the isolated system. This is illustrated in Fig. 8.7. We thus conclude that the larger the value of μ the larger the effect of retroactivity on the dynamical properties of the cycle, that is, the larger the response time, the phase lag, and the smaller the frequency bandwidth.

The bandwidth $\omega_{bandwidth,c}$ of the connected system can be increased by increasing α. One way to increase α is to equally (so not to alter the equilibrium of the system) increase the values of both E_{1T} and E_{2T}. The result is that the behavior of the connected system becomes closer to the one of the isolated system (Fig. 8.8). In the limit in which $A_c(0) = A(0)$, the behavior of the connected system approaches the one of the isolated system when both E_{1T} and E_{2T} are increased. That is, the system becomes *insulated* from retroactivity. This is in accordance with the principle for insulation based on time-scale separation, according to which faster system time-scales contribute to better retroactivity attenuation [19]. Note that if $\bar{\beta}$ is much smaller than β, that is, $A_c(0) \ll A(0)$, the dominant effect of retroactivity is on the steady state. In fact, increasing the frequency of the input stimulation will not result in a dramatic decrease of the connected system response compared to the isolated system response as these two responses are apart from each other already at zero frequency.

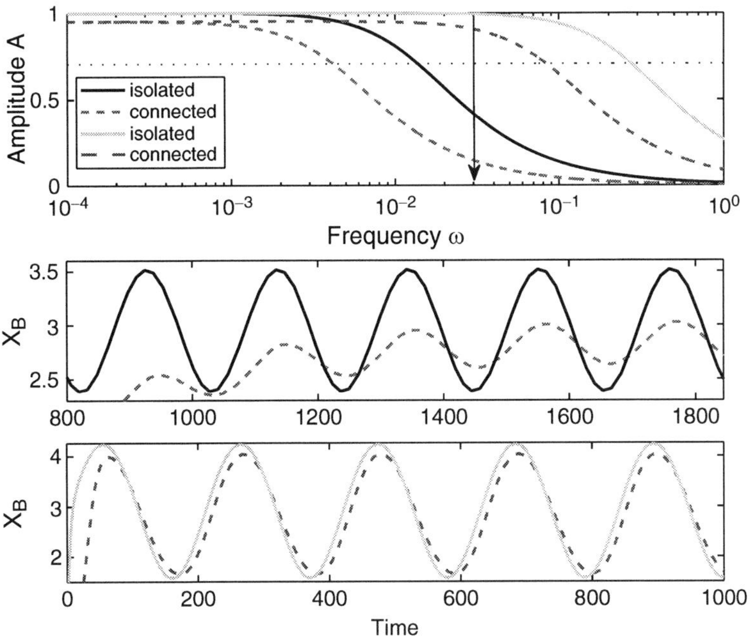

Fig. 8.8 Increasing the values of the enzymes E_{1T} and E_{2T} increases the bandwidth of the covalent modification cycle. As a result, the response of the connected system becomes closer to the one of the isolated system. The *thin solid* and *thin dashed plots* correspond to the isolated and connected system behavior, respectively, while the *bold solid* and *bold dashed plots* correspond to the isolated and connected system behavior, respectively, when the modification rates are increased by setting $E_{2T} = 30$ and $E_{1T} = 1.5$

Experimental Results

The prediction that retroactivity makes an ultrasensitive response into a graded one has been experimentally validated on a covalent modification cycle extracted from the nitrogen assimilation control system of *E. coli* and reconstituted *in vitro* [51]. Here, we briefly summarize these experimental results.

The instance of the covalent modification cycle of Fig. 8.5 employed in the experiments is highlighted in Grey in Fig. 8.9 [10, 20, 33]. This system was reconstituted *in vitro* to allow well controlled experimental conditions. Referring to Fig. 8.5, protein W is the PII signal transduction protein, protein W* is the active (uridylylated) protein PII-UMP, active enzyme E_1 is the UT activity of the UTase/UR bifunctional enzyme, while the active E_2 enzyme is the UR activity of the UTase/UR enzyme. The allosteric effector u is glutamine, which regulates both UT and UR activities by binding to a regulatory domain of the UTase/UR. The protein PII has one downstream signaling target, NRII.

The PII protein is a homotrimer, and can be uridylylated on each of its subunits (Fig. 8.10a). Hence, comparing Fig. 8.5 and Fig. 8.10a, we have that the modified

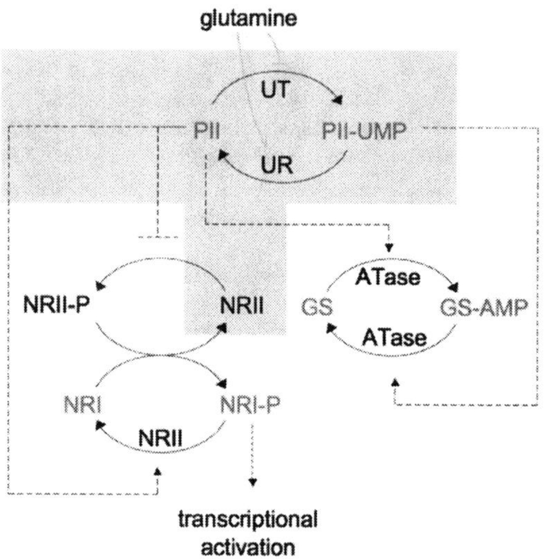

Fig. 8.9 Experimental system employed in [51]. The part used in the experiments is highlighted in Grey. Reprinted with permission from PNAS

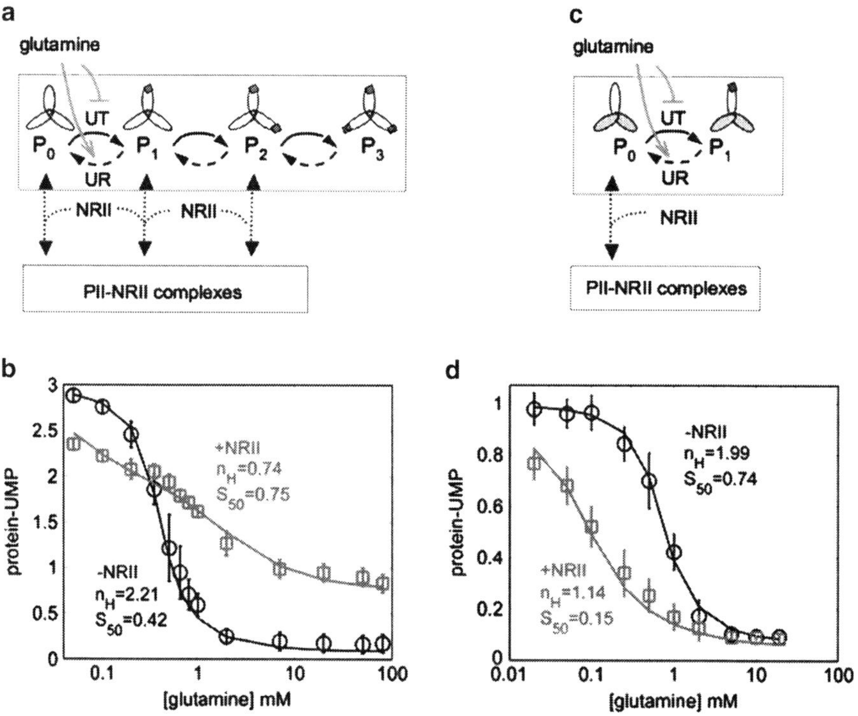

Fig. 8.10 Experimental results from [51]. (**a** and **b**) Using the trimeric PII protein. (**c** and **d**) Using a monovalent version of the PII protein

protein W* comprises all of the modified forms of PII (P_1, P_2, and P_3 of Fig. 8.10a). Also, partially modified forms of PII (P_1 and P_2, 8.10a) can bind to NRII. As a consequence, we have that the downstream targets L and N are the same and are given by the NRII protein. Thus, the use of the trimeric PII protein results in a cycle with 'double load' as depicted in Fig. 8.5, in which both the active and inactive protein species have downstream targets. In order to study the effects of applying the load on one side only of the cycle, which is a configuration often found in natural systems, we employed a monovalent version of the PII protein (Fig. 8.10c), which is obtained by proper mutation of two PII subunits.

Figure 8.10 illustrates how retroactivity makes an ultrasensitive input/output static response into a more graded response independently of where the load is applied. Also, it illustrates how the value of S_{50} decreases when the load is applied only on the inactive protein.

Discussion and Conclusion

In this work, we have summarized some recent results that illustrate how retroactivity impacts the behavior of biomolecular systems. Retroactivity by downstream targets slows down the dynamic response by decreasing the effective bandwidth and reduces the sensitivity of the steady state input/output characteristics. These effects, which are more dramatic as the amounts and affinity of downstream targets increase, indicate that the behavior of a biomolecular system cannot be understood in isolation. This is especially the case in signaling systems, in which covalent modification cycles have several downstream targets. What is the role of retroactivity in these systems? Signaling systems have been selected by nature for effective signal transduction. Hence, retroactivity must have a clear evolutionary advantage, or there must be insulation mechanisms to attenuate undesirable retroactivity effects. From a design point of view, the results summarized in this chapter indicate that retroactivity must be taken into account when engineering biomolecular circuits and that suitable insulation mechanisms should be designed in order to buffer connected components from each other [6, 7, 19].

References

1. Alon U (2007) An introduction to systems biology. Design principles of biological circuits. Chapman & Hall
2. Andrianantoandro E, Basu S, Karig DK, Weiss R (2006) Synthetic biology: new engineering rules for an emerging discipline. Mol Syst Biol 2:1–14
3. Asthagiri AR, Lauffenburger DA (2000) Bioengineering models of cell signaling. Annu Rev Biomed Eng 2:31–53
4. Atkinson MR, Savageau MA, Meyers JT, Ninfa AJ (2003) Development of genetic circuitry exhibiting toggle switch or oscillatory behavior in *Escherichia coli*. Cell 113(5):597–607
5. Cárdenas ML, Cornish-Bowden A (1989) Characteristics necessary for an interconvertible enzyme cascade to generate a highly sensitive response to an effector. Biochem J 257(2):339–345
6. Del Vecchio D, Jayanthi S (2008) Retroactivity attenuation in transcriptional networks: design and analysis of an insulation device. In: Proceedings conference on decision and control. Cancun pp 774–780

7. Del Vecchio D, Ninfa AJ, Sontag ED (2008) Modular cell biology: retroactivity and insulation. Nature/EMBO Mol Syst Biol 4:161
8. Del Vecchio D, Ninfa AJ, Sontag ED (2008) A systems theory with retroactivity: application to transcriptional modules. In: Proceedings of American control conference. Seattle, WA pp 1368–1373
9. Elowitz MB, Liebler S (2000) A synthetic oscillatory network of transcriptional regulators. Nature 403(6767):335–338
10. Fell D (1997) Understanding the control of metabolism. Portland Press Ltd, London
11. Gardiner CW (1996) Handbook of stochastic methods: for physics, chemistry and the natural sciences. Springer Berlin, Heidelberg, New York
12. Gardner TS, Cantor CR, Collins JJ (2000) Construction of the genetic toggle switch in *Escherichia Coli*. Nature 403(6767):339–342
13. Goldbeter A, Koshland DE (1981) An amplified sensitivity arising from covalent modification in biological systems. Proc Natl Acad Sci USA 78(11):6840–6844
14. Goldbeter A, Koshland DE Jr (1984) Ultrasensitivity in biochemical systems controlled by covalent modification. interplay between zero-order and multistep effects. J Biol Chem 259(23):14441–14447
15. Gomez-Uribe C, Verghese GC, Mirny LA (2007) Operating regimes of signaling cycles: statics, dynamics, and noise filtering. PLoS Comput Biol 3(12):2487–2497
16. Hartwell LH, Hopfield JJ, Leibler S, Murray AW (1999) From molecular to modular cell biology. Nature 402:47–52
17. Heinrich R, Neel BG, Rapoport TA (2002) Mathematical models of protein kinase signal transduction. Mol Cell 9:957–970
18. Jayanthi S, Del Vecchio D (2009) On the compromise between retroactivity attenuation and noise amplification in gene regulatory networks. In: Proceedings conference on decision and control. Shangai, China, pp 4565–4571
19. Jayanthi S, Del Vecchio D (2010) Retroactivity attenuation in biomolecular systems based on timescale separation. IEEE Trans Automatic Control, DOI: 10.1109/TAC.2010.2069631
20. Jiang P, Ninfa AJ (2007) *Escherichia coli* PII signal transduction protein controlling nitrogen assimilation acts as a sensor of adenylylate energy charge in vitro. Biochemistry 46:12979–12996
21. Jiang P, Mayo AE, Ninfa AJ (2007) *Escherichia coli* glutamine synthetase adenylyltransferase (ATase, EC 2.7.7.49): kinetic characterization of regulation by PII, PII-UMP, glutamine, and alpha-ketoglutarate. Biochemistry 46(13):4133–46
22. Kholodenko BN (2006) Cell signaling dynamics in time and space. Nat Rev Mol Cell Biol 7(3):165–176
23. Kholodenko BN, Kiyatkin A, Bruggeman FJ, Sontag E, Westerhoff HV, Hoek JB (2002) Untangling the wires: a strategy to trace functional interactions in signaling and gene networks. Proc Natl Acad Sci USA 99(20):12841–12846
24. Kirschner MW, Gerhart JC (2005) The plausibility of life: resolving Darwin's Dilemma. Yale University Press, New Haven and London
25. Klipp E, Herwig R, Kowald A, Wierling C, Lehrach H (2005) Systems biology in practice. Wiley, Weinheim
26. Kokotovic P, Khalil HK, O'Reilly J (1999) Singular perturbation methods in control. SIAM
27. Kremling A, Saez-Rodriguez J (2007) Systems biology - An engineering perspective. J Biotechnol 129:329–351
28. Krstić M, Kanellakopoulos I, Kokotović PV (1995) Nonlinear and Adaptive Control Design. Wiley, New York
29. Lauffenburger DA (May 2000) Cell signaling pathways as control modules: complexity for simplicity? Proc Natl Acad Sci USA 97(10):5031–5033
30. Mason O, Verwoerd M (Apr 2006) Graph theory and networks in biology. Technical report. http://arxiv.org/abs/q-bio.MN/0604006
31. Michel AN, Miller RK (1977) Qualitative Analysis of Large Scale Dynamical Systems. Academic Press, New York

32. Moylan PJ, Hill DJ (1978) Stability criteria for large-scale systems. IEEE Trans Automat Contr 23(2):143–149
33. Ninfa AJ, Jiang P (2005) PII signal transduction proteins: sensors of α-ketoglutarate that regulate nitrogen metabolism. Curr Opinion Microbiol 8:168–173
34. Paulsson J (2004) Summing up the noise in gene networks. Nature 427:415–418
35. Papin JA, Reed JL, Palsson BO (2004) Hierarchical thinking in network biology: the unbiased modularization of biochemical networks. Trends Biochem Sci 29:641–647
36. Pioszak AA, Jiang P, Ninfa AJ (2000) The *Escherichia coli* PII signal transduction protein regulates the activities of the two-component system transmitter protein NRII by direct interaction with the kinase domain of the transmitter module. Biochemistry 39(44):13450–13461
37. Polderman JW, Willems JC (2007) Introduction to Mathematical Systems Theory. A Behavioral Approach, 2nd edn. Springer-Verlag, New York
38. Rosenfeld N, Young JW, Alon U, Swain PS, Elowitz MB (2005) Gene regulation at the single-cell level. Science 307(5717):1962–1965, DOI: 10.1126/science.1106914
39. Rubinfeld H, Seger R (2005) The ERK cascade: a prototype of MAPK signaling. Mol Biotechnol 31(2):151–174
40. Sauro HM (2004) The computational versatility of proteomic signaling networks. Curr Proteomics 1(1):67–81
41. Saez-Rodriguez J, Kremling A, Gilles ED (2005) Dissecting the puzzle of life: modularization of signal transduction networks. Comput Chem Eng 29:619–629
42. Seger R, Krebs EG (1995) The MAPK signaling cascade. FASEB J 9:726–735
43. Sepulchre R, Janković M, Kokotović P (1997) Constructive nonlinear control. Springer-Verlag, New York
44. Snel B, Bork P, Huynen MA (2002) The identification of functional modules from the genomic association of genes. Proc Natl Acad Sci USA 99(9):5890–5895
45. Sontag ED (1998) Mathematical control theory. Springer-Verlag, New York
46. Sontag ED (2007) Input to state stability: Basic concepts and results. In: Nistri P, Stefani G (eds) Nonlinear and optimal control theory. Springer-Verlag, Berlin, pp 163–220
47. Stadtman ER, Chock PB (1977) Superiority of interconvertible enzyme cascades in metabolic regulation: analysis of monocyclic systems. Proc Natl Acad Sci USA 74(7):2761–2765
48. Thattai M, van Oudenaarden A (2001) Intrinsic noise in gene regulatory networks. Proc Natl Acad Sci USA pp 8614–8619
49. van der Schaft AJ (2000) $\mathcal{L}_2$-gain and passivity techniques in nonlinear control, 2nd edn. Springer-Verlag, Berlin
50. van Kampen NG (2007) Stochastic processes in physics and chemistry, 3rd edn. Elsevier, North Holland
51. Ventura AC, Jiang P, van Wassenhove L, Del Vecchio D, Merajver SD, Ninfa AJ (2010) Signaling properties of a covalent modification cycle are altered by a downstream target. Proc Natl Acad Sci USA 107(22):10032–10037
52. Ventura AC, Jiang P, van Wassenhove L, Ninfa AJ, Merajver SD, Del Vecchio D (2010) The impact of retroactivity on the input/output characteristic of a signaling component. In: Dynamic systems and control conference (in press)
53. Vidyasagar M (1981) Input-Output analysis of large scale interconnected systems. Springer-Verlag, Berlin
54. Šiljak DD (1978) Large-Scale systems: stability and structure. Dover, New York
55. Willems J (1972) Dissipative dynamical systems, Part I: General theory. Arch Ration Mech An 45:321–351
56. Willems JC (1972) Dissipative dynamical systems Part I: General theory; Part II: Linear systems with quadratic supply rates. Arch Ration Mech An 45:321–393

Chapter 9
Modularity, Retroactivity, and Structural Identification

Eduardo D. Sontag

Abstract Many reverse-engineering techniques in systems biology rely upon data on steady-state (or dynamic) perturbations – obtained from siRNA, gene knock-down or overexpression, kinase and phosphatase inhibitors, or other interventions – in order to understand the interactions between different 'modules' in a network. This paper first reviews one popular such technique, introduced by the author and collaborators, and also discusses why conclusions drawn from its (mis-)use may be misleading due to 'retroactivity' (impedance or load) effects. A theoretical result characterizing stoichiometric-induced steady-state retroactivity effects is given for a class of biochemical networks.

Keywords Retroactivity · Modularity · Reverse engineering

Introduction

The 'reverse engineering problem' in systems biology concerns itself with the discovery of the networks of interactions among the components of biomolecular networks, including signaling, gene regulatory, and metabolic control networks. The objective is to map out the direct or 'local' interactions among components, which capture the topology of the functional network, with the ultimate goal of elucidating the mechanisms underlying observed behavior (phenotype).

Typically, the analysis is based upon data gathered from steady-state perturbation experiments. Perturbations are done to particular gene or signaling components by means of traditional genetic experiments, RNA interference, hormones, growth factors, or pharmacological interventions. Observed are steady-state changes in concentrations of active proteins, mRNA levels, transcription rates, and so forth. A graph is used to summarize the deduced interactions. For example, if there are

E.D. Sontag (✉)
Department of Mathematics, Rutgers University, Piscataway, NJ
e-mail: sontag@math.rutgers.edu

H. Koeppl et al. (eds.), *Design and Analysis of Biomolecular Circuits: Engineering Approaches to Systems and Synthetic Biology*, DOI 10.1007/978-1-4419-6766-4_9,

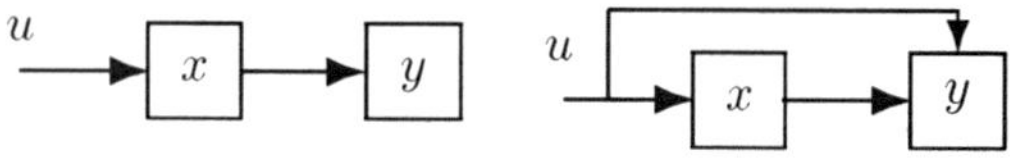

Fig. 9.1 Cascade and feedforward architectures

two components, labeled A and B, one may perform an up-perturbation in A. If this leads to an increased value of B, a directed edge $A \to B$ labeled 'activation' is introduced. If it leads to a decreased level of B, an edge labeled 'repression' is drawn. If there is no effect on B, no edge is put in. A major difficulty with such steady-state (or even time-resolved) experiments is that perturbations propagate, sometimes rapidly, throughout the network, thus causing 'global' changes which cannot be easily distinguished from direct effects. To illustrate this difficulty, consider the two graphs shown in Fig. 9.1 (arrows are supposed to be activating). In both instances, up-perturbations of the external signal u or of the block labeled x results in up-perturbations of the block y, but there is no obvious way to distinguish the two architectures. A major goal in reverse engineering is to unravel the local interactions among individual nodes from these observed global responses.

The 'unraveling', or 'Modular Response Analysis' (MRA) method proposed in [7] and further elaborated upon in [1–3, 12], (see [4, 13] for reviews) provides one approach to solving this global-to-local problem. The MRA experimental design compares those steady states which result after performing independent perturbations to each 'modular component' of a network. These perturbations might be genetic or biochemical. For example, in eukaryotes they might be achieved through the down-regulation of mRNA, and therefore protein, levels by means of RNAi, as done in [10]. That work employed MRA in order to quantify positive and negative feedback effects in the Raf/Mek/Erk MAPK network in rat adrenal pheochromocytoma (PC-12) cells; using the algorithms from [12] and [1], the authors of [10] uncovered connectivity differences depending on whether the cells are stimulated with epidermal growth factor (EGF) or instead with neuronal growth factor (NGF).

Let us illustrate the underlying idea with the simplest non-trivial example. Suppose that we are faced with the problem of distinguishing between the two possible architectures schematically shown in Fig. 9.1. In general, components may be described by single variables or by many variables. For instance, a gene expression component might be described at various levels of resolution: by just one variable (resulting protein levels), or by a far more complicated mechanism (including binding and unbinding of transcription factors, transcription initiation and mRNA polymerase dynamics, ribosome binding and translation dynamics, etc.). For simplicity, let us discuss a simple model in which each component is described by a scalar linear system. Thus both possible architectures are special cases of:

$$\dot{x} = -ax + bu$$
$$\dot{y} = cx - dy + pu$$

where all parameters are positive but otherwise unknown ($a, d > 0$, so the model is stable), and the question that we are interested in is that of deciding whether

$p = 0$ or $p \neq 0$. (Obviously, it would be difficult to distinguish a small $p \neq 0$ from $p = 0$, if measurements are noisy. We assume for this introductory discussion that measurements are exact.) The available data are the steady states for both x and y, for a constant (but unknown) input u, under these three scenarios: (1) $a = a_0$, $u = u_0$; (2) $a = a_0$, $u = u_1$; (3) $a = a_1$, $u = u_0$. Once again, we emphasize that u and a are not known. All we know is that they have changed, one at a time, in experiments (2) and (3), which represent a change in the concentration of u and a change in the degradation rate of x (e.g., due to a protease concentration being changed) respectively. In general, the steady state, obtained by setting $\dot{x} = \dot{y} = 0$, is (for constant u) given by $x(\infty) = (b/a)u$ and $y(\infty) = (cb/a + p)u/d$. Let us write $x_{\Delta u} = (b/a_0)u_1 - (b/a_0)u_0 = (b/a_0)\Delta u$, the difference between the measured steady state of x for experiments (2) and (1), and the corresponding quantity $y_{\Delta u} = (cb/a + p)\Delta u/d$ for y. Similarly, subtracting the data from experiments (3) and (1) provides the measured quantities $x_{\Delta a} = (1/a_1 - 1/a_0)bu_0$ and $y_{\Delta a} = (1/a_1 - 1/a_0)cbu_0/d$. Thus, we can compute from the data: $y_{\Delta u}/x_{\Delta u} - y_{\Delta a}/x_{\Delta a} = a_0 p/(bd)$. If this last number is zero, then $p = 0$ (cascade architecture), and if it is nonzero, then $p \neq 0$ (feedforward architecture). Our objective of distinguishing between the two structures has been achieved. (Moreover, we can even recover the numerical value $y_{\Delta a}/x_{\Delta a} = c/d$. And, if u_0 or u_1 were also known, then we would be able to compute b/a_0 from the steady state value of x, and hence we would also obtain the value of p/d, as $(b/a_0)\cdot(a_0 p/(bd))$. Therefore, the relative strengths of all the terms in the equation for $\dot{y}$ have been computed. Note that this is the best that one can do: the actual values of all three constants can never be obtained from purely steady-state data, because multiplying all constants by the same number doesn't affect the steady states.)

The MRA method generalizes the procedure shown for the above example, and is reviewed in section "Modular Response Analysis" together with an application and an extension to quasi-steady state data.

The name 'modular' arises from the fact that, in MRA, only communicating intermediaries in-between 'modules' are measured. When applying MRA in a modular fashion, only perturbation data on these communicating signals are collected. The connectivity strength among a pair of such intermediary signals (such as levels of activated signaling proteins) is estimated, even if this apparent connectivity is not due to a 'directed' biochemical interaction. In principle, an obvious advantage of the modular approach is that it can be applied regardless the degree of internal complexity of the nodes, since 'hidden' variables (such as non-activated forms of a signaling protein) only affect connectivity in an indirect fashion, Thus, functional interactions among communicating variables can be deduced without requiring detailed knowledge of all the components involved.

Unfortunately, this analysis may be misleading, due to 'impedance' or 'load' effects. Following work by Saez-Rodriguez and others [9], we generically called such effects *retroactivity* in [5]. In this paper, we wish to discuss how stoichiometric constraints (conservation laws) might lead to erroneous conclusions when using the MRA methodology. Let us illustrate this phenomenon with one of the simplest possible examples. Suppose that we want to study a system in which we postulate that there are two 'modules' involving enzymes X and Y, the 'active forms' of which are the 'communicating variables'. The active form X is reversibly produced from an inactive form X_0. We assume that Y is formed when X reversibly binds to a substrate S producing a complex C, which may dissociate into X and S or into X and Y (this is a standard Michaelis-Menten type of reaction). We also assume that Y can revert to S in one step; a more complicated model could be used as well, by modeling the phosphotase action in a Michaelis-Menten form, or by modeling mechanistically its binding and unbinding to Y, but the principle is the same.

The network of reactions is as follows:

$$X_0 \underset{1}{\overset{1}{\rightleftharpoons}} X\,, \quad X + S \underset{1}{\overset{1/2}{\rightleftharpoons}} C \overset{1/2}{\longrightarrow} X + Y\,, \quad Y \overset{\alpha/2}{\longrightarrow} S$$

and we consider experiments in which $X_0(0) = 3, X(0) = 0, C(0) = Y(0) = 0$, and $S(0) = \beta$, and either α or β is to be perturbed experimentally. We think of X_0 and X as constituting the first 'module' and S, C, Y as the second one. The two parameters α, β are usually viewed as affecting only the second module. The unique positive steady state (X, S, C, Y) is then obtained by solving:

$$2X + C = 3, \quad C = XS, \quad C = \alpha Y, \quad S + C + Y = \beta$$

(and $X_0 = X$).

We will consider perturbations around $\alpha = 1$ and $\beta = 3$. For these nominal parameter values, $(X, S, C, Y) = (1, 1, 1, 1)$. Taking implicit derivatives with respect to α, evaluating at $X = S = C = Y = 1, \alpha = 1, \beta = 3$, and denoting $x = \partial X/\partial\alpha$, $u = \partial S/\partial\alpha$, $v = \partial C/\partial\alpha$, $y = \partial Y/\partial\alpha$, we have that:

$$2x + v = 0, \quad v = x + u, \quad v = 1 + y, \quad u + v + y = 0$$

which solves to:

$$x = -1/7, \quad u = -3x, \quad v = -2x, \quad y = 5x$$

and thus 'dX/dY' computed as $\frac{\partial X/\partial\alpha}{\partial Y/\partial\alpha}$ equals $1/5 > 0$.

The MRA method, or any other sensitivity-based approach, applied to the phenomenological model in which only active X and Y are viewed as 'communicating intermediaries' will lead us to include an edge $Y \to X$ labeled 'activating'. But *such an edge does not represent a true feedback effect*: for example, it is not possible to delete this edge with a 'mutation' in the system that does not affect the forward edge. The edge merely reflects a 'loading' or impedance effect. In fact, the situation is even more confusing. Taking implicit derivatives with respect to β, evaluating at $X = S = C = Y = 1, \alpha = 1, \beta = 3$, and denoting $x = \partial X/\partial\beta$, $u = \partial S/\partial\beta$, $v = \partial C/\partial\beta$, $y = \partial Y/\partial\beta$, we now have that:

$$2x + v = 0, \quad v = x + u, \quad v = y, \quad u + v + y = 1$$

which solves to:

$$x = -1/7, \quad u = -3x, \quad v = -2x, \quad y = -2x$$

and thus 'dX/dY' computed as $\frac{\partial X/\partial \beta}{\partial Y/\partial \beta}$ now equals $-1/2 < 0$. Now the (false) effect is inhibition. (The intuition is that when we increase α, the substrate for X increases, sequestering more of X, and also Y is smaller. If instead we over-express S, then both X is sequestered more and Y is larger. But intuition is not enough: for some parameters, $dX/dY < 0$ for both experiments.)

Experimentally, it is often the case that one measures $X + C$ and Y, instead of X and Y, so that one would be interested in the relative variations of $\hat{x} = x + v$ and y. Since $2x + v = 0$, it follows that $\hat{x} = -x$. Thus, $d(X + C)/dY = -dX/dY$, so the signs are reversed, but are, again, ambiguous.

Of course, there is a simple explanation for the problem: the parameter α affects the differential equation for X, and the variables S and C in fact enter that differential equation. Thus, the conditions for applicability of MRA have been violated. The point, however, is that a naive application of sensitivity analysis (as usually done in practice) that does not account for these subtle dependencies is wrong. One way to avoid this potential pitfall is to insure that the postulated mechanism (without additional feedback loops) does not exhibit such 'load' effects. We will present an algorithm to detect such effects (at steady state).

Modular Response Analysis

Precise Problem Formulation

We consider systems

$$\dot{x} = f(x, p)$$

where $x = (x_1, \ldots, x_n)$ is the state and $p = (p_1, \ldots, p_m)$ is a vector of parameters. Parameters can be manipulated, but, once changed, they remain constant for the duration of the experiment. We will assume that $m \geq n$. In biological applications, the variables x_i might correspond to the levels of protein products corresponding to n genes in a network, and the parameters to translation rates, controlled by RNAi. Another example would be that in which the parameters represent total levels of proteins, whose half-lives are long compared to the time scale of the processes (such as phosphorylation modifications of these proteins in a signaling pathway) described by the variables x_i. Yet another example would be one in which the parameters represent concentrations of enzymes that control the reactions, and whose turnover is slow. The goal is to obtain, for each pair of variables x_i and x_j, the relative signs and magnitudes of the partial derivatives $\frac{\partial f_i}{\partial x_j}$, which quantify the *direct* effects of each variable x_j upon each variable x_i. The entries of $\partial f_i/\partial x_j$ of the Jacobian F of f with respect to x are functions of x and p. The steady-state version of MRA attempts to estimate this Jacobian when $x = \bar{x}$ is an 'unperturbed' steady state attained when the vector of parameters has an 'unperturbed' value $p = \bar{p}$. The steady-state condition means that $f(\bar{x}, \bar{p}) = 0$. Ideally, one would want to find the matrix F, since this matrix completely describes the influence of each variable x_j upon the rate of change of each other variable x_i. Unfortunately, such an objective is impossible to achieve from only steady-state data, because, for any parameter vector p and associated steady-state x, $f(x, p) = 0$ implies that $\Lambda f(x, p) = 0$, for any diagonal matrix $\Lambda = \text{diag}\,(\lambda_1, \dots, \lambda_n)$. In other words, the best that one could hope for is for steady state data to uniquely determine each of the rows

$$F_i = (F_{i1}, \dots, F_{in}), \quad i = 1, \dots, n$$

of F *only up to a scalar multiple*. For example, if we impose the realistic condition that $F_{ii} \neq 0$ for every i (these diagonal Jacobian terms typically represent degradation and/or dilution effects, and are in fact negative), one could hope to have enough data to estimate the ratios a_{ij}/a_{ii} for each $i \neq j$. Note that F_i is the same as the gradient ∇f_i of the ith coordinate f_i of f, evaluated at steady states.

The critical assumption for MRA, and indeed the main point of [7, 8, 12], is that, while one may not know the detailed form of the vector field f, often one does know which parameters p_j directly affect which variables x_i. For example, x_i may be the level of activity of a particular protein, and p_i might be the total amount (active plus inactive) of that particular protein; in that case, we might postulate that p_i only directly affects x_i, and only indirectly affects the remaining variables.

Under the above assumptions, the steady-state MRA experimental design consists of the following steps:

1. Measure a steady state $\bar{x}$ corresponding to the unperturbed vector of parameters $\bar{p}$;
2. Separately perform a perturbation to each entry of $\bar{p}$, and measure a new steady state.

The 'perturbations' are assumed to be small, in the sense that the theoretical analysis will be based on the computation of derivatives. Under mild technical conditions, this means that a perturbed steady state can be found near $\bar{x}$. Note that there are $m + 1$ experiments, and n numbers (coordinates of the corresponding steady state) are measured in each. In practice, of course, this protocol is repeated several times, so as to average out noise and obtain error estimates, as we discuss later. For our theoretical analysis, however, we assume ideal, noise-free measurements, and so we may assume that each perturbation is done only once.

Using these data (and assuming that a certain independence condition, which we review later, is satisfied), it is possible to calculate, at least in the ideal noise-free case, the Jacobian of f, evaluated at $(\bar{x}, \bar{p})$, except for the unavoidable scalar multiplicative factor uncertainty on each row.

The obtained results typically look as shown in Fig. 9.2, which is reproduced from [10]. The authors of that paper used MRA on their experimental data in order to infer positive and negative feedback effects in the Raf/Mek/Erk MAPK network

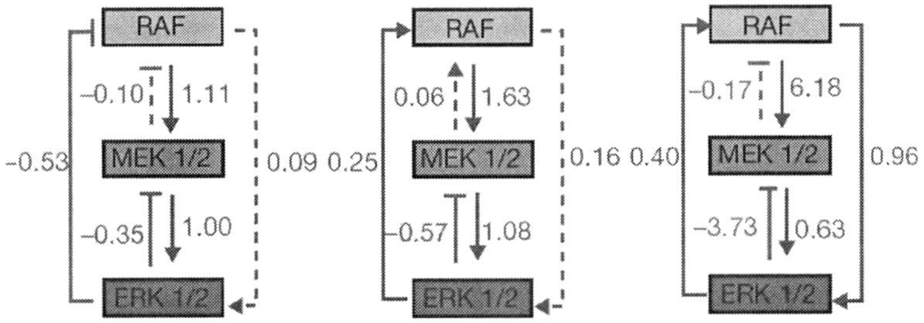

Fig. 9.2 Three reconstructed local interaction maps, in MRA experiments from [10]. Topologies derived from data obtained after stimulation by EGF (*left panel*, 5′) or NGF (*middle panel*, 5′, and *right panel*, 15′)

in PC-12 cells, employing perturbations in which total mRNA, and thus protein, levels are down-regulated by means of RNAi. The numbers in the arrows in Fig. 9.2 have been normalized to -1's in the diagonal of the Jacobian.

Mathematical Details

We assume given a parameter vector $\bar{p}$ and state $\bar{x}$ such that $f(\bar{x}, \bar{p}) = 0$ and so that the following generic condition holds for the Jacobian of f: $\det F(\bar{x}, \bar{p}) = \det \frac{\partial f}{\partial x}(\bar{x}, \bar{p}) \neq 0$. Therefore, we may apply the implicit function theorem and conclude the existence of a mapping φ, defined on a neighborhood of $\bar{p}$, with the property that, for each row i,

$$f_i(\varphi(p), p) = 0 \qquad \text{for all} \quad p \approx \bar{p}, \tag{9.1}$$

and $\varphi(\bar{p}) = \bar{x}$ (and, in fact, $x = \varphi(p)$ is the unique state x near $\bar{x}$ such that $f(x, p) = 0$).

We next discuss how one reconstructs the gradient $\nabla f_i(\bar{x}, \bar{p})$, up to a constant multiple. (The index i is fixed from now on, and the procedure must be repeated for each row f_i.) We do this under the assumption that it is possible to apply $n - 1$ independent parameter perturbations. Mathematically, the assumption is that there are $n - 1$ indices $j_1, j_2, \ldots, j_{n-1}$ with the following two properties:

(a) f_i does not depend directly on any p_j: $\partial f_i / \partial p_j \equiv 0$, for $j \in \{j_1, j_2, \ldots, j_{n-1}\}$, and
(b) the vectors $v_j = (\partial \varphi / \partial p_j)(\bar{p})$, for these j's, are linearly independent.

Assumption (a) is structural, and is key to the method and nontrivial, but assumption (b) is a weaker genericity assumption.

We then have, taking total derivatives in (9.1):

$$\nabla f_i(\bar{x}, \bar{p})\, v_j = 0, \qquad j \in \{j_1, j_2, \ldots, j_{n-1}\}.$$

Thus, the vector $\nabla f_i(\bar{x}, \bar{p})$ which we wish to estimate, and which we will denote simply as F_i, is known to be orthogonal to the $n-1$ dimensional subspace spanned by $\{v_1, \dots, v_{n-1}\}$. Therefore, it is uniquely determined, up to multiplication by a positive scalar. The row vector F_i satisfies

$$F_i \Sigma = 0 \tag{9.2}$$

where Σ is defined as the $n \times (n-1)$ matrix whose columns are the v_i's. Generically, we assume that there is no degeneracy, and the rank of Σ is $n-1$. Thus, F_i can be computed by using Gaussian elimination, as any vector which is orthogonal to the span of the columns of Σ. Another way to phrase this is to say that F_i is in the (one-dimensional) left nullspace of the matrix Σ. Of course, the sensitivities represented by the vectors v_i (entries of the matrix Σ, or $\Sigma^{\#}$ in the noisy case) cannot be directly obtained from typical experimental data. However, approximating the vectors v_j by finite differences, one has that $\nabla f_i(\bar{x}, \bar{p})$ is approximately orthogonal to these differences as well.

Handling Noise We next briefly discuss how to modify the algorithm to account for repeated but noisy measurements. In principle, such noise may be due to combinations of internal sources, such as stochasticity in gene expression, external sources affecting the process being studied, or measurement errors. Our discussion is tailored to measurement noise, although in an approximate way may apply to internal noise; however, the effect of internal noise on MRA has not been studied in any detail.

In practice, one would estimate not merely the results of just $n-1$ perturbation experiments, but many repetitions, collecting the data into a matrix $\Sigma^{\#}$ whose columns are derived from the different experiments. We will think of each column of $\Sigma^{\#}$ as having the form $v+e$, where v is a vector $(\partial\varphi/\partial p_j)(\bar{p})$, for some parameter p_j for which f_i does not depend directly on p_j, and where e is an 'error' vector. In matrix notation, $\Sigma^{\#} = \Sigma + E$, where E denotes an error matrix. Note that Eq. 9.2 implies that Σ has rank $n-1$. On the other hand, because of noise in measurements, $\Sigma^{\#}$ will have full rank n, which means that there is no possible nonzero solution F_i to Eq. 9.2 with the data matrix $\Sigma^{\#}$ used in place of the (unknown) Σ. So, we proceed as follows. Assuming that the signal to noise ratio is not too large, the experimental matrix $\Sigma^{\#}$ should be close to the ideal (noise-free) matrix, Σ. The best least-squares estimate of Σ, in the sense of minimization of the norm of E, is obtained by a singular value decomposition $\Sigma^{\#} = UMV^T$: the matrix Σ of rank $n-1$ for which $\|E\|$ is minimized is $\Sigma = UM_{n-1}V^T$, where M_{n-1} is the matrix obtained from M by setting the smallest singular value σ_n to zero. We now replace Eq. 9.2 by $F_i\Sigma^{\#} = 0$, which, because V is nonsingular, is the same as $F_iUM_{n-1} = 0$. Under the generic assumption that $\sigma_1, \dots, \sigma_{n-1}$ are nonzero, this means that $F_iU = \alpha e_n^T$, where α is a scalar and $e_n^T = (0, 0, \dots, 0, 1)$. We then conclude that, up to a constant multiple, $F_i^T = Ue_n$ is the right singular vector corresponding to the smallest singular value σ_n.

This procedure can also be interpreted as follows (see [1] for details). If we normalize F_i to have its ith entry as '-1' (in other words, we normalize the diagonal of the Jacobian to -1's), then the equation $F_i \Sigma^{\#} = 0$ can also be written as '$Az = b$' where z represents the unknown $n - 1$ remaining entries of F_i, b is the ith column of $\Sigma^{\#}$, and A is the matrix in which this column has been removed from $\Sigma^{\#}$. The estimation method outlined above is the 'total least squares' or 'errors in variables' procedure. Statistically, the method is justified if the elements of the noise matrix E are independent and identically distributed normal (Gaussian) random variables. If these entries are normal and independent but have different variances, then one must modify the above procedure to add an appropriate weighting, but in the general non-Gaussian case nonlinear SVD techniques are required.

Modular Approach Let us suppose that the entire network consists of an interconnection of n subsystems or 'modules', each of which is described by a set of differential equations such as:

$$\begin{aligned} \dot{x}_j &= g_j(\mathbf{y}_j, x_1, \ldots, x_n, p_1, \ldots, p_m), \quad j = 1, \ldots, n \\ \dot{\mathbf{y}}_j &= G_j(\mathbf{y}_j, x_1, \ldots, x_n, p_1, \ldots, p_m), \quad j = 1, \ldots, n, \end{aligned}$$

where the variables x_j represent 'communicating' or 'connecting' intermediaries of module j that transmit information to other modules, whereas the *vector variables* $\mathbf{y}_j$ represent chemical species that interact within module j. Each vector $\mathbf{y}_j$ has dimension ℓ_j. The integers ℓ_j, $j = 1, \ldots, n$ are in general different for each of the n modules, and they represent one less than the number of chemical species in the jth module respectively. Observe that, for each j, the rate of change of the communicating variable depends only on the remaining communicating variables $x_i, i \neq j$, and on the variables $\mathbf{y}_j$ in its own block, but does not directly depend on the internal variables of other blocks. In that sense, we think of the variables $\mathbf{y}_j$ as 'hidden' (except from the communicating variable in the same block).

We will assume, for each fixed module, that the Jacobian of G_j with respect to the vector variable $\mathbf{y}_j$, evaluated at the steady state corresponding to $\bar{p}$ (assumed to exist, as before) is nonsingular. The Implicit Mapping Theorem then implies that one may, in a neighborhood of this steady state, solve $G_j(\mathbf{y}_j, x, p) = 0$ (x denotes the vector $x_1, \ldots, x_n$, and similarly for p) for the vector variable $\mathbf{y}_j$, as a function of x, p, the solution being given locally by a function $\mathbf{y}_j = \mathbf{M}_j(x, p)$. Those steady states that are obtained by small perturbations of $\bar{p}$ are the same as the steady states of the 'virtual' system $\dot{x}_j = h_j(x_1, \ldots, x_n, p_1, \ldots, p_m) = g_j(\mathbf{M}_j(x, p), x, p)$, $j = 1, \ldots, n$. From here on, the analysis then proceeds as before, using the h_j's instead of the f_j's. A generalization to the case of more than one communicating intermediate in a module, namely a vector $(x_{j,1}, \ldots, x_{j,k_j})$, is easy.

Using Quasi-Steady State Data An example of the experimental data used to derive the diagrams in Fig. 9.2 is provided by Fig. 9.3, which shows the level of active (doubly phosphorylated) Erk1/2 when PC-12 cells have been stimulated by EGF and NGF. (The Figure shows only responses in the unperturbed case. Similar plots, not shown, can be derived from the data for the perturbation experiments given in [10].)

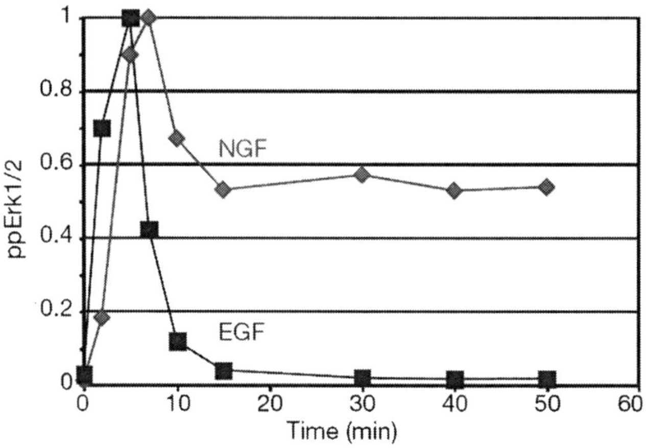

Fig. 9.3 Active form of Erk1/2, in MRA experiments from [10]. Data shown only for unperturbed case

The response to NGF stimulation allows the application of the steady-state MRA method, and leads to the results shown in the right-most panel in Fig. 9.2.

However, the plots in Fig. 9.3 indicate that, in certain problems, steady-state data cannot be expected to provide enough information, even for only finding the Jacobian rows up to multiplicative factors. Such a situation occurs when the system *adapts* to perturbations. In Fig. 9.3, notice that the steady state response to EGF stimulation is (near) zero (this holds for perturbed parameters as well, not shown). Thus, measuring steady-state level of activity of Erk1/2 after parameter perturbations, in the EGF-stimulated cells, will not provide nontrivial information. One needs more than steady-state data.

A variant of MRA, which allows for the use of general non-steady-state, time-series data was developed in [12]. However, that method requires one to compute second-order time derivatives, and hence is especially hard to apply when time measurements are spaced far apart and/or are noisy. In addition, as shown for $5'$ and $15'$ NGF stimulation by the middle and rightmost panels in Fig. 9.2, the relative strengths of functional interactions may change over time, so that a time-varying Jacobian may not be very informative from a biological standpoint. An appealing intermediate possibility is to use quasi-steady state data, meaning that one employs data collected at those times at which a variable has been observed to attain a local maximum (peak of activity) or a local minimum. Indeed, this is the approach taken in [10], which, for EGF stimulation, measured network responses at the time of peak Erk activity (approximately 5 min), and not at steady state. The left-most and middle panels in Fig. 9.2 represent, respectively, the networks reconstructed in [10] when using quasi steady-state data (at approximately 5 min) for EGF and NGF stimulation.

We next describe the extension to quasi-steady state MRA. We consider the following scenario. For any fixed variable, let us say the ith component x_i of x, we consider some time instant $\bar{t}_i$ at which $\dot{x}_i(t)$ is zero. Under the same independence hypothesis as in the steady-state case, plus the non-degeneracy assumption that the second time derivative $\ddot{x}_i(\bar{t}_i)$ is not zero (so that we have a true local minimum or local maximum, but not an inflection point), we show here that the MRA approach applies in exactly the same manner as in the steady-state case. Specifically, the ith row of the Jacobian of f, evaluated at the vector $(\bar{x}, \bar{p})$, is recovered up to a constant multiple, where $\bar{x} = x(\bar{t}_i)$ is the full state x at time $\bar{t}_i$. The main difference with the steady-state case is that different rows of f are estimated at different pairs $(\bar{x}, \bar{p})$, since the considered times $\bar{t}_i$ at which each individual $\dot{x}_i(t)$ vanishes are in general different for different indices i, and so the state $\bar{x}$ is different for different i's.

We fix an index $i \in \{1, \ldots, n\}$, and an initial condition $x(0)$, and assume that the solution $x(t)$ with this initial condition and a given parameter vector $\bar{p}$ has the property that, for some time $\bar{t} = \bar{t}_i$, we have that both $\dot{x}_i(\bar{t}) = 0$ and $\ddot{x}_i(\bar{t}) \neq 0$. At the instant $t = \bar{t}$, x_i achieves a local minimum or a local maximum as a function of t. We describe the reconstruction of the ith row of the Jacobian of f, which that is, the gradient ∇f_i, where f_i is the ith coordinate of f, evaluated at $x = \bar{x}$ and $p = \bar{p}$, where $\bar{x} = x(\bar{t})$.

To emphasize the dependence of the solution on the parameters (the initial condition $x(0)$ will remain fixed), we will denote the solution of the differential equation $\dot{x} = f(x, p)$ by $x(t, p)$. The function $x(t, p)$ is jointly continuously differentiable in x and p, if the vector field f is continuously differentiable. Note that, with this notation, the left-hand side of the differential equation can also be written as $\partial x/\partial t$, and that $x(\bar{t}, \bar{p}) = \bar{x}$.

Consider $\alpha(t, p) = \frac{\partial x_i}{\partial t}(t, p) = f_i(x(t, p), p)$. Thus,

$$\frac{\partial \alpha}{\partial t}(t, p) = \frac{\partial^2 x_i}{\partial t^2}(t, p) = \nabla f_i(x(t, p), p)\, f(x(t, p), p)$$

and $\alpha(\bar{t}, \bar{p}) = 0$. The assumption that $\ddot{x}_i(\bar{t}) \neq 0$ when $p = \bar{p}$ means that $\frac{\partial \alpha}{\partial t}(\bar{t}, \bar{p}) \neq 0$. Therefore, we may apply the implicit function theorem and conclude the existence of a mapping τ, defined on a neighborhood of $\bar{p}$, with the property that $\alpha(\tau(p), p) = 0$ for $p \approx \bar{p}$ and $\tau(\bar{p}) = \bar{t}$ (and, in fact, $t = \tau(p)$ is the unique value of t near $\bar{t}$ such that $(\partial x_i/\partial t)(t, p) = \alpha(t, p) = 0$). Finally, we define, also in a neighborhood of $\bar{p}$, the differentiable function $\varphi(p) = x(\tau(p), p)$ and note that $\varphi(\bar{p}) = \bar{x}$. Observe that, from the definition of α, we have that Eq. 9.1 holds, exactly as in the steady-state case. From here, the reconstruction of $\nabla f_i(\bar{x}, \bar{p})$ up to a constant multiple proceeds as in the steady-state case, again under the assumption that it is possible to apply $n - 1$ independent parameter perturbations. A noise analysis similar to that in the steady state case can be done here. However, there are now many more potential sources of numerical and experimental error, since measurements at different times are involved. In addition, internal (thermal) noise may introduce additional error, since, in the quasi-steady state case, the state probability distributions (solutions of the Chemical Master Equation) have not converged to steady state.

Retroactivity at Steady States

In this section, we analyze the *retroactivity* phenomenon of interconnections at steady states. As discussed in the introduction, such an analysis is required in order to understand the possible pitfalls of MRA. We will not define the term 'retroactivity' as such, but instead use it only informally; the results to be given provide a precise content to the term under slightly different contexts. The main question is, in any event, to understand what is the relation between the steady states of individual systems (described by chemical reactions) and the steady states of their interconnection.

Intuitively, one expects that retroactivity at steady state arises only when there are more conservation laws imposed by an interconnection, in addition to those that hold for each of the interconnected systems separately. Making this intuition precise is not completely trivial. In fact, unless certain properties are imposed on interconnections, the intuition is not even correct.

Our main results, Theorems 9.5 and 9.7, give sufficient conditions for retroactivity to exist or not, respectively. Neither is necessary. However, we will define a 'consistency' property for interconnections, under which Theorems 9.5 and 9.7 constitute a dichotomy.

See the Appendix for basic notations from chemical network theory. From now on, we will assume that the vector S of species has $N = n + m$ components, which we partition into two vectors $x \in \mathbb{R}^n$ and $z \in \mathbb{R}^m$: $S = (x', z')'$ (we use primes to indicate transpose). Corresponding to these coordinates, the reaction vector is partitioned into two vectors $R_1(x)$ and $R_2(x, z)$ of dimensions r_1 and $r_2 = r - r_1$ respectively: $R(x, z) = (R_1(x)', R_2(x, z)')'$. We also assume that, in terms of this partition, the stoichiometry matrix looks as follows:

$$\Gamma = \begin{pmatrix} P \\ Q \end{pmatrix} = \begin{pmatrix} A & B \\ 0 & C \end{pmatrix}$$

where $P \in \mathbb{R}^{n\times r}$, $Q \in \mathbb{R}^{m\times r}$, $A \in \mathbb{R}^{n\times r_1}$, $B \in \mathbb{R}^{n\times r_2}$, $C \in \mathbb{R}^{m\times r_2}$. (In some contexts, it will be convenient to use the '(P, Q)' form, while for other contexts the '(A, B, C)' form will be more useful.) The equations for the system take the following partitioned form:

$$\begin{aligned} \dot{x} &= AR_1(x) + BR_2(x, z) \\ \dot{z} &= CR_2(x, z) \end{aligned}$$

When there are no reactions involving x alone, we write $A = 0$, thought of as an $n \times 1$ matrix. Observe that, of course, the actual reactions entering x and z need not be the same, since B and C may multiply different elements of the vector $R_2(x, z)$ by zero coefficients.

We think of the overall system as an interconnection of the 'upstream' subsystem described by the x-variables, that feeds a signal to the 'downstream' subsystem described by the z variables. The 'x' appearing in $CR_2(x, z)$ is seen, in that sense, as an input signal to the second system. The role of $BR_2(x,z)$ is different. This term represents the 'retroactivity to the output', denoted by the letter 'r' in [5], and is interpreted as a 'load' effect that arises due to the physical interconnection. Of course, these interpretations are subjective, and partitioning a system into an interconnection can be done in non-unique ways. However, the questions to be posed depend on one such partition.

In this context, we call the system $\dot{x} = AR_1(x)$ the *isolated system*, and the full system $\dot{S} = \Gamma R(S)$ the *interconnected system*. We use the notation $\Delta_1(x)$ for the stoichiometry class of a state x of the isolated system:

$$\Delta_1(x) \;=\; (x + \Delta_1)\bigcap \mathbb{R}^n_{\geq 0}\,,$$

where Δ_1 is the span of the columns of A.

Example 9.1. Our first example is this network:

$$X_0 \underset{k_2}{\overset{k_1}{\rightleftharpoons}} X\,, \quad X + S \underset{k_4}{\overset{k_3}{\rightleftharpoons}} C \overset{k_5}{\longrightarrow} X + Y\,, \quad Y \overset{k_6}{\longrightarrow} S$$

which represents the interaction of five species: an kinase which exists in inactive (X_0) or active (X, typically phosphorylated) form, a substrate S for the active kinase, a complex C that is a dimmer of X and S, and a 'product' Y of the enzymatic reaction. (For simplicity, we assume that the reverse transformation of Y back to S happens at a constant rate; more complicated models can be studied in exactly the same way.)

We wish to think of this system of chemical reactions as consisting of the upstream system described by the vector $x = (X_0, X)'$ which drives the downstream system described by the vector $z = (S, C, Y)'$. Thus, with $n = 2$, $m = 3$, $r_1 = 2$, and $r_2 = 3$, we take, using mass-action kinetics, $R_1(x) = (k_1X_0, k_2X)'$ and $R_2(x,z) = (k_3XS, k_4C, k_5C, k_6Y)$. Note that

$$A = \begin{pmatrix} -1 & 1 \\ 1 & -1 \end{pmatrix}, \quad B = \begin{pmatrix} 0 & 0 & 0 & 0 \\ -1 & 1 & 1 & 0 \end{pmatrix}$$

$$C = \begin{pmatrix} -1 & 1 & 0 & 1 \\ 1 & -1 & -1 & 0 \\ 0 & 0 & 1 & 1 \end{pmatrix}.$$

There are conservation laws in this system which tie together the isolated (x) system to the downstream (z) system, and one may expect that retroactivity effects appear. Indeed, this system will satisfy the sufficient condition for retroactivity given in Theorem 9.5 below. □

Example 9.2. Consider these reactions:

$$1 \overset{u}{\rightarrow} X\,, \quad X \overset{\delta}{\rightarrow} 0\,, \quad X + P \underset{k_2}{\overset{k_1}{\rightleftharpoons}} C.$$

Then, with $x = X$ and $z = (P, C)'$, and listing reactions in the obvious order:

$$A = \begin{pmatrix} 1 & -1 \end{pmatrix}, \; B = \begin{pmatrix} -1 & 1 \end{pmatrix}, \; C = \begin{pmatrix} -1 & 1 \\ 1 & -1 \end{pmatrix}.$$

Because of the production and/or decay of X, there are no conservation laws tying together the X and the P, C systems, and there is no retroactivity effect. Indeed this system will satisfy the sufficient condition in Theorem 9.7 for non-retroactivity. □

Example 9.3. Consider the following reaction:

$$X \xrightarrow{k_1} Z \,, \quad Z \xrightarrow{k_2} 0$$

with $x = X$ and $z = Z$. Here $B = (-1\ 0)$ and $C = (1\ -1)$, and $A = 0$. This example is one in which there are no conservation laws whatsoever, yet retroactivity holds. Neither Theorem 9.5 nor Theorem 9.7 applies to this example, showing the gap between the conditions. However, this example is somewhat pathological, as it represents an 'inconsistent' interconnection in the sense defined below. □

Main Results Consider the following property:

$$\operatorname{rank} \begin{pmatrix} P \\ Q \end{pmatrix} = \operatorname{rank} P + \operatorname{rank} Q \qquad (*)$$

Remark 1. Since the weak inequality '$\leq$' is always true, the negation of (*) is equivalent to:

$$\operatorname{rank} \begin{pmatrix} P \\ Q \end{pmatrix} < \operatorname{rank} P + \operatorname{rank} Q \qquad (\not\ast)$$

or, equivalently, the requirement that the row spaces of P and Q have a nonzero intersection. □

If property ($\not\ast$) holds, then there is retroactivity at steady state. The precise statement is as follows:

Lemma 9.4. *Suppose that Property* $(*)$ *does not hold. Then, for each positive state* $\bar{S} = (\bar{x}', \bar{z}')'$ *of the interconnected system, there exists a state* $S_0 = (x_0', z_0')'$ *such that*

$$\Delta(S_0) = \Delta(\bar{S}) \quad \textit{but} \quad \Delta_1(x_0) \neq \Delta_1(\bar{x})\,.$$

Moreover, S_0 *can be picked arbitrarily close to* $\bar{S}$.

Proof. Suppose that ($\not\ast$) holds, and pick any positive state $\bar{S}$. By ($\not\ast$), there is some nonzero row vector θ which is in the row spaces of P and Q, that is to say, there are two row vectors μ_0 and ν_0 such that $\theta = \mu_0 P = \nu_0 Q \neq 0$. Replacing ν_0 by $-\nu_0$, we will assume that $\mu_0 P = -\nu_0 Q \neq 0$. Let r be any vector such that $\mu_0 P r \neq 0$ (for example, one may pick $r = P'\mu_0'$), and let $u := Pr$ and $v := Qr$. Let, as earlier, $\Pi := \Delta^{\perp}$. Note that, for each $\pi = (\mu, \nu) \in \Pi$, $\mu P + \nu Q = 0$, by definition of Π, and therefore also $\mu u + \nu v = (\mu P + \nu Q)r = 0$. In particular, $(\mu_0, \nu_0) \in \Pi$ satisfies that $\mu_0 u = \mu_0 Pr \neq 0$ and also $\nu_0 v = \nu_0 Qr = -\mu_0 Pr \neq 0$. Since $P = (A\ B)$ and $Q = (0\ C)$, every element $(\mu, \nu) \in \Pi$ has the property that, in particular, $\mu A = 0$.

Notice that one could pick u and v as close to zero as wanted (multiplying, if necessary, u and v by a common small positive factor). So, without loss of generality, we assume that both $x_0 := \bar{x} + u$ and $z_0 := \bar{z} + v$ are non-negative, and write $S_0 := (x_0', z_0')$. We claim that S_0 and $\bar{S}$ are in the same stoichiometry class. Indeed, for any $(\mu, \nu) \in \Pi$: $\mu\bar{x} + \nu\bar{z} = \mu\bar{x} + \nu\bar{z} + 0 = \mu\bar{x} + \nu\bar{z} + \mu u + \nu v = \mu(\bar{x} + u) + \nu(\bar{z} + v) = \mu x_0 + \nu z_0$.

Finally, we claim that x_0 and $\bar{x}$ are not in the same stoichiometry class for the isolated system. Since $\mu_0 A = 0$, μ_0 is a conservation law for the isolated system. So it will be enough to show that $\mu_0\bar{x} \neq \mu_0\xi$. Indeed, $\mu_0\bar{x} = \mu_0(x_0 - u) = \mu_0\xi - \mu_0 u$, and $\mu_0 u \neq 0$. ■

Lemma 9.4 implies a steady-state retroactivity effect, in the following sense. Suppose that $\bar{S}$ is an attractor for points near it and in $\Delta(\bar{S})$. If x_0 is taken as the initial state of a trajectory $x(t)$ for the isolated system, then every limit point ξ of this trajectory is in $\Delta_1(x_0)$. On the other hand, if the composite system is initialized at this same state x_0 for the x-subsystem, and at z_0 for the z-subsystem, then the ensuing trajectory converges to the steady state $\bar{S}$, with x-component $\bar{x}$. But $\xi \neq \bar{x}$, because $\bar{x} \notin \Delta_1(x_0)$. The following result formalizes this fact.

Theorem 9.5. *Suppose that there is some positive steady state $\bar{S} = (\bar{x}', \bar{z}')'$ of the interconnected system which is a local attractor relative to its stoichiometry class. If Property $(*)$ is false, then there exist x_0 and z_0 such that, with the initial condition $S_0 = (x_0', z_0')'$:*

1. *$\varphi(t, S_0) \to \bar{S}$ as $t \to +\infty$, but*
2. *for the solution $x(t)$ of the isolated system $\dot{x} = AR_1(x)$ with $x(0) = x_0$, $\bar{x} \notin$ clos$\{x(t), t \geq 0\}$.*

Proof. We use Lemma 9.4. Let S_0 be as there. Since S_0 can be picked arbitrarily close to $\bar{S}$ and in $\Delta(\bar{S})$, we may assume that S_0 belongs to the domain of attraction of the steady state $\bar{S}$. Property (1) in the Theorem statement is therefore satisfied. Finally, we consider the solution $x(t)$ of the isolated system $\dot{x} = AR_1(x)$ with initial condition $x(0) = x_0$, and pick any state $\xi \in$ clos$\{x(t), t \geq 0\}$. As $\Delta_0(\xi) = \Delta_0(x_0) \neq \Delta_0(\bar{x})$, it follows that $\xi \neq \bar{x}$. ■

Next, consider the following property:

$$\text{rank}(A\ B) = \text{rank}A \qquad (**)$$

i.e., the column space of B is included in that of A. Note that if this condition holds, then $(*)$ holds too.

Lemma 9.6. *Suppose that $(**)$ holds. Pick any two states $\bar{S} = (\bar{x}', \bar{z}')'$ and $S_0 = (x_0', z_0')'$ of the interconnected system. Then*

$$\Delta(\bar{S}) = \Delta(S_0) \;\Rightarrow\; \Delta_1(\bar{x}) = \Delta_1(x_0)\,. \qquad (9.3)$$

Proof. As $\bar{S} - S_0$ belongs to the column space Δ of Γ, in particular, $\bar{x} - x_0$ is in the column space of $(A\ B)$. Since the latter equals the column space of A, it follows that $\bar{x} - x_0$ is in the column space of A, which means that x_0 and $\bar{x}$ are in the same stoichiometry class in the isolated system. ∎

Lemma 9.4 implies a steady-state retroactivity effect, in the following sense. Suppose that there is a unique steady state in each stoichiometry class in the isolated system, and that this steady state is a global attractor relative to its class. Then, every omega-limit point of the composite system has the property that its x-component equals this same steady state of the isolated system. The following result formalizes this discussion.

Theorem 9.7. *Suppose that* $(**)$ *holds. For any initial condition* $S_0 = (x_0', z_0')'$, *if a state* $\bar{S} = (\bar{x}', \bar{z}')'$ *of the interconnected system is in the omega-limit set of* S_0, *then* x_0 *and* $\bar{x}$ *are in the same stoichiometry class relative to the isolated system.*

Proof. If $\bar{S} = (\bar{x}', \bar{z}')'$ is in the omega-limit set of S_0 then $\Delta(S_0) = \Delta(\bar{S})$. The conclusion thus follows from Lemma 9.6. ∎

There is a gap between the negation of Property $(*)$ in Theorem 9.5 and Property $(**)$ in Theorem 9.7. In order to bridge this gap, we introduce the following property:

$$\ker C \subseteq \ker B \qquad (\mathcal{C})$$

which we call *consistency*.

An interpretation of property $(\mathcal{C})$ is as follows. Suppose that $S = (\bar{x}', \bar{z}')'$ is a steady state of the interconnected system. That is to say, $AR_1(\bar{x}) + BR_2(\bar{x}, \bar{z}) = 0$ and $CR_2(\bar{x}, \bar{z}) = 0$. Since then $R_2(\bar{x}, \bar{z}) \in \ker C \subseteq \ker B$, this means that also $BR_2(\bar{x}, \bar{z}) = 0$, and therefore we can conclude that $AR_1(\bar{x}) = 0$. In summary, *under consistency, the x-component of every steady state of the interconnected system is a steady state of the isolated system.* Moreover, the 'retroactivity' signal $BR_2(x, z)$ also vanishes at steady state. This property is satisfied in most interesting interconnections.

Property $(\mathcal{C})$ is equivalent to the requirement that the row space of B be a subspace of the row space of C. Under this property, $\operatorname{rank}\Gamma = \operatorname{rank} A + \operatorname{rank} C$, and therefore Property $(*)$, i.e. $\operatorname{rank}\Gamma = \operatorname{rank}(A\ B) + \operatorname{rank} C$ is equivalent to Property $(**)$. In other words, for consistent interconnections, the two Theorems provide a dichotomy. Summarizing this discussion and consequences of the two technical lemmas:

Corollary 9.8. *Suppose that Property* $(\mathcal{C})$ *holds. Then, the following statements are equivalent:*

(a) Property $(*)$ *holds.*
(b) Property $(**)$ *holds.*
(c) Property (9.3) holds for any two states.

Example 9.1 fails Property $(*)$: the ranks of P and Q are 2 and 3 respectively, but the composite matrix has rank $4 < 5$. Thus this example exhibits retroactivity, by Theorem 9.5. Note that this example is consistent.

Example 9.2 does not exhibit any retroactivity effects, as is easy to see directly, or appealing to Theorem 9.7, since Property $(**)$ is satisfied. Note that this example is consistent.

Example 9.3 satisfies Property $(*)$, but nonetheless exhibits a retroactivity effect, in the sense that every state of the isolated system is a steady state, but for the interconnected system $\dot{x} = -x$, $\dot{y} = x - y$ every solution converges to $x = y = 0$. However, Property $(**)$ cannot be used to show retroactivity, since this property also fails. Intuitively, this is a system that has no conservation laws, yet retroactivity fails. However, this system is 'inconsistent' in the sense that property $(\mathcal{C})$ does not hold.

Appendix: Chemical Reaction Network Formalism

The differential equations for the evolution of the concentrations of the reactants in a chemical reaction system are written in the following standard 'chemical reaction network' formalism. Suppose that there are N species $S_1, \ldots, S_N$ taking part in a reaction system, where each $S_i = S_i(t)$ is a non-negative function of time that lists the concentration of species i at time $t \geq 0$. (We use the same letter for a chemical species and for its concentration.) Collecting all entries into an N-dimensional column vector S, one writes the evolution equations as follows: $\dot{S} = \Gamma R(S)$. The matrix $\Gamma \in \mathbb{R}^{N \times r}$ is the *stoichiometry matrix*, and $R(S) \in \mathbb{R}^r$ is the vector of *reactions*: $R(S(t))$ indicates the values of the reaction rates when the species concentrations are $S(t)$. A technical assumption is that solutions that start non-negative remain so. This property is automatically satisfied for all the usual chemical reaction rate forms, including mass-action kinetics. Mathematically, what is required is that, for each $i \in \{1, \ldots, N\}$, the ith entry of $\Gamma R(S)$ is non-negative whenever $S_i = 0$. We will also assume that, for each initial condition $S_0 \in \mathbb{R}^N_{\geq 0}$, the solution $\varphi(t, S_0)$ of $\dot{S} = \Gamma R(S)$ with $S(0) = S_0$ is defined for all times $t \geq 0$.

For any chemical reaction system $\dot{S} = \Gamma R(S)$, and any state S_0, the *stoichiometry equivalence class of* S_0, denoted here as $\Delta(S_0)$, is the intersection of the affine manifold $S_0 + \Delta$ with $R^N_{\geq 0}$, where Δ is the span of the columns of Γ. Thus, two states S_0 and S_1 are in the same stoichiometry class if and only if $S_0 - S_1 \in \Delta$, or equivalently if $\Delta(S_0) = \Delta(S_1)$. Observe that $\varphi(t, S_0) \in \Delta(S_0)$ for all $t \geq 0$. Moreover, since $\Delta(S_0)$ is a closed set, any S in the closure of the forward orbit $\mathcal{O}^+(S_0) = \{\varphi(t, S_0), t \geq 0\}$ is also in $\Delta(S_0)$.

We also introduce the vector space of 'conservation laws'. This is the set of all vectors perpendicular to the stoichiometry space, written as rows: $\Pi := \Delta^\perp = \{\pi \in \mathbb{R}^{1 \times N} | \pi \Gamma = 0\}$ Observe that a state S_1 is in the stoichiometry class of a state S_0 (that is, $S_1 - S_0 \in \Delta$) iff $\pi(S_1 - S_0) = 0$ for all $\pi \in \Pi$.

For any chemical reaction system $\dot{S} = \Gamma R(S)$, and any steady state $\bar{S}$ (that is, $\Gamma R(\bar{S}) = 0$) we say that $\bar{S}$ is *a local attractor relative to its stoichiometry class* if there is some neighborhood $\mathcal{U}$ of $\bar{S}$ in $R^N_{\geq 0}$ such that, for each $S_0 \in \mathcal{U} \bigcap \Delta(\bar{S})$, $\varphi(t, S_0) \to \bar{S}$. A *positive* state S is one for which all components are strictly positive, that is, $S \in \mathbb{R}^N_{>0}$. Under certain hypotheses on the structure of the chemical reaction network, one may insure that in each stoichiometry class there is at least one positive steady state that is a local attractor relative to the class. Moreover, this steady state is often unique, and is a global attractor relative to the class; see for example [6, 11].

References

1. Andrec M, Kholodenko BN, Levy RM, Sontag ED (2005) Inference of signaling and gene regulatory networks by steady-state perturbation experiments: structure and accuracy. J Theor Biol 232(3):427–441
2. Berman P, Dasgupta B, Sontag ED (2007a) Algorithmic issues in reverse engineering of protein and gene networks via the modular response analysis method. Ann N Y Acad Sci 1115:132–141
3. Berman P, Dasgupta B, Sontag ED (2007b) Randomized approximation algorithms for set multicover problems with applications to reverse engineering of protein and gene networks. Discrete Appl Math Spec Ser Comput Mol Biol 155:733–749
4. Crampin EJ, Schnell S, McSharry PE (2004) Mathematical and computational techniques to deduce complex biochemical reaction mechanisms. Prog Biophys Mol Biol 86(1):77–112
5. Del Vecchio D, Ninfa AJ, Sontag ED (2008) Modular cell biology: retroactivity and insulation. Mol Syst Biol 4:161
6. Feinberg M (1987) Chemical reaction network structure and the stability of complex isothermal reactors - i. the deficiency zero and deficiency one theorems. Chem Eng Sci 42:2229–2268
7. Kholodenko BN, Kiyatkin A, Bruggeman F, Sontag ED, Westerhoff H, Hoek J (2002) Untangling the wires: a novel strategy to trace functional interactions in signaling and gene networks. Proc Natl Acad Sci USA 99:12841–12846
8. Kholodenko BN, Sontag ED (2002) Determination of functional network structure from local parameter dependence data. Technical report. arXiv physics/0205003, May 2002
9. Saez-Rodriguez J, Kremling A, Gilles ED (2005) Dissecting the puzzle of life: modularization of signal transduction networks. Comput Chem Eng 29:619–629
10. Santos SDM, Verveer PJ, Bastiaens, PIH (2007) Growth factor induced MAPK network topology shapes Erk response determining PC-12 cell fate. Nat Cell Biol 9:324–330
11. Sontag ED (2001) Structure and stability of certain chemical networks and applications to the kinetic proofreading model of T-cell receptor signal transduction. IEEE Trans Autom Control 46(7):1028–1047
12. Sontag ED, Kiyatkin A, Kholodenko BN (2004) Inferring dynamic architecture of cellular networks using time series of gene expression, protein and metabolite data. Bioinformatics 20(12):1877–1886
13. Stark J, Callard R, Hubank M (2003) From the top down: towards a predictive biology of signalling networks. Trends Biotechnol 21(7):290–293

Part III
Design and Standardization

Chapter 10
Computer-Aided Design for Synthetic Biology

Deepak Chandran, Frank T. Bergmann, Herbert M. Sauro, and Douglas Densmore

Abstract Computer-aided design (CAD) for synthetic biology has been proposed to parallel similar efforts in other engineering disciplines, such as electrical engineering or mechanical engineering. However, there is an important distinction between the fields, which is that the mechanisms by which biological systems function are not currently fully understood in sufficient detail to make completely predictive tools. Computational models of biological systems provide, at best, a qualitative understanding of the system under investigation. Quantitative models are limited by the large number of unknown parameters in any given biological system as well the lack of understanding of the detailed mechanisms. It is difficult to determine how much detail is required for predictable design of biological systems. Even assembling individual DNA sequences has shown to be unpredictable due to secondary DNA structures. As a result, the phrase 'computer-aided design' takes a very different meaning in synthetic biology: designing biological systems is as much an exploratory process as it is a rational design process. Through design and experimentation, the science of engineering biology is furthered, and that knowledge must be explicitly fed back into the design process itself. Due to its complexity, the challenge of predictably designing biological systems has become a community effort rather than a competitive effort. Consequently, several software developers in synthetic biology have recognized that supporting a community is a necessary component in synthetic biology design applications. Existing software tools in synthetic biology can be categorized into a three broad categories. First, there are software tools for mathematical analysis of biological systems. This category also includes tools from the field of systems biology. Secondly, there are software tools for assembling DNA sequences and analyzing the structure of the resulting composition. This category builds on concepts from genetic engineering for manipulating DNA sequences. The third category of tools are for database access. Synthetic biologists need a catalog of biological components, or 'parts', from which systems can be built; therefore, databases, whether local or distributed, are integral for synthetic biology research. This chapter will cover these categories of tools and

D. Chandran (✉)
University of Washington, Seattle, Washington
e-mail: deepakc@u.washington.edu

H. Koeppl et al. (eds.), *Design and Analysis of Biomolecular Circuits: Engineering Approaches to Systems and Synthetic Biology*, DOI 10.1007/978-1-4419-6766-4_10,

how they contribute to synthetic biology. We also consider design by combinatorial optimization, which may work well in biological engineering due to properties of DNA replication.

Keywords Synthetic biology · Software · Computer-aided design · CAD · Systems biology · Design · Specification · Assembly · Analysis · Modeling

Introduction

The purpose of computer-aided design (CAD) in synthetic biology is to assist an engineer in the process of designing a system with a desired behavior (**specification and design**) and understanding the system in sufficient detail to construct the physical realization of the system (**assembly**). Presumably the system being designed is at a level of complexity that a manual design process is either too lengthy, error prone, or costly. The specification and design process should encourage reuse, reduce design ambiguity, remove potential human error, and have the ability to simulate the proposed design. A system's desired behavior may be represented using mathematical formalism which is amendable to analysis by computer software. In synthetic biology, the assembly of a system often is a cell with the specific DNA sequence that encodes the designed system function. The following are some examples of desired behaviors:

- Optimized production of a specific metabolite; [28]
- A specific input-output response curve, where the input might be a specific molecule or environmental signal; [13]
- A specific temporal or spatial pattern; [1]
- A defined distribution of a cell phenotypes in a population; [4]

In biological engineering, there is no established design methodology for moving from a specification, or desired behavior, to the end result, the living cell. Synthetic biologists often attempt to mimic the procedures from other engineering disciplines [9, 27, 42]; however, due to the fact that the underlying physical mechanisms in biological systems are different, it is impractical to assume that the procedures from other engineering disciplines will directly transition into biology. The term, 'biological circuit' for example, often misleads scientists in thinking of biology in the same way as electronic circuits. It is important to realize that the notion of 'circuit' can mean various things even in established engineering fields; hence, it is possible that the term may restrict biological engineers to a specific perspective. Similarly, Boolean algebra is an abstraction applied to only specific digital electronic systems; synthetic biologists often assume that biological systems can be abstracted using Boolean logic, which can be arguably correct in some cases [53] but not in other cases where molecular concentration-dependent response is important [13]. While there is no right or wrong approach at the current stage of synthetic biology, it is important to understand that biological systems are inherently different,

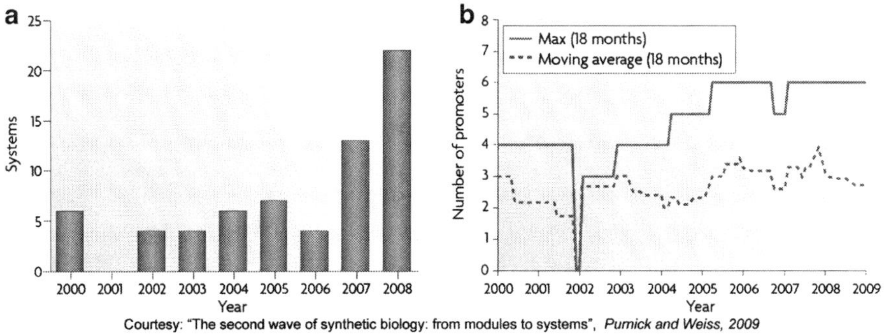

Fig. 10.1 Progression of synthetic biology: (**a**) illustrates the design growth in synthetic biology and (**b**) illustrates synthetic biology system complexity as measured by regulatory regions. From [39]

and therefore, the rules, abstractions, and techniques from other engineering disciplines cannot simply be applied broadly. Nonetheless, this does not preclude CAD from being part of synthetic biology. Figure 10.1 illustrates that in fact synthetic biology designs have grown in number but have not increased in complexity. This just underscores the need for design methodologies utilizing CAD to help cope with increased system complexity. We make the case that synthetic biology will not live up to its true potential without developing such a design methodology. The fact that biology is inherently different from mechanical systems or electronics simply implies that novel methods, techniques, and interdisciplinary research will be required for making CAD successful in synthetic biology. This is precisely why this field is so fertile.

This chapter will illustrate how a rigorous, complete CAD methodology can assist a synthetic biologist and where currently available software applications fit in our vision of this design methodology.

Methodology

A CAD methodology in synthetic biology should start with a specification, which is defined by the human engineer, and end with the manufacturing of the biological system. The manufacturing process itself will be a mixture of automation and human experimentation, but the design methodology should take a design up to the point where it can directly enter the manufacturing protocol and process. The ideal design methodology can be described using six stages: *specification, design, analysis, composition of parts, and assembly*. In the first stage of the methodology, the requirements and constraints of the design are formally captured. In the second stage, CAD assists an engineer transform the requirements into a conceptual design of a biological network that can potentially satisfy the objectives under consideration. Mathematical analysis will be used in the third stage to assess the design and may reveal potential issues with the design; the design may require refinements after the analysis step. Once the conceptual design is satisfactory, it must be converted to the set of real biological parts that represent the design in the fourth stage. Biological parts are encoded as DNA sequence information, which will then be optimized for manufacturing purposes in the fifth stage. In this last stage, the design process will use standard manufacturing protocols so that the final design can be directly assembled with minimal human labor. Of note is the role that feedback and iteration play in the design process. Notice that the results of assembly feedback in the analysis stage for use in future designs. Also notice that stages may require iteration in order to proceed to the next stage. Figure 10.2 illustrates this design methodology.

As a note, an alternative method we will discuss is to use 'directed evolution' to generate a working design. Using directed evolution would shift the design and analysis steps to the end of the design flow. CAD for directed evolution may become necessary in synthetic biology as directed evolution becomes more commonplace. However, using directed evolution does not remove the requirement for analysis of design, because any results of directed evolution must be analyzed so that the

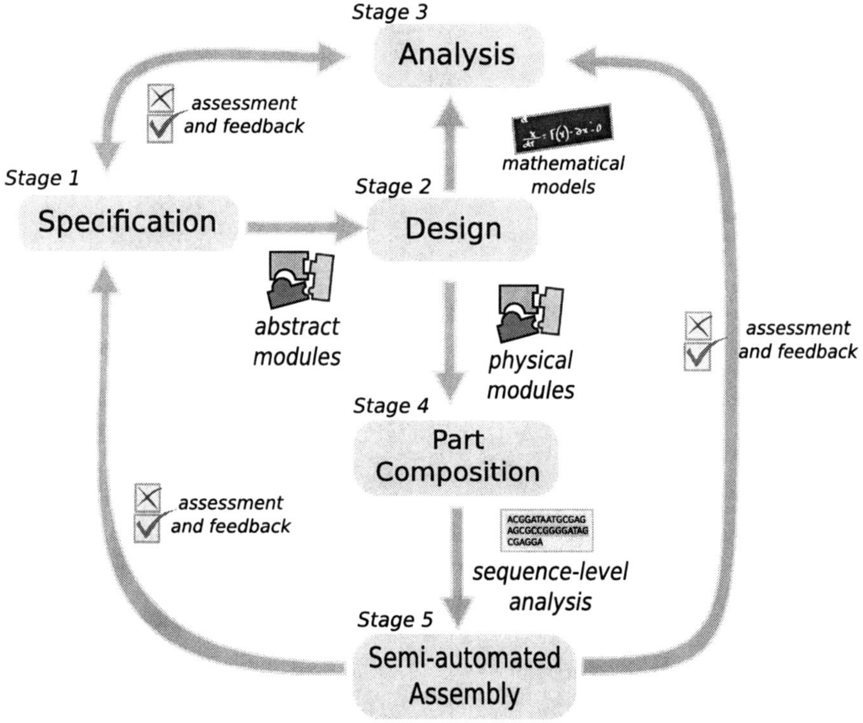

Fig. 10.2 A CAD methodology for synthetic biology. CAD will play the central role in an efficient design process in synthetic biology. The process includes specification, design, analysis, composition, and assembly of the system. Software CAD tools would take the process up to the construction step; in the ideal case, standard protocols will be used to automate much of the construction as well

ideas can be reused in future. Next sections of this chapter will cover the individual stages of this design process and how existing software applications can be used in each stage. After each section, a table is provided to summarize the list of software tools that are useful for each stage of the design process (See Tables 10.1, 10.2, 10.3, and 10.4).

Table 10.1 Synthetic biology: design tools

Tool	Description
Antimony[a]	A modular model definition language where a system is defined as a set of reactions or basic DNA parts. The modules can be connected to each other, and the final system can be converted to a mathematical model for analysis.
BioJADE[b]	A visual design application where symbols from electrical engineering are used to represent biological networks. With the support of a parts database, BioJADE is able to associate DNA sequence information with the visual design.
GEC[c]	A language for describing biochemical reactions in terms of events and rules that govern interactions. DNA components such as promoters are be used to to specify the types of components in a model. GEC also supports modularity.
GenoCAD[d]	A web-based program where a sequence of parts are placed in accordance to a set of grammatical rules. These motifs are later 'mapped' to physical implementations. The graphical environment restricts the way in which the functional motifs are composed to encourage a functioning final product.
ProMoT[e]	A modular model definition language and a visual tool where a system is defined as a set of reactions. The modules can be connected to each other, and the final system can be converted to a mathematical model for analysis.
Spectacles[f]	A functional 'schematic' based approach to biological device design where modules are individual functional motifs. These motifs are later 'mapped' to physical implementations. The graphical environment restricts the way in which the functional motifs are composed to encourage a functioning final product.
SynBioSS[g]	A web-based and desktop-based tool that allows users to connect biological parts, such as promoters and coding regions, and build a consistent mathematical model
TinkerCell[h]	A visual drawing tool for constructing modular and semantically structured diagrams. Due to semantic descriptions, the network diagrams can be automatically mapped to mathematical models or biological parts. TinkerCell has extensive support for plug-ins.

[a]http://antimony.sourceforge.net/
[b]http://web.mit.edu/jagoler/www/biojade/
[c]http://research.microsoft.com/en-us/projects/gec/
[d]http://www.genocad.org
[e]http://www.mpi-magdeburg.mpg.de/projects/promot
[f]Spectacles is a plugin for the Clotho design environment.
[g]http://synbioss.sourceforge.net
[h]http://www.tinkercell.com

Table 10.2 Systems biology: mathematical analysis tools

Tool	Description
JDesigner	A visual model construction program where reaction networks are constructed using nodes and reactions. In addition to basic simulation, the SBW suite is used to perform various types of analyses.
Jarnac	A text-based model construction program where simple scripts are used to describe reactions and rate equations. Programming language features such as loops and matrix operations are available for analysis of models.
CellDesigner	A visual model construction program where reaction networks are constructed using molecules such as RNA, proteins, and genes. Simulation and other analyses are made available through simulation libraries and SBW.
COPASI	A graphical interface for analysis of models. Some of the types of analyses supported include simulations, optimization, and parameter scans.
PySCeS	A program that is built on Scientific Python and provides all the programming flexibility of Python. PySCeS uses a simple text-based language to describe reactions and nodes that comprise a model. Available analyses include simulations, metabolic control analysis, bifurcation analyses, and any function available in Scientific Python.

Table 10.3 Synthetic biology: part composition and management tools

Tool	Description
BioMotar[a]	BioMortar is an application to facilitate the construction of BioBricks by consolidating information is an easy to access database and provide Standard Assembly-driven lab protocols.
Clotho[b]	Connects users to repositories of biological parts. Plugin tools then define the various functions that can be performed. Clotho makes it easier to share data and get it in a data model useful to synthetic biologists.
GenoCAD	Web based tool for the design of biological devices using an attribute grammar which defines the legal composition of parts.
j5[c]	Designs assembly strategies for a variety of different assembly protocols given a list of initial parts. Works for SLIC, Gibson, CPEC, and Golden-Gate based assemblies.
MIT Registry of Standard Parts[d]	A large repository of standard biological parts.
Viz-a-brick[e]	A visual environment for navigating registries of biological parts.

[a] http://igem.uwaterloo.ca/biomortar/
[b] www.clothocad.org
[c] http://jbei-exwebapp.lbl.gov/j5
[d] http://partsregistry.org/
[e] http://gcat.davidson.edu/VizABrick

Table 10.4 Genetic engineering: sequence refinement tools

Tool	Description
GeneDesign[a]	An open-source web application that provides functions such as removal of unwanted restriction sites, codon optimization, and annotation.
GeneDesigner[b]	A proprietary desktop application for primer-design *in silico* cloning, codon optimization, and visualization of DNA sequences.
Sequence Refiner[c]	A package for modifying natural DNA sequences so that they conform to specific synthetic biology standard. Part of the BIOFAB code collection.
Vector NTI[d]	A proprietary desktop application for primer-design *in silico* cloning, codon optimization, finding restriction sites, finding markers and key features, and visualization of DNA sequences.

[a]http://www.genedesign.org
[b]https://www.dna20.com/genedesigner2/
[c]Sequence Refiner is a plugin for the Clotho design environment and a standalone application created by researchers at the BIOFAB (http://www.biofab.org).
[d]A product of Life Technologies.

Specification: *Stage 1*

The first step in the design process is to have a well defined set of objectives that the final product should implement. A well designed objective is one that can be tested and described unambiguously. However that objective should not tie itself too closely with a physical implementation so that alternate designs can be explored from the same specification. For example, creating cells that are able to excrete insulin at a specified rate during their optimal growth phase would be a testable objective. An objective to create cells with a higher growth rate might be difficult to test because growth rate depends on several factors such as nutrition and population density. In such cases, a measurable aspect of growth needs to be specified along with the conditions for performing the measurement.

In addition to the objectives, the specification should also detail any constraints or requirements that system must adhere to. Using the previous example, perhaps a certain volume of insulin must be produced. Constraints can detail which biological building blocks should or should not be used in combinations. The Eugene programming language[1] provides constraints on part compositions explicitly in the form of *<operator operand operator>*, for example *promoter1 NOTWITH rbs2*. These constraints should be modular so that the types of constraints and their parameters can be swapped out to represent different operating conditions and requirements.

Design: *Stage 2*

An ideal CAD flow should provide a framework for an engineer to construct a design from the specification. This process should provide the engineer with a set of

[1] http://www.eugenecad.org

building blocks that ease the task of making complex systems while ensuring that the specification is adhered to. In mechanical engineering, a CAD program would provide the engineer with building components such as dampers and gears. The properties of these mechanical components are well defined, and therefore, there are established ways in which each component should be used to achieve particular design objectives as well as rules for their composition. An analogous set of components in biology would include small networks motifs with known properties. A phosphorylation-dephosphorylation cycle can be an example of such a biological motif. This particular motif has a sharp sigmoid response curve, where the input is the phosphorylating enzyme [15]. Another motif can be an expression cassette with a positive feedback, which can either be used to provide a sensitive response or memory [51]. Conversely, negative autoregulation expression cassettes can be used to make the response more linear and less abrupt [36, 43]. Since the dynamics of these small network motif have been investigated in the literature, they can be used by engineers to satisfy particular aspects of the specification. However, in comparison to mechanical components such as dampers or gears, network motifs in biology are poorly studied for the purpose of design and composition. Specifically, network motifs have been studied individually, but the consequence of connecting one motif to another is not a well studied phenomenon. Hence, constructing a network composed of several biological motifs may or may not have the anticipated behavior. Subtle details, such as time scale separation, can play an integral role in the interface of two biological motifs. Such minute details have been well studies in physical systems, allowing mechanical and electrical engineers to accurately predict the response of a system composed of multiple components. In order to reach the same level of sophistication is biology, it is important to build models composed of smaller building blocks and understand how the individual building blocks contribute to the whole.

Figure 10.3 shows an example from the software application called TinkerCell (www.tinkercell.com) where 'biological modules' are used to simplify a design challenge [7]. In this example, the specification is a feedback oscillator using two genes. The feedback needs to be strong in order to trigger oscillations, so the engineer needs some method of increasing the strength of the feedback. A CAD program such as TinkerCell with previously designed modules might contain a phosphorylation-dephosphorylation cycle, a functional module that can convert a linear input into a sigmoid output. This function is ideal for the increasing the strength of the feedback. Without the help of CAD, the engineer would be required to rediscover this network design. The design in Fig. 10.3 shows an oscillator constructed using three modules, where the phosphorylation module in the middle is used to provide the necessary amplification for a genetic oscillator.

The Role of Modularity in Design

Ideally, CAD should automatically convert between biological parts and mathematical models, as shown in Fig. 10.2. This is a difficult challenge for a number

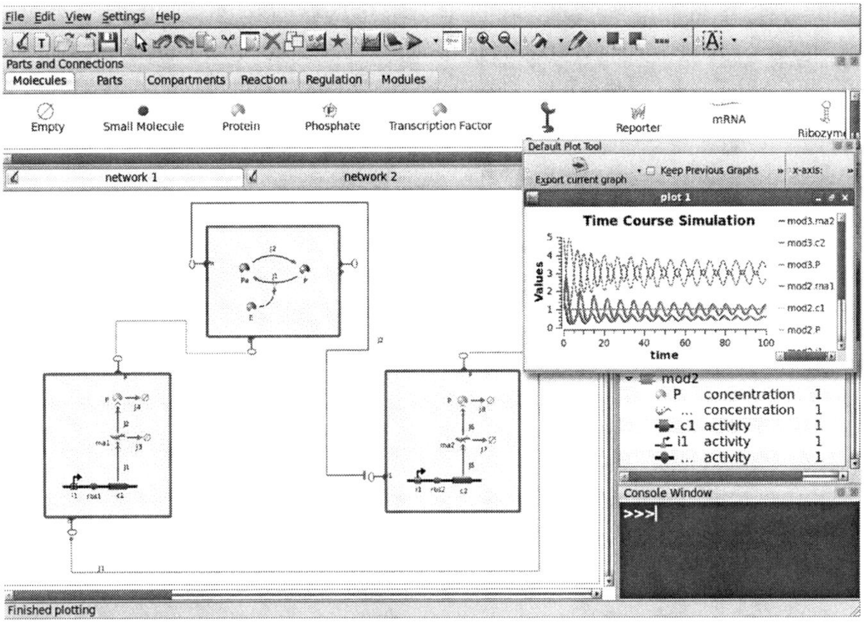

Fig. 10.3 Modules in TinkerCell. TinkerCell (www.tinkercell.com) allows users to define modules, or networks with interfaces, that can be connected to each other. TinkerCell checks whether connections between modules are valid by looking at the type of components that are connected, thus preventing connections that do not have any biological meaning

of reasons, including lack of knowledge of the underlying biological processes and technical difficulties in constructing arbitrary systems. One solution used by software developers has been to introduce some form of modularity in the design process. For example, at the simplest level, individual components such as promoters can be treated as modules. The software application SynBioSS [20] allows users to construct a DNA sequence of functional components such as promoters, ribosomal binding sites, and protein coding regions. From the sequence of components, a model is generated, with the assumption that the components behave relatively independent of each other. Similarly, mathematical models for individual biological components have been constructed with the hope that these models can be used to map between annotated DNA sequences and mathematical models [41].

For systems with several interacting components, it may be impossible to compose the mathematical model of the entire system from models of individual components. For example, concentrations of metabolites might be part of the system dynamics, and these processes may not be captured by the proteins coded in the DNA. For such cases, the concept of biological modules may be required, such as the one shown in Fig. 10.3. A few software applications support the concept of computational biological modules. From the field of systems biology, ProMoT [35] and Antimony [48] are two examples where the user is allowed to describe chemical reactions using scripts or visual diagrams. GEC [38] is another language for defining biochemical modules using rules of chemical interaction. The modules in these applications can be linked to one another by declaring molecules that are shared between modules. While such modules capture the chemical kinetic, they do not contain the necessary information to be translated into a physical realization, i.e. a DNA sequence, except perhaps GEC due to its use of semantics. BioJADE [16] was one of the first applications to describe synthetic biology systems using modules similar to electrical devices such as inverters. Each module, or device, in BioJADE could be modeled computationally and had a corresponding DNA sequence stored in a local database. The similar application, TinkerCell, uses semantic annotations to describe modules, which also bridges the gap between the mathematics and the physical system. Clotho [10] plugins, such as Eugene Scripter and Spectacles,[2] allows for the composition of functional concepts (e.g. promoters, ribosome binding sites, coding regions) which can be mapped to actual DNA sequence information at a later time. Eugene supports the concept of biological 'devices', or modules defined by their individual components encoded in DNA. The device definition describes the physical realization of a system as opposed to describing the mathematics of the system. With ample annotations, it should be possible to map such physical device descriptions to modules describing system dynamics, but that is active area of research at the moment.

Mathematical Analysis: *Stage 3*

An integral step in CAD is to perform analysis on a given design to quantitatively investigate how well it satisfies the specification. This step requires formulating a given design as a mathematical model. Mathematical models of a biological system can be constructed in different ways [8]. The choice of modeling method will depend on the specific objective. The most common type of models are differential equation models, where the concentration change of each molecular species in the design is captured using an ordinary differential equation (ODE). The set of ODEs can be understood as a matrix multiplication between the stoichiometry matrix and the reaction rates. The stoichiometry matrix and the rate equations are necessary for generating stochastic models of the biological system, which is a modeling method that can capture random fluctuations in concentration values that may impact the overall behavior of the system. More detailed modeling methods can include simulations where the spatial distribution of molecules is also taken into account. However, the more detailed mathematical models require more parameters describing the biological system. Biological parameters are generally unknown and difficult to measure, for which reason many researchers attempt to simplify

[2] http://2009.igem.org/Team:Berkeley_Software/Spectacles

the models with reduced parameter sets. One of the simplest types of models are Boolean models, which abstracts each molecules has being in a 'on' or 'off' state [50]. A CAD program needs to explore the possible modeling paradigms and select the one that is most fitting for assessing a given design and a given objective. For example, if the objective specifies something about noise levels in the system, then it would be mandatory to use some form of stochastic modeling procedure that can capture noise levels. On the contrary, if the objective is highly abstract, then perhaps a simple Boolean model might be sufficient to test the design. For metabolic control analysis, the stoichiometry matrix becomes highly important. For example, optimal enzyme concentrations can be determined using flux balance analysis, which is essentially a linear programming method that uses the stoichiometry matrix to define the optimization problem. In summary, there are numerous methods available for analysis of a given design, and the choice depends on the questions that are being investigated. The correct type of analysis is an integral part of CAD. For this reason, we envision a horizontal integration of various approaches at the analysis stage, each offering their own approach which the designer can select from.

Programming Languages for Mathematical Analysis

Once a biological system has been described mathematically, then the procedure for analyzing the biological system is generally the same as analysis of any other dynamical system. Numerous software packages exist for mathematical analysis of dynamical systems. The commonly used commercial packages are scripting languages such as MATLAB (www.mathworks.com) and Mathematica (www.wolfram.com). Open-source software packages such as R [25], Octave [12], SciLab [17], SciPy (www.scipy.org), and Maxima (maxima.sourceforge.net) also provide much of the functionality available in the commercial packages. All of these software tools require the user to perform some amount of computer programming to construct and analyze the model. However, the benefit of programming is that it allows a great degree of flexibility. Further, many of these general-purpose scripting languages contain useful packages from fields such as statistics and control theory, which increase the types of analysis that can be done.

One weakness of using programming languages to encode mathematical models is that the models will be represented using custom code. In most cases, only the author of the code would be able to interpret the biological significance of each variable in the code. Therefore, the models are difficult to share. The field of systems biology introduced a more standard method of representing a computational model [23], allowing different software tools to exchange mathematical models without requiring interpretation from the original author. As a result, numerous packages are available from the field of systems biology for performing mathematical analysis, a few of which are described next.

Systems Biology Software Applications

The Systems Biology community has developed a large toolset to aid in mathematical analysis of computational models. Most of these tools are suitable for the analysis stage of synthetic biology. Most software tools in systems biology support the Systems Biology Markup Language (SBML) [23]. Thus to make it easy to use one of these tools the synthetic model should be restated in SBML. This can be done either directly, or via the aforementioned Antimony language.

Applications for Systems Biology fall into one of the following categories:

- Model Editors – used to construct mathematical models
- Simulators – used to simulate mathematical models
- Analysis Tools – used to perform other mathematical operations, such as optimization or parameter scan
- Translators – used to convert one model format to another

Some software packages try to fulfill several categories. In the following we briefly review a selection of popular software applications.

Systems Biology Workbench (SBW) The Systems Biology Workbench (SBW) [2, 3, 22, 46] consists of a collection of model editors, simulators and analysis tools. While the model editors are targeted towards metabolic or signaling networks, they can be used to model the environment of the synthetic networks. In any case the tools allow transcribing the mathematical model of a synthetic network. Specifically the model editors are JDesigner and Jarnac:

JDesigner JDesigner [2, 46], a software application developed for the Windows platform, was one of the first visual modeling applications available. It is tailored for modeling generic biochemical reaction networks but the nodes in the network can be used implicitly to modeling metabolic, signaling or gene-regulatory networks. Via integration with the Systems Biology Workbench (SBW), JDesigner also provides advanced capabilities for simulation and analysis of the active model.

Jarnac Jarnac [44] was developed for the Windows platform and is a successor to SCAMP [45]. It is also one of the first simulation applications for Windows. Jarnac features a simple model description language as well as a powerful control language that operates on the model description. This makes even advanced simulation and analysis tasks possible through the programming of looping constructs and manipulation of matrices. Jarnac offers a very fast model development time and is thus a good tool for prototyping models. The recent language called Antimony [48] is a software library that can parse Jarnac-like scripts and has additional powerful features such as modularity and capability to define genetic networks.

Apart from these model editors SBW also includes with RoadRunner a state of the art deterministic simulator. A general simulation environment, the 'Simulation Tool' (Fig. 10.4) allows users to experiment with the model by visually defining a variety of simulation experiments.

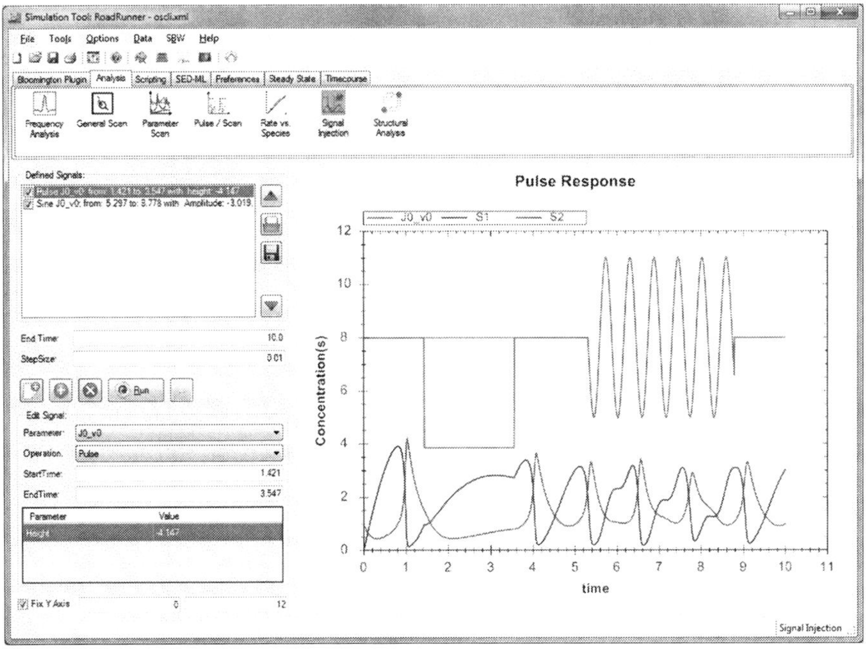

Fig. 10.4 Simulation tool 'Signal Injection' pulse/sine signal. Additional Simulation experiments include Frequency Analysis, 1D and 2D parameter scans as well as network analysis

Finally the SBW comes with a selection of translators that translate models defined in SBML into a variety of other formats, among them Matlab and Mathematica.

CellDesigner CellDesigner [14] is widely known in the systems biology community and used by many research groups. Similar to JDesigner, a user of CellDesigner would place one of several nodes onto the graphics canvas in order to draw a pathway. While JDesigner gives the user the freedom to adjust the graphics of a given node freely, CellDesigner provides a fixed set of symbols, each with a given meaning (e.g. forward-slanted parallelogram for RNA, backward-slanted parallelogram for Antisense RNA and such). The notation using those symbols has been dubbed a Process Diagram. In future versions CellDesigner will adopt the Systems Biology Graphical Notation (SBGN). CellDesigner integrates the SBML ode Solver [33] library to provide basic simulation capabilities for the currently active model. CellDesigner is SBW enabled and so further analysis is possible through the SBW menu, which allows the model to be sent to SBW capable applications. The CellDesigner team is also experimenting with a plug-in system, which will allow third parties to interface with CellDesigner directly.

COPASI COPASI [21] is the successor of Gepasi [34], one of the first popular simulation applications for Windows. COPASI is a state of the art modeling, simulation and optimization environment for computational models. Since 2008, COPASI has been SBW enabled. This enhances COPASI by an SBW menu, allowing models to be exchanged between COPASI and other SBW enabled application. COPASI also has the capabilities of exporting models to C code, Berkeley Madonna or XPP AUTH.

PySCeS PySCeS [37] is a scripting environment built on top of the popular script language Python . PySCeS model definition language looks rather similar to that of Jarnac and the example given above would be written as follows in PySCeS:

```
X0 > S1
k1*X0
```

PySCeS uses SciPy at its core and Matplotlib [24] to generate a variety of plots. Arguably, the biggest strength of PySCeS is that researchers have the full power of the Python programming language at their disposal.

Biological Part Composition: *Stage 4*

The fourth stage our CAD flow takes a 'satisfactory' design and identifies the final DNA sequence that represents the design. A DNA sequence can be considered as a linear chain of biological 'parts', where each part is a DNA sequence that encodes a specific function in the designed system. The CAD program needs to link the mathematical analysis with the parts that comprise the designed system. This is perhaps the most difficult challenge faced by current applications, because the mathematics is usually abstract and not directly connected to the parts. Moreover, the dynamics of the system results from proteins and other molecules in the system, not the DNA that encodes them. In the mathematics, the biological parts encoded in the DNA are often hidden inside parameters or rate equations. For example, a promoter part that is present in the DNA sequence might be reflected in the model as a parameter in the mRNA production rate equation. In a different modeling scheme, the same promoter might be reflected in two parameters. At this point, there are no established methods for deriving the mathematics from the parts alone. It is possible that there are multiple models that can be derived from the same set of parts. The choice of model may change depending on the circumstances and the questions being addressed by the modeler.

The difficulties of interlinking mathematical modeling and biological parts is often shadowed by difficulties in the physical construction step. Once a synthetic biologist has a working design in terms of biological parts, the next stage is to construct the DNA sequence where all the parts are placed in the correct order. Protocols for efficiently obtaining parts and assembling multiple parts into a single construct are still in active development [11]. DNA synthesis of the entire sequence is an alternative, but that option is not affordable by many synthetic biology labs at this time. As a result, software efforts have been placed in organizing parts so that reuse of existing parts would be easier. j5[3] is a software package developed at Lawrence Berkeley National Lab which specifically creates assembly plans for a variety of different assembly protocols and chemistries. Organizing parts means more than simply placing the available set of parts in a database. Usually when a synthetic biologist wishes to construct a new design, they would like to find specific patterns, such as promoters that are regulated by two transcription factors or expression cassettes with fluorescent proteins. In order to find these patterns, it is important to represent a sequence of parts in a well structured format. The web-based application called GenoCAD [5] is one of the first tools to propose a structured representation for defining sequences of biological parts. GenoCAD (Fig. 10.5) defines a grammar for interpreting linear sequences of biological parts much in the same way the English language defines grammatical rules for forming sentences from words. Similarly, the domain specific language Eugene uses specific syntax to interpret a linear chain of biological parts. The parser uses a flexible set of constructs where new rules and terms can be defined. Such software tools provide a systematic way for interpreting a sequence of biological parts. The systematic representation has two major benefits. One, the parts that are constructed by synthetic biologists can be stored in an organized format, allowing other researchers to search for specific patterns of parts. Second, the structured representation opens possibilities for automatically linking mathematical models with parts. For example, a chain of parts that is composed of a promoter, ribosomal binding site, and protein coding region can be mapped to

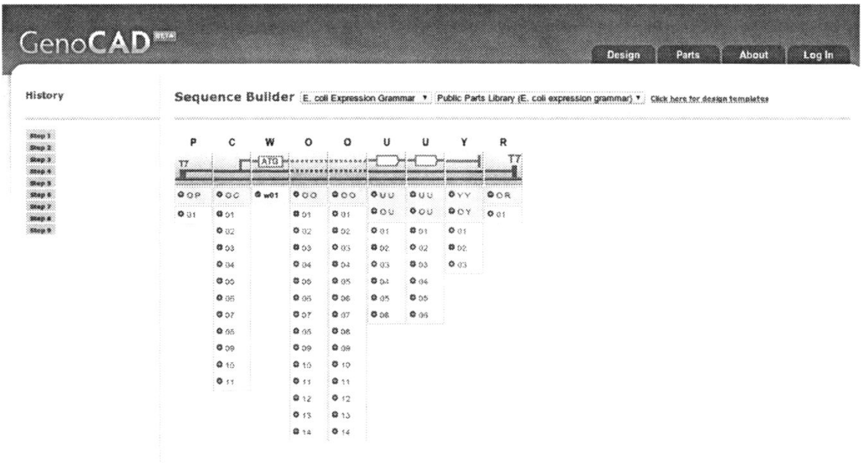

Fig. 10.5 Composing a sequence of parts using GenoCAD. GenoCAD allows users to construct sequence of parts similar to the way a sequence of English words are used to form a sentence. The underlying grammar ensures that the sequence will be biologically meaningful

[3] http://jbei-exwebapp.lbl.gov/j5

a simple model for protein production. An ongoing effort in this direction is the Synthetic Biology Open Language[4] semantic, which represents parts and their relationships to each other using semantic connections.

Representing a system as a set of biological parts acts as a link between the mathematical model and the final DNA sequence. The parts imply function, which can be represented as mathematical models, and the parts consist of a unique DNA sequence, which can be used to construct the final DNA sequence. Therefore, correctly defining biological parts is possibly the key step in building the complete pipeline from design to construction of the DNA sequence. One of the ambitions of the TinkerCell application was to represent synthetic biology diagrams in a way that can be converted to mathematical models or DNA parts. TinkerCell adopted semantic descriptions, similar to the Synthetic Biology Open Language semantics, as the method of reaching this goal together.

Sequence Refinement

Once a design has been reduced to a set of parts, the DNA sequence can be constructed and refined based on specific criteria. Examples include avoiding specific restriction sites or optimizing the codon usage. The sequence might also be optimized to avoid repeated regions that may be prone to recombination or secondary structures. Once the DNA sequence has been refined, the sequence can either be assembled using genetic engineering techniques or synthesized using DNA synthesis technology. The resulting DNA can be placed in a living cell and tested. The results of the testing will be used to refine the original design.

Many sequence refinement tools are available from classical genetic engineering. Some examples of commercial and free software applications that can be used to perform sequence edits such as codon optimization or removal of restriction sites include Gene Designer [52], Vector NTI [32], and GeneDesign [40]. The same tools are used in synthetic biology at present. As synthetic biology matures, specialized functions will be needed, particularly for standardization purposes. As a community, synthetic biologists have begun to agree on certain standard procedures, such as use of restriction enzymes. Such standard procedures allow better exchange or parts between research labs. As a result, such standards would need to be supported by software tools. Software efforts such as Clotho [10] that already make use of such exchange standards will provide features for validating parts against the community standards and offering options to edit the sequence to conform to the standards.

Further, while computer programs often treat DNA as a string of information, the fact is that it is a three-dimensional structure, and certain sequences can influence the structure of the DNA (or RNA) to produce unexpected results. For instance, the region upstream of the protein coding region in the mRNA can often form secondary

[4] http://www.sbolstandard.org

structures, preventing ribosomes from translating the mRNA into proteins. Therefore, checking for undesired secondary structures is another analysis that must be done at the sequence refinement stage.

Assembly: *Stage 5*

One major goal of sequence refinement is to make the conversion from parts to the final product very simple and predictable. For instance, if all parts are stored in DNA sequences with compatible sticky ends, then assembling those parts might be much simpler. Similarly, if the possibility of any DNA or RNA secondary structures are eliminated, then the parts may behave much closer to what is expected. If the process of assembling parts can be streamlined in such manner, it may be possible to establish standard protocols for assembling parts, some of which are already being practiced [47]. Efficient ways of assembling DNA fragments is an ongoing area of research [30]. When this technology is perfected, it will undoubtedly boost the progress of synthetic biology, because much of the time in any synthetic biology experiment is spent in the physical construction of the system. DNA synthesis is another area of research that will greatly enhance the development cycle in synthetic biology. It is arguable that DNA synthesis will entirely replace DNA parts assembly as the means of creating the final DNA encoding the designed system [6]. At the same time, parts assembly will generally be cheaper and perhaps more efficient than DNA synthesis due to the fact that individual parts have already been constructed. In either case, the benefit of standard procedures for constructing synthetic systems will eliminate the most time-consuming step in synthetic biology at present. In addition to assembly of DNA, it may even be possible to automatically test newly constructed systems using standard protocols. Such automation has been demonstrated for molecular biology experiments [49], but the cost of maintaining such automated systems makes them impractical for synthetic biology labs at present.

Design by Evolution

While the five steps (Fig. 10.2) in CAD describe a rational design approach, it is also possible to take a combinatorial approach to design. Combinatorial design is especially advantageous in biological systems because of the natural machinery for generating variety. Novel enzymes can be generated by creating trillions of different versions of an initial design and screening for the enzyme's desired activity. Similarly, one can imagine generating variants of a network design and screening for the network that satisfies an objective. This is a valid design process for biological systems. However, there are no CAD software applications in synthetic biology that support such design methodology at present. If design by selection becomes a common method in synthetic biology, CAD applications may emerge to support this methodology (Fig. 10.6).

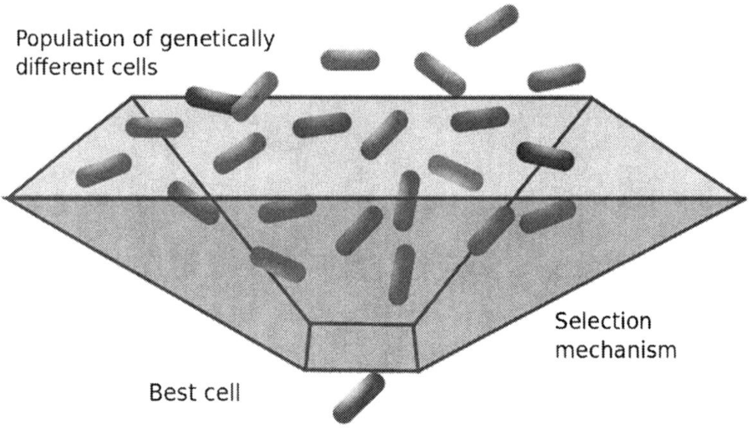

Fig. 10.6 Directed evolution. Direction evolution involves selecting for the few wanted cells from a population of billions. The process will often involve repeated cycles of selection, where each cycle produces slightly better cells. The challenge in directed evolution is designing a screening process that correctly selects for cells with the desired behavior

Nature invents new systems by combining existing ones. For example, new proteins are created by shuffling existing protein domains, and new organisms are created by rewiring the regulatory or signaling networks of existing organisms. Homologous recombination is often used by nature as a means of optimizing functions; the immune system uses such an optimization method to generate antibodies that target specific foreign particles [29]. There is evidence suggesting that homologous recombination is an important tool for evolution of viruses and microbes [19, 31]. While protein engineers and synthetic biologists have used random mutations as a means of optimizing the function of a proteins or adjusting a promoter site [18, 26], random mutations cannot be used to design an entire system. However, homologous recombination can be used to create new systems, because they can rewire existing regulation sites or even protein binding domains, creating new networks. The recombination sites can be planned ahead of time, much similar to the way immune system uses specific recombination sites to evolve antibodies. Recombination sites can be planted in the DNA sequence such that only certain recombination events are possible. This process can be summarized as a planned combinatorial optimization of synthetic network.

The role of computer-aided design in such combinatorial optimization would be to assist the engineer in designing the DNA sequence. The repeats in the DNA sequence would control the recombination events, and therefore the possible networks that would be generated by the process can be modeled by a CAD program.

While directed evolution can be used to create the end product, it does not remove the requirement for further analysis. Evolution can often discover design strategies that human engineers might have overlooked. When such strategies are discovered through evolution, those designs should be thoroughly analyzed and incorporated into the repertoire of design patterns for synthetic biology. Therefore, design by evolution is not disjoint from rational design. Rather, they are two approaches that should be used in conjunction with one another.

Education

The success of CAD in synthetic biology will require expanding on knowledge from established fields such as systems biology, electronic design automation, control theory, and genetic engineering as well as a community of researchers who can understand concepts from multiple disciplines. One important aspect of software development that is often missed is its impact on education. Using visual software allows students as well as researchers understand some of the details of a foreign discipline. For example, a student using Clotho can learn about how synthetic biology data is related and create a plug-in to address a biological design problem using that data; a student using TinkerCell would realize the details that are required to transition from a cartoon diagram to a mathematical model. For researchers, well designed software tools can bridge gaps in knowledge by automating many standard procedures. For example, TinkerCell allows users to draw abstract concepts such as transcription regulation and automatically generates a default model to represent that abstract concept, which the user can opt to modify. As 'best practices' for modeling and wet-lab protocols are establishes, such automatic features in software will become more commonplace.

There are two excellent opportunities for designers of CAD software for synthetic biology to engage the community:

- The International Genetically Engineered Machine Competition (iGEM) – The premiere undergraduate Synthetic Biology competition. Student teams are given a kit of biological parts at the beginning of the summer from the Registry of Standard Biological Parts. Working at their own schools over the summer, they use these parts and new parts of their own design to build biological systems and operate them in living cells.
- The International Workshop on Bio-Design Automation (IWBDA) – Brings together researchers from the synthetic biology, systems biology, and design automation communities. The focus is on concepts, methodologies and software tools for the computational analysis of biological systems and the synthesis of novel biological systems.

Summary

The goal of a CAD methodology in synthetic biology is to provide an efficient design flow for creating and analyzing new biological systems. The difficulty in building this design flow is due to lack of established methods in biology coupled with our relative lack of understanding regarding biological behavior. For instance, modeling of biological systems is still a developing area of research, so there are no 'best practices' for building predictive models of a given system. The difficulties in modeling are amplified by the need to represent a model in terms of biological parts that can be encoded as DNA. Even if the modeling aspect of synthetic biology is resolved, no established procedures exist for describing a 'biological part' and constructing a system using multiple parts. The construction of DNA from individual parts is an active area of research, where a few of the unresolved issues include DNA secondary structure formation or interference with cell growth, both of which can influence a system in unexpected ways.

Due to the numerous challenges in building an efficient CAD system, the synthetic biology community has recognized obvious advantageous in collaborative efforts to resolve common issues. The drive for collaboration is reflected in many of the new software tools. Applications such as TinkerCell and Clotho are designed so that different collaborators can add plug-ins that add new features and integrate information. Web-based applications such as a GenoCAD and SynBioSS utilize databases of parts to support reuse of biological parts, which was a feature first seen in BioJADE [16].

We have presented a flow which we feel outlines the required design activities that CAD can help introduce in the synthetic biology community. While the individual algorithms and tools which participate in this flow will change, the general structure provides a robust foundation to push forward the level of complexity, robustness, and reproducibility possible in synthetic biological designs.

Acknowledgements The authors of this chapter would like to acknowledge the National Science Foundation (NSF 0527023-FIBR) and the National Institute of Health (NIH GM081070 and NIH/NIBIB BE08407) for their support.

References

1. Basu S, Gerchman Y, Collins CH, Arnold FH, Weiss R (2005) A synthetic multicellular system for programmed pattern formation. Nature 434(7037):1130–1134
2. Bergmann FT, Sauro HM (2006) SBW-a modular framework for systems biology. Proceedings of the 38th conference on winter simulation. Winter simulation conference, pp 1637–1645
3. Bergmann FT, Vallabhajosyula RR, Sauro HM (2006) Computational tools for modeling protein networks. Curr Proteomics 3(3):181–197
4. Çagatay T, Turcotte M, Elowitz MB, Garcia-Ojalvo J, Süel GM (2009) Architecture-dependent noise discriminates functionally analogous differentiation circuits. Cell 139(3):512–522
5. Cai Y (2010) GenoCAD: linguistic approaches to synthetic biology. PhD thesis
6. Carlson R (2009) The changing economics of DNA synthesis. Nat Biotechnol 27(12): 1091–1094
7. Chandran D, Bergmann FT, Sauro HM (2009a) TinkerCell: modular CAD tool for synthetic biology. J Biol Eng 3(1):19
8. Chandran D, Copeland WB, Sleight SC, Sauro HM (2009b) Mathematical modeling and synthetic biology. Drug Discov Today Dis Models 5(4):299–309

9. Chinnery DG, Keutzer K (2000) Closing the gap between asic and custom: an asic perspective. DAC '00: proceedings of the 37th annual design automation conference, ACM, New York, pp 637–642
10. Densmore D, Van Devender A, Johnson M, Sritanyaratana N (2009) A platform-based design environment for synthetic biological systems. The fifth Richard Tapia celebration of diversity in computing conference: intellect, initiatives, insight, and innovations, ACM, pp 24–29
11. Densmore D, Hsiau THC, Kittleson JT, DeLoache W, Batten C, Anderson JC (2010) Algorithms for automated dna assembly. Nucleic Acids Res 38(8):2607–2616
12. Eaton JW, Bateman D, Hauberg S (1997) Gnu octave. Citeseer
13. Entus R, Aufderheide B, Sauro HM (2007) Design and implementation of three incoherent feed-forward motif based biological concentration sensors. Syst Synth Biol 1(3):119–128
14. Funahashi A, Matsuoka Y, Jouraku A, Morohashi M, Kikuchi N, Kitano H (2008) CellDesigner 3.5: a versatile modeling tool for biochemical networks. Proc IEEE 96(8):1254–1265
15. Goldbeter A, Koshland DE (1981) An amplified sensitivity arising from covalent modification in biological systems. Proc Natl Acad Sci USA 78(11):6840
16. Goler JA (2004) BioJADE: A Design and Simulation Tool for Synthetic Biological Systems. PhD thesis
17. Gomez C (1999) Engineering and scientific computing with Scilab. Birkhauser
18. Haseltine EL, Arnold FH (2007) Synthetic gene circuits: design with directed evolution. Annu Rev Biophys Biomol Struct 36:1–19
19. Hendrix RW, Lawrence JG, Hatfull GF, Casjens S (2000) The origins and ongoing evolution of viruses. Trends Microbiol 8(11):504–508
20. Hill AD, Tomshine JR, Weeding E, Sotiropoulos V, Kaznessis YN, (2008) SynBioSS. Bioinformatics 24(21):2551–2553
21. Hoops S, Sahle S, Gauges R, Lee C, Pahle J, Simus N, Singhal M, Xu L, Mendes P, Kummer U (2006) COPASI–a COmplex PAthway SImulator. Bioinformatics 22(24):3067
22. Hucka M, Finney A, Sauro HM, Bolouri H, Doyle J, Kitano H (2002) The ERATO systems biology workbench: enabling interaction and exchange between software tools for computational biology. In: Pac Symp Biocomput, vol 1. Citeseer, pp 450–461
23. Hucka M, Finney A, Sauro HM, Bolouri H, Doyle JC, Kitano H et al (2003) The systems biology markup language (SBML): a medium for representation and exchange of biochemical network models. Bioinformatics 19(4):524
24. Hunter JD (2007) Matplotlib: a 2D graphics environment. Comput Sci Eng 9:90–95
25. Ihaka R, Gentleman R (1996) R: a language for data analysis and graphics. J Comput Graph Stat 5(3):299–314
26. Kaur J, Sharma R (2006) Directed evolution: an approach to engineer enzymes. Crit Rev Biotechnol 26(3):165–199
27. Keutzer K, Newton AR, Rabaey JM, Sangiovanni-Vincentelli A (2000) System-level design: orthogonalization of concerns and platform-based design. IEEE Trans Comput-Aided Des Integr Circuits Syst 19(12):1523–1543
28. Lee SK, Chou H, Ham TS, Lee TS, Keasling JD (2008) Metabolic engineering of microorganisms for biofuels production: from bugs to synthetic biology to fuels. Curr Opin Biotechnol 19(6):556–563
29. Lieber MR (1991) Site-specific recombination in the immune system. FASEB J 5(14):2934
30. Linshiz G, Yehezkel TB, Kaplan S, Gronau I, Ravid S, Adar R, Shapiro E (2008) Recursive construction of perfect DNA molecules from imperfect oligonucleotides. Mol Syst Biol doi:10.1038/msb.2008.26
31. Lonsdale DM, Brears T, Hodge TP, Melville SE, Rottmann WH (1988) The plant mitochondrial genome: homologous recombination as a mechanism for generating heterogeneity. Philos Trans R Soc London Ser B, Biol Sci 319(1193):149–163
32. Lu G, Moriyama EN (2004) Vector NTI, a balanced all-in-one sequence analysis suite. Brief Bioinform 5(4):378
33. Machné R, Finney A, Muller S, Lu J, Widder S, Flamm C (2006) The SBML ODE solver library: a native API for symbolic and fast numerical analysis of reaction networks. Bioinformatics 22(11):1406

34. Mendes P (1993) GEPASI: a software package for modelling the dynamics, steady states and control of biochemical and other systems. Bioinformatics 9(5):563
35. Mirschel S, Steinmetz K, Rempel M, Ginkel M, Gilles ED (2009) PROMOT: modular modeling for systems biology. Bioinformatics 25(5):687
36. Nevozhay D, Adams RM, Murphy KF, Josić K, Balázsi G (2009) Negative autoregulation linearizes the dose–response and suppresses the heterogeneity of gene expression. Proc Natl Acad Sci USA 106(13):5123
37. Olivier BG, Rohwer JM, Hofmeyr JHS (2005) Modelling cellular systems with PySCeS. Bioinformatics 21(4):560–561
38. Pedersen M, Plotkin GD (2010) A language for biochemical systems: design and formal specification. Trans Comput Syst Biol XII: Spec Issue Model Methodol 5945:77
39. Purnick PEM, Weiss R (2009) The second wave of synthetic biology: from modules to systems. Nat Rev Mol Cell Biol 10(6):410–422
40. Richardson SM, Wheelan SJ, Yarrington RM, Boeke JD (2006) GeneDesign: rapid, automated design of multikilobase synthetic genes. Genome Res 16(4):550
41. Rouilly V, Canton B, Nielsen P, Kitney R (2007) Registry of BioBricks models using CellML. BMC Syst Biol 1(Suppl 1):P79
42. Sangiovanni-Vincentelli A (2007) Quo vadis, sld? reasoning about the trends and challenges of system level design. Proc IEEE 95(3):467–506
43. Sauro HM, Ingalls B (2007) Mapk cascades as feedback amplifiers. Arxiv preprint arXiv: 0710.5195
44. Sauro HM (2000) Jarnac: a system for interactive metabolic analysis. Animating the cellular map: proceedings of the 9th international meeting on bioThermoKinetics, Stellenbosch University Press, pp 221–228
45. Sauro HM, Fell D (1991) SCAMP: a metabolic simulator and control analysis program. Math Comp Model 15(12):15–28
46. Sauro HM, Hucka M, Finney A, Wellock C, Bolouri H, Doyle J, Kitano H (2003) Next generation simulation tools: the systems biology workbench and BioSPICE integration. Omics J Integr Biol 7(4):355–372
47. Shetty RP, Endy D, Knight TF Jr (2008) Engineering BioBrick vectors from BioBrick parts. J Biol Eng 2:5
48. Smith LP, Bergmann FT, Chandran D, Sauro HM (2009) Antimony: a modular model definition language. Bioinformatics 25(18):2452
49. Sparkes A, Aubrey W, Byrne E, Clare A, Khan M, Liakata M, Markham M, Rowland J, Soldatova L, Whelan K, Young M, King R (2010) Towards robot scientists for autonomous scientific discovery. Autom Exp 2(1):1
50. Thomas R (1973) Boolean formalization of genetic control circuits* 1. J Theor Biol 42(3): 563–585
51. Thomas R, D'Ari R (1990) Biological feedback. CRC
52. Villalobos A, Ness JE, Gustafsson C, Minshull J, Govindarajan S (2006) Gene designer: a synthetic biology tool for constructing artificial DNA segments. BMC Bioinformatics 7(1):285
53. Yuh CH, Bolouri H, Davidson EH (1998) Genomic cis-regulatory logic: experimental and computational analysis of a sea urchin gene. Science 279(5358):1896

Chapter 11
High-Level Programming Languages for Biomolecular Systems

Jacob Beal, Andrew Phillips, Douglas Densmore, and Yizhi Cai

Abstract In electronic computing, high-level languages hide much of the details, allowing non-experts and sometimes even children to program and create systems. High level languages for biomolecular systems aim to achieve a similar level of abstraction, so that a system might be designed on the basis of the behaviors that are desired, rather than the particulars of the genetic code that will be used to implement these behaviors. The drawback to this sort of high-level approach is that it generally means giving up control over some aspects of the system and having decreased efficiency relative to hand-tuned designs. Different languages make different tradeoffs in which aspects of design they emphasize and which they automate, so we expect that for biology, there will be no single 'right language', just as there is not for electronic computing. Because synthetic biology is a new area, no mature languages have yet emerged. In this chapter, we present an in-depth survey of four representative languages currently in development – GenoCAD, Eugene, GEC, and Proto – as well as a brief overview of other related high-level design tools.

Keywords Synthetic biology · Abstraction · High level languages · GenoCAD · GEC · Proto · Eugene · Modeling · Design · XOR

Overview

A 'high-level' programming language is one that abstracts many of the details of how a computation will actually be implemented. The programmer writes down a simple description, capturing the essence of the computation, and this description is automatically expanded to produce a complete implementation that can be executed on the available computational substrate.

J. Beal (✉)
BBN Technologies, Cambridge, Massachusetts
e-mail: jakebeal@bbn.com

H. Koeppl et al. (eds.), *Design and Analysis of Biomolecular Circuits: Engineering Approaches to Systems and Synthetic Biology*, DOI 10.1007/978-1-4419-6766-4_11,

On modern digital computers, this process can go through many different layers. Consider, for example, an entry in a Microsoft Excel spreadsheet that adds up a column of figures. The expression itself, something like '= **SUM(C1:C10)**,' is a transparently simple statement in an arithmetic-centric high-level language. Within Excel, this statement is interpreted into a set of calls to various functions within Excel, in the process adding implicit behaviors like error handling. These Excel functions were themselves written in some high-level language, and then compiled into machine code that can execute on the computer where Excel is running, in the process making routine decisions like how to implement each mathematical operation using the resources of the machine's processor. Even that machine code goes through another layer of interpretation, as the processor itself restructures the code to operate more efficiently given the current state of the processor.

The essence of the idea behind high-level languages is this: as an engineering field matures, finding good-enough solutions to sub-problems of design becomes routine. Highly routine problem solving can then be automated, reifying the knowledge of skilled engineers into a piece of software. The software solutions to individual parts of the design process can then be connected together to form a complete tool-chain, translating from high-level descriptions down to working implementations without any need for human intervention.

Separating the programmer from the implementation details has three important benefits:

- **Accessibility:** less knowledge is required to build a system, since much of the required knowledge has been captured in software.
- **Scalability:** since routine design work is automated, it is possible to build larger and more complex systems, and to re-use the same programs on different platforms.
- **Reliability:** aspects of design that are automated are no longer subject to programmer error; software can also check for common errors in the programmer's high-level design.

On electronic computers, high-level languages have become so successful that few people ever use anything besides a high-level language. In the programming of biomolecular systems, high-level languages are just beginning to emerge.

For the purposes of this chapter, we will define a high-level language for biomolecular systems as any system description language where the choice of implementing biological parts may routinely be left unspecified. We will focus primarily on programming languages for *in vivo* biomolecular computation, reviewing four representative languages: GenoCAD, Eugene, GEC, and Proto in rough order from lower to higher levels of abstraction. To aid comparison and understanding, we apply each language to a simple example problem:

> Express green fluorescent protein (GFP) when either of the small-molecule signals aTc or IPTG is present, but not when both are present.

At the end of the chapter, we also review the scope of other related high-level design tools for biomolecular computing systems.

GenoCAD

GenoCAD (www.genocad.org) is one of the earliest CAD tools for synthetic biology, built upon the foundation of formal grammars. In this section, we summarize the basics of grammars, the theoretical foundation underneath GenoCAD and also a brief tutorial on how programs are constructed using the GenoCAD web service.

Formal Language & Syntactic Model

A formal language is a set of (possibly infinite) strings derived from an alphabet Σ, which encodes information for communication purposes. There are several kinds of languages, including natural languages (e.g. English and Chinese), computer languages (e.g. C and HTML), and mathematical languages (e.g. first-order logic). However not all the strings over a language's alphabet actually belong to that language, only those which follow its rules. A grammar is a finite set of rules that specifies the syntax (permissible structure) of a language. A grammar G contains four components:

- A finite set N of non-terminal symbols.
- A finite set Σ of terminal symbols that is disjoint from N.
- A finite set P of rewriting rules, each rule is in the form of $\alpha \rightarrow \beta$, where α and β are both strings of symbols, and α contains at least one symbol from N. More formally, a rewriting rule can be represented as $(\Sigma \cup N)^* N (\Sigma \cup N)^* \rightarrow (\Sigma \cup N)^*$, where * is the Kleene star operation (meaning zero or more copies of the preceding statement) and $\cup$ is the set union operation.
- A distinguished symbol $S \in N$ that is the start symbol.

In the 1950s, Chomsky classified grammatical models into four classes based on the forms of their production rules, which reflect their expressive power [13]. In a nutshell, selecting a class of grammatical model as the representation of biological sequences is a tradeoff between the expressivity and the compilation complexity. Since GenoCAD uses a Context-Free Grammar (CFG), we will only give the mathematical definition of CFG. A good general introduction to formal languages and the Chomsky hierarchy may be found in [37].
A **Context-Free Grammar** allows any production rule of the form $A \rightarrow \alpha$. The left-hand side only consists of a single non-terminal symbol A, and the right hand side can be any string α, where $A \in N$, and $\alpha \in (N \cup \Sigma)^*$. The corresponding automaton for a context free grammar is a push-down automaton. The computational complexity to recognize a context free grammar is polynomial.

GenoCAD formalizes many generic design principles of molecule biology in the form of a context free grammar. The biological parts are the terminals, while the devices/systems composing multiple parts are categorized as non-terminals in

Table 11.1 GenoCAD small grammar set of terminal and non-terminal symbols

Non-terminals	Terminals
S	–
Operon	–
Cistron	–
Promoter	$prom1, prom2, prom3$
RBS	$rbs1, rbs2, rbs3$
Gene	$lacI, tetR, gfp$
Terminator	$b0012, b0015$

Table 11.2 GenoCAD small context free grammar of gene expression

Number	Rule
$P1$	$S \rightarrow Operon$
$P2$	$Operon \rightarrow Operon, Operon$
$P3$	$Operon \rightarrow Promoter, Cistron, Terminator$
$P4$	$Cistron \rightarrow RBS, Gene$
$P5$	$Cistron \rightarrow Cistron, Cistron$
$P6$	$Terminator \rightarrow Terminator, Terminator$
$P7 \cdots P9$	$Promoter \rightarrow prom1 \mid prom2 \mid prom3$
$P10 \cdots P12$	$RBS \rightarrow rbs1 \mid rbs2 \mid rbs3$
$P13 \cdots P15$	$Gene \rightarrow lacI \mid tetR \mid gfp$
$P16, P17$	$Terminator \rightarrow b0012 \mid b0015$

the grammar. In this review, only a small grammar will be presented: two more comprehensive grammars are published elsewhere [9, 11].

Table 11.1 summarizes the non-terminals and terminals used in this small grammar. S is a special non-terminal which is used as the start symbol of the grammar. *Operon* and *Cistron* are complex devices, which are composed of multiple basic parts (terminals). In the category of *Promoter*, there are three terminals, namely $prom1, prom2$ and $prom3$. Similarly, a ribosome binding site *RBS* can be chosen from $rbs1$, $rbs2$ and $rbs3$, while a *Gene* could be *lacI* or *tetR* or *gfp*. Finally, there are two terminals $b0012$ and $b0015$ belong to the non-terminal *Terminator*. Table 11.2 presents a context free grammar for designing gene expression cassettes. The whole grammar can be divided into two sections: rules $P1 - P6$ transform the structure of a design, while rules $P7 - P17$ are used to select a particular terminal for each non-terminal category. The design starts with $P1$, where the start symbol S becomes an expression *Operon*. Multiple *Operon*s are allowed by applying rule $P2$ multiple times: for a design with n cassettes, $P2$ is applies $n-1$ times. Rule $P3$ specifies the structure of an *Operon* to be a *Promoter*, followed by a *Cistron* and a *Terminator*. A *Cistron* can be broken down by rule $P4$ as an *RBS* and a *Gene*. Multiple *Cistron*s and *Terminators* are allowed in a design by rules $P5$ and $P6$, respectively. After the structure of a design is defined, rules $P7 - P17$ are used to transform each non-terminal to a specific biological part (terminal). For instance, rules $P7$, $P8$ and $P9$ specify $prom1, prom2$ and $prom3$ respectively to replace non-terminal *Promoter* (the '|' sign indicates OR relationship).

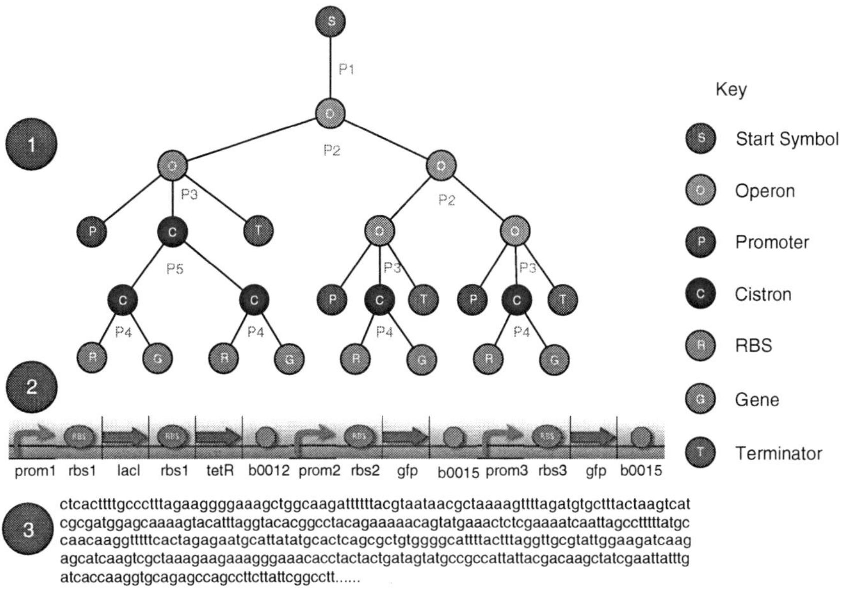

Fig. 11.1 Grammatical design of a DNA sequence. *Panel 1*: A parsing tree showing the step-by-step application of rules to generate the sequence (excepting terminal selection). Each step is labeled with the rule applied. *Panel 2*: Representation of the generated DNA part sequence, using a standard set of synthetic biology icons. *Panel 3*: The designed DNA sequence

Figure 11.1 shows how this simple syntactic model can be applied to generate a sequence structurally consistent with the XOR gates developed below in our presentation of Eugene (Fig. 11.3) and GEC (Fig. 11.4). The design process starts with applying $P1$ to the start symbol S to transform the design into a single *Operon*. After applying $P3$ twice, the design becomes three *Operon*s. In the next step, rule $P3$ defines the structure of each *Operon* as a *Promoter*, a *Cistron* and a *Terminator*. In order to express *lacI* and *tetR* under control of the same constitutive promoter, $P5$ is applied to allow two *Cistron*s in the leftmost *Operon*. Finally, rule $P4$ breaks down each *Cistron* into an *RBS* followed by a *Gene*. Once the structure of the design is decided, a part is selected for each category (panel 2 in Fig. 11.1) and mapped to a DNA sequence that can be exported for synthesis (panel 3 in Fig. 11.1)

If we operate the process in Fig. 11.1 in reverse, then rather than generating a DNA sequence, we can validate whether a specified DNA sequence is consistent with the syntactic model. This is carried out with an automated process known as 'parsing' in computer science. The parser operates in the reverse order of the design process: the GenoCAD parser takes the DNA sequence (panel 3 in Fig. 11.1) as input and breaks it into a series of biological parts (panel 2 in Fig. 11.1). It then checks for the existence of at least one rule application tree that can generate this series of parts using the context-free grammar. Realizing that we can build parsers from the syntactic model opens up the possibility of viewing DNA sequences as a programming language. One can make changes to a DNA sequence just like writing source code, and use the parser to check whether the new DNA sequence is still consistent with the syntactic model (which formalizes the biological knowledge).

It should be noted that the syntactic model only checks the structure, but not the meaning of the design. A syntactically correct sentence is not necessarily meaningful. In the context of synthetic biology, this means the syntactic model only controls the order of putting biological parts together to ensure a successful gene expression, but the function of the DNA sequence (i.e., what does this sequence do?) remains unknown. Recently, GenoCAD has been extended to address this area with the introduction of an attribute grammar to develop semantic models of DNA sequences [10]. By associating biological attributes with parts, and coupling semantic actions with each production rule, the semantic models are capable of translating a class of DNA sequences to mathematical models that describe the encoded phenotypic behavior.

GenoCAD Web Service

Based on the syntactic model originally described in [9], an open-source web application (www.genocad.org) has been implemented. GenoCAD constrains the design space using the underlying syntactic models, and guides the user through the design process in a 'point and click' fashion. This has been extended recently with a second syntactic model, designed specifically for BioBrick^{TM}-based constructs [11].

The GenoCAD web tool applies these syntactic models to support both design and validation of sequences [15] (though at the moment when this chapter was written, the validation section was offline for development). The design space (Fig. 11.2.a) has two distinct sections: on the left hand side is the 'History Record' which keeps track of each design step, while the right hand side shows the current design. On top of the right hand side is an icon representation of the current design, which will evolve as the design proceeds. Underneath the icon representation is the main design space, where a user can point and click on a grammar rule to transform the design or to decide on a specific biological part for a category. After the design is finished (i.e., the structure is finalized, and all parts are selected), GenoCAD will offer the user an option to export the DNA sequence being designed.

If a DNA sequence is designed outside GenoCAD, it can be taken into the validation section (Fig. 11.2.b) to check whether the composition of biological parts is consistent with GenoCAD grammars. It should be noted that if a sequence fails in the validation, it does not necessarily mean this sequence is non-functional. Rather, it means that the GenoCAD grammar could not find a parsing tree to generate this sequence, and that sequence requires a closer inspection by human experts. On entering the validation page, a user firsts select a grammar to validate against, then pastes the DNA sequence into GenoCAD. The tool will then interpret the DNA sequence into a series of parts and (if successful) report whether this sequence has a correct structure as defined by the selected grammar.

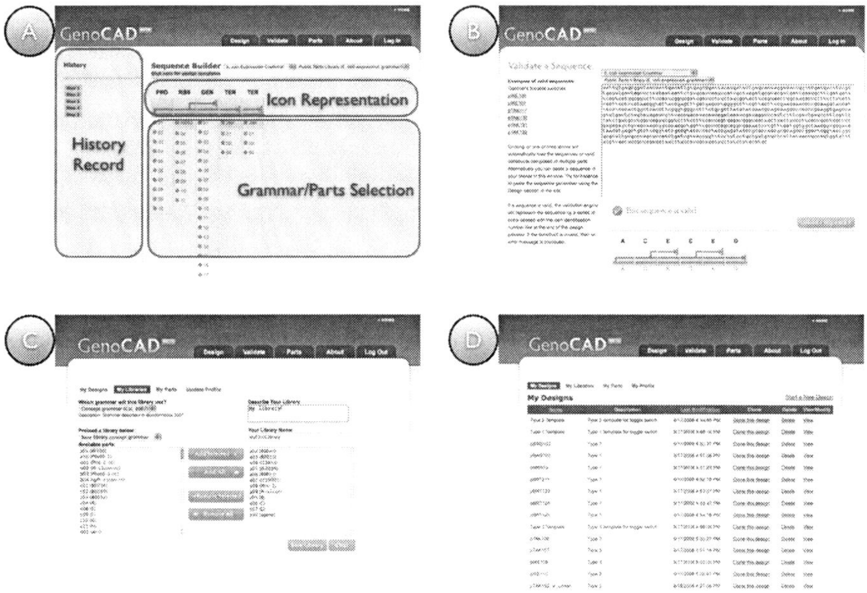

Fig. 11.2 Screenshots of the GenoCAD.org web service showing pages for sequence design (**a**), sequence validation (**b**), part library customization (**c**), and the user's workspace (**d**)

Finally, users who elect to register an account with the GenoCAD web tool have more privileges in customizing their design space in GenoCAD. A registered user can create new libraries (Fig. 11.2.d), add new parts, and save intermediate and final designs for later use (Fig. 11.2.d).

Eugene

Eugene (www.eugenecad.org) [22, 23] is a human readable, executable specification, which reflects the creation of biological systems by defining, specifying, and combining collections of biological parts. Eugene is inspired by the languages of the Electronic Design Automation (EDA) industry (e.g. Verilog, VHDL) in terms of its ability to provide a biological design netlist (a collection of abstract components and their connections) which can be synthesized (automatically transformed) into collections of physical implementations in a design library.

Eugene bridges the synthetic biology 'part' and 'device' (composite of multiple basic parts) hierarchy levels by explicitly addressing the components in different levels of the hierarchy. These relationships are explicitly reflected in Eugene's data types: **Device** and **Part** declarations abstract low-level implementation details (captured by **Property** statements), while still providing the capability to capture the lower level information through the encapsulation of specific design information with **Part Instance** objects composed of specific **Properties**. These features address the need for flexibility in biological part and device specification. Moreover, Eugene can directly interface with design tools like Clotho [8, 17] which extract information from repositories of biological parts and encapsulate that information as Eugene 'header files.' These files define specific instances of Parts and their Properties for a given 'design library.' These header files are modular and allow changes from one design library to the other with the inclusion of different files without modifying the Eugene Device declarations.

Eugene is also an executable specification since it is an interpreted language. At runtime, the Eugene interpreter can create collections of **Devices** based on conditional execution statements (e.g. **if**) coupled with specific functions to create new **Devices** at runtime. These features address the need for the combinatorial exploration of devices from a wide variety of different biological parts. For example, if a particular Part's Property does not meet a specific threshold, the body of the conditional statement can be used to swap that Part out with one that does meet the requirements.

Finally, Eugene allows for the creation and assertion of design rules. A **Rule** directly applies to the relationship between various **Parts** in a **Device** and provides the validation mechanisms needed to ensure the successful creation of a construct. These rules are not predefined in the language but rather created by the user from a rich set of rule primitives. Such flexibility allows users to define and assert numerous combinations of rules.

Eugene Constructs

The language supports five predefined primitive types. These are **txt**, **num**, **boolean**, **txt[]** (a list of text sequences), and **num[]** (a list of numbers). **Properties** represent characteristics of interest and are defined by primitives and associated with parts. The data type **Part** represents a standard biological part, such as a BioBrick^{TM}in the MIT Registry. **Part** definitions do not construct any parts, but rather specify which parts can be constructed. Declarations of those parts create instances of predefined **Parts** and assigns values to their properties. **Device** statements represent an ordered composite of standard biological parts and/or other devices. Below are examples of these constructs:

```
//Eugene Primitives
txt[] listOfSequences = ["ATG", "TCG", "ATCG"];
txt specificSequence = listOfSequences[2];
num[] listOfNumbers = [2.5, 10, 3.4, 6];
num ten = listOfNumbers[1];

//Eugene Properties
Property Sequence(txt);
Property RelativeStrength(num);
```

```
//Eugene Part Declarations
Part Promoter(ID, Sequence, Orientation);
Part ORF(ID, Sequence, Orientation);
Part RBS(ID, Sequence, Orientation);

//Eugene Part Instances
RBS rbs (.Sequence("gatcttaattgcggagacttt"),
 .Orientation("Forward"));
ORF orf (.Sequence("gatcttaattgcggagacttt"),
 .Orientation("Forward"));

//Eugene Device
Device BBa_K112234(rbs, orf);
```

Eugene Rules

The specification of rules provides the ability to validate **Device** declarations. **Rule** declarations themselves do not perform the validation. They have to be 'noted,' 'asserted' or used as expressions inside an if-statement to affect program operation. **Rule** declarations are single statements consisting of a left and right operand and one rule operator. The rule operators **BEFORE**, **AFTER**, **WITH**, **NOTWITH**, **NEXTTO**, **NOTCONTAINS**, **CONTAINS**, and **NOTMORETHAN** can be applied to **Part** instances or **Device** instances. These operators also have been defined with specific semantics as well (e.g. their commutative properties). Property values of **Part/Device** instances or primitives in relation with one **Part/Device** can be operators in rule declarations when using the relational operators $<$, $<=$, $>$, $>=$, $!=$, $==$. These operators are overloaded when evaluating text and the text is compared according to alphabetical precedence. The following are examples of rules in Eugene:

```
Rule r1(rbs BEFORE orf);
Rule r2(rbs WITH promoter);
Rule r3(promoter NEXTTO rbs);

Rule r4(rbs.Sequence != orf.Sequence);

Rule r5(rbsStrong.RelativeStrength > rbsWeak.
 RelativeStrength);

num relativeStr = rbsStrong.RelativeStrength;
Rule r6(p.RelativeStrength > relativeStr);

Assert(r6); // Strong enforcement of the rule
 (stop compilation)
Note(r4);  // Weak enforcement of the rule (warning)
```

Currently rules must be defined explicitly in the body of the Eugene program or in a header file. However work is in progress to examine ways to associate rules with Parts types and instances as well as generate constraints in response to experimental work done in laboratories which is fed back to Eugene at runtime. In addition, the automated assembly system j5 [29] uses Eugene constraints as part of its combinatorial exploration of alternate devices.

Eugene Functionality

The use of conditional statements breaks up the flow of execution and allows selected blocks of code to be executed. Eugene supports two kinds of **if** statements to achieve this: rule validating **if** statements and standard **if** statements. The three logical operators **AND**, **OR**, and **NOT** can combine statements of each type but cannot mix them together.

```
Rule r7(rbs BEFORE orf);
if(on (BBa_K112234) r7) {
  Block statement, in case of true evaluation
} else {
  Block statement, in case of false evaluation
}

boolean test = true;
if(test) {
  Assert(r7);
} else {
  Assert(NOT r7);
}
```

The **permute** function automates the specification of many **Devices** that share the same basic structure. Applying **permute** generates a **Device** for every combination of predefined **Parts**, maintaining the **Part** type of each component in the original **Device**. For example, the following code will result in eight devices at the completion of the permute operation.

```
Promoter p1(.Sequence("atc"));
Promoter p2(.Sequence("gcta"));
RBS rbs1(.Sequence("gatct"),.Orientation("Forward"));
RBS rbs2(.Sequence("gatcttaatt"), .Orientation("Forward"));
Device d2(p2, d1, rbs2);
permute(d2);
```

Permute also can be given additional parameters that limit the number of Devices created or force the Devices to adhere to the rules currently defined. The latter provides an intelligent design space exploration process. For example, *Permute*($d2, 4,$ *strict*) will create four Devices which adhere to the rules currently defined while maintaining the overall structure of Device d2.

XOR Design Example

We now show how Eugene can be applied to design our example XOR system, which only produces a green fluorescent protein (GFP) with either aTc or IPTG

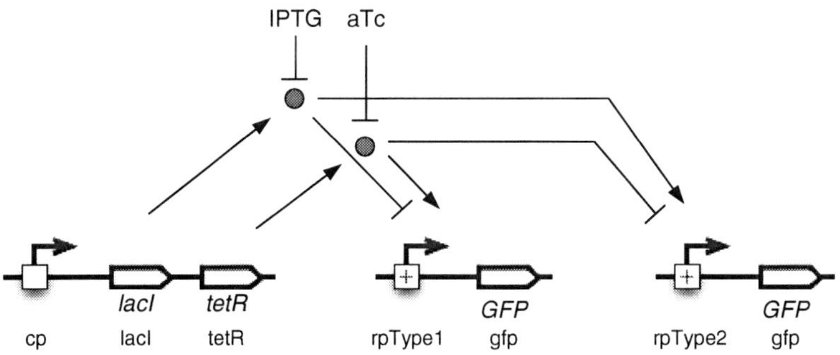

Fig. 11.3 XOR Gate design example designed with Eugene

(but not both) present. Figure 11.3 shows the proposed network of parts and the regulation present. The code snippet below shows what this design would look like in Eugene. Of note is that since Eugene is based around the specification of devices from individual parts, there is not a natural way to express small molecular interactions. These manifest themselves as properties of the parts. Control statements could then check these properties to create alternate networks reflecting the presence or absence of these molecules. If the proper DNA sequences are provided for all the parts, these interactions themselves would occur naturally in the physical device. The provided design merely captures the topology of the XOR device as an ordered collection of parts.

```
Property sequence(txt);
Property smallMoleculeInteraction(txt);

Property type(num);
//1 - neg regulated by lacI, pos regulated by tetR
//2 - neg regulated by tetR, pos regulated by lacI

Part ConstitutivePromoter(sequence);
Part RegulatedPromoter(sequence, type);

Part ORF(sequence, smallMoleculeInteraction);

ConstitutivePromoter cp("ACGT...");
RegulatedPromoter rpType1("ACGT...", 1);
RegulatedPromoter rpType2("ACGT...", 2);

ORF gfp("ACGT...", "none");
ORF lacI("ACGT...", "IPTG");
ORF tetR("ACGT...", "aTc");

Device xor(cp, lacI, tetR, rpType1, gfp, rpType2, gfp);
```

Notice that here no rules are actually specified. However, were this design actually given to a downstream tool chain for automated assembly, one would want to create many potential devices in case the provided device either fails to function or assemble. Potential rules could be:

```
//This places the ConstitutivePromoter before lacI and tetR
Rule cpLocation1(cp BEFORE lacI);
Rule cpLocation2(cp BEFORE tetR);
Assert(cpLocation1 AND cpLocation2);

//Ensures that only two RegulatedPromoters are in the system
Rule UniquePromoter1(rpType1 NOTMORETHAN 1);
Rule UniquePromoter2(rpType2 NOTMORETHAN 1);
Assert (UniquePromoter1 AND UniquePromoter2);
```

For the sake of space, all the rules are not listed here but one should specify the relationship between the gfp ORF part and the RegulatedPromoters as well. This would be followed by a *Permute(xor, strict)* function call which would create a variety of devices (e.g. with the position of the lacI and tetR parts swapped). These devices would then be given to an automated assembly program [4] for downstream use with laboratory automation.

GEC

This section describes a programming language for Genetic Engineering of Cells (GEC), initially presented in [31] and available at http://research.microsoft.com/gec. The main goal of GEC is to facilitate the design, analysis and implementation of biological devices inside living cells. GEC builds on previous research in the field of synthetic biology, including a registry of standard parts (http://partsregistry.org) together with experimental techniques for combining these parts into higher-level devices. More recently, a range of software tools have been developed for designing and simulating biological devices, as discussed for example in [31, 33]. The main innovation behind GEC is to take the design process a step further, by allowing biological devices to be designed with little or no knowledge of the specific parts available. The user needs only a basic knowledge of the available *part types*, namely *promoters*, *ribosome bindings sites*, *protein coding regions* and *terminators*. These elementary part types can be composed and the *properties* of the desired parts can be expressed as constraints in the GEC language. Once a biological device has been designed in this way, the GEC compiler automatically determines the set of actual parts that satisfy the design constraints. In most cases, multiple solutions are possible for a given design. GEC can compile each of the solutions to a set of chemical reactions, which can then be simulated or analyzed by the user. The solutions that exhibit the desired behavior can then be synthesized and put to work in living cells. Although there is no guarantee that a solution which produces the desired simulation results will function correctly inside a living cell, analyzing the design on a computer is an effective way to rapidly detect design errors prior to building the physical device – a process which can take several days and for which even small errors can prove very costly.

a

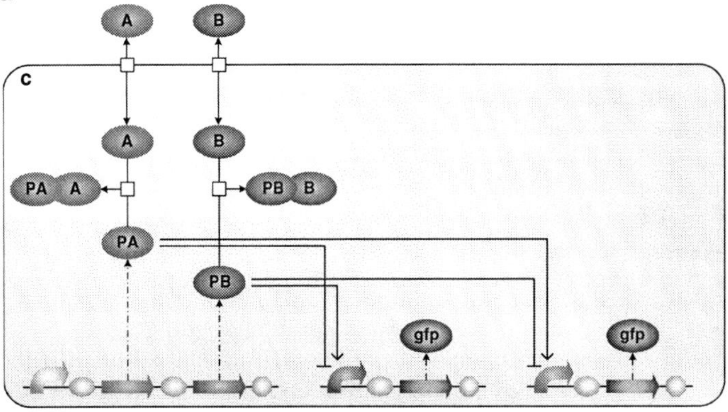


```
directive sample 2000.0 1000
directive plot c[gfp]
initPop A 10000.0 | initPop B 10000.0 |
c[
prom<con(RT)>;rbs;pcr<codes(PA)>;rbs;pcr<codes(PB)>;ter ;
prom<neg(PA),pos(PB)>;rbs;pcr<codes(gfp)>;ter ;
prom<neg(PB),pos(PA)>;rbs;pcr<codes(gfp)>;ter |
PA + A -> PA-A | PB + B -> PB-B | RT > 0.1
] | A -> c[A] | B -> c[B]
```

b

[("A", "iptg"); ("B", "aTc"); ("PA", "lacI"); ("PB", "tetR")]

[[r0051; b0034; c0012; b0034; c0040; b0015;
rU2; b0034; e0040; b0015;
rU1; b0034; e0040; b0015]]

Fig. 11.4 Designing an exclusive OR (XOR) logic gate in GEC. (**a**) GEC code for the XOR gate, together with its graphical representation, expressed in terms of part types, part properties and logical variables. Note that none of the part identifiers are specified explicitly. The design yields a number of possible solutions. (**b**) One of the solutions proposed by the GEC tool, expressed as a mapping from logical variables to molecules, together with a list of the part identifiers that make up the design

We illustrate the design approach of the GEC language on a simple exclusive OR (XOR) logic gate (Fig. 11.4). The system is specified as a collection of three transcriptional units, where each unit consists of a sequences of part types. The first transcriptional unit consists of a promoter (prom), a ribosome binding site (rbs) a protein coding region (pcr), followed by another ribosome binding site and protein coding region, followed by a terminator (ter):

```
prom; rbs; pcr; rbs; pcr; ter
```

Additional constraints on part types are specified in the form of part properties. In the first transcriptional unit, the part prom<con(RT)> denotes a promoter with a constitutive transcription rate RT, the part pcr<codes(PA)> denotes a protein coding region that codes for protein PA, and the part pcr<codes(PB)> denotes a protein coding region that codes for protein PB:

```
prom<con(RT)>; rbs; pcr<codes(PA)>; rbs; pcr<codes(PB)>; ter
```

The transcription rate RT and the proteins PA and PB start with an upper case letter, which means that they are *logical variables* representing an unknown rate and unknown proteins. Although the values of these variables are not known in advance, the GEC compiler takes into account the full set of design constraints in order to find suitable values that satisfy the desired properties. For example, the property RT > 0.1 states that the constitutive transcription rate of the promoter must be above a certain threshold. In the second transcriptional unit the part prom<neg(PA),pos(PB)> denotes a promoter region that is negatively regulated by protein PA and positively regulated by protein PB:

```
prom<neg(PA),pos(PB)>; rbs; pcr<codes(gfp)>; ter
```

This places additional constraints on the proteins PA and PB, which must act as a positive and negative regulator, respectively. The third transcriptional unit places further constraints on the proteins PA and PB, which must now act as both positive and negative regulators simultaneously:

```
prom<neg(PB),pos(PA)>; rbs; pcr<codes(gfp)>; ter
```

Note that the protein gfp starts with a lower case letter, meaning that it represents a known protein.

In order to map logical variables and design constraints to physical parts, GEC includes a database of parts. Each of the parts in the database is associated with a part identifier together with zero or more part properties. A subset of a GEC parts database is shown in Table 11.3. The part properties are also associated with rate constants, which are used to simulate the design solutions. For example, the

Table 11.3 A subset of the GEC parts database, which can be defined and extended by the user. Each of the parts in the database is associated with a part identifier together with zero or more part properties

ID	Type	Properties
e0040	pcr	codes(gfp, 0.01)
c0012	pcr	codes(lacI,0.01)
c0040	pcr	codes(tetR, 0.01)
b0034	rbs	rate(0.1)
b0015	ter	
r0051	prom	neg(cI, 1.0, 0.5, 0.00005), con(0.12)
r0040	prom	neg(tetR, 1.0, 0.5, 0.00005), con(0.09)
rU1	prom	neg(tetR,1.0,0.01,0.0), pos(lacI,1.0,0.5,0.1), con(0.0)
rU2	prom	neg(lacI,1.0,0.01,0.0), pos(tetR,1.0,0.5,0.1), con(0.0)
rU3	prom	neg(tetR,1.0,0.5,0.0), pos(lacI,1.0,0.5,0.1), con(0.0)

database entry (c0040↦ pcr,codes(tetR, 0.01)) denotes a protein coding region c0040, which codes for the protein tetR with degradation rate 0.01. The entry (r0051↦prom,neg(cI,1.0,0.5, 0.00005),con(0.12)) denotes a promoter r0051 that is negatively regulated by the protein cI, which binds to the promoter at rate 1.0 and unbinds at rate 0.5, where the repressed transcription rate is 0.00005 and the constitutive transcription rate is 0.12.

The design of the XOR gate in Figure 11.4 on page 237 also includes interactions between proteins and transport reactions across the cell membrane. The following constraints require that the protein `A` binds to `PA` and forms a complex `PA-A`, and that the protein `B` binds to `PB` and forms a complex `PB-B`. A vertical bar is used to separate multiple constraints:

```
PA + A -> PA-A | PB + B -> PB-B
```

This effectively specifies that the inputs `A` and `B` to the XOR gate can inhibit the activity of the transcription factors `PA` and `PB` by forming inert complexes with these transcription factors. Finally, the following properties require that both `A` and `B` are able to cross the cell wall:

```
A -> c[A] | B -> c[B]
```

These properties are essential in order for the input signals of the XOR gate to be read by the cell. In order to map these reaction constraints to physical parts, the GEC system includes a database of reactions. Each of the reactions in the database consists of a set of reactants, a set of products and a corresponding reaction rate. A subset of a GEC reactions database is shown in Table 11.4. For example, the reaction (lacI + iptg → {1.0} lacI-iptg) denotes the formation of a complex between lacI and iptg. In many cases accurate rate information for these reactions is missing, and approximate rate constants are used instead.

The above design constraints for the XOR gate are solved by the GEC compiler in order to find an appropriate solution. For example, the first protein coding region of the first transcriptional unit must produce a protein `PA` that can both inhibit the promoter of the second transcriptional unit, activate the promoter of the third transcriptional unit and also form a complex with a compound that is capable of crossing the cell membrane. In the general case multiple solutions are possible for a given design. One of the possible solutions is shown in Figure 11.4. The solution maps the inputs `A` and `B` to iptg and aTc respectively, and the transcription factors `PA` and `PB` to lacI and tetR, respectively. The corresponding part identifiers are also listed, which denote specific nucleotide sequences that could potentially be inserted inside a bacterium in order to program an XOR gate.

Table 11.4 A subset of the GEC reactions database, which can be defined and extended by the user

Reactants	Rate	Products
lacI + iptg	1.0	lacI-iptg
tetR + aTc	1.0	tetR-aTc
iptg	1.0	c[iptg]
aTc	1.0	c[aTc]

The main characteristic of the XOR gate is that green fluorescent protein (GFP) is only produced when one of the input signals A or B is present, but not both. When the user compiles the XOR gate design in GEC, they are presented with a set of possible solutions that satisfy the design constraints. The user can then simulate each of the solutions in order to choose the most desirable one. The design can be further refined by specifying that certain rates such as transcription, translation or transcription factor binding must lie within a specified range. This helps to reduce the initial set of possible solutions. In the case of the XOR design, one of the solutions represents a condition whereby GFP is produced even in absence of both inputs A and B. This occurs because the rate of repression of one of the promoters by transcription factor PA is less than its rate of activation by transcription factor PB, meaning that activation out-competes inhibition. This unwanted solution can be eliminated by adding the constraint that the inhibitor transcription factors bind more tightly than the activator transcription factors.

In order to simulate a given design, GEC automatically compiles the design to a set of chemical reactions, using the rates associated with the part properties and reactions in the GEC databases. The set of reactions for the XOR gate design is summarized in Fig. 11.7, and simulation of the reactions is shown in Fig. 11.5. Additional details about the compilation to reactions are provided in [31], and a screen shot of the tool is shown in Figure 11.6.

In this section we have illustrated the design of genetic devices in GEC using a simple XOR gate as an example. In order to effectively design more complex devices, however, further work is needed to characterize the properties of individual parts. At present only a few parts are well-characterized and many reaction rates are unknown, so the part and reaction databases described here do not yet exist on a large scale. As one potential consumer of such databases, GEC may help guide how these are designed and populated with information about biological devices.

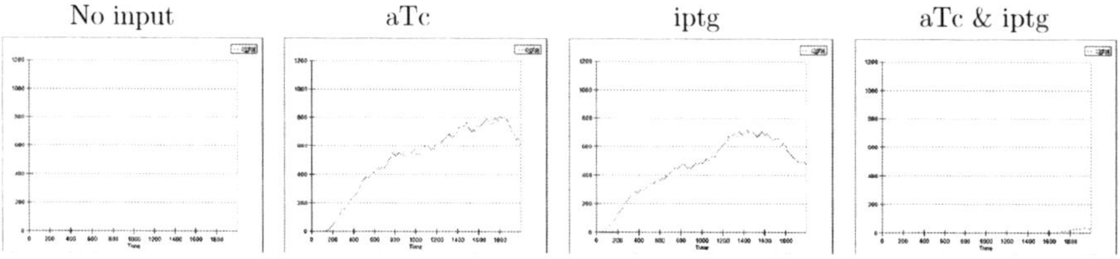

Fig. 11.5 Simulation of gfp concentration over time for an exclusive OR gate in GEC, with four combinations of inputs. The simulation uses the chemical reactions of Fig. 11.7, which were automatically generated from the chosen solution of Figure 11.4 on page 237. The solution exhibits the desired behavior and is a candidate for synthesis

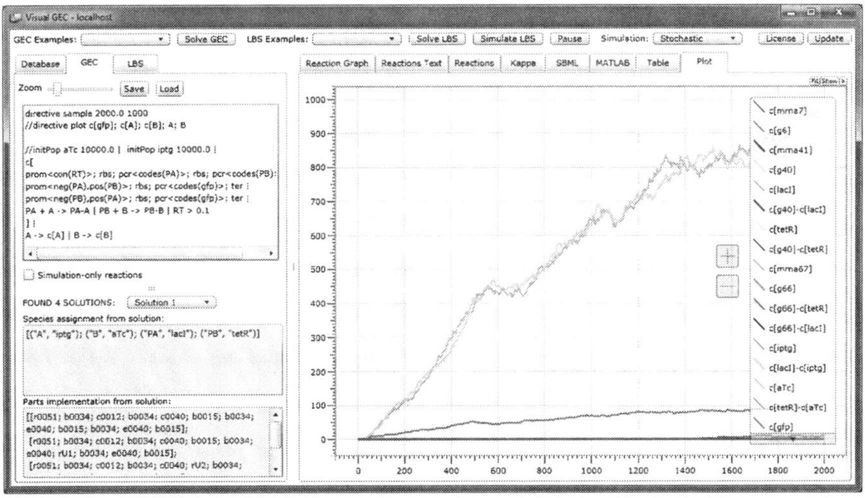

Fig. 11.6 Screen shot of the GEC tool in action. The GEC program is entered on the left as a collection of part types, part properties and logical variables. The design is then compiled to a set of solutions, which can be individually selected. A given solution can then be simulated by the tool in order to observe the expected evolution of the molecular species over time

Proto

Proto is a truly high-level language for synthetic biology, in the sense that a programmer specifies the computation they wish to execute, but the implementation of that computation as a genetic regulatory network is entirely automated. This greatly increases the power of the programmer, at the cost of programs that typically consume more resources than hand-tuned systems. The same sort of optimization techniques that apply to conventional processors, however, can be applied to the genetic regulatory networks generated by Proto, making this a reasonable approach to designing complex synthetic biology systems.

Amorphous Medium and Proto

The original focus of Proto [4] was not synthetic biology, and synthetic biology is still not its primary focus. Rather, it was designed for programming spatial computers – potentially large aggregates of locally communicating computing devices distributed to fill a physical space, such as sensor networks, robotic swarms, smart materials, or FPGAs. A colony of cells is also a spatial computer – albeit one that may have billions or trillions of devices, rather than the paltry dozens in many sensor networks. Proto's continuous space-time abstraction lets it scale gracefully to such large numbers and its functional dataflow semantics match well with genetic regulatory networks, particularly for describing the spatial differentiation necessary to construct complex multicellular systems like biofilms or tissues.

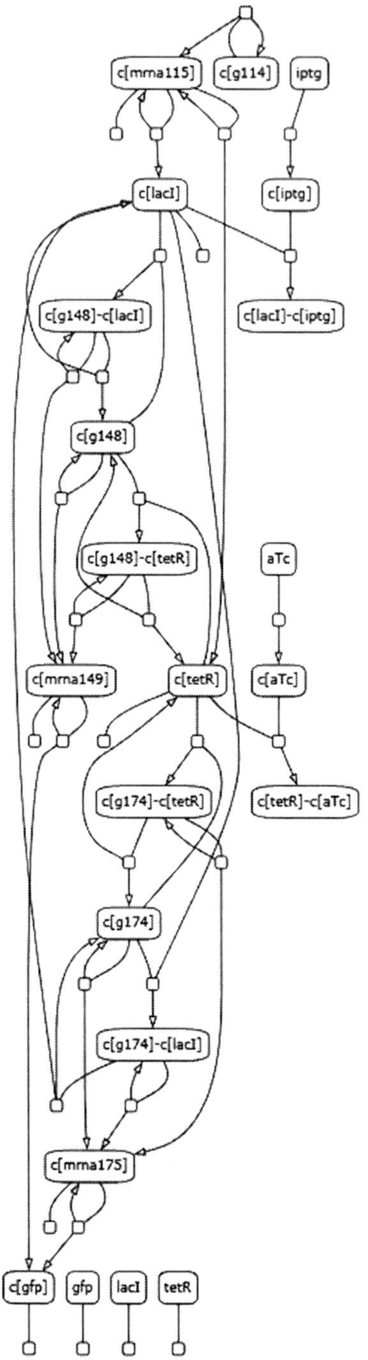


```
directive sample 2000.0 1000
directive plot c[gfp]
rate mDeg = 0.001;
init aTc 10000 |
init iptg 10000 |
c [
init glacI_tetR 1 |
mlacI_tetR ->{mDeg} |
glacI_tetR->{0.12} glacI_tetR +
mlacI_tetR |
init ggfp1 1 |
mgfp1 ->{mDeg} |
ggfp1 ->{0.0} ggfp1 + mgfp1 |
ggfp1+ lacI ->{1.0} ggfp1-lacI |
ggfp1-lacI ->{0.01} ggfp1 + lacI |
ggfp1-lacI ->{0.0} ggfp1-lacI +
mgfp1 |
ggfp1+ tetR ->{1.0} ggfp1-tetR |
ggfp1-tetR ->{0.5} ggfp1 + tetR |
ggfp1-tetR ->{0.1} ggfp1-tetR +
mgfp1 |
init ggfp2 1 |
mgfp2 ->{mDeg} |
ggfp2 ->{0.0} ggfp2 + mggfp2 |
ggfp2+ tetR ->{1.0} ggfp2-tetR |
ggfp2-tetR ->{0.01} ggfp2 + tetR |
ggfp2-tetR ->{0.0} ggfp2-tetR +
mgfp2 |
ggfp2+ lacI ->{1.0} ggfp2-lacI |
ggfp2-lacI ->{0.5} ggfp2 + lacI |
ggfp2-lacI ->{0.1} ggfp2-lacI +
mgfp2 |
lacI + iptg ->{1.0} lacI-iptg | tetR
+ aTc ->{1.0} tetR-aTc |
mgfp2 ->{0.1} mgfp2 + gfp |
mgfp1 ->{0.1} mgfp1 + gfp |
mlacI_tetR ->{0.1} mlacI_tetR +
lacI | mlacI_tetR ->{0.1}
mlacI_tetR + tetR
] |
iptg->{1.0} c[iptg] |
aTc->{1.0} c[aTc] |
c [
gfp ->{0.01} |
lacI ->{0.01} |
tetR ->{0.01}
]
```

Fig. 11.7 Network of reactions generated from the design of Figure 11.4 on page 237. The graphical representation on the left was also generated by the GEC tool, and is equivalent to the textual representation on the right

Fig. 11.8 An amorphous medium is a manifold where every point is a general computational device that knows its neighbors' recent past state

Proto's approach to the challenges of spatial computing is to focus not on the network of devices, but on the continuous space that they occupy, using the *amorphous medium* abstraction. An amorphous medium [3] is a manifold with a general computational device at every point, where each device knows the recent past state of all other devices in its neighborhood (Fig. 11.8). While an amorphous medium cannot, of course, be constructed, it can be approximated on the discrete network of a spatial computer.

Proto uses the amorphous medium abstraction to factor programming a spatial computer into three loosely coupled subproblems: global descriptions of programs, compilation from global to local execution on an amorphous medium, and discrete approximation of an amorphous medium by a real network.

Proto is a functional language that is interpreted to produce a dataflow graph of operations on fields. This program is then evaluated against a manifold to produce a field with values that evolve over time. Proto uses four families of operations: pointwise operations like + that involve neither space nor time, restriction operations that limit execution to a subspace, feedback operations that establish state and evolve it in continuous time, and neighborhood operations that compute over neighbor state and space-time measures and summarize the computed values in the neighborhood with a set operation like integral or minimum.

With appropriate operators, compilation and discrete approximation are straightforward. Thus, Proto makes it easy for a programmer to carry out complicated spatial computations using simple geometric programs that are robust to changes in the network and self-scale to networks with different shape, diameter, density of nodes, and execution and communication properties [1].

For example, Weiss' band detector [2] uses diffusing AHL to detect intermediate distance from a high aTc concentration. This can be implemented using the Proto program:

```
(def band-detect (signal lo hi)
  (and (> signal lo) (< signal hi))))

(let ((signal (diffuse (aTc) 0.8 0.05)))
  (green (band-detect signal 0.2 1)))
```

where **aTc** is a function for sensing aTc and **green** is an actuator that sets the level of GFP expression. Figure 11.9 (from [5]) shows the Proto band detector program interpreted to produce a dataflow graph, then evaluated against an irregularly shaped space. Executing the Proto band detector in simulation produces results

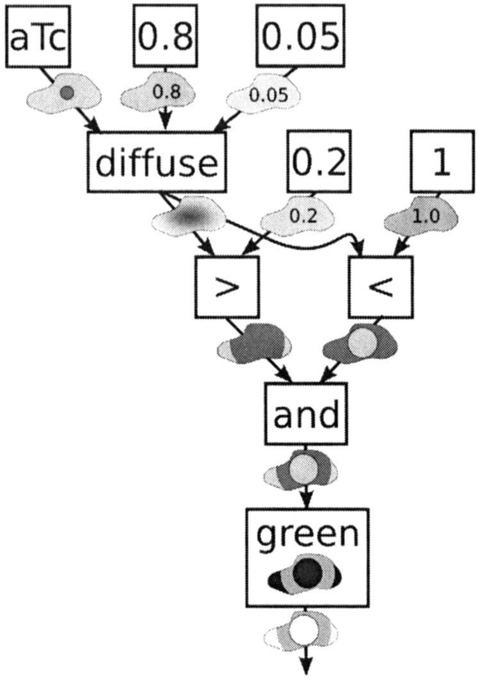

Fig. 11.9 A Proto program specifies a dataflow graph of operations on fields. When evaluated on a space, each operation produces a field of values over that space. Here the band detector program is shown evaluated on an irregularly shaped space, with values shown as varying levels of grey. The actuation produced by **green** is shown inside that operation

equivalent to Weiss's band detector. Figure 11.10 (from [5]) compares execution on a network of 2,000 simulated wireless devices distributed randomly through a 100 by 100 unit region with a 10 unit communication radius to Weiss' original results.

Motif-Based Compilation and Optimization

Given a library of devices and standards to compile to, Proto programs can be transformed into genetic regulatory network designs by a process of motif-based compilation [5, 6]. The resulting design can then be optimized using adapted forms of standard code optimization techniques.

The basis of this compilation are associations of each Proto primitive to be compiled with a genetic regulatory network fragment. These are declared as annotations on primitives. For example, the logical **not** operator is associated with a biological inverter motif by the statement shown in Fig. 11.11. The first line declares the **not** operator as a primitive that takes a boolean as input and returns a boolean as output. The second line annotates this declaration with a description of a genetic regulatory region – in this case, a strong promoter repressed by whatever protein will represent the **not** operator's input, followed by coding regions for the proteins representing its outputs (each of which is implicitly headed by the necessary ribosome binding site), then finally a terminator.

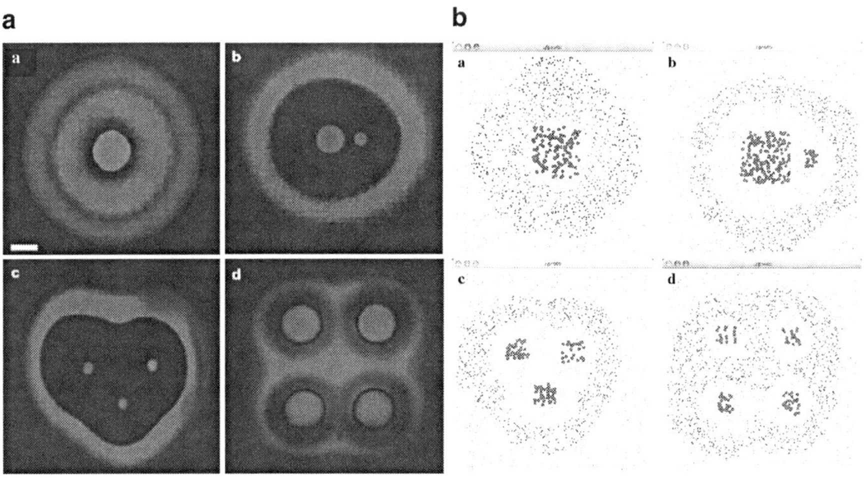

Fig. 11.10 Examples of the Weiss lab band detector in use (**a**), reprinted by permission from Macmillan Publishers Ltd: Nature [2], copyright (2005). The circular regions in the center are active sender bacteria, while the fuzzy areas around them are receiver bacteria expressing fluorescent protein. A Proto implementation produces equivalent results (**b**) on a network of 2,000 simulated devices

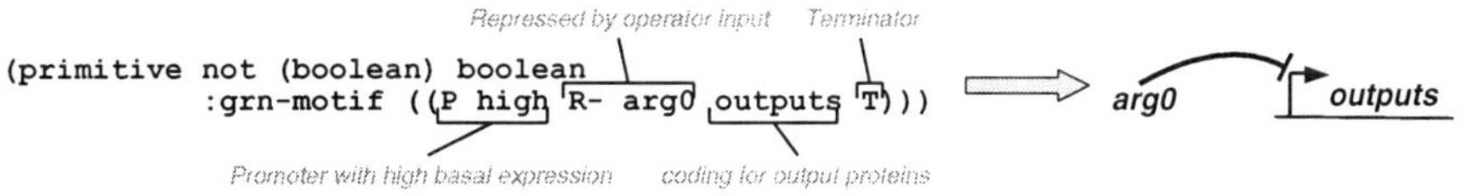

Fig. 11.11 Motif declaration for logical **not** operator

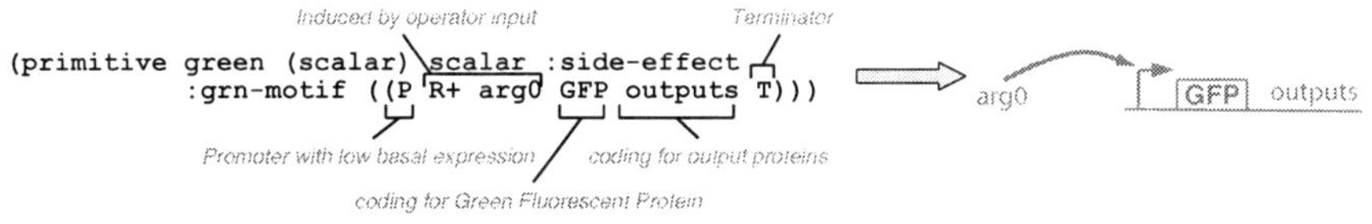

Fig. 11.12 Motif declaration for green fluorescence actuator

Motifs can include many other elements as well. For example, a motif can specify particular chemicals to be used, as in the case of the **green** actuator shown in Fig. 11.12, whose green fluorescence side effect is implemented by the inclusion of a GFP coding region in the motif. Motifs can also include chemical reactions, as in the case of the **IPTG** sensor shown in Fig. 11.13, which uses repression of LacI to detect the presence of the small-molecule signal IPTG. They may even declare internal signaling variables, to be filled in by the compiler, as in the case of the **and** operator shown in Fig. 11.14, which implements a non-branching logical AND using inverter input to a NOR gate.

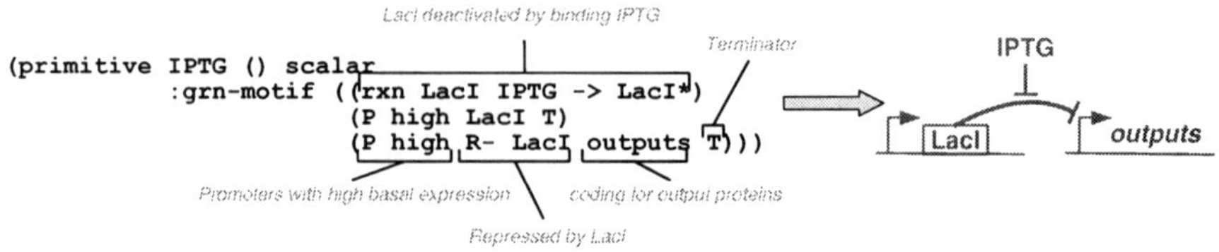

Fig. 11.13 Motif declaration for IPTG sensor. An aTc sensor uses the same motif, except that aTc replaces IPTG and TetR replaces LacI

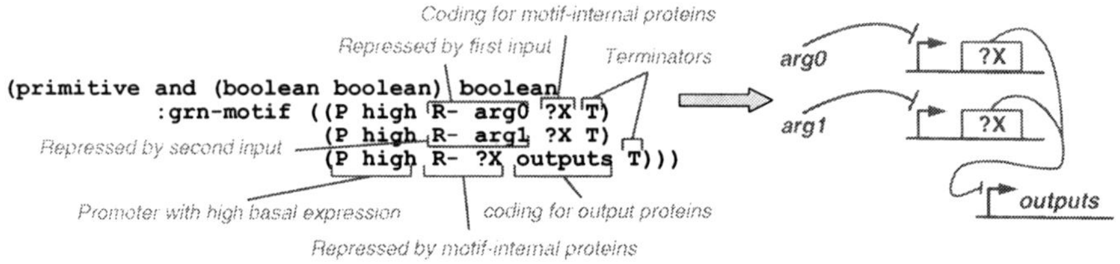

Fig. 11.14 Motif declaration for a non-branching logical AND operator. A logical OR uses the same motif, except that all repressors are switched to activators and promoters have low base activity

In order to transform a Proto dataflow computation into an abstract genetic regulatory network, the compiler maps each operator to its associated motif and maps each dataflow edge and internal motif variable to a regulatory protein. These motifs and proteins are then linked together, using the structure of the dataflow graph, to form an abstract genetic regulatory network. The particular choice of chemicals and sequences to implement this network is not fully determined, but left for a later stage of compilation, such as might be provided by systems like GEC [31] or Eugene [7]. An initial set of target chemical rate constants for the network (to be modified as the implementation is determined) are filled in from the motifs where specified and filled in by a default set-point in the standards family where not specified.

Consider, for example, the following declaration and use of logical XOR to implement our example program:

```
(def xor (a b)
  (or (and a (not b))
      (and b (not a))))

(green (xor (aTc) (IPTG)))
```

This program should create cells that fluoresce green when precisely one of IPTG or aTc is present at high concentration.

This program is first interpreted to produce the dataflow computation shown in Fig. 11.15. Each operator is then mapped to the motifs specified by the declarations shown above, producing the network shown in Fig. 11.16. The dataflow edges are assigned to arbitrary regulatory proteins *A*, *B*, etc. The consuming motifs set the type of protein, such that *A* and *B* are activators, *C* is a repressor, etc.

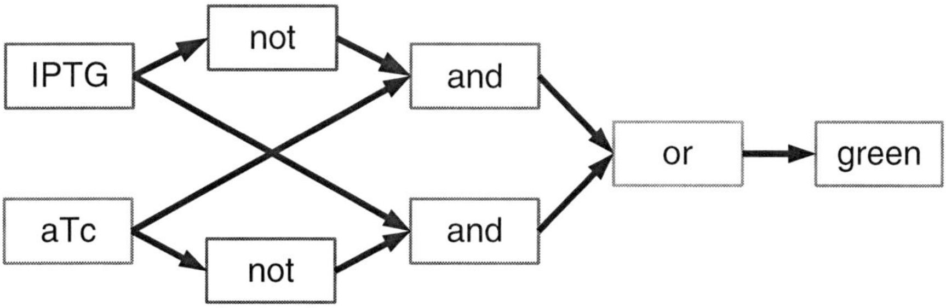

Fig. 11.15 A Proto dataflow computation implementing the XOR example program

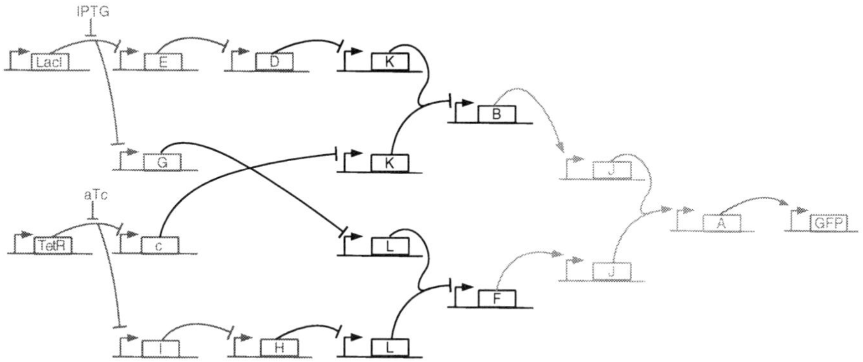

Fig. 11.16 A Proto dataflow computation is compiled to an abstract genetic regulatory network in two stages. First, each operator is mapped to a motif and each dataflow edge is mapped to a regulatory protein. These elements are then linked together, using the structure of the dataflow graph, to form an abstract genetic regulatory network

We now have a genetic regulatory network design that implements our high-level computation, though as yet it is still unoptimized and may be extremely inefficient. As we have demonstrated in [5], standard code optimization techniques such as copy propagation, dead code elimination, and common subexpression elimination, can be adapted to operate on genetic regulatory networks.

For example, copy propagation tests whether a protein is being used only to copy a value; if so, the original input may be used directly rather than the copy. In this case of this XOR program, copy propagation changes the input of the GFP-expressing element from A to J. This then leaves protein A not regulating anything. Similarly, copy propagation switches the regulation of J from B to K and from F to L.

Dead code elimination deletes proteins that are not regulating anything, network elements with no products, and proteins that can never be expressed. Since protein A is no longer regulating anything, it is deleted, along with all of the protein coding sequences that can produce it. Since A was the only product of one of the network elements regulated by J, that whole network element is deleted. Likewise, B and F and their producing elements are deleted by dead code elimination.

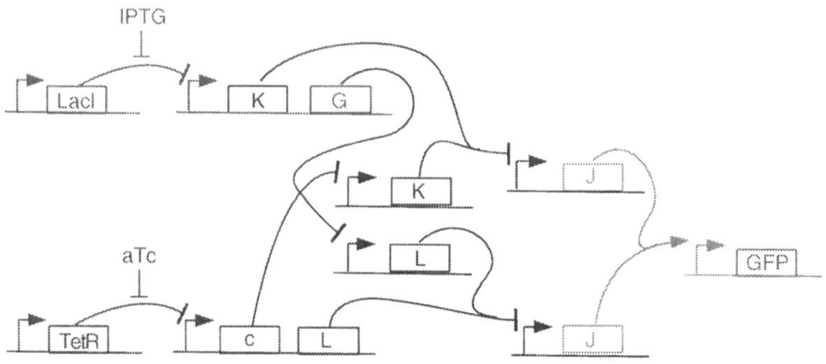

Fig. 11.17 Optimized genetic regulatory network for XOR example

Another example of optimization is double negative elimination, which looks for sequences of two inverters and snips them out of the network. In the case of this XOR program, this results in changing the production of E to production of K, since E's only use is to repress D, which in turn represses K. Similarly production of I is changed to production of L. This leaves I and E produced nowhere and D and H unable to be expressed, so dead code elimination deletes another piece of unneeded genetic regulatory network.

These optimizations and more are all applied automatically by the compiler, eventually resulting in the network shown in Fig. 11.17. All told, the complexity of the generated network is reduced by approximately 50% in every measure of complexity: from 15 to 8 proteins, from 18 to 9 network elements, and from 7 to 4 stages of propagation delay.

We thus see that high-level computations specified in Proto can be automatically transformed into an abstract genetic regulatory network through a strategy of motif-based compilation. The resulting genetic regulatory network can be optimized using adapted forms of standard code optimization techniques, and might then be mapped onto particular parts from a database using lower-level languages like Eugene or GEC. Although the network is more complex than a hand-optimized design like those encoded in the other tools above, stronger optimizations will likely be able to continue to close the gap, as they have for electronic computers.

Other High-Level Design Tools for Biological Computation

We have chosen to focus this chapter on high-level programming languages for *in vivo* biomolecular computation, where the metaphor of cell as computer holds most strongly. We have covered most of the existing tools in this space, omitting only a few, such as [21], that are quite similar to those described. There are a number of related areas outside of this scope, however, in which high-level design tools for biomolecular systems have been developed, which we now briefly survey.

Macroizing CAD Tools. A number of synthetic biology design tools, such as TinkerCell [12] and SynBioSS [20], use biological rules to aid the programmer in designing reaction networks. For example, SynBioSS (the Synthetic Biology Software Suite) is a software suite for the generation, storage, and quantitative simulation of synthetic biological networks. One component of this software suite, called SynBioSS Designer, uses biological rules to create a reaction network given a series of biological parts, such as promoters and ribosome binding sites, and the spatial and temporal connectivity of these parts.

These systems also frequently include the ability to abstract a portion of the network being designed. This type of 'macroization' is a step toward a high-level language: the details of the abstracted portion are hidden and it can be given a name that describes its overall function. The programmer must still be aware of the details, however, since the set of parts in the abstracted sub-network are fixed and can interfere with other portions of the design.

Specialized Automated Design Tools. Complementary to Macroizing CAD tools are specialized automated design tools, which might be thought of as limited high-level languages. An example is the boolean circuit design tool recently described in [25]. Given a truth table mapping inputs to desired outputs, this tool applies the Karnaugh map method from electronic circuit design to find a minimal set of boolean formulas, then maps these formulas onto a library of established biomolecular boolean gates.

Cell-Free Biomolecular Computation. A number of biomolecular computation systems have been constructed to operate in cell-free *in vitro* environments, and the design challenges for many of these systems are being addressed with high-level design tools. For example, the VERB compiler [35] transforms circuit designs written in Verilog into a biochemical reaction network, and CAD tools have been written to generate DNA origami structures [34].

Bio-Inspired Languages. There are a number of biologically-inspired languages that have been designed to mimic the behavior of engineered biological systems. For the most part, these are at a level of abstraction too high to currently be able to map to a biomolecular systems implementation, though Weiss' Microbial Colony Language [38] is close. Many of these languages are focused on pattern formation, such as the Origami Shape Language [28], which develops geometric structure through folding, and the Growing Point Language [14], which develops topological structure through tropisms. Yet others either model high-level biological development without connection to the details necessary to implement it, as in the case of L-systems [32] and MGS [19], or use biological metaphors for decidedly non-biological programming, as in the case of membrane computing [30].

Modeling Languages. Biological modeling languages such as Antimony [36], ProMoT [26], iBioSim [27] and little b [24] raise the level of abstraction for constructing models of biomolecular reactions, but do not directly address the problem of designing computations. For example, Antimony is a modular model definition language that allows scientists to define and use reaction networks. It is designed to be human-writable and acts as an extension to other tools by translating the model to SBML [18]. Antimony models composable DNA parts and also allows reaction networks to be abstracted and parameterized, but does not provide any design automation for its user.

Summary

In this chapter, we have examined four high-level languages for the design of biomolecular computing systems. Although the philosophy and the level of abstraction varies between systems, all are fulfilling the same basic goal of hiding complexity from the programmer. Each thus allows a programmer to specify the computing system they wish to create without the full details of how it will be implemented, then automatically generates the remaining details.

At present, none of the available high-level languages can be considered mature. They are, however, an important and rapidly developing research area. Major challenges in the near future for this area include:

- Development of concise high-level abstractions that map well to efficient biomolecular implementations of a broad range of goals.
- Enhancing the range and quality of automation.
- Integration with other simulation, design, and assembly tools to form complete tool-chains.
- Transitioning from research software to production quality software.

Assuming that progress continues in these areas, however, the advent of high-level programming languages for biomolecular systems is likely to fundamentally transform the field, much as they have done in computer science, by enabling much more complex biomolecular systems to be designed more reliably by a vastly larger number of practitioners.

References

1. Bachrach J, Beal J, Fujiwara, T (2007) Continuous space-time semantics allow adaptive program execution. In proceedings of IEEE International conference on Self-Adaptive and Self-Organizing Systems (SASO), Cambridge, MA
2. Basu S, Gerchman Y, Collins CH, Arnold FH, Weiss R (2005) A synthetic multicellular systems for programmed pattern formation. Nature 434:1130–1134
3. Beal J (2005) Programming an amorphous computational medium. In: Banâtre J-P, Fradet P, Giavitto J-L, Michel O (eds) Unconventional programming paradigms international workshop: lecture notes in computer science, vol 3566. Springer, Berlin/Heidelberg, p 97
4. Beal J, Bachrach J (2006) Infrastructure for engineered emergence in sensor/actuator networks. IEEE Intell Syst 21(2):10–19
5. Beal J, Bachrach J (2008) Cells are plausible targets for high-level spatial languages. Spatial computing workshop. Venice, Italy

6. Beal J, Lu T, Weiss R (2010) Automatic compilation from high-level languages to genetic regulatory networks. In: Proceedings of 2nd International Workshop on Bio-Design Automation, Anaheim, CA
7. Berkeley Software 2009 iGem Team (October 2009) Eugene. http://2009.igem.org/Team:Berkeley_Software/Eugene. Accessed 10 May 2010
8. Bing X, Swapnil B, Ben B, Maisam D, Douglas D, Anderson JC (2011) Clotho: a software platform for the creation of synthetic biological systems, A developer's and user's guide for clotho v2.0, Methods in enzymology, vol 498 (in press)
9. Cai Y, Hartnett B, Gustafsson C, Peccoud J (2007) A syntactic model to design and verify synthetic genetic constructs derived from standard biological parts. Bioinformatics 23(20):2760–2767
10. Cai Y, Lux MW, Adam L, Peccoud J (2009) Modeling structure-function relationships in synthetic dna sequences using attribute grammars. PLoS Comput Biol 5(10):e1000529
11. Cai Y, Wilson ML, Peccoud J (2010) Genocad for igem: a grammatical approach to the design of standard-compliant constructs. Nucleic Acids Res 38(8):2637–2644
12. Chandran D, Bergmann F, Sauro H (2009) Tinkercell: modular cad tool for synthetic biology. J Biol Eng 3(1):19
13. Chomsky N (1956) Three models for the description of language. IRE Trans Inf Theory 2(3):113–124
14. Coore D (1999) Botanical computing: a developmental approach to generating inter connect topologies on an amorphous computer. PhD thesis, MIT, Cambridge, MA, USA
15. Czar MJ, Cai Y, Peccoud J (2009) Writing dna with genocad. Nucleic Acids Res 37(Web Server issue):W40–W47
16. Densmore D, Hsiau TH-C, Kittleson JT, DeLoache W, Batten C, Anderson JC (2010) Algorithms for automated dna assembly. Nucleic Acids Res 38(8):2607–2616
17. Densmore D, Van Devender A, Johnson M, Sritanyaratana N (2009) A platform-based design environment for synthetic biological systems. TAPIA '09: the fifth Richard Tapia celebration of diversity in computing conference, ACM, New York, pp 24–29
18. Finney A, Hucka M, Bornstein BJ, Keating SM, Shapiro BM, Matthews J, Kovitz BK, Schilstra MJ, Funahashi A, Doyle J, Kitano H (2006) Software infrastructure for effective communication and reuse of computational models. In: Szallasi Z, Stelling J, Periwal V (eds) System modeling in cell biology: from concepts to nuts and bolts. MIT Press, Cambridge, MA, USA
19. Giavitto J-L, Godin C, Michel O, zemyslaw Prusinkiewicz Pr (2002) Computational models for integrative and developmental biology. Technical report 72-2002, Univerite d'Evry, LaMI
20. Hill AD, Tomshine JR, Weeding EMB, Sotiropoulos V, Kaznessis YN (2008) Synbioss: the synthetic biology modeling suite. Bioinformatics 24(21):2551–2553
21. Umesh P, Naveen F, Rao C, Nair A (2011) Programming languages for synthetic biology. Springer, Netherlands, pp 1–5
22. Lesia B, Adam L, and Douglas D (2011) The eugene language for synthetic biology, Methods in enzymology, vol. 498 (in press)
23. Lesia B, Adam L, Sherine C, Emma W, Bing X, Mariana L, Anderson JC, Douglas D Eugene – a domain specific language for specifying and constraining synthetic biological parts, devices, and systems, PLoS ONE (in press)
24. Mallavarapu A, Thomson M, Ullian B, Gunawardena J (2009) Programming with models: modularity and abstraction provide powerful capabilities for systems biology. J R Soc Interface 6(32):257–270
25. Marchisio MA, Stelling J (2010) Automatic design of digital synthetic gene circuits. 2nd international workshop on bio-design automation, San Diego, USA
26. Mirschel S, Steinmetz K, Rempel M, Ginkel M, Gilles ED (2009) Promot: modular modeling for systems biology. Bioinformatics 25(5):687–689
27. Myers CJ, Barker N, Jones K, Kuwahara H, Madsen C, Nguyen NP (2009) Ibiosim: a tool for the analysis and design of genetic circuits. Bioinformatics 25:2848–2849
28. Nagpal R (2001) Programmable self-assembly: constructing global shape using biologically-inspired local interactions and origami mathematics. PhD thesis, MIT, Cambridge, MA, USA

29. Nathan JH (2010) j5 automated dna assembly software. http://jbei-exwebapp.lbl.gov/j5. Accessed 28 Sept. 2010
30. Paun G (2002) Membrane computing: an introduction. Springer-Verlag New York, Inc., Secaucus, NJ, USA
31. Pedersen M, Phillips A (2009a) Towards programming languages for genetic engineering of living cells. J R Soc Interface 6(Suppl 4):S437–S450
32. Prusinkiewicz P, Lindenmayer A (1990) The algorithmic beauty of plants. Springer-Verlag, New York
33. Purnick PEM, Weiss R (2009) The second wave of synthetic biology: from modules to systems. Nat Rev Mol Cell Biol 10(6):410–422
34. Rothemund PWK (2005) Design of DNA origami. In: International Conference on Computer Aided Design (ICCAD'05). San Jose, CA, pp 471–478
35. Shea A, Riedel M, Fett B, Parhi K (2009) Synthesizing sequential register-based computation with biochemistry. In: Computer Aided Design – Digest of Technical Papers (ICCAD'09). IEEE/ACM International Conference on 2–5 Nov. San Jose, CA, USA, pp 136–143
36. Smith LP, Bergmann FT, Chandran D, Sauro HM (2009) Antimony: a modular model definition language. Bioinformatics 25(18):2452–2454
37. Sudkamp TA (2006) Languages and machines: an introduction to the theory of computer science, 3rd edn. Pearson Addison-Wesley, Boston
38. Weiss R (2001) Cellular computation and communications using engineered genetic regulatory networks. PhD thesis, MIT, Cambridge, MA, USA

Chapter 12
Rational Design of Robust Biomolecular Circuits: from Specification to Parameters

Marc Hafner, Tatjana Petrov, James Lu, and Heinz Koeppl

Abstract Despite the early success stories synthetic biology, the development of larger, more complex synthetic systems necessitates the use of appropriate design methodologies. In particular, the integration of smaller circuits in order to perform complex tasks remains one of the most important challenges faced in synthetic biology. We propose here a methodology to determine the region in the parameter space where a given dynamical model works as desired. It is based on the inverse problem of finding parameter sets that exhibit the specified behavior for a defined topology. The main issue we face is that such inverse mapping is highly expansive and suffers from instability: small changes in the specified dynamic property could lead to large deviations in the parameters for the identified models. To solve this issue, we discuss regularized maps complemented by local analysis. With a stabilized inversion map, small neighborhoods in the property space are mapped to small neighborhoods in the parameter space, thereby finding parameter vectors that are robust to the problem specification. To specify dynamic circuit properties we discuss Linear Temporal Logic (LTL). We apply these concepts to two models of the cyanobacterial circadian oscillation.

Keywords Robustness · Inverse problems · Robust control · Optimal control · Circuit design · Formal verification

Introduction

In recent years, a variety of synthetic circuits have been successfully implemented and characterized [52]. These range from oscillators [21, 59, 61, 63], to toggle switches [26, 35] and intercellular senders/receivers or quorum sensing communication systems [5, 66, 67]. In spite of the wide range of behaviors exhibited by these

H. Koeppl (✉)
Swiss Federal Institute of Technology Zurich (ETHZ), Zurich, Switzerland
e-mail: koeppl@ethz.ch

H. Koeppl et al. (eds.), *Design and Analysis of Biomolecular Circuits: Engineering Approaches to Systems and Synthetic Biology*, DOI 10.1007/978-1-4419-6766-4_12,

circuits, they were implemented from a small number of components configured in well-known network topologies, with the appropriate choices of parameter values found by trial-and-error.

Despite the early success of these synthetic circuits, the development of larger, more complex synthetic systems necessitates the use of appropriate design methodologies. In particular, the integration of smaller circuits in order to perform complex tasks remains one of the most important challenges facing synthetic biology [52]. *In silico* analyses can provide significant insights into the construction of complex synthetic systems, but due to the poor understanding and quantification of biological environments, the predictive capability of *in silico* models for *in vitro* implementations remains limited [59]. One way to remedy this shortcoming is to use robustness as a design criterion, which has already been successfully utilized in engineering as a way to deal with uncertainty. Due to the noisy and less controlled environment encountered in biology, this should be an even more important criterion to consider. In naturally occurring biological systems, robustness has been widely observed as an intrinsic property [16, 20, 43, 53, 56, 57, 68] and there is strong evidence that natural selection favors robust biological systems [65]. This suggests that robustness, alongside with performance, can be chosen as an important design principle in the construction of synthetic systems.

In this chapter, a methodology based on robustness is presented as a way to analyze and design synthetic circuits. In contrast to the task of model simulation (Fig. 12.1a), robustness analysis with respect to parameter fluctuations can be viewed as an inverse problem of mapping design constraints to the choice of model parameters (Fig. 12.1c); this aspect is discussed in section "Robustness of Biological Systems". In section "Formalism for Robustness Analysis", a formalism based on linear temporal logic is introduced as a specific way to define the design constraints, which is followed by discussions on parameter sampling and regularization of the reverse map. Finally, an application of robustness analysis is presented in section "Application on Circadian Clock Models".

Robustness of Biological Systems

Robustness may be broadly defined as the ability of a system to maintain its function in spite of external perturbations and internal fluctuations [34]. The term robustness should be understood in a relative sense: in qualifying a system as being robust, it has to be stated which of the system properties are being referred to, as well as the types of perturbations under consideration. For biomolecular circuits, the function of a system is context dependent: for instance, due to the role that circadian oscillators play in orchestrating the daily rhythm of biological processes, their function is that of a clock exhibiting stable oscillations with a near-constant period as well as an amplitude within a predetermined range. The perturbations that a system may face include: changes in the network structure [47, 64] or

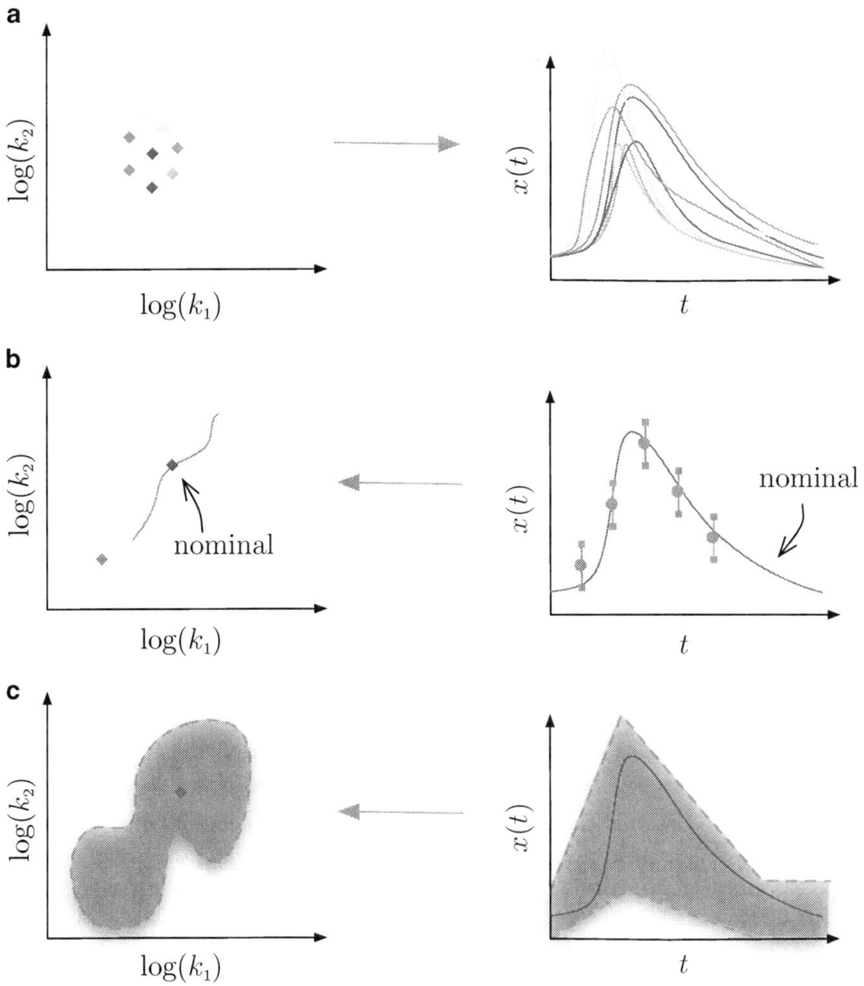

Fig. 12.1 Illustration of forward and reverse problems: (**a**) simulation as a mapping from parameter vectors to trajectories; (**b**) parameter identification as a mapping from measured data to parameter vectors; (**c**) robustness analysis as a mapping from specifications to consistent parameter regions

kinetics values due to mutations; molecular noise resulting from low copy numbers [28,38,60]; fluctuating external concentrations or parameters values due to variability in environmental factors (including temperature [54], pH-value, ionic strength and so on [9]). As this example indicates, the function of a system can be derived from the constancy of multiple system characteristics under a number of variable conditions.

Methods for Robustness Analysis

Although there is evidence that robustness is prevalent in biochemical systems, few methods have been developed to quantify it [4, 45, 53, 57]. In control engineering, the notion of robustness already exists but its definition is too restrictive for it to be applicable to biological systems: in parametric robust control, the property to be maintained is typically stability, while for biology one would like to maintain system functions of a higher-order nature. For instance, the heat shock (HS) system in *E. coli* is able to adapt its fraction of mis-folded proteins to a temperature change by producing specific HS proteins [20], while at the same time being able to cope with molecular noise [19]. Therefore, the property that should be maintained in a robust manner is homeostasis. As this example shows, robustness in biology is arguably more complex to quantify than the existing robust stability measures.

Amongst the different robustness analysis methods are those that can be classified as local methods [15, 28, 56, 57]. These methods try to answer the underlying question: *'how sensitive are the system's functions to small perturbations?'* The classification 'local' means that the analysis is performed on a given model with a specific choice of parameter vector. An example of these methods is sensitivity analysis, which is used to study the effect of parameter variations on different system's functions [1,56,57]. Another example is the quantification of the effect of molecular noise on either a steady state value [19,62] or on the period of an oscillator [27,28]. In this local approach, the robustness of a system is measured as the insensitivity of the property of interest.

The second category of methods that quantify robustness tries to answer the following question: *'how many parameter vectors allow the system to function as described?'* As opposed to local methods, these 'global' methods aim to explore large regions in the parameter space and try to characterize the geometry of the *consistent* region, defined as the region(s) in the parameter space for which the system shows the specified behavior [12, 18, 30]. The lowest-order geometric characterization is the volume: a small volume of the consistent region forces a precise tuning of the parameters. On the other hand, a consistent region of large volume allows a system to successfully face changes in environmental conditions, because its parameters can adapt, sometimes by orders of magnitude, without impairing its systemic properties. Hence, larger consistent volumes correspond to higher robustness. The second characterization which plays another important role in a system's robustness is the geometrical shape of the consistent region. Strong correlations in the consistent region have been widely observed in biological systems, a property called sloppiness [10]. Consistent volumes with skewed geometries are more prone to be left with little variations in the stiff directions. On the other hand, geometries close to a sphere in normalized quantities are more robust as they permit moderate fluctuations in arbitrary parameter directions without leaving the consistent region.

Existing global analysis methods take the *forward* approach, namely: many parameter vectors are generated, each of which is evaluated for the systemic properties and those that are consistent are selected [12, 18, 30]; a method for parameter sampling based on this approach is presented later in this chapter. Note that these approaches differ from parameter identification (Fig. 12.1b) as they yield a region in the parameter space (Fig. 12.1c) rather than a single parameter vector. An alternative approach to the forward sampling is to view the task as an inverse problem: in the third part of section "Formalism for Robustness Analysis", we discuss the issue of stability that arises in this formulation and propose a way to *regularize* it.

Inverse Problems

In contrast to the task of simulating a given model, the task of design for robustness is a problem of the inverse type: in particular, the specified system properties, as codified in some set of specifications, play the role of input, to which the goal is to find the consistent range of parameters. Inverse problems typically violate one or more of the following conditions for well-posedness as defined by Hadamard [22]:

1. For all admissible data, a solution exists;
2. For all admissible data, the solution is unique;
3. The solution depends continuously on the data.

There are various ways to remedy the ill-posedness of a problem. First of all, by appropriately enlarging the solution space or relaxing the specified properties, as for instance by seeking the solutions of minimum-deviation, one may reformulate the problem such that the first condition can be satisfied. In the context of design for robustness, the second condition for uniqueness is usually not of particular concern: in fact, one may wish to identify not only one but the set of all models that satisfy a particular specification. The violation of the third condition merits the most concern, as this implies that small changes in the specified properties could lead to vastly different geometries in the consistent parameter regions. In order to solve such inverse problems, regularization strategies (as discussed in section "Formalism for Robustness Analysis") can be employed to replace the original unstable problems with nearby, more stable approximations.

We remark that, while in this chapter only the parameter values are taken as variables, in general the task of circuit design involves solving inverse problems in the space of network topologies together with the associated parameters. The issues of stability and appropriate regularization strategies for sparsity [11,70] would become important factors to consider.

Formalism for Robustness Analysis

Given an *in vivo* biomolecular reaction system, involving a set of genes, RNAs, proteins, signaling molecules and so forth, we summarize its environmental operating conditions together with its kinetic rate constants into an m-dimensional parameter vector $\mathbf{k}$. Biophysical constraints restrict the parameter space to some subspace $\mathcal{K} \subset \mathbb{R}^m$ (applying a logarithmic representation). For every $\mathbf{k} \in \mathcal{K}$, we denote by

$\varphi(\mathbf{k}) \in L_2([0,T];\mathbb{R}^n)$ the corresponding system trajectory, upon which the observation operator ψ is applied to give the systemic properties $\psi \circ \varphi(\mathbf{k}) \in \mathbb{R}^q$; see (12.1). We denote by $\mathcal{X} \equiv \varphi(\mathcal{K})$ the space of all (admissible) trajectories that are mapped from the parameter subspace $\mathcal{K}$. Depending on the choice of modeling paradigm, we will be looking at different subspaces of $L_2([0,T];\mathbb{R}^n)$: for the case where ordinary differential equations (i.e. reaction-rate equations) are used, under appropriate smoothness conditions on the vector field (i.e. kinetic rate laws) the trajectories are at least continuously differentiable and hence $\mathcal{X} \subset C^1([0,T];\mathbb{R}^n)$; for stochastic models, the set of trajectories lie in the Skorokhod space of right-continuous functions with left limits, $\mathcal{X} \subset D([0,T];\mathbb{R}^n)$. In this chapter, we consider applying the formalism to ordinary differential equation (ODE) models of biological circuits, where at each time point the trajectory takes on values from the state space lying in the positive orthant, $\mathcal{S} \subseteq \mathbb{R}^n_{\geq 0}$. Hence, a given trajectory $\mathbf{x} \in \mathcal{X} \subset C^1([0,T];\mathcal{S})$ is the function $\mathbf{x} : [0,T] \to \mathcal{S}$. In terms of mapping between spaces, the evaluation of systemic properties of a given model can be expressed as follows:

$$\underbrace{\mathcal{K} \subseteq \mathbb{R}^m}_{\text{parameters}} \xrightarrow{\varphi} \underbrace{\mathcal{X} \subseteq L_2([0,T];\mathbb{R}^n)}_{\text{trajectories}} \xrightarrow{\psi} \underbrace{\Pi \subseteq \mathbb{R}^q}_{\substack{\text{systemic}\\ \text{properties}}}, \tag{12.1}$$

where $\Pi \equiv \psi \circ \varphi(\mathcal{K})$ is the corresponding space of systemic properties.

Depending on the application, systemic properties take on various forms. For instance, in calibrating a model to experimental time-course data, one may wish to determine parameter regions exhibiting trajectories that stay within a predefined interval around the experimental time-course, reflecting the uncertainty in the acquired data due to error sources. In the section "Application on Circadian Clock Models" of this chapter, in an application of robustness analysis to circadian models for the cyanobacteria, elements of the property space $\boldsymbol{\pi} \in \Pi$ contain the components $\boldsymbol{\pi} = (\pi_T, \pi_M, \pi_A)$, i.e. the period, peak value (maximum) and amplitude for the circadian oscillation in the protein of interest, respectively.

We now look at the inverse problem of specifying constraints on the systemic properties and derive the corresponding parameter intervals, which we refer to as the consistent parameter region $\mathcal{C}$. Let us denote by $\tilde{\Pi}$ the space of specified properties that a circuit designer may wish to achieve. In general, the specified properties may not be attainable with the given model, in which case $\tilde{\Pi} \nsubseteq \Pi$ and there might be even no solution to the specification, i.e. $\mathcal{C} = \emptyset$. The goal is to find parameter regions, if non-empty, that lead to the satisfaction of specified properties. That is, for $\boldsymbol{\pi} \in \tilde{\Pi}$ one would like to infer the corresponding consistent parameter region, as given by the preimage $\varphi^{-1} \circ \psi^{-1}(\boldsymbol{\pi})$:

$$\underbrace{\mathcal{C} \subseteq \emptyset \cup \mathcal{K} \subseteq \mathbb{R}^m}_{\substack{\text{consistent}\\ \text{parameters}}} \xleftarrow{\varphi^{-1}} \underbrace{\tilde{\mathcal{X}} \subseteq \emptyset \cup L_2([0,T];\mathbb{R}^n)}_{\text{trajectories}} \xleftarrow{\psi^{-1}} \underbrace{\tilde{\Pi} \subseteq \mathbb{R}^q}_{\substack{\text{specified}\\ \text{properties}}} \tag{12.2}$$

More specifically, the preimages are set-valued, that is: for $\boldsymbol{\pi} \in \tilde{\Pi}$, $\psi^{-1}(\boldsymbol{\pi}) = \{x \in \mathcal{X} | \psi(x) = \boldsymbol{\pi}\}$. For non-attainable design specifications, the preimages are therefore empty sets : $\psi^{-1}(\boldsymbol{\pi}) = \emptyset$, $\forall \boldsymbol{\pi} \in \tilde{\Pi} \setminus \Pi$.

The input–output relationship from parameters to properties is captured in the *forward* map, consisting of the composition $F = \psi \circ \varphi$. The inverse problem is, in turn, encapsulated by the map, $F^{-1} = \varphi^{-1} \circ \psi^{-1}$. Once the space of properties Π has been endowed with an appropriate metric, one can study the stability properties of the inverse problems.

We now focus on the second map shown in (12.1) – the map ψ from trajectories to specified properties. In order to consider inverse problems, we first need to describe the space of specified properties $\tilde{\Pi} \subseteq \mathbb{R}^q$ in a mathematical form.

Formal Specification Languages

Formal verification is the area of computer science which deals mainly with the generally undecidable problem of how to prove that a system is behaving correctly with respect to the formal specification [17, 29]. It also allows for the 'bottom–up' approach in systems design – one could build systems which are 'correct by construction', by performing the provably correct refinement steps to transform a specification into a design, and the actual implementation (e.g. [2, 31]). One convenient way to formally specify systemic properties is to use *temporal logic*, which appears as an extension of the classical propositional reasoning, where the propositional variables may change their truth values over time. One such temporal logic is the *linear temporal logic* (LTL), when the truth value of the propositions is interpreted over a time *line* [50]. Time is viewed as a linear sequence of events, and properties are specified over a single path. We can assert not only property such as 'proposition p holds at current time', but also for example 'p holds at next time step' (denoted $\bigcirc\, p$), 'p holds globally in time' (denoted $\Box\, p$), 'p will hold eventually in the future' (denoted $\Diamond\, p$), or 'p holds continuously until the time point where q is satisfied' (denoted $p\mathbf{U}q$). The combination of these properties allows for defining safety properties ('something bad never happens'), liveness ('something good eventually happens'), fairness ('if something is requested, it eventually gets granted'), or strong fairness ('if something is repeatedly requested, it repeatedly gets granted'). The other most widely used temporal logic is the *computational tree logic* (CTL) [7], where the truth value of the propositions is interpreted over more possible branches of a time line. CTL formulae are interpreted over a transition graph, i.e. more paths may be quantified at the same time, allowing one to express statements such as 'there exists a path, such that a property p globally holds'. The linear and branching temporal logics have strictly different expressive powers, in the sense that there are formulae which can be expressed in LTL and not in CTL, and vice versa – neither is superior to the other.

In order to consider the inverse problem (12.2), we have to describe the space of specified properties $\tilde{\Pi} \subseteq \mathbb{R}^q$ in a mathematically precise fashion. Inspired by [23], we define the robustness of the systemic properties expressed in LTL with respect to a single sampled trajectory, which is then used to define the robustness with respect to a set of sampled trajectories.

Linear Temporal Logic (LTL). Let $\mathcal{P}$ be the set of atomic propositions which may be assigned to a time point. For the purposes of this text, these propositions will refer to the sampled trajectory and they will be defined by statements such as 'the current value is equal to **s**'. Let us observe the logical expressions generated by the following grammar (Backus Normal Form [6])

$$\phi := p \in \mathcal{P} \mid \top \mid \bot \mid \neg\phi \mid \phi\wedge\psi \mid \phi\vee\psi \mid \bigcirc\phi \mid \Box\phi \mid \Diamond\phi \mid \phi\mathbf{U}\psi.$$

We denote the set of such formulae by **FORM** and we interpret the formulae in a discrete, linear model of time $\mathcal{M} = (\mathbb{N}, I)$, where $I : \mathbb{N} \to \wp(\mathcal{P})$ with $\wp$ denoting the powerset, maps each moment in time to a set of propositions from $\mathcal{P}$. In this way, the model is interpreted as assigning to each time point a set of propositions which hold true. Note that, if the propositions are described by 'the current value is equal to **s**', then exactly one proposition can be valid at one time point, i.e. the interpretation sets will contain exactly one proposition. The truth value of an LTL formula is interpreted over a particular model and time point by a satisfaction relation $\models \subseteq \mathcal{M} \times \mathbb{N} \times \mathbf{FORM}$, where $(\mathcal{M}, i, \phi) \in \models$ is implicitly written as $(\mathcal{M}, i) \models \phi$, which is to say that the model $\mathcal{M}$ at time point i *satisfies* the LTL formula ϕ. The satisfaction relation is interpreted in the following way: the model $\mathcal{M}$ at time point i *satisfies* the atomic proposition p, if the proposition p belongs to the interpretation set of the time point i in model $\mathcal{M}$,

$$(\mathcal{M}, i) \models p \text{ iff } p \in I(i). \tag{12.3}$$

The composed formulae are interpreted with the following equivalences:

$$(\mathcal{M}, i) \models \neg\phi \text{ iff } (\mathcal{M}, i) \not\models \phi$$
$$(\mathcal{M}, i) \models \phi\wedge\psi \text{ iff } (\mathcal{M}, i) \models \phi\wedge(\mathcal{M}, i) \models \psi$$
$$(\mathcal{M}, i) \models \bigcirc\phi \text{ iff } (\mathcal{M}, i+1) \models \phi$$
$$(\mathcal{M}, i) \models \phi\mathbf{U}\psi \text{ iff exists } j.((j \geq i).(\mathcal{M}, j) \models \psi \wedge \forall k \in [i, j].(\mathcal{M}, k) \models \phi).$$

The rest of the operators may be defined as compositions of the previously defined ones:

$$(\mathcal{M}, i) \models \top(= \phi \vee \neg\phi)$$
$$(\mathcal{M}, i) \not\models \bot(= \phi \wedge \neg\phi)$$
$$(\mathcal{M}, i) \models \phi\vee\psi \text{ iff } (\mathcal{M}, i) \models \phi\vee(\mathcal{M}, i) \models \psi(= \neg(\neg\phi \wedge \neg\psi))$$
$$(\mathcal{M}, i) \models \Diamond\phi \text{ iff there exists } j.(j \geq i).(\mathcal{M}, j) \models \phi(= \top\mathbf{U}\phi)$$
$$(\mathcal{M}, i) \models \Box\phi \text{ iff for all } j.j \geq i \Rightarrow (\mathcal{M}, j) \models \phi(= \neg\Diamond\neg\phi).$$

Finally, we say that $\mathcal{M}$ satisfies a formula ϕ, if it satisfies it at the initial time point: $\mathcal{M} \models \phi$ iff $(\mathcal{M}, 1) \models \phi$.

For the application of LTL on models described by ODEs, we consider the trajectory space consisting of continuously differentiable functions, $\tilde{\mathcal{X}} \subseteq C^1([0,T];\mathcal{S})$. For every $\mathbf{x} \in \tilde{\mathcal{X}}$ we take samples at times t_i with $i \in \mathcal{T} = \{1,\ldots,s\}$ and $t_s \leq T$. We denote the space of sampled trajectories as $\hat{\mathcal{X}} \subseteq \ell_2(\mathcal{T};\mathcal{S})$, i.e. the space of square-integrable sequences. The map $\Delta : \tilde{\mathcal{X}} \to \hat{\mathcal{X}}$ is referred to as the sampling operator and can be realized by a convolution with a comb of Dirac distributions. Practically, this means that given $\mathbf{x} \in \tilde{\mathcal{X}}$, the sampled trajectory $\hat{\mathbf{x}} \in \hat{\mathcal{X}}$ is given by $\hat{\mathbf{x}}(i) = \mathbf{x}(t_i)$. The atomic propositions are parametrized by the values $\mathbf{s} \in \mathcal{S}$ – we denote by $p_\mathbf{s}$ the proposition 'the current value is $\mathbf{s}$'. The trajectory $\hat{\mathbf{x}}$ is considered as a model $\mathcal{M} = (\mathcal{T}, I)$, such that $I(i) = \{p_\mathbf{s}\}$. In other words, the satisfaction relation in (12.3) may now be written as

$$(\hat{\mathbf{x}}, i) \models p_\mathbf{s} \text{ iff } (\hat{\mathbf{x}}(i) = \mathbf{s}).$$

Since we have assumed that the state space $\mathcal{S}$ is finite, for any $a, b \in \mathbb{R}$, we may introduce predicates in the form of inequalities over the projections of the state vector to its components – the truth value of inequality '$\hat{x}_j(i) \geq a$' ($\hat{x}_j(i)$ denotes the j-th component of the vector $\hat{\mathbf{x}}(i)$) is equivalent to that of a disjunction '$\bigvee_{\{\mathbf{s}\in\mathcal{S}.s_j\geq a\}} p_\mathbf{s}$'. Now we may write

$$\begin{aligned}(\hat{\mathbf{x}}, i) &\models (s_j \geq a) \text{ iff } (\hat{x}_j(i) \geq a), \text{ or}\\ (\hat{\mathbf{x}}, i) &\models (s_j \leq b) \text{ iff } (\hat{x}_j(i) \leq b),\end{aligned} \tag{12.4}$$

whose conjunction defines a predicate '$s_j \in [a,b]$'. Readers interested in more technical discussion on testing interval temporal logic (MITL) specifications on continuous time signals using only discrete time analysis may refer to [24].

As a one-dimensional state space example, $(\hat{\mathbf{x}}, i) \models \bigcirc(s = 5)$ stands for $(\hat{\mathbf{x}}, i) \models \bigcirc p_5$ and this means that at time point $i+1$ the sampled value is 5. Moreover, writing $\hat{\mathbf{x}} \models \Box(s \in [10,12] \Rightarrow \Diamond s \leq 1)$ expresses that 'it always holds that if the value of s is between 10 and 12, then it will eventually in the future fall below value 1'. Note that LTL formulae do not allow for 'counting' steps in the sense of specifying something like 'the property p holds at every second time step'.

Distance Measure for Systemic Properties. The use of LTL allows for a succinct and quantitative description for a given trajectory in terms of the systemic properties of interest, as well as providing a distance measure to the specification. Each formula $\phi \in$ **FORM** contains a particular composition of temporal and Boolean operators, and propositions in the form of inequalities as in (12.4). The values which are present in the propositions are referred to as the *quantitative* part of the formula. For example, the quantitative part of the formula $\phi = \Box(s \in [2,6] \Rightarrow \Diamond(s \leq 3))$ consists of the values 2, 6 and 3. For this example, these values define the specification vector, denoted by $\boldsymbol{\pi} \in \mathbb{R}^q$, as $\hat{\boldsymbol{\pi}} = (2, 6, 3)$ (we write $\hat{\boldsymbol{\pi}}$ to denote a particular instance of a vector of specifications). By ϕ we denote a formula whose quantitative part can be any vector $\boldsymbol{\pi} \in \mathbb{R}^q$.

Putting this into the context of the general robustness formalism introduced earlier, the observation operator ψ can be viewed as a composition of $\Lambda \circ \Delta$, with $\Lambda : \hat{\mathcal{X}} \to \Pi \subseteq \mathbb{R}^q$. Informally, the functional Λ maps a sequence $\hat{\mathbf{x}}$ to a vector $\boldsymbol{\pi}$, that satisfies the formula ϕ with equalities instead of inequalities. For instance, in considering the formula $\Box\, s \in [\hat{\pi}_1, \hat{\pi}_2]$ we choose Λ to map the one-dimensional sequence $\hat{x} \in \hat{\mathcal{X}}$ to $\boldsymbol{\pi} \equiv (\pi_1, \pi_2) \in \Pi$ with $\pi_1 = \min_{i \in \mathcal{T}} \hat{x}(i)$ and $\pi_2 = \max_{i \in \mathcal{T}} \hat{x}(i)$. In testing whether $\hat{\mathbf{x}}$ satisfies the specification $\phi(\hat{\boldsymbol{\pi}})$, the computed maximum and minimum of the trajectory are then compared to the quantitative part of the formula, i.e. the specification $\hat{\boldsymbol{\pi}} = (2, 6)$.

Once that the observation $\boldsymbol{\pi}$ is computed, we are able to define how well a particular trajectory $\hat{\mathbf{x}}$ satisfies the specification $\phi(\hat{\boldsymbol{\pi}})$. The *safisfaction degree* is a measure of distance between a sampled trajectory $\hat{\mathbf{x}}$ and a formula $\phi(\hat{\boldsymbol{\pi}})$,

$$\mathrm{sd}(\hat{\mathbf{x}}, \phi(\hat{\boldsymbol{\pi}})) = \frac{1}{1 + \mathrm{dist}(\phi(\hat{\boldsymbol{\pi}}), \mathcal{D}_{\hat{\mathbf{x}},\phi})} \in [0, 1], \tag{12.5}$$

where the *satisfaction domain* $\mathcal{D}_{\hat{\mathbf{x}},\phi}$ is defined as the subspace of systemic properties which are satisfied by the trajectory $\hat{\mathbf{x}}$,

$$\mathcal{D}_{\hat{\mathbf{x}},\phi} = \{\boldsymbol{\pi} \in \mathbb{R}^q \mid \hat{\mathbf{x}} \models \phi(\boldsymbol{\pi})\}. \tag{12.6}$$

The distance $\mathrm{dist}(\phi(\hat{\boldsymbol{\pi}}), \mathcal{D}_{\hat{\mathbf{x}},\phi})$ measures how much the quantitative part of the formula $\phi(\hat{\boldsymbol{\pi}})$ should be changed, so that the trajectory $\hat{\mathbf{x}}$ falls in its satisfaction region. If the trajectory $\hat{\mathbf{x}}$ satisfies the properties given by $\phi(\hat{\boldsymbol{\pi}})$, i.e. $\hat{\boldsymbol{\pi}} \in \mathcal{D}_{\hat{\mathbf{x}},\phi}$, then the distance equals to zero. Otherwise, the distance depends on which metric is chosen. We illustrate the introduced terminology in Fig. 12.2; the reader can refer to [23] for a more detailed explanation and analysis.

The above definition can be extended to the case where one has not a single, but instead, a set of trajectories $\hat{\mathcal{X}}$. The value which reflects the minimal change in the formula such that satisfaction holds under all trajectories in the set is proposed to be as follows:

$$\mathrm{Rsd}_{\phi(\hat{\pi}),\hat{\mathcal{X}}} = \frac{1}{1 + \mathrm{dist}(\phi(\hat{\boldsymbol{\pi}}), \bigcap_{\hat{\mathbf{x}} \in \hat{\mathcal{X}}} \mathcal{D}_{\hat{\mathbf{x}},\phi})}. \tag{12.7}$$

The effort involved in the computation of the satisfaction domains depends on the structure of the LTL formula, and on the number of its parameters. Moreover, the complexity grows with the size of the sampled trajectory. The authors of [23] have implemented the computation of the robust satisfaction degree of an LTL formula with respect to a single trajectory in BIOCHAM, which is a modeling tool for the analysis of biological systems [3].

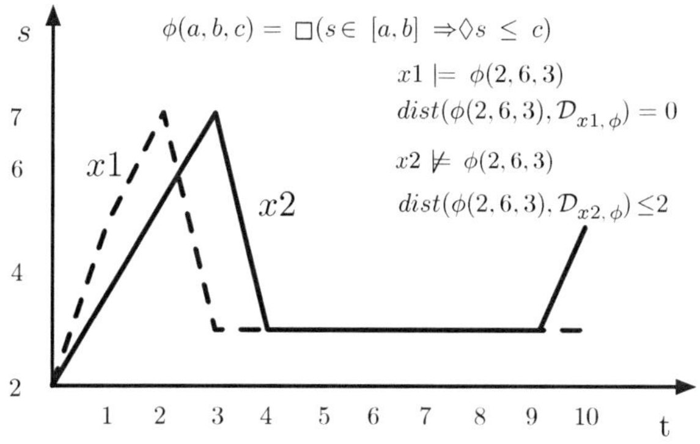

Fig. 12.2 An illustration of how the distance between the LTL formula and a trajectory is evaluated. We observe a one dimensional state space, and the LTL formula $\phi(a,b,c) = \Box(s \in [a,b] \Rightarrow \Diamond(s \leq c))$. If we instantiate the parameters of the systemic property by $\phi(2,6,3)$, then the trajectory $x1 = (2,4,5,7,3,3,3,3,3,3,3,3,3,3)$ satisfies the property, i.e. $x1 \models \phi(2,6,3)$. On the other hand, we have that the trajectory $x2 = (2,3,4,5,6,7,3,3,3,3,3,3,3,5)$ does not satisfy the formula $\phi(2,6,3)$, i.e. $x2 \not\models \phi(2,6,3)$. Since changing the parameter c to 5 gives an instance $\phi(2,6,5)$, such that $x2 \models \phi(2,6,5)$, using the Euclidian metrics we have that $dist(\phi(2,6,3), \mathcal{D}_{x2,\phi}) \leq 2$

We have used LTL to define the robustness of systemic properties with respect to a single trajectory, which is subsequently used in defining the robustness with respect to the set of sampled trajectories. In addition to this, each of the trajectories may be weighted by a certain probability. In the following we discuss an alternative approach to capturing systemic properties, where all sampled trajectories are captured in a model of a transition graph and branching temporal logic is used instead. All possible 'one-step' transitions from a state $\mathbf{s}$ to $\mathbf{s}'$ are specified, and the probability for a transition to happen is independent of the previously visited states on the trajectory (Markovian property). Then, instead of having a model $\mathcal{M}$ interpreted over a linear time line, we think of a transition graph where one state can have either multiple successors or none. Each state of the transition graph is assigned a set of propositions. Such a transition graph allows for the interpretation of CTL formulae at any of its nodes. These formulae reason about the subgraph whose root is in the observed node. Moreover, one can enrich such models by assigning weights (probabilities) to the transitions in the transition graph. These models enable one to ask questions such as 'the state where property p holds is reachable with probability larger than 0.5'. Verifying that a probabilistic system satisfies a property with a given likelihood is called *probabilistic model checking*. The logic used here are the probabilistic temporal logic (PCTL) [36], or even more general frameworks which involve continuous state spaces, such as modal logics on labelled transition systems [13].

Parameter Sampling

The major task of system design is to ensure that the resulting circuit fulfills the specified systemic properties. As discussed above, for a given model these properties depend on the choice of parameter values and the problem is reduced to one of finding the region in the parameter space for which the model yields the specified behavior, i.e. the subset of consistent parameter vectors

$$\mathcal{C} = \left\{\mathbf{k} \in \mathcal{K} | F(\mathbf{k}) \in \tilde{\Pi}\right\}. \tag{12.8}$$

This question relates to the global methods for robustness analysis as discussed previously, which can be reformulated as *'how large is the parameter region where the model functions according to the specification?'*. The answer to this question is relevant in system design, because parameter values in biology are never accurate, moreover they may fluctuate *in vivo*. This observation rises to the condition that a designed circuit should be insensitive to small parameter fluctuations. The problem of parameter tuning can be approached in two different ways: as a forward or inverse problem. In this section, we discuss the forward approach based on parameter sampling to find consistent parameter vectors and quantify global robustness.

Monte Carlo Integration. In this approach, a broad sampling is performed in the parameter space and the systemic properties of the model are measured for each vector. This gives, atop a large number of consistent parameter vectors, a quantification of the global robustness of the model, i.e. the volume $V = \text{Vol}(\mathcal{C})$ of the region where the model functions as specified. The larger the consistent region $\mathcal{C}$, the more robust the system will be against parameter fluctuations.

This volume, V, can be evaluated with a Monte Carlo integration, which is a numerical method used to obtain the integral value of a function [51]. For example, to integrate the function $f(x) > 0$ over a given range of $x \in [a, b]$ (see Fig. 12.3a),

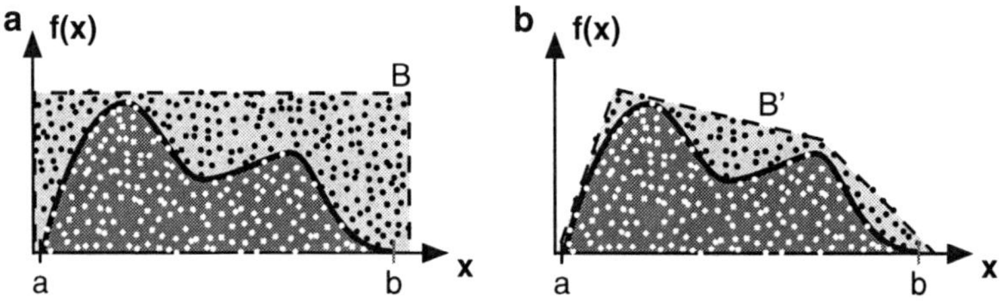

Fig. 12.3 Sketch of the Monte Carlo integration methods. (**a**) A uniform sampling is performed in the box B to evaluate the value of the integral of $f(x)$ in the range [a,b] (*red surface*). The estimated value is the surface of B multiplied by the fraction of points under $f(x)$ (*the white points*). (**b**) To optimize the sampling, a box B', tighter, could be defined for the sampling. It requires some *a priori* knowledge to fit the function without missing some part of it

one has to first define a box B that contains the function over the given range (e.g. the box defined by the boundaries $y = 0$, $y = \max_{x\in[a,b]}(f(x))$, $x = a$ and $x = b$). Secondly, the function $f(x^i)$ is evaluated for each element of the set $\{\boldsymbol{\xi}^i \in B|\boldsymbol{\xi}^i = (x^i, y^i),\ i = 1,\ldots,n\}$, uniformly distributed in B. An estimate of the integral $I = \int_a^b f(x)dx$ is given by

$$I \approx \frac{1}{n}\mathrm{Vol}(B)\sum_{i=1}^{n}\theta(f(x^i) - y^i), \tag{12.9}$$

where Vol(B) is the volume (or surface in this case) of the box B and the test function θ is the Heaviside function (i.e. equal to one if $y^i < f(x)$, zero otherwise). This expression has the interpretation that the integral is approximately equal to the surface of B multiplied by the fraction of points in B below the function $f(x)$.

This approach can be extended to the high-dimensional parameter space, where B is a hyperbox in the space $\mathcal{K}$ and the test function θ is equal to one if for the input parameter vector the model shows the specified systemic properties. This can be formalized as

$$V \approx \hat{V} = \frac{1}{n}\mathrm{Vol}(B)\sum_{i=1}^{n}\theta(\mathbf{k}^i), \tag{12.10}$$

where $\mathbf{k}^i$ are uniformly distributed parameter vectors in B and $\theta(\mathbf{k}^i) = 1$ if $F(\mathbf{k}^i) \in \tilde{\Pi}$ (implying $\mathbf{k}^i \in \mathcal{C}$), otherwise it is zero.

In order to compare models with different number of parameters, we define another robustness property: the normalized consistent volume $R = \sqrt[m]{V}$. The value R represents the average variation per parameter axis in the consistent space. Although it is almost always the case that the consistent range varies along different axes, R can still be thought of as a parameter robustness of a model.

An issue of the Monte Carlo integration is that a large fraction of tested parameter vectors can be rejected if the sampling box is too loose. This problem becomes more pronounced in higher dimensions and brute force approaches become unsuitable. One way to remedy this is to follow the ideas of Monte Carlo integration [51] with importance sampling, whereby one can adapt the sampling box B' to better fit the consistent parameter region as shown in Fig. 12.3b. The construction of a region enclosing the consistent space as tightly as possible prior to performing the Monte Carlo integration is critical for algorithm efficiency and precision. We now describe a method which helps in finding such an optimized box, B'.

PCA Sampling Method. This method is based on iterative Gaussian sampling and principal component analysis (PCA) to guide the sampling and obtain consistent parameter vectors more efficiently [30]. Briefly, at every iteration this method determines the mean value and the covariance matrix of the previously identified consistent vectors in parameter space in guiding the additional sampling. This algorithm is easy to implement and tune, but its efficiency depends on the convexity of the consistent region and the number of consistent vectors obtained at each iteration.

In more details, at each iterative step j the method generates a finite set $\hat{\mathcal{K}}^{(j)}$ of vectors in $\mathcal{K}$ and identifies the subset of consistent parameter vectors $\hat{\mathcal{C}}^{(j)}$. The first set $\hat{\mathcal{K}}^{(1)}$ is a Monte Carlo sample of the parameter space obtained via a large ($n > 10^4$) number of p-dimensional Gaussian random variates, centered on a known consistent parameter vector (Fig. 12.4a–b). The consistent subset $\hat{\mathcal{C}}^{(1)}$ of $\hat{\mathcal{K}}^{(1)}$, which should comprise of around 100 to 1,000 elements (the number depends on the dimension of the parameter space) is then determined. The subsequent step of the procedure consists of carrying out a PCA of $\hat{\mathcal{C}}^{(1)}$ [25], which is used to identify associations among consistent parameters, thereby guiding the sampling in subsequent iterations. More specifically, the set $\hat{\mathcal{K}}^{(2)}$ and subsequent sets are generated from previous parameter sets by

$$\hat{\mathcal{K}}^{(j)} = \left\{\mathbf{k}_i = \langle\hat{\mathcal{C}}^{(j-1)}\rangle + \lambda^{(j-1)}\,\boldsymbol{\xi}_i \mid i = 1,\ldots,n_s\right\}, \qquad (12.11)$$

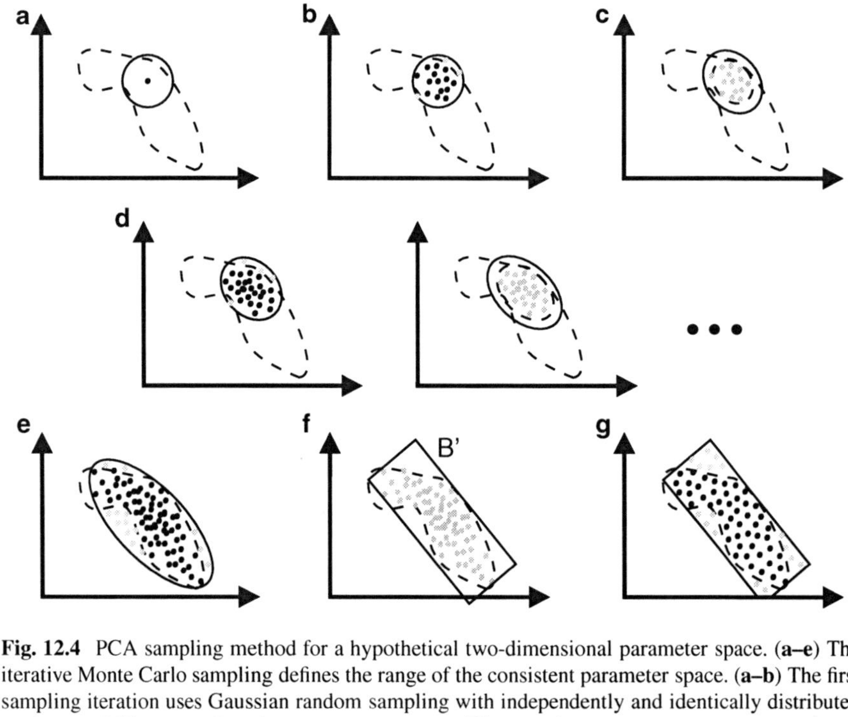

Fig. 12.4 PCA sampling method for a hypothetical two-dimensional parameter space. (**a–e**) The iterative Monte Carlo sampling defines the range of the consistent parameter space. (**a–b**) The first sampling iteration uses Gaussian random sampling with independently and identically distributed random variables around a given parameter vector. The tested parameter vectors are consistent (*black points*) or not (*light gray points*). (**c–d**) For subsequent iterative steps, sampling occurs according to the covariance matrix of consistent parameters estimated in the previous steps (*gray points*). The procedure is iterated to convergence or until a predefined number of iterations is reached (**e**). (**f–g**) Monte Carlo integration. To estimate the volume of the consistent region $\mathcal{C}$, a hyperbox B' (rectangle in **f**) that contains all the consistent parameters of the last iteration is defined, which is then uniformly sampled (**g**)

for all $j > 1$, where $\langle \hat{\mathcal{C}}^{(j-1)} \rangle$ stands for the element-wise mean of parameter vectors in the set $\hat{\mathcal{C}}^{(j-1)}$ and $\boldsymbol{\xi}_i$ is the i-th realization of a p-dimensional Gaussian process with zero mean and covariance matrix $\boldsymbol{\Sigma}^{(j-1)}$. The cardinality of $\hat{\mathcal{K}}^{(j)}$, n_s, could be adjusted such that the number of consistent vectors found at each iteration is in the order of 100 to 1,000. The entries $\Sigma_{nm}^{(j-1)}$ are then computed from the pair-wise covariances of parameters k_n and k_m in the set $\hat{\mathcal{C}}^{(j-1)}$, which assemble to give rise to a matrix whose eigenvectors are the principal axes of the set $\hat{\mathcal{C}}^{(j-1)}$. The real valued factor $\lambda^{(j-1)}$ determines the variance of the j-th Gaussian process by scaling the standard deviations of the distribution along the PCA directions of the $(j-1)$-th iteration (Fig. 12.4c–d). In this approach, the use of PCA helps to avoid wasting sampling effort in parameter regions where consistent parameter vectors are not likely to be found. The procedure is iterated either to convergence or until a predefined number of iterations is reached.

Finally, the Monte Carlo integration is then performed on the tight hyperbox B', constructed in the parameter space $\mathcal{K}$ with axes parallel to the PCA axes of the last iteration (Fig. 12.4f). In each dimension, the limits of this box are defined by the most extreme components of the consistent parameters found in all the previous samplings. Then, a set $\hat{\mathcal{K}}$ of a large number of parameter vectors sampled uniformly within B' is generated. The size of this set should be adjusted according to the desired precision (see the discussion below) and the dimensionality of the parameter space. In the application shown later, we found 10^5 to be an adequate number of the sampled vectors for the two models used. Finally, the set of uniformly distributed consistent points is defined as $\hat{\mathcal{C}} = \left\{ \mathbf{k} \in \hat{\mathcal{K}} | F(\mathbf{k}) \in \tilde{\Pi} \right\}$ (Fig. 12.4g). This Monte Carlo integration yields a global measure of robustness for any type of model: the estimated consistent volume

$$\hat{V} = \mathrm{Vol}(B') \frac{|\hat{\mathcal{C}}|}{|\hat{\mathcal{K}}|}, \tag{12.12}$$

where $|.|$ denotes the cardinality of the given set.

This method requires very little adjustments, the only potential limitation being the initialization of the iterative procedure that requires a consistent parameter vector. In the application presented below, published data were used to define the seed. However, even where such information is unavailable, random sampling and optimization techniques [41] are available to find such a vector. The drawback of this method is that its efficiency decreases when the consistent region differs strongly from an ellipsoidal shape as in the case of non-convex or poorly connected spaces for instance.

We cite here a second method that overcomes the limitation of the PCA sampling method [69]. It consists of two stages: (1) a coarse-grained sampling of the consistent space, which allows the identification of regions where consistent parameter vectors are found. This procedure delivers starting points for stage (2), where a more detailed subsequent local exploration consisting of various applications of the PCA method is applied. The sampled points define a domain for the subsequent Monte

Carlo integration in computing the volume. In this algorithm, in order to optimize the coverage of non-convex consistent regions a stratified sampling [51, p. 412] is implemented: the integration is performed on different sampling regions that cover the entire consistent region. The first stage requires a cost function that maps continuously and monotonically with respect to the degree of satisfaction of the specified property. The distance function described above in the LTL formalism (Fig. 12.2) could be used for this purpose.

Error in the Monte Carlo Integration. In deriving estimates for the sampling errors in the consistent fractions and volumes, we note that $|\hat{\mathcal{C}}|$ as obtained by Monte Carlo integration is a binomially distributed random variable [25, 51]. An estimate of its standard deviation is

$$\Delta(|\hat{\mathcal{C}}|) = \sqrt{\frac{|\hat{\mathcal{C}}|(|\hat{\mathcal{K}}| - |\hat{\mathcal{C}}|)}{|\hat{\mathcal{K}}|}}. \tag{12.13}$$

Of interest is the coefficient of variation, or relative error, defined as the standard deviation divided by the mean. The relative error of $R = \sqrt[m]{V}$ is equivalent to the relative error on $|\hat{\mathcal{C}}|$ (Eq. 12.13 divided by $|\hat{\mathcal{C}}|$) times m^{-1}

$$\frac{\Delta R}{R} = \left(\frac{1}{m}\right)\sqrt{\frac{|\hat{\mathcal{K}}| - |\hat{\mathcal{C}}|}{|\hat{\mathcal{C}}||\hat{\mathcal{K}}|}}, \tag{12.14}$$

which scales as $|\hat{\mathcal{K}}|^{-1/2}\, m^{-1}$.

Furthermore the necessary sample size $|\hat{\mathcal{K}}|$ for a given relative accuracy δ and confidence can be estimated. Applying Hoeffding's inequality [32] we obtain

$$\Pr\left\{\left|1 - \frac{\mathsf{E}(|\hat{\mathcal{C}}|)}{|\hat{\mathcal{C}}|}\right| \geq \delta\right\} \leq 2\, e^{-2\delta^2\left(\frac{|\hat{\mathcal{C}}|}{|\hat{\mathcal{K}}|}\right)^2 |\hat{\mathcal{K}}|},$$

where $\mathsf{E}(\cdot)$ denotes the expectation operator. Thus, estimating the sampling acceptance ratio $|\hat{\mathcal{C}}|/|\hat{\mathcal{K}}|$ from a sufficiently large ensemble and assuming it to be constant for the successive sampling, one can compute a lower bound for the necessary sample size. For example, asking for 10% accuracy with a confidence of 95% at an acceptance ratio of $1/20$, Hoeffding's bound requires the sample size to be $|\hat{\mathcal{K}}| > 60{,}000$.

We advice caution since in practice, one can never be certain that the entire consistent space is contained in the integration domain. The agreement between the actual consistent volume $V = \mathrm{Vol}(\mathcal{C})$ and the estimated consistent volume $\hat{V} = \mathrm{Vol}(B')|\hat{\mathcal{C}}|/|\hat{\mathcal{K}}|$, depends on the proportion of the consistent region $\mathcal{C}$ that is enclosed in sampling region B'. Due to the high dimensionality of the problem, in order to keep the computational time within reasonable bounds it is necessary to balance between a large and conservative sampling region (high accuracy) and a tight region (higher efficiency).

Regularizing the Design Problem

In the design for robustness, the issue of stability with respect to the specifications needs to be examined since the number of properties of interest is typically much smaller compared to the model dimensions, that is $q \ll m$. In particular, some subsets of the parameters could be non-influential for the properties of interest, hence the forward map $F : \mathcal{K} \rightarrow \Pi$ from parameters to systemic properties would be highly contractive along certain dimensions. This implies that the reverse map, F^{-1}, would be highly expansionary: small changes in the specified property $\pi \in \tilde{\Pi}$ could, via the intermediary state trajectories, lead to large deviations in the parameters for the identified models. The consideration of stability is important not only for problems involving inputs containing measurement noise but also when they are the given design specifications. In both situations, an important factor to consider is the inversion map φ^{-1} from the trajectory space $\mathcal{X}$ onto the model space, $\mathcal{K}$. If the stability of the reverse map is lost, one obtains parameter solutions that change in a discontinuous manner as one requests different quantitative values in the dynamic features; since specified criteria are often not fixed *a priori* but may change as the experimental data on the characterization of other circuit components accumulate, the sensitive dependence on the design goal is undesirable. We would like to stabilize the inversion map, φ^{-1}, such that small neighborhoods in the property space are mapped to small neighborhoods in the parameter space, thereby finding parameter vectors that are robust to the problem specification (see Fig. 12.5 for an illustration).

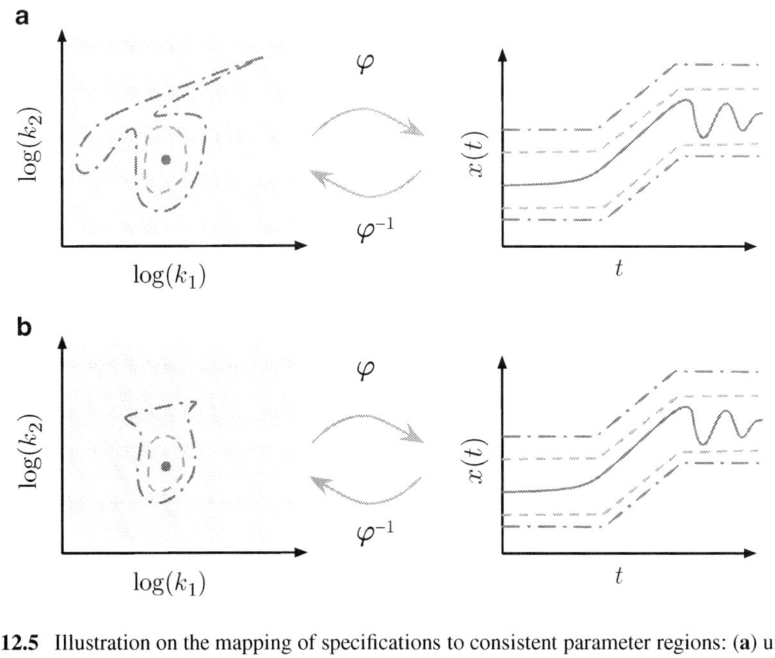

Fig. 12.5 Illustration on the mapping of specifications to consistent parameter regions: (**a**) unregularized case; (**b**) regularized case

One way to stabilize inverse problems is to employ *regularization strategies* to find the right compromise between accuracy to the specified design goal and its stability. In the following we assume attainable specifications, i.e. $\tilde{\Pi} \subseteq \Pi$. In finding parameter vectors $\mathbf{k}^* \in \mathcal{K}$ that approximate a given specification $\boldsymbol{\pi} \in \tilde{\Pi}$ via *variational regularization*, the following minimization problem is solved

$$\mathbf{k}^* \in \arg\min_{\mathbf{k}\in\mathcal{K}} \left\{ \| F(\mathbf{k}) - \boldsymbol{\pi} \| + \mu \varrho(\mathbf{k}, \mathbf{k}^0) \right\}, \quad \text{with } \boldsymbol{\pi} \in \tilde{\Pi}, \tag{12.15}$$

where $\varrho : \mathcal{K} \times \mathcal{K} \to \mathbb{R}_{\geq 0}$ is an appropriate *regularization function*, $\mathbf{k}^0$ an *a-priori* choice of parameter values and μ, the regularization parameter which trades off between the stability of the solution and the fidelity to the design goal. We note that in general, the minimizer may not be unique, in which case $\mathbf{k}^*$ is a set of vectors.

The above formulation provides an element-wise map from a single specification to a parameter vector (or an ensemble of vectors, in the case where there is non-uniqueness); this can be extended to the case where we wish to map a region in the design space $\tilde{\Pi}$ to the corresponding regularized parameter region $\mathcal{C}^\mu$, parametrized by μ

$$\mathcal{C}^\mu = \left\{ \mathbf{k} \in \mathcal{K} \,\middle|\, \exists \boldsymbol{\pi} \in \tilde{\Pi},\ \mathbf{k} \in \arg\min_{\mathbf{k}\in\mathcal{K}} \left\{ \| F(\mathbf{k}) - \boldsymbol{\pi} \| + \mu \varrho(\mathbf{k}, \mathbf{k}^0) \right\} \right\}.$$

That is, $\mathcal{C}^\mu$ consists of the set of parameter vectors that are minimizers of the regularized function corresponding to some point in the design space, $\tilde{\Pi}$. Finally, we remark that the choice of the regularization function ϱ would depend on the application of interest. For certain design applications, regularization terms can be used not only to stabilize the solution but also as a way to bind the cost of the identified design. For instance, if one wishes to redesign an existing component to perform a different function (for instance, from a switch to an oscillator), the size of the sparsity-based penalty terms [11, 70] can be related to the complexity involved in such a modification.

Optimizing Robustness

The determination of the parameter region that shows consistency with respect to a given specification yields information on the maximal parametric deviation. The size of these deviations does not need to coincide with the assumed perturbation scenario *in vivo*. Moreover, the perturbation scenario may be characterized by a probability distribution giving particular kinds of perturbations higher chances to occur. For instance, if we consider perturbations due to temperature variations one expects to observe perturbations in kinetic parameters that are strongly correlated due to the common biophysical principle of Arrhenius' law [37]. Taking the general, but computationally less tangible robustness notion of Kitano [34] and applying it to the formalism developed in this chapter we obtain

$$R(F, p, \sigma) = \int_{\mathbf{k}\in\mathcal{K}} p(\mathbf{k})\,\sigma(F(\mathbf{k}); \tilde{\Pi})\mathrm{d}\mathbf{k}, \tag{12.16}$$

where $p : \mathcal{K} \to [0, 1]$ is a probability density function representing the assumed *in vivo* perturbation model and $\sigma : \mathbb{R}^q \times \mathbb{R}^q \to \mathbb{R}_{\geq 0}$ a distance function in specification space. The robustness functional (12.16) thus shows the expected dependencies, namely on the particular model and its systemic properties characterized by the forward operator $F = \psi \circ \varphi$, the perturbation model p and the specification metric σ. Assuming a standard ℓ_p metric we can define (with a slight abuse of notation) the distance to the specification set $\tilde{\Pi}$ as

$$\sigma(\boldsymbol{\pi}; \tilde{\Pi}) = \begin{cases} 0 & \text{if } \boldsymbol{\pi} \in \tilde{\Pi}, \\ \min_{\tilde{\boldsymbol{\pi}}\in\tilde{\Pi}} \|\boldsymbol{\pi} - \tilde{\boldsymbol{\pi}}\|^p & \text{otherwise.} \end{cases} \tag{12.17}$$

The question of optimizing robustness comes up when one considers the forward design process. Although large uncertainties in the design process exist, the designer needs to aim for a nominal parameter vector (e.g. on/off rates) and decide on nominal values for external and internal conditions around which they will fluctuate. We encapsulate this nominal operating condition by $\mathbf{k}^0$ for which we naturally require $\psi \circ \varphi(\mathbf{k}^0) \in \tilde{\Pi}$ and obtain the following stochastic optimal control problem [58],

$$\begin{aligned} &\max_{\mathbf{k}^0\in\mathcal{C}} \mathsf{E}\left[\sigma(F(\mathbf{k}^0 + \delta\mathbf{k}); \tilde{\Pi})\right] \\ &\qquad \text{subject to:} \\ &\qquad \dot{\mathbf{x}} = \mathbf{f}(\mathbf{x}, \mathbf{k}^0 + \delta\mathbf{k}) \quad \text{with } \mathbf{x}(0) = \mathbf{x}_0, \end{aligned} \tag{12.18}$$

where the expectation operator E is taken with respect to the zero mean perturbation variable $\delta\mathbf{k}$. We have exemplified the dynamic constraint in terms of the reaction rate equations; however, the constraint can take on other forms, for instance as Markov jump processes or its diffusion approximation in terms of a system of stochastic differential equations. In practice, the constraint $\mathbf{k}^0 \in \mathcal{C}$ can be relaxed to $\mathbf{k}^0 \in \mathcal{K}$. Due to the functional form of (12.17) the solution of problem (12.18) should lie within $\mathcal{C}$.

Accounting for the particular specification map realized by LTL constraints, denoted by $\Lambda \circ \Delta$, one can define robustness in a similar manner to (12.16), using the satisfaction degree (12.5) [23]

$$\tilde{R}(\varphi, p, \mathrm{sd}) = \int_{k\in\mathcal{K}} p(\mathbf{k})\,\mathrm{sd}(\Delta \circ \varphi(\mathbf{k}); \phi)\mathrm{d}\mathbf{k}, \tag{12.19}$$

where we denote by $\Delta \circ \varphi(\mathbf{k})$ an element of the sequence space corresponding to parameter $\mathbf{k}$.

Application on Circadian Clock Models

As an application of the robustness analysis with a global sampling, we focus on two recent models of the cyanobacterial circadian oscillator [39, 55]. We chose this study system for the following reasons. Firstly, it is an area of very active recent model development, [8, 39, 42, 55], driven by recent insights into the molecular mechanisms of the oscillator [44]. Secondly, the behavior and function of circadian oscillators are well-characterized: they exhibit oscillations at substantial amplitudes with a period of approximately 24 h [14] as well as low sensitivity to non-periodic environmental perturbations. Thirdly, *in vitro* and *in vivo* experiments show that the cyanobacterial circadian clock is robust to many perturbations [33, 40].

Two Cyanobacterial Clock Models

The *in vitro* experiments are based on the mixing of the three key proteins KaiA, KaiB and KaiC with ATP [44]. We chose two models that capture the important empirical observations about the cyanobacterial circadian cycle: phosphorylation of KaiC with the help of KaiA [46], inhibition of this phosophorylation step by KaiB when bound to phosphorylated KaiC [46, 49], and finally dephosphorylation to complete the cycle [46]. However, the models are also fundamentally different in some key assumptions about the underlying mechanism. These significant differences may play a decisive role in the model robustness.

The first model [39] (Fig. 12.6a) involves the complex formation of KaiC with the other proteins, as well as the cyclic phosphorylation and desphosphorylation of KaiC. In this model, KaiA first binds to KaiC (the top reaction in Fig. 12.6a). The resulting complex KaiAC catalyzes the phosphorylation of KaiC, forming KaiAC*. A central element of this model is that KaiAC* then exerts a positive feedback on its own formation (denoted by the *gray arrow* in Fig. 12.6a). This autocatalysis is inhibited in a subsequent step, by the binding of KaiB to the complex KaiAC*. As the completion of the cycle, KaiA is released, followed by KaiB, and finally KaiC* is dephosphorylated. We will refer to this model as the 'autocatalytic model'. It contains 8 states variables and 7 reactions with 7 individual parameters [39].

The second model [55] makes a distinction between the two phosphorylation sites S and T of KaiC [46], resulting in three possible phosphorylated states: $KaiC^T$, $KaiC^S$ and $KaiC^{ST}$ (see Fig. 12.6b). KaiA catalyzes the phosphorylation of both S and T sites and inhibits the dephosphorylation of $KaiC^{ST}$ and $KaiC^S$. These actions of KaiA are inhibited by $KaiC^S$ (*gray bar* in Fig. 12.6b). Although $KaiC^S$ exerts its effects on KaiA jointly with KaiB [49], KaiB does not appear in the equations, because it is assumed to be at saturation level in this model. We will refer to this model as the 'two (phosphorylation) sites model'. It contains 4 state variables and 8 reactions with 12 parameters [55] since the concentration of KaiA is expressed as a function of $KaiC^S$ concentration.

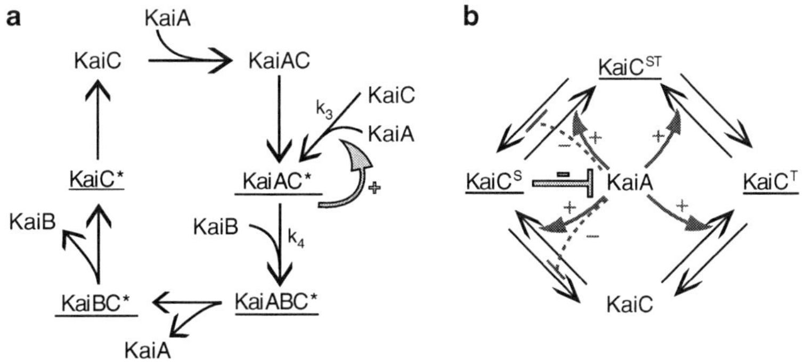

Fig. 12.6 Two models of the cyanobacterial circadian cycle. (**a**) Autocatalytic model from Mehra et al. [39]. 'C*' stands for phosphorylated KaiC. The cycle proceeds clockwise, starting from the upper left. The sum of concentrations of the KaiC*-containing complexes (*underlined*) form the output of the model. The *gray arrow* denotes the autocatalytic effect of KaiAC* on its synthesis. (**b**) Two phosphorylation sites model from Rust et al. [55]. There are three possible phosphorylated states for KaiC: KaiC^T, KaiC^S and KaiCST. The sum of concentrations of phosphorylated KaiC molecules (*underlined*) is the output of the system. KaiA catalyzes phosphorylation reactions (*curved arrows*) and inhibits some dephosphorylation reactions (*dashed bars*). KaiC^S (complexed with KaiB, not explicitly modeled) inhibits the action of KaiA (*gray bar*). Only relevant parameters are written, for additional details of the models refer to [30]

Specified Systemic Properties

In evaluating the global robustness of both models, we define $\mathcal{K}$ as a range of six orders of magnitude for each parameter, centered at published parameter values for both models [39, 55]. To obtain a uniform sampling of the consistent region $\mathcal{C}$, we used the PCA sampling method described above (see section "Formalism for Robustness Analysis). For the systemic properties $\boldsymbol{\pi}$, we chose the intervals in which the period π_T, the (maximum) peak value π_M and the amplitude π_A of phosphorylated KaiC concentration. The bounds of the specified range $\tilde{\Pi} = [\underline{\boldsymbol{\pi}}, \overline{\boldsymbol{\pi}}] = \tilde{\Pi}_T \times \tilde{\Pi}_M \times \tilde{\Pi}_A$ are chosen [48] to be 10% below and above the respective values of the properties π_T, π_M, and π_A for each model evaluated with the published parameter vector [39,55].

The specification is given in terms of three intervals: (1) the specified interval for period oscillations: $\tilde{\Pi}_T = [\underline{T}, \overline{T}]$; (2) the specified interval for the maximum value: $\tilde{\Pi}_M = [\underline{M}, \overline{M}]$; (3) the specified interval for the amplitude: $\tilde{\Pi}_A = [\underline{A}, \overline{A}]$. Given the maximum value π_M for the oscillations, the minimum should be in the interval $[\pi_M - \overline{A}, \pi_M - \underline{A}]$. We may express the three constraints on $\hat{\boldsymbol{\pi}} = (\underline{T}, \overline{T}, \underline{M}, \overline{M}, \underline{A}, \overline{A})$ using the LTL specifications (see section "Formalism for Robustness Analysis") by the formula

$$
\begin{aligned}
\phi(\hat{\boldsymbol{\pi}}) &= \phi(\underline{T}, \overline{T}, \underline{M}, \overline{M}, \underline{A}, \overline{A}) \\
&= \exists T \in [\underline{T}, \overline{T}].\Box\,(\bigcirc^T s_j \wedge s_j)
\end{aligned}
$$

$$\wedge \left[\exists \pi_M \in [\underline{M}, \overline{M}].(\Box\,(s_j \le \pi_M) \wedge \Diamond\,(s_j = \pi_M) \right.$$
$$\left. \wedge\, \Box\,(s_j > \pi_M - \overline{A}) \wedge \Diamond\,(s_j \le \pi_M - \underline{A}))\right],$$

with the j-th component of $\hat{\mathbf{x}}$ being the total concentration of phosphorylated KaiC, i.e. for the autocatalytic model, the sum of KaiAC*, KaiABC*, KaiBC* and KaiC* concentrations and, for the two-sites model, the sum of KaiC^T, KaiC^S and KaiC^{ST} concentrations. Note that the periodicity condition enforces the alternation between the maximal and minimal amplitudes.

Robustness Results

Figure 12.7a shows the normalized consistent volumes R for the two models. These volumes can be interpreted as the average allowable variation per parameter that leaves the circadian oscillations intact. The two-sites model is much more robust than the autocatalytic model. Specifically, the value $R = 0.718$ for the autocatalytic model means that the parameters can vary over 0.7 orders of magnitude, or 5.2-fold. For the two-sites model, the value of $R = 1.60$ is more than twice as large as for the autocatalytic model, corresponding to a 39-fold allowable variation.

We then asked what is responsible for the lower robustness of the autocatalytic model. One possibility is that strong associations exist between individual parameters in $\mathcal{C}$, such that some parameters cannot vary independently from others. Figure 12.7b shows the standard deviations of consistent parameters along the principal axes of both models. With one exception, the amount of variation along the principal components are similar over the two models. The single exception (indicated by the *arrow* in the Fig. 12.7b) is the lowest PCA standard deviation for the autocatalytic model.

The high level of constraint along this axis is caused by a strong positive correlation between the rate for the autocatalytic reaction, parameter k_3, and the rate for the formation of the complex KaiABC*, k_4 (see Fig. 12.7c). This strong association contributes to the lack of global robustness observed in the autocatalytic model. The implication is that a perturbation of parameter k_3 which is not be matched by a corresponding perturbation in parameter k_4 would prevent the model from preserving its systemic properties $\boldsymbol{\pi}$. Examining the structure of the autocatalytic model (Fig. 12.7a), we found that the mechanistic cause for this association lies in the dynamics of KaiAC*: on one hand, if k_3 is too large, the concentration of KaiAC* increases too fast and the autocatalytic effect is too strong; on the other hand, if k_4 is too large, the concentration of KaiAC* is too low and the autocatalytic effect is too weak. The parameters k_3 and k_4 need to be delicately balanced in order to have the correct concentration of KaiAC* and result in the appropriate feedback strength. Collapsing the highly correlated parameters k_3 and k_4 into one (assuming that k_3 and k_4 are linearly dependent) yields a global robustness estimate of $R = 1.09$, from this we conclude that the strong correlation accounts partially for the lower robustness of the autocatalytic model.

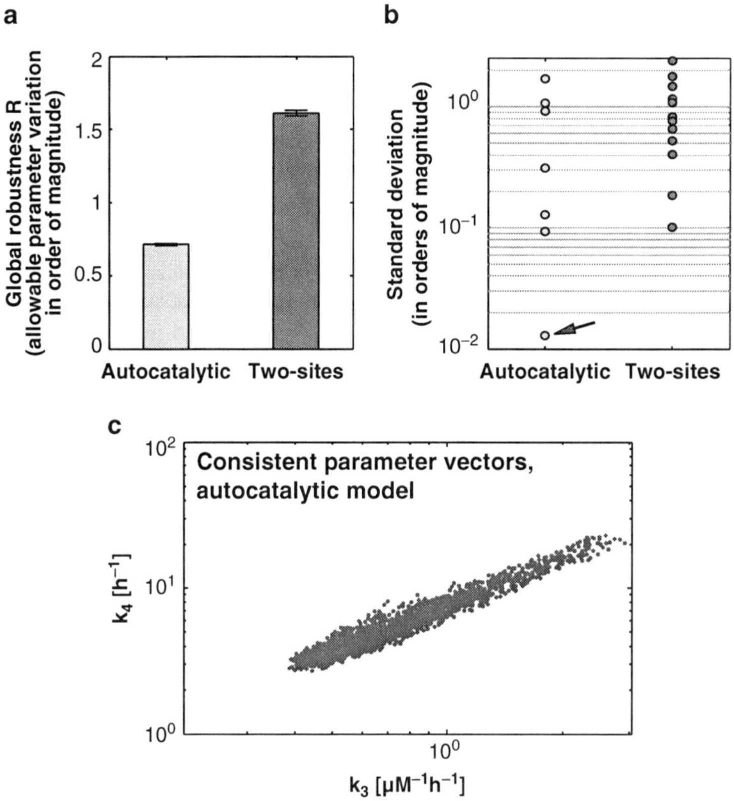

Fig. 12.7 Results of the global robustness analyses for both models. (**a**) The two-sites model (*right*) has significantly greater nomalized consistent volume than the autocatalytic model (*left*). Error bars ($<1\%$) correspond to standard deviations over five independent estimates. (**b**) Standard deviations along the principal axes of consistent parameters for the autocatalytic model and the two-sites model. Note the logarithmic scale. The autocatalytic model has a strongly constrained axis (*arrow*); amounts of variation along the other axes are overall smaller for the autocatalytic model. (**c**) Projection of the consistent vectors of the autocatalytic model after the MC integration on the plane (k_3, k_4). These two parameters are strongly correlated, resulting in the lowest standard deviation for the autocatalytic model (**b**)

Conclusion

With the advance of assembly technologies in synthetic biology, an increasing number of design choices for synthetic constructs are becoming available. In making a rational choice amongst them, one should be directed by performance constraints or behavioral specifications required for the particular synthetic circuit. Acknowledging the ubiquitous randomness and fluctuations in cellular environment we consider a circuit's robustness to those fluctuations to be a central design constraint.

The chapter presents approaches on how to determine a circuit's operating range so that it is in accordance with a predetermined specification. The larger the operating range, the more robust we consider a circuit to be. The outlined formalization allows one to assess objectively the robustness of different circuit architectures as well as different nominal parameter sets. Based on this analysis, circuits can be discriminated as well as optimized. We relate robustness analysis to inverse problems in mathematics and discuss the algorithmic challenges that arise in such. Linear temporal logic (LTL) is discussed as one framework to define general specifications for system behavior. Efficient sampling procedures to determine the parametric operating region that is consistent with a specification are presented. The proposed mathematical framework is general and can be extended to include additional environmental fluctuations and specifications with respect to which a circuit needs to conform.

Acknowledgment Marc Hafner, Tatjana Petrov and James Lu acknowledge the funding of SystemsX.ch (the Swiss initiative for systems biology) for their IPhD and BIP (JL) projects. Tatjana Petrov gratefully acknowledges Barbara Jobstmann for her helpful comments and suggestions to the LTL part. Heinz Koeppl acknowledges the support from the Swiss National Science Foundation, grant no. 200020-117975/1 and grant no. PP00P2_128503/1.

References

1. Bagheri N, Stelling J, Doyle FJ (2007) Quantitative performance metrics for robustness in circadian rhythms. Bioinformatics 23(3):358–364
2. Balzer R (1985) A 15 year perspective on automatic programming. IEEE Trans Software Eng SE-11(11):1257–1268
3. Calzone L, Fages F, Soliman S (2006) BIOCHAM: an environment for modeling biological systems and formalizing experimental knowledge. Bioinformatics 22(14):1805–1807
4. Carlson JM, Doyle J (2002) Complexity and robustness. Proc Natl Acad Sci U S A 99(Suppl 1): 2538–2545
5. Chen M-T, Weiss R (2005) Artificial cell-cell communication in yeast saccharomyces cerevisiae using signaling elements from arabidopsis thaliana. Nat Biotechnol 23(12):1551–1555
6. Chomsky N (1956) Three models for the description of language. IRE Trans Inform Theor 2:113–124
7. Clarke EM, Emerson EA, Sistla AP (1986) Automatic verification of finite-state concurrent systems using temporal logic specifications. ACM Trans Progr Lang Sys 8:244–263
8. Clodong S, Duhring U, Kronk L, Wilde A, Axmann I, Herzel H, Kollmann M (2007) Functioning and robustness of a bacterial circadian clock. Mol Syst Biol 3:90
9. Cornish-Bowden A (2004) Fundamentals of enzyme kinetics, 3rd edn. Portland Press, London
10. Daniels BC, Chen Y-JJ, Sethna JP, Gutenkunst RN, Myers CR (2008) Sloppiness, robustness, and evolvability in systems biology. Curr Opin Biotech 19(4):389–395
11. Daubechies I, Defrise M, De Mol C (2004) An iterative thresholding algorithm for linear inverse problems with a sparsity constraint. Comm Pur Appl Math 57(11):1413–1457
12. Dayarian A, Chaves M, Sontag ED, Sengupta AM (2009) Shape, size, and robustness: feasible regions in the parameter space of biochemical networks. PLoS Comput Biol 5(1):e1000256+
13. Desharnais J, Edalat A, Panangaden P (2002) Bisimulation for labelled Markov processes. Inf Comput 179(2):163–193
14. Ditty JL, Williams SB, Golden SS (2003) A cyanobacterial circadian timing mechanism. Annu Rev Genet 37(1):513–543

15. Doyle F, Gunawan R, Bagheri N, Mirsky H, To T (2006) Circadian rhythm: a natural, robust, multi-scale control system. Comput Chem Eng 30(10–12):1700–1711
16. Doyle J, Csete M (2005) Motifs, control, and stability. PLoS Biol 3(11):e392+
17. D'Silva V, Kroening D, Weissenbacher G (2008) A survey of automated techniques for formal software verification. IEEE Trans Comput Aid D 27(7):1165–1178
18. Eissing T, Allgöwer F, Bullinger E (2005) Robustness properties of apoptosis models with respect to parameter variations and intrinsic noise. IEEE Proc Syst Biol 152(4):221–228
19. El-Samad H, Khammash M (2006) Regulated degradation is a mechanism for suppressing stochastic fluctuations in gene regulatory networks. Biophys J 90(10):3749–3761
20. El-Samad H, Kurata H, Doyle JC, Gross CA, Khammash M (2005) Surviving heat shock: control strategies for robustness and performance. Proc Natl Acad Sci USA 102(8):2736–2741
21. Elowitz MB, Leibler S (2000) A synthetic oscillatory network of transcriptional regulators. Nature 403(6767):335–338
22. Engl HW, Hanke M, Neubauer A (1996) Regularization of inverse problems. Mathematics and its applications, vol 375. Kluwer Academic Publishers Group, Dordrecht
23. Fages F, Rizk A (2008) On temporal logic constraint solving for analyzing numerical data time series. Theor Comput Sci 408(1):55–65
24. Fainekos GE, Pappas GJ (2007) Robust sampling for MITL specifications. In: FORMATS 2007 Proceedings of the 5th international conference on Formal modeling and analysis of timed systems, Springer-Verlag Berlin, Heidelberg
25. Fukunaga K (1990) Introduction to statistical pattern recognition. Computer science and scientific computing, 2nd edn. Academic Press, San Diego CA
26. Gardner TS, Cantor CR, Collins JJ (2000) Construction of a genetic toggle switch in Escherichia coli. Nature 403(6767):339–342
27. Gonze D, Goldbeter A (2006) Circadian rhythms and molecular noise. Chaos 16(2):026110+
28. Gonze D, Halloy J, Goldbeter A (2002) Robustness of circadian rhythms with respect to molecular noise. Proc Natl Acad Sci U S A 99(2):673–678
29. Grumberg O, Veith H (eds) (2008) 25 Years of model checking - history, achievements, perspectives. Lecture notes in computer science, vol 5000. Springer, Berlin Heidelberg
30. Hafner M, Koeppl H, Hasler M, Wagner A (2009) 'Glocal' robustness analysis and model discrimination for circadian oscillators. PLoS Comput Biol 5(10):e1000534+
31. Hamza J, Jobstmann B, Kuncak V (2010) Synthesis for regular specifications over unbounded domains. In: Conference on formal methods in computer aided design (FMCAD), pp 101–110
32. Hoeffding W (1963) Probability inequalities for sums of bounded random variables. J Am Stat Assoc 58(301):13–30
33. Johnson CH (2004) Global orchestration of gene expression by the biological clock of cyanobacteria. Genome Biol 5(4):217
34. Kitano H (2007) Towards a theory of biological robustness. Mol Syst Biol 3:137
35. Kramer BP, Viretta AU, Baba MD, Aubel D, Weber W, Fussenegger M (2004) An engineered epigenetic transgene switch in mammalian cells. Nat Biotechnol 22(7):867–870
36. Kwiatkowska MZ, Norman G, Sproston J (2003) PCTL model checking of symbolic probabilistic systems. Technical Report CSR-03-2. University of Birmingham, School of Computer Science
37. Laidler KJ (1987) Chemical kinetics, 3rd edn. Prentice Hall, New York, NY
38. McAdams HH Arkin A (1997) Stochastic mechanisms in gene-expression. Proc Natl Acad Sci U S A 94(3):814–819
39. Mehra A, Hong CI, Shi M, Loros JJ, Dunlap JC, Ruoff P (2006) Circadian rhythmicity by autocatalysis. PLoS Comput Biol 2(7):e96
40. Mihalcescu I, Hsing W, Leibler S (2004) Resilient circadian oscillator revealed in individual cyanobacteria. Nature 430(6995):81–85
41. Moles CG, Mendes P, Banga JR (2003) Parameter estimation in biochemical pathways: a comparison of global optimization methods. Genome Res 13(11):2467–2474
42. Mori T, Williams DR, Byrne MO, Qin X, Egli M, Mchaourab HS, Stewart PL, Johnson CH (2007) Elucidating the ticking of an *in vitro* circadian clockwork. PLoS Biol 5(4):e93

43. Morohashi M, Winn AE, Borisuk MT, Bolouri H, Doyle J, Kitano H (2002) Robustness as a measure of plausibility in models of biochemical networks. J Theor Biol 216(1):19–30
44. Nakajima M, Imai K, Ito H, Nishiwaki T, Murayama Y, Iwasaki H, Oyama T, Kondo T (2005) Reconstitution of circadian oscillation of cyanobacterial KaiC phosphorylation in vitro. Science 308(5720):414–415
45. Newman MEJ, Girvan M, Farmer JD (2002) Optimal design, robustness, and risk aversion. Phys Rev Lett 89(2):028301+
46. Nishiwaki T, Satomi Y, Kitayama Y, Terauchi K, Kiyohara R, Takao T, Kondo T (2007) A sequential program of dual phosphorylation of KaiC as a basis for circadian rhythm in cyanobacteria. EMBO J 26(17):4029–4037
47. Nowak MA, Boerlijst MC, Cooke J, Smith JM (1997) Evolution of genetic redundancy. Nature 388(6638):167–171
48. Ouyang Y, Andersson CR, Kondo T, Golden SS, Johnson CH (1998) Resonating circadian clocks enhance fitness in cyanobacteria. Proc Natl Acad Sci U S A 95(15):8660–8664
49. Pattanayek R, Williams DR, Pattanayek S, Mori T, Johnson CH, Stewart PL, Egli M (2008) Structural model of the circadian clock KaiB-KaiC complex and mechanism for modulation of KaiC phosphorylation. EMBO J 27(12):1767–1778
50. Pnueli A (1977) The temporal logic of programs. In Ann IEEE Simp Found 46–57, IEEE, New York, NY
51. Press WH, Flannery BP, Teukolsky SA, Vetterling WT (1992) Numerical recipes in C: the art of scientific computing, 2nd edn. Cambridge University Press, Cambridge UK
52. Purnick PEM, Weiss R (2009) The second wave of synthetic biology: from modules to systems. Nat Rev Mol Cell Biol 10(6):410–422
53. Rand D, Shulgin B, Salazar J, Millar A (2006) Uncovering the design principles of circadian clocks: mathematical analysis of flexibility and evolutionary goals. J Theor Biol 238(3):616–635
54. Ruoff P (1992) Introducing temperature-compensation in any reaction kinetic oscillator model. J Interdiscipl Cycle Res 23(2):92–99
55. Rust MJ, Markson JS, Lane WS, Fisher DS, O'Shea EK (2007) Ordered phosphorylation governs oscillation of a three-protein circadian clock. Science 318(5851):809–812
56. Stelling J, Gilles ED, Doyle FJ (2004a) Robustness properties of circadian clock architectures. Proc Natl Acad Sci U S A 101(36):13210–13215
57. Stelling J, Sauer U, Szallasi Z, Doyle FJ, Doyle J (2004b) Robustness of cellular functions. Cell 118(6):675–685
58. Stengel RF (1986) Stochastic optimal control. Wiley, New York. Theory and application
59. Stricker J, Cookson S, Bennett MR, Mather WH, Tsimring LS, Hasty J (2008) A fast, robust and tunable synthetic gene oscillator. Nature 456(7221):516–519
60. Swain PS, Elowitz MB, Siggia ED (2002) Intrinsic and extrinsic contributions to stochasticity in gene expression. Proc Natl Acad Sci U S A 99(20):12795–12800
61. Swinburne IA, Miguez DG, Landgraf D, Silver PA (2008) Intron length increases oscillatory periods of gene expression in animal cells. Gene Dev 22(17):2342–2346
62. Thattai M, van Oudenaarden A (2001) Intrinsic noise in gene regulatory networks. Proc Natl Acad Sci U S A 98(15):8614–8619
63. Tigges M, Marquez-Lago TT, Stelling J, Fussenegger M (2009) A tunable synthetic mammalian oscillator. Nature 457(7227):309–312
64. Wagner A (2000) Robustness against mutations in genetic networks of yeast. Nat Genet 24(4):355–361
65. Wagner A (2007) Robustness and evolvability in living systems: (princeton studies in complexity), 1st edn. Princeton University Press, Princeton NJ
66. Weber W, Schuetz M, Dénervaud N, Fussenegger M (2009) A synthetic metabolite-based mammalian inter-cell signaling system. Mol Biosyst 5(7):757–763
67. Weiss R, Knight T (2001) Engineered communications for microbial robotics. In: Condon A, Rozenberg G, (eds) DNA Computing. Lecture Notes in Computer Science, vol 2054. chapter 1. Springer, Berlin/Heidelberg, pp 1–16

68. Wolf J, Becker-Weimann S, Heinrich R (2005) Analysing the robustness of cellular rhythms. Syst Biol 2(1):35–41
69. Zamora Sillero E, Hafner M, Ibig A, Stelling J, Wagner A (2010) Efficient characterization of high-dimensional parameter spaces for systems biology. Submitted to BMC Systems Biology
70. Zarzer C (2009) On Tikhonov regularization with non-convex sparsity constraints. Inverse Probl 25(2):025006

Chapter 13
Data Model Standardization for Synthetic Biomolecular Circuits and Systems

Michal Galdzicki, Deepak Chandran, John H. Gennari, and Herbert M. Sauro

Abstract While biological engineers strive to capture the biophysical theory essential in predicting how a newly designed synthetic organism will behave, the current state of this knowledge is far from ideal. To facilitate the research towards this goal, specifically through the application of computational tools, the data required to engineer biological systems should be electronically accessible and interpretable. The challenge to represent such information computationally is complicated by the enormous diversity and size of biological data. There is a plethora of biological components, interacting physically and chemically, with implications for behavior at multiple time and spatial scales. The many scientists working to move the synthetic biology field forward have to communicate their research findings and should understand each other despite their diverse academic backgrounds. The challenge and demand for data standardization arises from the need to collaborate in order to engineer ever more complex biomolecular circuits and to understand and control biological systems. The bioinformatics field provides us with a history of experience in its attempts to facilitate collaboration in the biomedical research community. We draw on the lessons from the application of information technology solutions to inform and inspire the new efforts in synthetic biology. Furthermore, we acknowledge fundamental differences in the nature of the two fields and discuss the need to standardize data models for the purpose of engineering and design of novel biomolecular circuits and systems.

Keywords Bioinformatics · BioModelsDB · CellML language · Community · Data model · Data standarization · EDIF · GenoCAD · Information sharing · LibSBML · MIAME · Nomenclature · Ontology · OWL · PDB · PoBoL · Policy · Publication · Registry of standard biological parts · SBGN · SBML · SBOL · SBOL-sematic · SBOL-visual · Software data models · Software tools · Standard format · Standard symbols · Standardized data · Synthetic biology data exchange group · TinkerCell · Visual standards

H.M. Sauro (✉)
Department of Bioengineering, University of Washington, Seattle, Washington
e-mail: hsauro@uw.edu

H. Koeppl et al. (eds.), *Design and Analysis of Biomolecular Circuits: Engineering Approaches to Systems and Synthetic Biology*, DOI 10.1007/978-1-4419-6766-4_13,

Introduction

The need for data standards in the design and analysis of biomolecular circuits is posited on the experience that once we move from individual examples, with often neatly hidden complexity, the number of components and their relationships grows astronomically large. Applying abstraction is a powerful tool in overcoming challenges in design due to complexity. Abstraction helps reduce such problems to essential conceptualizations of relevant facts. Within the paradigm for complexity engineering there is also the eventual goal to realize the design or to include further details previously hidden. When working towards a goal of limited complexity these functions maybe achievable by a 'manual' process. Today, it is likely that most of that 'manual' process will be performed by using computer support. If synthetic biologists are to move beyond designs limited to 5–20 genes, computational assistance is necessary. To allow computer applications to work with previously described data and information, there need to be standards which define the structure and meaning of that data. A call for the need to standardize data in life sciences research [7] was heard within the field of bioinformatics when the research focus moved beyond the one or a few genes to thousands of genes. Bioinformatics has concerned itself with the understanding, analysis, and management of life sciences information for decades [39], lessons from success and failures within this related field can serve to inform efforts in synthetic biology. While the goal of synthetic biology is the design and implementation of new biological systems the physical substrate with which we are concerned is the same as the other life sciences. The ability to manipulate and interrogate molecular level components, especially DNA, is what enables synthetic biologists to realize biological circuit designs. Software tools which aid in planning, performing molecular techniques, and interpreting the results require a diversity of up-to-date information. Standardized data from multiple sources and the capability to manipulate those data structures allows for an improvement in the efficiency of research and in some cases offers new possibilities. If computational analyses are to generate meaningful results there need to be established conventions for naming and describing biological objects in terms relevant to the goal of the analyses. Additionally, the information provided will need to be provided in a format which can be parsed computationally, its descriptions will have to conform to a structure, and a constrained terminology. Once it is possible to interpret these vast information resources computationally, the novel insights gained can be leveraged to improve new designs. These questions of how to manage an ever increasing body of knowledge about biological systems remain unanswered by either field. Challenges exist both technically and socially to provide an information technology infrastructure that meets the needs of both individual scientists and the broader community. However, significant progress is being made in understanding how to represent such information and the successes of grass roots and institutional efforts to adopt standards offers hope in changing the research culture toward taking advantage of data standards.

Early Examples of Success: PDB

The Protein Data Bank (PDB), is the oldest electronic repository of biological data. It contains standardized computational representation of structures of macromolecules, such as proteins and nucleic acids. These 3D structures are obtained by methods such as X-ray crystallography or NMR spectroscopy. Not only are these molecular structures an important source of knowledge to use in engineering novel proteins and interaction, but its *pourquoi* story provides synthetic biologists with a history of a successful standardized data model. The PDB began as a grassroots effort around 1971, since then it has grown tremendously, as can be illustrated by the number of structures archived, a dozen at the beginning, to now more than 68,000 entries (http://www.pdb.org). This success has created the authoritative source for structural biology information and can be attributed to the responsiveness of the PDB to the evolution of the field, technology and attitudes about data sharing [5]. Throughout the 1970s the PDB founders focused on personal communication with the community. For example, writing letters to the authors of articles inviting them to submit reported structures to the collection. Driven by the increased appreciation of the value of structural biology, advances in the methods rapidly sped up the pace of structure determination in the 1980s. The growth of the field, and a definitive source of knowledge for the molecular basis of biology and medicine, created the impetus for establishing a policy that would require data deposition into the PDB. By 1989, a formal recommendation specifying requirements for data deposition was published (International Union of Crystallography, 1989). Such policies are premised on the future value that disclosing the detailed structures of macro-moecules will provide tremendous value for downstream researcher. The recognition of this value was echoed by major journals, hence requiring PDB submissions concurrently with manuscripts. Furthermore, the National Institute for General Medical Sciences made research funding dependent on such open sharing of data. To support such sharing of structural data and management PDB researchers developed an information infrastructure and new data representation methods. Now, the PDB coordinates international efforts to integrate, or link, PDB information to related information sources, for example GenBank [4], UniProt [1], etc. Understanding the experiences of the PDB, especially during the early part of its history, is illustrative of the kinds of challenges and possible solutions that the synthetic biology standards community should learn from, to increase the likelihood of their own success.

Data Requirement Success – Microarrays

Another success story that inspires synthetic biologists, is the request of a checklist of variables that should be included in every microarray publication [8]. This bioinformatics effort organized by the MGED society has become a standard known as the Minimum Information About a Micro Array Experiment (MIAME) [8]. This request was backed by not only the international repositories of microarray data

Gene Expression Omnibus (GEO) [3, 18], ArrayExpress [35], and CIBEX [26], but also by the support of editorial policy at major journals [2]. These helped establish the needed incentive which led to high adoption across the field. Throughout the last decade the request has found broad support, compliance, and now MIAME is required by editors for successful publication. The standard begets uniform information which then allows both meta-analysis and interpretation by any software designed to read such a format. Today, the majority of microarray software is capable of reading and writing such standardized data files. The outcome of the MIAME effort is that results of gene expression studies are now easily accessible for downstream analysis via the web. Synthetic biologists who hope to design ever more sophisticated biological systems can draw upon this example to inform the process of standardization of experimental data exchange.

Standards for Models – SBML and CellML

The systems biology field is known for the development of dynamic models of cellular systems. Researchers in this field use a variety of formalisms and computational methodologies as appropriate to model the great diversity of biological dynamic behaviors. For example, some of the mathematical techniques used to represent how biological components change over time are ordinary differential equations (ODEs), deterministic hybrid models, differential-algebraic equations (DAEs), partial differential equations (PDEs), and stochastic modeling [38]. It should be noted that to describe genetic regulatory systems, a common type of synthetic biomolecular circuits, alternative formalisms can also be used such as directed graphs, bayesian networks, boolean networks, and rule-based formalisms [16]. Among this great diversity of computational methods, quantitative models based on ordinary differential equations (ODEs) are the commonly used form [38]. In order to facilitate the exchange of such models two XML-based exchange formats for computational models in systems biology have been developed, the Systems Biology Markup Language (SBML) [25] and the CellML language [34]. SBML relies on the definition of fundamental concepts for dynamic biochemical models: the Species, a chemical or other participant of a reaction; and Reaction, a statement describing change to the quantity of species (reaction definitions link the product and reactant species with their kinetic laws), other fundamental concepts such as Compartment, Parameters, Unit definitions, and Rules are also included [25]. This model structure allows for a relatively comprehensive representation of biochemical systems and it is consistent with the well established biochemistry perspective that chemical reactions have reactants and products. CellML, on the other hand, represents cellular models using a mathematical description, more closely following the structure of the mathematical equations of the model. This view is capable of representing almost arbitrary mathematical models, providing greater generalizability, at the cost of complexity of the representation. Both CellML and SBML use standard XML based metadata (using RDF) as described by the MIRIAM requirements [32].

One of the barriers to adoption by system biologists is the ease of encoding models into the standard format created in the many specialized software applications for quantitative modeling. To enable the use of SBML its authors developed a software library, libSBML [6], which could be used within existing software to translate the software's internal representation into SBML. The result of making the source code of libSBML freely available, has been overwhelming success as indicated by its adoption in more than 180 software systems such as simulators, model editors, and databases (http://sbml.org/SBML_Software_Guide).

Both SBML and its development process are far from perfect, as not all kinetic model formalisms are supported and individual software projects create models with varied quality. However, the syntactic standardization, enforced by libSBML produces a base line level of interoperability 'good enough' to have gained the considerable buy-in from an active community of researchers. Additionally, the success of SBML can be attributed to the initial effort of a small number of collaborators who adopted the open-innovation model and encouraged community participation. Through support from an international community of interested researchers and participants, it has grown into the *de facto* standard format in its field. Ongoing development is helping to expand the utility of SBML. For example, a new software library, libAnnotationSBML, links SBML ontology annotations to the web services that describe these ontological terms [41]. The growth of capabilities in creating models and the ability to unambiguously annotate the concepts through a curatorial process led to the creation of the BioModels database [31, Li 2010]. BioModelsDB now holds 249 models, which have been validated by a professional team of curators, and 224 additional models unverified by human inspection (As of Sept 2010). Not only is the SBML standards approach an important example of a successful standardization effort, but the systems biology community offers direct benefit to synthetic biologists designing new systems. The software applications for simulation of quantitative models provide ready to use tools or at least an advanced starting point for developing new tools that purposely serve the design-build-test engineering process for synthetic biologists. The library of the models in BioModelsDB includes many genetic regulatory, metabolic, and signaling pathways, thoroughly described, ready for download into SBML compatible tools, to serve as biological inspiration for new designs.

Standards for Synthetic Biology

Synthetic Biologists deal with DNA which encodes biochemical systems of interest. The tools they use are molecular techniques which manipulate the DNA sequence and the mathematical models which predict their behavior. Both require software which reads and then helps the researcher interpret the sequence or model. The challenge is to facilitate the work process of engineering biological circuits in a unified computational framework without limiting the ability of these researchers to apply the latest tools available. In the engineering field such approaches have led to wild success, far exceeding the efforts in the life sciences to date. For example, VLSI CAD applications use object models to distinguish between design objects with a common interface but different implementations. One example, of a standard in electrical engineering is the Electronic Design Interchange Format (EDIF) [27]. Motivated by prior success in both life sciences and engineering, synthetic biologists attempting to create the infrastructure for engineering biological systems have begun to standardize the biological substrates and the information about them.

The Registry of Standard Biological Parts

The Registry of Standard Biological Parts (partsregistry.org), is a pioneering effort to store and distribute BioBrick parts [37], standardized biological parts for synthetic biologists [19]. The Registry, is the only publicly accessible resource of information about BioBrick parts, plasmid DNA whose sequence conforms to physical assembly standards [29,40]. There are more than 13,000 BioBrick part records within the parts registry (As of Sept 2010). Furthermore, as in other biological research, synthetic biology faces a rapidly growing body of literature and molecular data. Exacerbating the data deluge, is the BIOFAB (biofab.org), a facility for the fabrication and functional characterization of standard biological parts on a large scale [12, 28]. Whilst immensely valuable in the pursuit of predicable biological design, it will in the near future generate immense amounts of quantitative information as part of its effort. A standard electronic form of such information would allow synthetic biologists to effectively exploit the data within computational tools. To make this data available to synthetic biologists there is now a need to standardize the electronic form of the knowledge about biological components.

Visual Representation Standards

Visual standards are important in any field where diagrams are used to exchange information. Electronics is the most prominent field where standard symbols such as resistors, capacitors, and inducers are used to unambiguously represent electronic circuits. However, fields such as mechanical engineering have a less obvious symbol set. Biological systems will require a large symbol set due to the enormous variety; however, since synthetic biology deals with a subset of the possible biological components, developing standard symbols may be convenient. There are two approaches for generating the library of standard symbols: (1) adopt the commonly used symbols from the community or (2) construct a new set of symbols. The first option will be easier to get community acceptance, but the symbol set may not be well organized. The second option will produce a more organized and systematic set of symbols, but getting community acceptance of a new symbol set may be challenging. The Systems Biology Graphical Notion (SBGN) [30], which is a suggested set of standard symbols for biological networks, favors the second option. *See SBOLv for standard symbols in synthetic biology.*

Software Data Models

Synthetic biologists designing new biological systems need software tools to aid in managing the complexity inherent to these systems [14]. New software tools are being actively developed to support the work of biological engineers. These applications range from those that will help the engineer at the laboratory bench to those that help simulate model systems to predict possible behaviors. Custom data models are almost a requirement for most software projects. This is because most software tools will have custom data that is specific for the software application's data model and may not contain corresponding data in a standard data model. As a result, software developers generally provide import and export functions for supporting standard formats rather than using the standard data model as the default. This general rule can be an exception if the standard format has an option allowing software applications to add customized information. This option is available in standard file formats such as GenBank and SBML, which is why many software programs use these formats as the default format. Nevertheless, even this extra freedom provided by the standard format can be limiting if the general structure of a software application's data model is different. An example is the data model of the TinkerCell application [13]. While TinkerCell is essentially a program for generating models of biological systems, it is not able to store the model in the standard SBML format because of subtle differences in the underlying structure. Additionally, all connections between molecular species in SBML represent reactions, whereas connections between nodes in TinkerCell can mean multiple things. The difference between TinkerCell and the SBML data model is due to the fact that TinkerCell's model is designed to represent a diagram rather than a dynamical system, which is what SBML describes. Therefore, the basic perspective on the model is different. This problem may exist for other software applications that have a different perspective on what a model of a biological system is. Applications such as GenoCAD define a model using a special grammar describing the relationships between biological parts [9–11,15]. This perspective of a model is not compatible with TinkerCell's or SBML's view of a model. Developing standards that bridges different perspectives can be a challenging task due to these fundamental differences.

SBOL – An Emerging Standard

Taking inspiration from the success of prior efforts such as MIAME, PDB, SBML, and CellML we have had the privilege to lead an effort to establish the Synthetic

Biology Data Exchange Group. This group aims to develop standards and technologies to facilitate the electronic exchange of synthetic biology information. The overall goal is to describe data in the domain using a defined but extensible scheme to enable electronic exchange and unambiguous communication of the information. To address these goals two complementary projects emerged to define the Synthetic Biology Open Language (SBOL). One is to develop an ontology, SBOL-semantic, which serves both as an organizing structure for information and as a standard exchange format through its use of RDF/OWL language. The second project is the definition of a set of graphical symbols SBOL-visual (SBOLv) which assigns a preferred icon for commonly used concepts, thereby reducing the ambiguity of diagrams used informally, within graphical user interfaces, and published.

Community The Synthetic Biology Data Exchange group is following a grass roots model, a community of researchers motivated to improve informatics capabilities in synthetic biology. The goals for this group where established in a spirited discussion at the Standards and Specifications in Synthetic Biology workshop on April 2008 in Seattle, WA. The result of the meeting was the submission of a Request For Comments documents to the BioBrick Foundation [21], which specified a core data standard for information about BioBrick parts. Emphasizing the preliminary nature the format was named the Provisional BioBrick Language, (PoBoL). Following this meeting, other researchers interested in participating joined and continued to meet annually as the Synthetic Biology Data Exchange Group (sbolstandard.org). The members of this group represent stakeholders from the synthetic biology community, especially those developing software. Following the next meeting at Stanford University in 2009, the name of the main effort was changed to better reflect its broader ambition to Synthetic Biology Open Language (SBOL). This group of researchers is attempting to forge a consensus on terminology and the technical requirements needed to standardize the computational representation of information used by synthetic biologists.

Standard Symbols The Synthetic Biology Open Language visual standard (SBOLv) symbols are based on many symbols that are already in use today. The key contribution is to limit the number of different symbols which correspond to the same concept. For example, symbols to indicate a promoter, ribosome binding sequence (RBS), coding sequence (CDS), and transcription terminator the symbols have been proposed (Fig. 13.1). (www.sbolstandard.org/sbol-visual).

A standardized nomenclature is needed to reduce ambiguity. The experience of the HUGO nomenclature committee (HGNC) in human gene names and symbols illustrates difficulties that can face an organization in attempting to establish unique and meaningful names [42]. While attempting to reduce ambiguity such as synonymy, multiple terms for the same concept, and homonymy, one symbol used for multiple concepts, the HGNC has struggled to increase the adoption of the official gene symbols throughout the broader community. While supported by journal policies, actual usage of the official gene symbols in publications has not gained broad adoption. The increase of mentioned official symbols over other aliases only rose from 35% in 1994 to 44% in 2004 [42]. The success of a standard is measured

Fig. 13.1 Examples of proposed standard symbols to represent types of biological parts used in synthetic biology. This example depicts a transcriptional unit starting with a promoter regulatory region through its termination signal sequence. Such symbols defined by SBOL-visual help diagram composition of biological parts in an unambiguous visual form. The SBOLv project aims to define the many symbols needed to clearly communicate biological part designs (http://www.sbolstandard.org/sbol-visual)

by how well the community adopts it. Due to its early introduction and the small size of the synthetic biology software developer community, who needed symbols for visual representation, SBOL-visual has been readily adopted. For example, the software tools Spectacles/Clotho [17], TinkerCell [13], SynBioSS [24, 43], DeviceEditor (http://jbei-exwebapp.ibl.gov/j5), and GenoCAD [11] use SBOL-visual symbols.

Standard for data and information SBOL-semantic is an ontology which describes common concepts used in synthetic biology. It aims to standardize and facilitate information exchange for synthetic biologists using recommended information technologies for data on the web. At the heart of SBOL-semantic is a core ontology, a set of fundamental synthetic biology concepts and their relationships. SBOL defines these concepts such as Parts, Sequence Features, and Assembly Standards (Table 13.1) as a hierarchy and specifies how they connect to each other (sbolstandard.org). These core concepts are the result of an approximate consensus reached by the Synthetic Biology Data Exchange Group. To easily allow for further expansion of the standard the group follows an open process for the evolution of SBOL [23]. The SBOL implementation conforms to W3C recommended technology for the Semantic Web. Use of formats such as RDF/OWL, allows data to be read, manipulated, and interpreted using generic Semantic Web tools. Encoding SBOL in these languages allows for compatible software tools to read and interpret data annotated using SBOL. The ontology is written in the Web Ontology Language (OWL), the W3C standard language for the definition of ontologies, chosen for its capabilities in modeling knowledge using a computable and standardized format. For example, enabling query based information retrieval of information from Parts Registry information translated into SBOL [22]. The overall objective of the SBOL effort is to represent and manipulate data that spans scales from plasmids, to cells, to tissues. Tackling the challenge to organize the vast and inherently complex biological systems information needs a robust and systematic solution for knowledge management.

Policy In the attempt to provide a long term solution to aid the vision of engineering biology, there is also the complementary, top down model to consider. The top-down model would involve, funding organizations to enact policies which require sharing of data. Historically, policies mandating data sharing in a standardized form were

Table 13.1 SBOL-semantic root classes which represent generalized concepts, with examples of a subclass (italicized) below its parent class as a hierarchy. Terms correspond to commonly used concepts in synthetic biology. Each class is used to describe sets of individual data elements as categories of common information objects used in the synthetic biology engineering process. SBOL terms also include a simple definition to clarify their intended use. As the SBOL ontology is expanded it will provide a richer vocabulary for the description of synthetic biology constructs

Class	Definition
Sample	Aliquot of Cells or DNA material in a physical container
Cell	Basic functional unit of life
Physical DNA	Continuous DNA molecule
Plasmid	Extra-chromosomal DNA capable of replicating apart from chromosomal DNA
Part	A standardized DNA component for synthetic biology
Vector Backbone	A kind of Part into used as a carrier for a construct of interest
Assembly Standard	Set of Sequence Features which designate a physical composition standard
Sequence Annotation	Position and direction describing the region for a Sequence Feature of a Part
Sequence Feature	Description of primary Annotations of nucleic acid sequence
BioBrick Scar	Sequence between adjacent Parts, byproduct of BBF Assembly Standard 10
Terminator	Transcriptional terminator sequence

found in the environmental and social sciences, where studies can last 30 years and require long term information management plans [20]. Field et al make a case for the need to enact such policies in the 'omics' or high throughput data fields which are generating massive amounts of data. Such an approach may be eventually needed for synthetic biology to incentivize participation in submission of standardized data, a process which places a significant cost on the individual researcher in terms of time and consequently funds. In creating and maintaining institutional infrastructure to manage the information, centralizing such an effort does provide economies of scale, although with a substantial direct cost. Additionally, regulatory agencies have a strong interest to encourage participation in order to review outcomes of synthetic biology efforts as necessary. Such policies can be enforced by grant application data sharing plans, specified time periods, and in a accordance with international standards. Journal referees and editors can uphold and extend these policies analogous to the accession number for DNA sequences. Once consensus is reached on the value and need for an information sharing, a policy mandating timely and public release of data will be needed [20]. Such policies, which obligate the researchers to submit information in a standardized form would serve a common aspiration in synthetic biology, to make biological systems easier to engineer [36]. Standardized data aids in the gathering, preservation, and amalgamation of research output by greatly reducing the barrier to accessing the knowledge created.

Taking advantage of well supported technology developed for information management on the web will provide long term benefits for the synthetic biology community. Specifically, standardization and dissemination of synthetic biology knowledge resources will greatly increase the potential for its re-use by downstream researchers and engineers.

Acknowledgements The data model standardization work was carried out in collaboration with members of the Synthetic Biology Data Exchange Group. We would like to especially thank Cesar A. Rodriguez, Laura Adam, J. Christopher Anderson, Douglas Densmore, Drew Endy, Raik Gruenberg, Timothy Ham, Matthew Lux, Akshay Maheshwari, Barry Moore, Chris J. Myers, Jean Peccoud, Cesar A. Rodriguez, Nicholas Roehner, Guy-Bart Stan, and Mandy Wilson for their ideas and support. This work was partially supported by grants from the National Library of Medicine (R41 LM010745, T15 LM007442), and the National Institute of Biomedical Imaging and Bioengineering (BE08407).

References

1. Apweiler R, Bairoch A, Wu CH, Barker WC, Boeckmann B, Ferro S, Gasteiger E, Huang H, Lopez R, Magrane M (2004) UniProt: the universal protein knowledgebase. Nucleic Acids Res 32(suppl 1):D115
2. Ball CA, Sherlock G, Parkinson H, Rocca-Sera P, Brooksbank C, Causton HC, Cavalieri D, Gaasterland T, Hingamp P, Holstege F, et al (2002) Standards for microarray data. Science 298:539
3. Barrett T, Troup DB, Wilhite SE, Ledoux P, Rudnev D, Evangelista C, Kim IF, Soboleva A, Tomashevsky M, Marshall Ka, Phillippy KH, Sherman PM, Muertter RN, Edgar R (2009) NCBI GEO: archive for high-throughput functional genomic data. Nucleic Acids Res 37:D885–D890
4. Benson DA, Karsch-Mizrachi I, Lipman DJ, Ostell J, Wheeler DL (2006) GenBank. Nucleic Acids Res 34:D16–D20
5. Berman HM (2008) The Protein Data Bank: a historical perspective. Acta crystallogr A 64: 88–95
6. Bornstein BJ, Keating SM, Jouraku A, Hucka M (2008) LibSBML: an API library for SBML. Bioinformatics 24:880–881
7. Brazma A (2001) On the importance of standardisation in life sciences. Bioinformatics 17: 113–114
8. Brazma A, Hingamp P, Quackenbush J, Sherlock G, Spellman P, Stoeckert C, Aach J, Ansorge W, Ball CA, Causton HC, Gaasterland T, Glenisson P, Holstege FC, Kim IF, Markowitz V, Matese JC, Parkinson H, Robinson A, Sarkans U, Schulze-Kremer S, Stewart J, Taylor R, Vilo J, Vingron M (2001) Minimum information about a microarray experiment (MIAME)-toward standards for microarray data. Nat Genet 29:365–371
9. Cai Y, Hartnett B, Gustafsson C, Peccoud J (2007) A syntactic model to design and verify synthetic genetic constructs derived from standard biological parts. Bioinformatics 23(20):2760–2767
10. Cai Y, Lux MW, Adam L, Peccoud J (2009) Modeling structure-function relationships in synthetic DNA sequences using attribute grammars. PLoS Comput Biol 5(10):e1000529
11. Cai Y, Wilson ML, Peccoud J (2010) GenoCAD for iGEM: a grammatical approach to the design of standard-compliant constructs. Nucleic Acids Res 38(8):2637–2644
12. Canton B, Labno A, Endy D (2008) Refinement and standardization of synthetic biological parts and devices. Nat Biotechnol 26:787–793
13. Chandran D, Bergmann FT, Sauro HM (2009) TinkerCell: modular CAD tool for synthetic biology. J Biol Eng 3(19)
14. Clancy K, Voigt CA (2010) Programming cells: towards an automated 'Genetic Compiler'. Curr Opin Biotechnol 21(4):572–581

15. Czar MJ, Cai Y, Peccoud J (2009) Writing DNA with GenoCAD. Nucleic Acids Res 37(suppl 2):W40–W47
16. de Jong H (2002) Modeling and simulation of genetic regulatory systems: a literature review. J Comput Biol 9:67–103
17. Densmore D, Van Devender A, Johnson M, Sritanyaratana N (2009) A platform-based design environment for synthetic biological systems. In: The fifth Richard Tapia celebration of diversity in computing conference: intellect, initiatives, insight, and innovations. ACM, pp 24–29
18. Edgar R, Domrachev M, Lash AE (2002) Gene Expression Omnibus: NCBI gene expression and hybridization array data repository. Nucleic Acids Res 30:207–210
19. Endy D (2005) Foundations for engineering biology. Nature 438:449–453
20. Field D, Sansone S-A, Collis A, Booth T, Dukes P, Gregurick SK, Kennedy K, Kolar P, Kolker E, Maxon M, Millard S, Mugabushaka A-M, Perrin N, Remacle JE, Remington K, Rocca-Serra P, Taylor CF, Thorley M, Tiwari B, Wilbanks J (2009) OMICS: 2009, 2010, and beyond. OMICS 326:234–236
21. Galdzicki M, Chandran D, Nielsen A, Morrison J, Cowell M, Grunberg R, Sleight S, Sauro H (2009) BBF RFC 31: Provisional BioBrick Language (PoBoL). doi:1721.1/45537
22. Galdzicki M, Rodriguez C, Chandran D, Sauro HM, Gennari JH (2011) Standard biological parts knowledgebase. PLoS ONE 6(2):e17005
23. Grunberg R (2009) BBF RFC 30: Draft of an RDF-based framework for the exchange and integration of synthetic biology data. doi: 1721.1/45143
24. Hill AD, Tomshine JR, Weeding EMB, Sotiropoulos V, Kaznessis YN (2008) SynBioSS: the synthetic biology modeling suite. Bioinformatics 24(21):2551–2553
25. Hucka M, Finney A, Sauro HM, Bolouri H, Doyle JC, Kitano H, et al (2003) The systems biology markup language (SBML): a medium for representation and exchange of biochemical network models. Bioinformatics 19:524
26. Ikeo K, Ishi-i J, Tamura T, Gojobori T, Tateno Y (2003) CIBEX: center for information biology gene expression database. C R Biol 326:1079–82
27. Kahn HJ (1995) EDIF version 350/400 and information modelling. In: Design automation conference, 1995. Proceedings of the ASP-DAC '95/CHDL '95/VLSI '95, IFIP international conference on hardware description languages; IFIP international conference on very large scale integration, Asian and South Pacific
28. Kelly JR, Rubin AJ, Davis JH, Ajo-Franklin CM, Cumbers J, Czar MJ, de Mora K, Glieberman AL, Monie DD, Endy D (2009) Measuring the activity of BioBrick promoters using an in vivo reference standard. J Biol Eng 3:1–13
29. Knight TF (2003) Idempotent vector design for standard assembly of BioBricks. Technical report, Massachusetts Institute of Technology Artificial Intelligence Lab, Cambridge, Massachusetts
30. Le Novere N, Hucka M, Mi H, Moodie S, Schreiber F, Sorokin A, Demir E, Wegner K, Aladjem MI, Wimalaratne SM, et al (2009) The systems biology graphical notation. Nat Biotechnol 27(8):735–741
31. Le Novere N, Bornstein B, Broicher A, Courtot M, Donizelli M, Dharuri H, Li L, Sauro H, Schilstra M, Shapiro B, Snoep JL, Hucka M (2006) BioModels Database: a free, centralized database of curated, published, quantitative kinetic models of biochemical and cellular systems. Nucleic Acids Res 34:D689–D691
32. Le Novere N, Finney A, Hucka M, Bhalla US, Campagne F, Collado-Vides J, Crampin EJ, Halstead M, Klipp E, Mendes P, Nielsen P, Sauro H, Shapiro B, Snoep JL, Spence HD, Wanner BL (2005) Minimum information requested in the annotation of biochemical models (MIRIAM). Nat Biotechnol 23:1509–1515
33. Li C, Donizelli M, Rodriguez N, Dharuri H, Endler L, Chelliah V, Li L, He E, Henry A, Stefan MI, Snoep JL, Hucka M, Le Novere N, Laibe C (2010) Biomodels database: an enhanced, curated and annotated resource for published quantitative kinetic models. BMC Syst Biol 4:92
34. Lloyd CM, Halstead MDB, Nielsen PF (2004) CellML: its future, present and past. Prog Biophys Mol Biol 85:433–450

35. Parkinson H, Kapushesky M, Kolesnikov N, Rustici G, Shojatalab M, Abeygunawardena N, Berube H, Dylag M, Emam I, Farne A, Holloway E, Lukk M, Malone J, Mani R, Pilicheva E, Rayner TF, Rezwan F, Sharma A, Williams E, Bradley XZ, Adamusiak T, Brandizi M, Burdett T, Coulson R, Krestyaninova M, Kurnosov P, Maguire E, Neogi SG, Rocca-Serra P, Sansone S-A, Sklyar N, Zhao M, Sarkans U, Brazma A (2009) ArrayExpress update–from an archive of functional genomics experiments to the atlas of gene expression. Nucleic Acids Res 37: D868–D872
36. Peccoud J, Anderson JC, Chandran D, Densmore D, Galdzicki M, Lux MW, Rodriguez CA, Stan GB, Sauro HM (2011) Essential information for synthetic DNA sequences. Nat Biotechnol 29(1):22
37. Peccoud J, Blauvelt MF, Cai Y, Cooper KL, Crasta O, DeLalla EC, Evans C, Folkerts O, Lyons BM, Mane SP, Shelton R, Sweede MA, Waldon SA (2008) Targeted development of registries of biological parts. PLoS ONE 3:7
38. Sauro HM (2006) Standards, platforms, and applications. In: Kriete A, Eils R (ed) Computational systems biology. Elsevier Academic Press, Burlington, Massachusetts, p 103
39. Searls DB (2010) The roots of bioinformatics. PLoS Comput Biol 6:e1000809
40. Shetty RP, Endy D, Knight TF (2008) Engineering BioBrick vectors from BioBrick parts. J Biol Eng 2:5
41. Swainston N, Mendes P (2009) libAnnotationSBML: a library for exploiting SBML annotations. Bioinformatics 25:2292–2293
42. Tamames J, Valencia A (2006) The success (or not) of HUGO nomenclature. Genome Biol 7:402
43. Weeding E, Houle J, Kaznessis YN (2010) Synbioss designer: a web-based tool for the automated generation of kinetic models for synthetic biological constructs. Brief Bioinform 11(4):394–402

Chapter 14
DNA Assembly Method Standardization for Synthetic Biomolecular Circuits and Systems

Nathan J. Hillson

Abstract As molecular biology tasks progress from single gene expression to the implementation of entire metabolic pathways and behavioral genetic circuitry, DNA assembly, the process of cloning/constructing a contiguous piece of DNA from a set of composite parts, poses an increasingly formidable challenge. Standardized DNA assembly methodologies have recently emerged that enable and facilitate part re-use, assembly design automation, and high-throughput physical assembly protocols. This chapter reviews the BioBrick, SLIC, Gibson, CPEC and Golden Gate methods, and compares and contrasts their respective strengths, limitations and extents of standardization.

Keywords Synthetic biology · DNA assembly · BioBricks · SLIC · Gibson · CPEC · Golden Gate

The DNA Assembly Challenge

Simply put, the DNA assembly challenge is to take a set of double-stranded DNA fragments, and physically (as well as informatically) stitch them together in the proper order and proper orientation to yield a single, potentially circular, assembled DNA sequence. These DNA sequence fragments are often referred to as 'parts' in the synthetic biology lexicon, especially when the fragments are each associated with a particular biological function, such as a promoter, a coding sequence, a terminator, etc.

Figure 14.1 depicts a representative DNA assembly. We start with 8 non-degenerate (non-repeated) composite biological parts (shown at the bottom of the figure): a vector backbone, 5 protein coding sequences (*orfA* to *orfE*) with upstream ribosome binding sites (RBS), a terminator, and a promoter. We assemble

N.J. Hillson (✉)
Joint BioEnergy Institute, Emeryville, California
e-mail: njhillson@lbl.gov

H. Koeppl et al. (eds.), *Design and Analysis of Biomolecular Circuits: Engineering Approaches to Systems and Synthetic Biology*, DOI 10.1007/978-1-4419-6766-4_14,

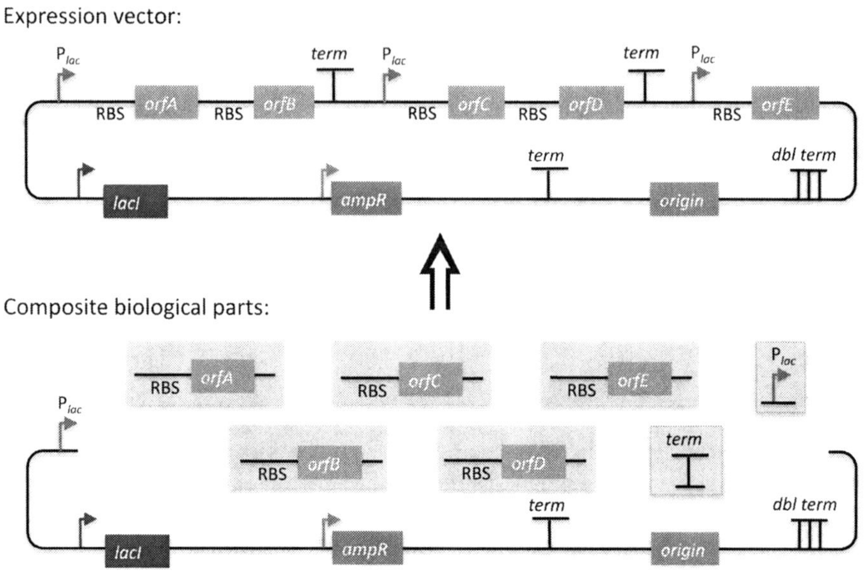

Fig. 14.1 The DNA assembly challenge

these 8 parts into the target expression vector (shown at the top of the figure). Note that in this particular example, we used the same terminator and promoter parts twice to achieve the target expression vector.

The Traditional Multiple Cloning Site Approach

Figure 14.2 shows an expression destination vector designed with the traditional approach in mind. In this specific example, a multiple cloning site (MCS) follows a T7 promoter, and is in turn followed by a T7 terminator. If a researcher wants to integrate a protein coding sequence of interest into this expression vector, he or she: (1) identifies two restriction sites present in the MCS, but absent from the coding sequence of interest, (2) PCR amplifies the coding sequence of interest with DNA oligo primers flanked with the selected restriction sites, (3) digests the PCR product as well as the destination vector with the corresponding restriction enzymes, (4) gel-purifies the digested PCR product and the destination vector backbone, (5) ligates the purified digested PCR product and destination vector, and (6) transforms the resulting ligation reaction into competent *E. coli*.

This approach works well for integrating a single coding sequence into the MCS of the destination expression vector. The large number of multiple cloning sites (11 in the vector shown in Fig. 14.2) results in a high likelihood that at least two of the sites will be absent from the coding sequence of interest. However, as soon as we would like to incorporate multiple DNA sequence fragments into the same destination vector, such as an entire metabolic pathway or genetic circuit, as shown in

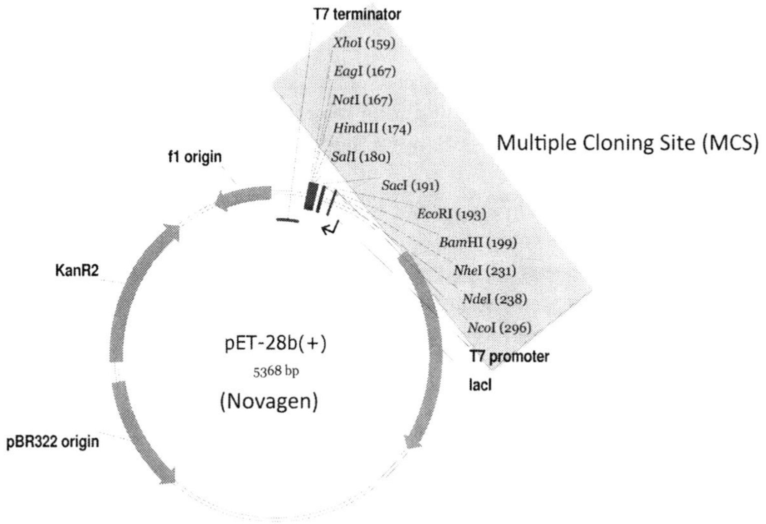

Fig. 14.2 A typical multiple cloning site expression vector

the previous DNA assembly example, Fig. 14.1, the odds are less on our side. Now, we must find as many distinct restriction sites (or resulting overhang sequences, to be more precise; with sequential sites absent from the DNA fragment they flank) as the number of DNA fragments to be assembled (including the destination vector), and two of these (flanking the linearized destination vector) must be present in the MCS. In Fig. 14.1, with ten fragments total (the terminator and promoter were each used twice), we would need ten restriction sites with distinct resulting overhang sequences, including two from the MCS, with the corresponding requirement that each is absent from its flanking assembly fragments. Generally speaking, it becomes increasingly unlikely that this constraint will be met with each additional sequence fragment to be assembled.

Molecular biologists have tackled this recurring obstacle with various strategies. A non-exhaustive set of examples includes: adding (silent) point mutations to DNA fragments to disrupt restriction site sequences, splicing together two or more fragments with PCR (e.g. splicing by overlap extension (SOE) [9]), using cohesive single-stranded overhangs that (when ligated) do not result in a recognizable/recleavable restriction site, partial DNA digests, annealing single stranded DNA fragments to yield double stranded DNA with the desired single stranded overhangs, site specifically protecting a methyl-sensitive restriction enzyme site from methylation with a DNA oligo/RecA complex [7], sequentially performing the assembly hierarchically (so that the same restriction site may potentially be used more than once; however, this makes downstream cloning and re-use more difficult), and the list goes on and on. It should be explicitly pointed out here that direct DNA synthesis, while perhaps cost-prohibitive at the moment (although certainly less so in the near future), is a very viable alternative to DNA assembly in general, and has the capacity to make many of these obstacles and concerns obsolete.

The BioBrick [13], SLIC [11], Gibson [8], CPEC [12] and Golden Gate [6] DNA assembly methods utilize, or are derived from, many of these modifications to the multiple cloning site method. What sets these methods apart from the traditional approach is 'standardization'. In traditional cloning, the set of selected restriction enzymes (as well as the point mutations made to disrupt undesired replicate restriction sites) is entirely dependent on the number, sequences and order of the fragments to be assembled. Thus, every different assembly might require a different combination of restriction enzymes, point mutations, reaction temperature and buffer conditions. Furthermore, a given sequence fragment may have to be re-cloned entirely for each new assembly, precluding re-use. While restriction enzyme companies (such as NEB and Fermentas), have made much progress in ensuring that all of their restriction enzymes can operate under a single reaction condition (temperature, buffer, etc.), in general, it is very unlikely that a single enzymatic 'master mix' can be applied across independent traditional assemblies, making the process less amenable to parallelization and automation (especially via high-throughput liquid handling robotics platforms). The BioBrick, SLIC/Gibson/CPEC and Golden Gate methods, in contrast, use the same (standardized) set of enzymes and reaction conditions for every assembly. When point mutations are required (as is potentially the case for BioBrick and Golden Gate assembly, which utilize restriction enzyme(s)), the same mutations are required for every assembly, and thereby each sequence fragment only needs to be cloned once, facilitating re-use. Thus, these standardized methods are much more amenable to parallelization and automation than the traditional approach.

The BioBrick Approach

BioBricks standardize the DNA assembly process, facilitating automation and part re-use. There are several BioBrick assembly standards, such as that originally developed at MIT [13], as well as the UC Berkeley BglBrick standard [1], which is depicted in Fig. 14.3.

In the BglBrick standard, a part (or DNA sequence fragment that is nominally associated with a biological function) is flanked with two restriction enzyme sites at its 5′ end, namely *Eco*RI and *Bgl*II, and is flanked with *Bam*HI and *Xho*I at its 3′ terminus. To comply with the BglBrick standard, these four restriction sites must be absent from the sequence of the part itself. The 'BglBrick', then, spans from the *Eco*RI to the *Xho*I site, and the BglBrick-bearing vector backbone makes up the residual plasmid sequence, which should also be devoid of the four BglBrick restriction sites. To assemble *partA* followed by *partB*, followed by the *partA*-bearing vector backbone, the *partA* BglBrick vector is digested with *Bam*HI and *Xho*I, and the *partB* vector is digested with *Bgl*II and *Xho*I. The resulting digest fragments containing *partA* and *partB* are then ligated together, resulting in the desired plasmid. The overhang sequences resulting from *Bam*HI and *Bgl*II digest are complementary (base-pair/anneal perfectly and are cohesive with one-another), but the resulting

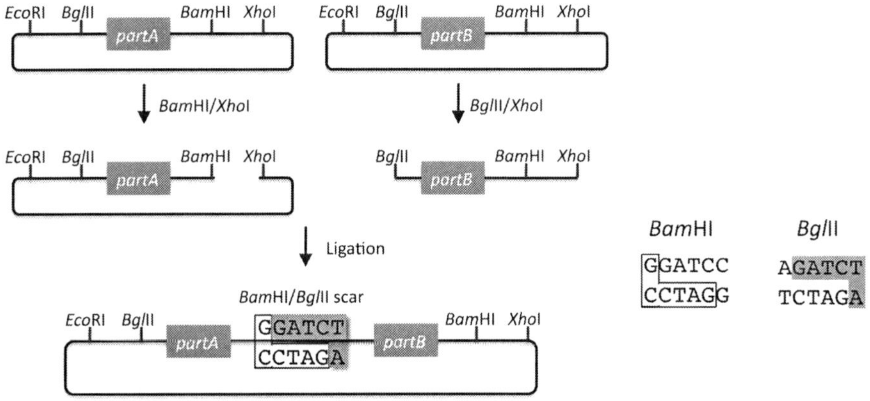

Fig. 14.3 BglBrick assembly of *partA*, *partB*, and the *partA*-bearing vector backbone

ligation product sequence is not recognized/recleaved by either *Bam*HI or *Bgl*II. Thus, the assembly results in a new BglBrick, containing *partA* followed by a 6 bp scar sequence, followed by *partB*. Since assembling two BglBrick'd parts results in a new BglBrick, this process can be iterated successively to assemble an arbitrary number of parts together, using the same protocol repetitively. It is possible to assemble *partB* in front of *partA*, and/or to select the *partA* or *partB*-bearing vector backbone for the resulting construct, by using different combinations of the four BglBrick restriction enzymes. Other BioBrick standards are completely analogous to BglBricks, and simply use alternate sets of the four restriction enzymes.

Contrasting with the traditional approach, there are several advantages to using BioBricks: (1) only four restriction enzymes are utilized, (2) once a part is BioBrick'd, it is never necessary to re-clone it (or even re-PCR amplify it, reducing the probability of PCR-derived mutations), and (3) assembling an arbitrary number of parts (in any desired arrangement) is no more difficult than putting two together (plasmid size considerations aside). It should be highlighted that, in contrast with the SLIC, Gibson, CPEC and Golden Gate methods, BioBrick assembly not only standardizes the assembly process (e.g. the set of four restriction enzymes, protocols, etc.), but also physically standardizes the BioBrick'd parts themselves, as they all have the same 5′ and 3′ terminal sequences, and internally share the same 6 bp scar vestiges of prior assemblies. There are burgeoning repositories of these standardized parts (physical and/or informatic), such as the MIT Registry of Standard Biological Parts [10] and the Joint BioEnergy Institute (JBEI) Registry, and supporting organizations, such as the BioBricks Foundation, that allow and facilitate researcher re-use of characterized and validated parts, preempting wasteful redundant efforts.

Returning to the DNA assembly challenge presented in Fig. 14.1, Fig. 14.4 shows how we could use BioBrick assembly to put together the pathway. Note that there are many different possible routes (assembly trees) to put together this pathway using BioBricks. Some of the intermediate parts, such as the terminator fused to the promoter, need only be made once, and can be re-used multiple times. Recently,

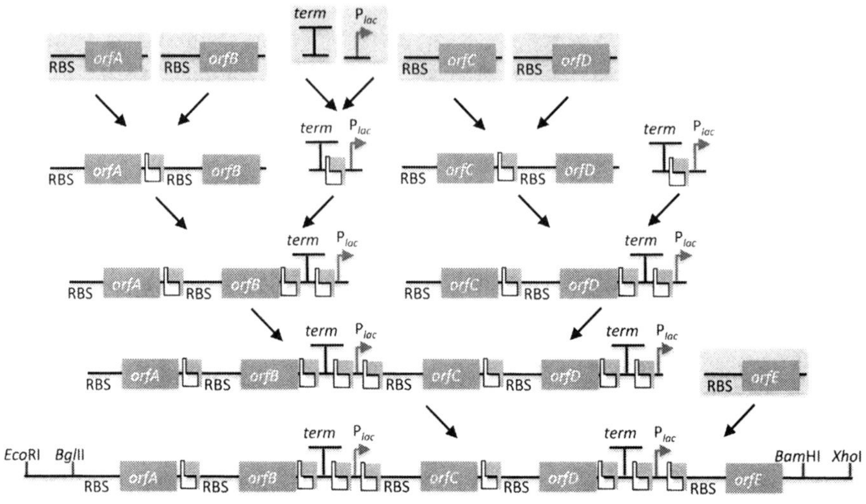

Fig. 14.4 BioBrick pathway assembly

algorithms have been developed [4] to optimize the design of binary BioBrick assembly trees, and the development of an automated in vivo BglBrick assembly process utilizing liquid-handling robotics is currently underway (J.C. Anderson, personal communication).

BioBrick Limitations and Obstacles

Despite the many merits of the BioBrick approach, there are some drawbacks. First, there is no control over the existence and sequence of the 6 bp scars resulting from each binary BioBrick assembly. While in many cases, these scars will not prove problematic, there are scenarios where the scar sequences, affecting coding sequences or mRNA secondary structure, can adversely perturb the desired protein, RBS, terminator, etc. function. Second, unless every intermediate part is archived along the binary assembly tree, it is necessary to repeat the entire process from scratch in order to replace a composite part (e.g. *orfC* in the example shown in Fig. 14.4) in the assembled BioBrick; even if all intermediate parts are archived, many of the steps must still be repeated. Third, combinatorial library diversity generation is potentially at odds with the BioBrick assembly process, because diversity must be recaptured after each and every binary assembly step (which requires aggregating approximately 5 times as many post-transformation colonies per binary assembly as the sequence diversity to be retained). Fourth, BioBrick assembly only works with previously BioBrick'd parts, and another cloning method must be used to first create the BioBricks to be assembled.

The SLIC, Gibson and CPEC Assembly Methods

SLIC, Gibson, and CPEC are related methods that offer standardized, scarless, (largely) sequence–independent, multi-part DNA assembly. Some discussion of the advantages of each method over the others is provided below.

SLIC

SLIC, or sequence and ligase independent cloning [11], as its name implies, does not utilize restriction enzymes or ligase. A DNA sequence fragment to be cloned into a destination vector is PCR amplified with oligos whose 5′ termini contain about 25 bp of sequence homology to the ends of the destination vector, linearized either by restriction digest or PCR amplification. Sequence homology regions are depicted in white and grey in Fig. 14.5.

The linearized destination vector and the PCR product containing *partA* are mixed together with T4 DNA polymerase in the absence of dNTPs. In the absence of dNTPs, T4 DNA polymerase has 3′ exonuclease activity, which begins to chew-back the linearized destination vector and the PCR product from 3′ to 5′. Once the termini of the linearized destination vector and the PCR product have sufficient complementary single-stranded 5′ overhangs exposed, they will be able anneal to each other once mixed. With the addition of dCTP, the T4 DNA polymerase changes activity

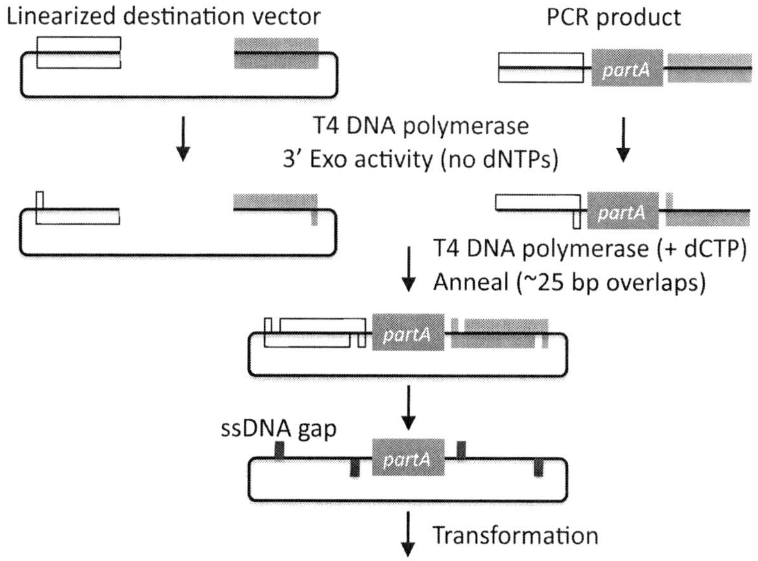

Fig. 14.5 SLIC assembly of *partA* with a linearized destination vector

from 3′ exonuclease to polymerase, but stalls because not all dNTPs are present, retaining most, if not the entirety, of each chewed-back overhang. Alternatives to the 3′ chew-back with T4 DNA polymerase in the absence of dNTPs include the use of mixed or incomplete PCR products (so this does not apply to the linearized vector backbone if it is derived from a restriction enzyme digest), which can also result in the desired 5′ overhangs, as described in the original SLIC publication [11]. The chewed-back linearized destination vector and PCR product are mixed together, and annealed to each other. Since there is no ligase in the reaction, this results in a plasmid with four single stranded gaps or nicks. Once transformed into competent *E. coli*, the gaps are repaired. Note that SLIC assembly is standardized, in that it always uses the same reaction components and conditions, scar-less, since there is no sequence in the resulting assembly that is not user-designed, and sequence–independent, as the method is not (at least to a large extent, but see below) sensitive to the sequences of either the destination vector or the part to be incorporated.

Gibson

Gibson DNA assembly, given its name after the developer of the method [8], is analogous to SLIC, except that it uses a dedicated exonuclease (no dCTP addition step), and uses a ligase to seal the single stranded nicks, as shown in Fig. 14.6.

The linearized destination vector and the PCR product containing *partA* are mixed together with T5 exonuclease, which chews-back the linearized destination vector and the PCR product from 5′ to 3′, Phusion polymerase, which (with the annealed linearized destination vector and PCR product effectively priming each

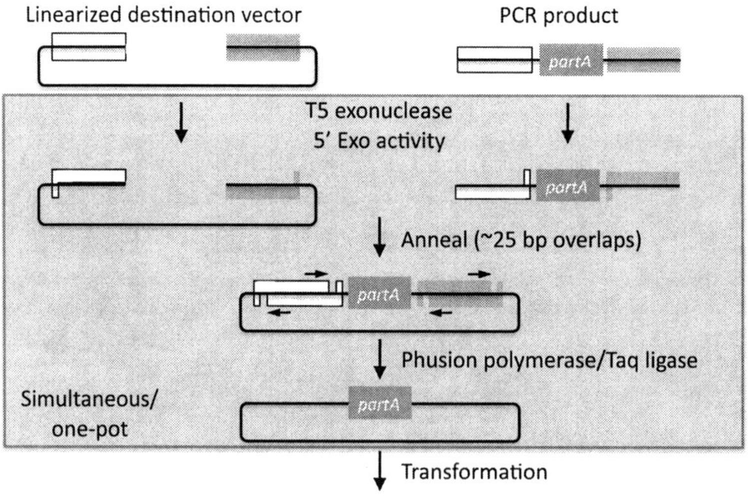

Fig. 14.6 Gibson assembly of *partA* with a linearized destination vector

other) fills in the gaps, and ligase, which seals the four single stranded nicks. The polymerase chases the exonuclease around the plasmid, with the polymerase eventually overtaking, as the exonuclease is gradually heat-inactivated (and Phusion is extremely fast). Like SLIC, Gibson assembly is standardized, scar-less, and largely sequence–independent. Gibson is advantageous over SLIC in that it is a simultaneous one pot reaction (the two-step addition of dCTP is not required), the presence of ligase may boost assembly efficiency, and since the assembly reaction occurs at an elevated temperature relative to SLIC, there may be fewer problems when somewhat stable secondary structures occur at the ends of assembly pieces; the disadvantage of the Gibson method is that the T5 exonuclease, Phusion polymerase, and Taq ligase cocktail is more expensive than that required for SLIC (only T4 DNA polymerase, or none at all if mixed or incomplete PCR products are used). An anecdotal/empirical limitation of the Gibson method is that it works best to assemble DNA fragments that are at least 250 bp in length or longer; this is perhaps due to the likelihood that the T5 exonuclease would entirely chew through a short DNA fragment before it has a chance to anneal and prime the Phusion polymerase for extension. While the same could be said for SLIC, the timing of dCTP addition provides some control in switching from the exonuclease to the polymerase activity of T4 DNA polymerase (the use of mixed or incomplete PCR products can prevent this problem all together), although caution should be applied when using SLIC to assemble small DNA fragments. Prior to Gibson (or SLIC) assembly, it is recommended to SOE (splice by overlap extension) together neighboring assembly fragments until their cumulative size is larger than 250 bp. Fortunately, the very same PCR products designed for Gibson (and SLIC) assembly, already contain the flanking homology sequences required for SOEing.

CPEC

CPEC, or circular polymerase extension cloning [12], is analogous to SOEing together the fragments to be assembled, except that no external amplification oligos are utilized (the linearized destination vector and PCR product prime each other, as in Gibson assembly) and there are typically only a few thermo-cycles required, as shown in Fig. 14.7.

Since there are no (or very few) re-amplifications of a given template sequence, PCR-derived mutations are not propagated to the same extent as one would anticipate for standard SOEing reactions. Like SLIC and Gibson assembly, CPEC is standardized, scar-less, and largely sequence–independent. CPEC is advantageous in that, since there is no exonuclease chew-back, small sequence fragments can be assembled directly without a preliminary SOEing step, there is no dCTP addition step (unlike SLIC), there is only a single enzyme (polymerase) required (unlike Gibson), and since the CPEC assembly reaction occurs at higher temperatures than either SLIC or Gibson, stable secondary structures at the ends of assembly pieces are relatively less of a concern. The main disadvantages of CPEC is that it is more

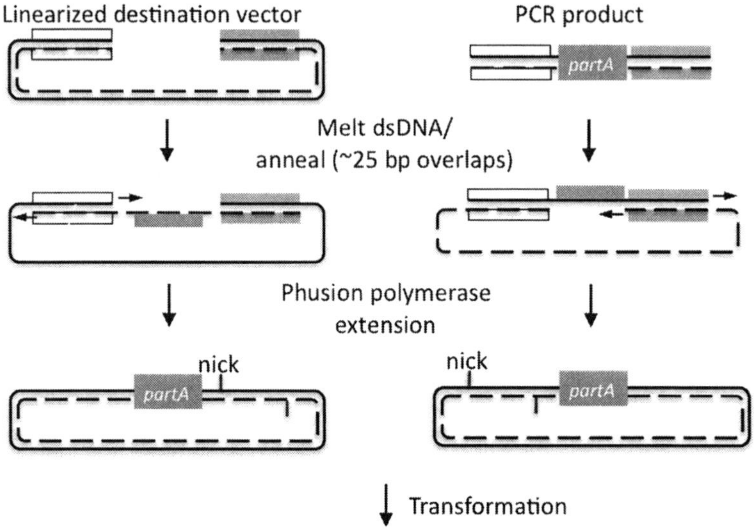

Fig. 14.7 CPEC assembly of *partA* with a linearized destination vector

likely to result in polymerase-derived mutations than SLIC or Gibson, mis-priming events are now possible anywhere along the sequences of the fragments to be assembled (as opposed to only at the termini of the fragments), and parts that are difficult to PCR amplify (such as those that contain direct sequence repeats) are also difficult to assemble with CPEC, although the Gibson method, depending on how much of a head start the T5 exonuclease has, could suffer from similar drawbacks.

SLIC, Gibson, and CPEC Similarities

Despite their differences in implementation, the SLIC, Gibson, and CPEC assembly methods all start with the same input materials and result in the same final products, as shown in Fig. 14.8. Thus, an assembly designed for CPEC will be equally applicable to SLIC or Gibson assembly. Returning to the previous DNA assembly challenge presented in Figs. 14.1 and 14.9 shows how we could use SLIC/Gibson/CPEC assembly to put together the pathway.

In Fig. 14.9, each homology region is distinctly color-coded, from light to dark and in various gradient patterns, in an analogous fashion to the white and grey homology region coloring in the single part example in Fig. 14.8. It should be noted that with SLIC/Gibson/CPEC, unlike BioBrick assembly, we can put together many parts at the same time in the same pot (multi-part assembly). Consequences of multi-part, in contrast with hierarchical binary BioBrick, assembly is that we have immediate access to each and every part to be assembled, and with only one transformation step, combinatorially generated diversity is captured a single

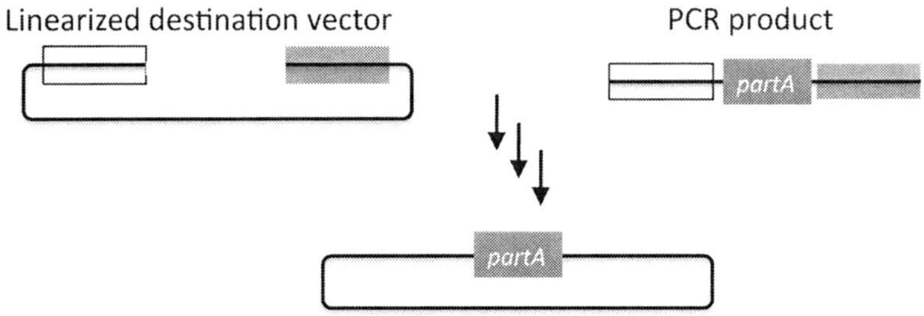

Fig. 14.8 SLIC, Gibson and CPEC assembly similarities

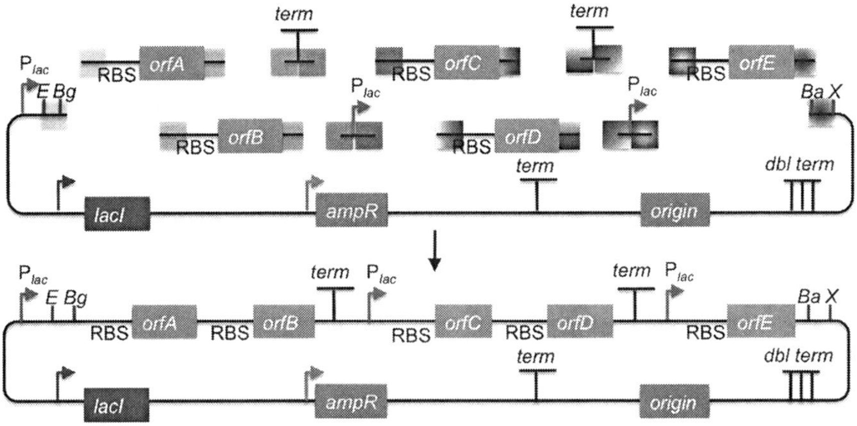

Fig. 14.9 SLIC/Gibson/CPEC pathway assembly

time. As shown in Fig. 14.9, we can use SLIC/Gibson/CPEC assembly to generate a BioBrick (BglBrick) vector, although since we didn't use BioBrick assembly during the construction process, we did not introduce any undesirable scar sequences. The downside of SLIC/Gibson/CPEC assembly is that we must now design the 5′ flanking homology sequence of each oligo specifically for each assembly junction, a process that can be tedious, laborious, and error-prone.

SLIC, Gibson, and CPEC Limitations and Obstacles

A major limitation to SLIC/Gibson/CPEC assembly is that the termini of the DNA sequence fragments to be assembled should not have stable single stranded DNA secondary structure, such as a hairpin or a stem loop (as might be anticipated to occur within a terminator sequence), as this would directly compete with the required single-stranded annealing/priming of neighboring assembly fragments. To some extent, it may be possible to mitigate this by padding these problematic termini with sequence from their neighboring assembly fragments. Repeated sequences (such as the repeated terminators and promoters in Fig. 14.9) are often obstacles to SLIC/Gibson/CPEC assembly, since assembly is directed by sequence homology, and if two distinct assembly fragments are identical at one terminus (such as the 3′ termini of the terminators in Fig. 14.9), this can lead to assemblies that do not contain all of the desired parts, or may contain parts arranged in the wrong order. To circumvent these obstacles, it is often necessary to perform a sequential hierarchical assembly so as not to place assembly fragments with identical termini in the same assembly reaction at the same time. When ever possible, it is highly preferred to substitute repeated sequences with sequence pairs that are not identical, yet encode comparable biological function; this provides a benefit not only to the DNA assembly process, but will also enhance the DNA stability of the resulting construct. Finally, SLIC/Gibson/CPEC might not be the optimal choice for combinatorial assembly if sequence diversity occurs at the very ends of the sequence fragments to be assembled (within about 15 bps of the termini), since this will preclude the reuse of the same homology sequences throughout all of the combinations. These limitations, which assert that the SLIC/Gibson/CPEC assembly methods are not completely sequence–independent, are largely addressed by the Golden Gate assembly method.

The Golden Gate Assembly Method

The Golden Gate method [5, 6] offers standardized, quasi-scarless, multi-part DNA assembly, and is an excellent choice for combinatorial library construction. The Golden Gate method relies upon the use of type IIs endonucleases, whose recognition sites are distal from their cut sites. Although there are several different type IIs endonucleases to choose from, the example in Fig. 14.10 uses *Bsa*I (equivalent to *Eco*31I) (the Golden Gate method only uses a single type IIs endonuclease at at time).

The *Bsa*I recognition sequence 'GGTCTC' is separated from its 4 bp overhang by a single bp, and *Bsa*I activity is independent of the sequences of the single bp spacer and the 4 bp overhang. The recognition site for *Bsa*I is not palindromic, and is therefore directional. In the notation used here, the recognition site is abstractly represented by a rectangle below the dsDNA line (with an arrowhead on the bottom segment of the rectangle pointing to the cut site), and the 4 bp overhang sequence is represented by a colored box (with different colors indicating different 4 bp sequences). Using this notation, the PCR product containing *partA* in Fig. 14.10 is flanked by two *Bsa*I recognition sites, both pointing inward towards *partA*, with a gray overhang at its 5′ terminus and a dark gray overhang at its 3′ end. The linearized destination vector is similarly depicted. If the PCR product is mixed with *Bsa*I and ligase, the PCR product is (reversibly) digested, resulting in three DNA fragments (the squiggly line abstractly representing the double-stranded cut), and ligated back together again. The same is true of the linearized destination vector.

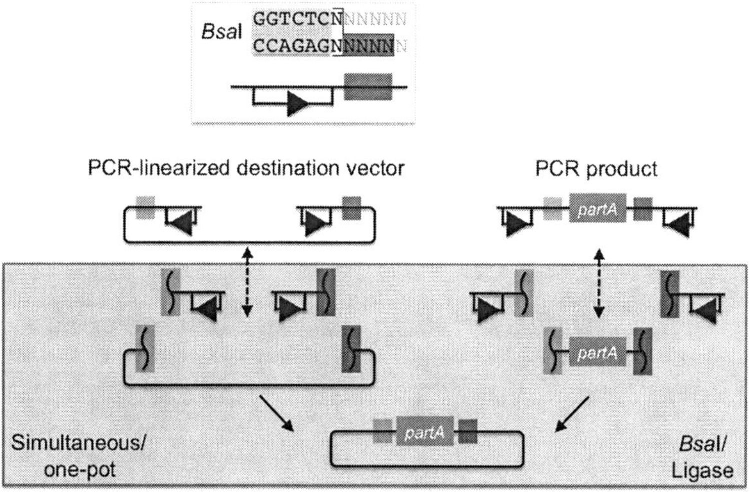

Fig. 14.10 A variant of Golden Gate assembly of *partA* with a linearized destination vector

However, if the PCR product and the linearized destination vector (each of which contains one gray and one dark gray 4 bp overhang) are both mixed together with *Bsa*I and ligase, as shown, the cut linearized destination vector will irreversibly ligate (dead-end reaction product) with the cut PCR product containing *partA*. This particular ligation is irreversible, because the ligation product no longer contains any *Bsa*I recognition sequences. Thus, over time, all reactions will tend towards the desired assembly product. It should be pointed out that the sequences of the of gray and dark gray 4 bp overhangs are (almost) entirely user-specifiable. In this regard, Golden Gate assembly is scar-less, since we have complete control over the sequence of the resulting assembly product. There are some exceptions to this (such as the overhang sequences themselves must not be palindromic (or they would be self-complimentary), and any two (e.g. the gray and the dark gray) overhang sequences must differ by at least one and preferably 2 bps so that the different overhangs are not cross-complimentary), but in general this is not an issue, because for each assembly junction there are multiple overhang sequences to choose from, within a region spanning from several bp to the 5′ to several bp to the 3′ of the junction, that still result in a scar-less assembly. It should be pointed out that the original Golden Gate method calls for the assembly using uncut plasmids, in contrast with the PCR product and the PCR-linearized destination vector shown in Fig. 14.10 [5, 6]. The proposed benefit of using uncut plasmids as the source material is that it is easier to control the assembly stoichiometry, and with each of the plasmid substrates sequence verified and without the use of PCR, accumulating PCR-derived point mutations is not a concern. The limitation of using uncut plasmids as the source material is that the destination vector, and all of the parts to be assembled, must already be cloned into a Golden Gate format plasmid system with fixed overhang sequences, which is very analogous to the physical composition standardization enforced by BioBrick assembly. While PCR amplifying the destination vector backbone and the

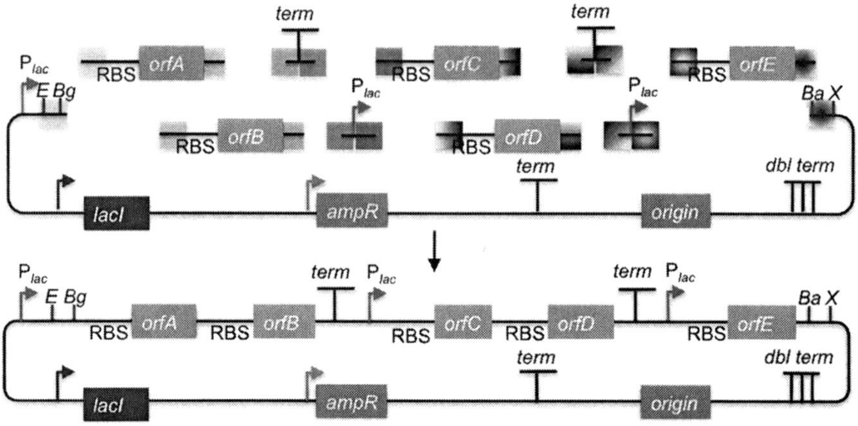

Fig. 14.11 A variant of Golden Gate pathway assembly

parts to assemble may result in PCR-derived point mutations, using PCR products as the Golden Gate assembly source material provides the freedom to use any destination vector, and any parts to be assembled into it, without an initial round of cloning that locks in the overhang sequences. One additional point is that for optimal performance of Golden Gate assembly, the linearized destination vector and the part to be incorporated should lack any additional *Bsa*I recognition sites, other than those explicitly depicted in the example above. Since the digestion/ligation reaction is reversible for any internal *Bsa*I recognition sites, it is generally not obligatory to make (silent) point mutations to remove them, however it is usually preferable to do so to maximize efficiency, and to assure that the internal overhang sequences will not anneal to the designed overhangs, and lead to incorrect assemblies.

Returning to the previous DNA assembly challenge presented in Figs. 14.1 and 14.11 shows how we could use Golden Gate assembly to put together the pathway. In Fig. 14.11, each 4 bp overhang is distinctly color-coded, from light to dark and in various gradient patterns, in an analogous fashion to the gray and dark gray overhang coloring in the previous single part example in Fig. 14.10 (the *Bsa*I recognition sites, while present and inwardly facing in all of the sequence fragments to be assembled, are not explicitly depicted in Fig. 14.11). As is true of SLIC/Gibson/CPEC assembly, we can put together many parts at the same time in the same pot (multi-part assembly), and consequently Golden Gate assembly provides immediate access to each and every part to be assembled, and with only one transformation step, combinatorially generated diversity is captured a single time. As shown in Fig. 14.11, and as is the case for SLIC/Gibson/CPEC assembly, we can use Golden Gate assembly to generate a BioBrick (BglBrick) vector, and since we didn't use BioBrick assembly during the construction process, we did not introduce any undesirable scar sequences. The downside of Golden Gate assembly (as for SLIC/Gibson/CPEC) is that we must now design the 4 bp overhang sequences for each assembly junction and incorporate them into the 5′ flanking sequence of each oligo, a process that can be tedious, laborious, and error-prone.

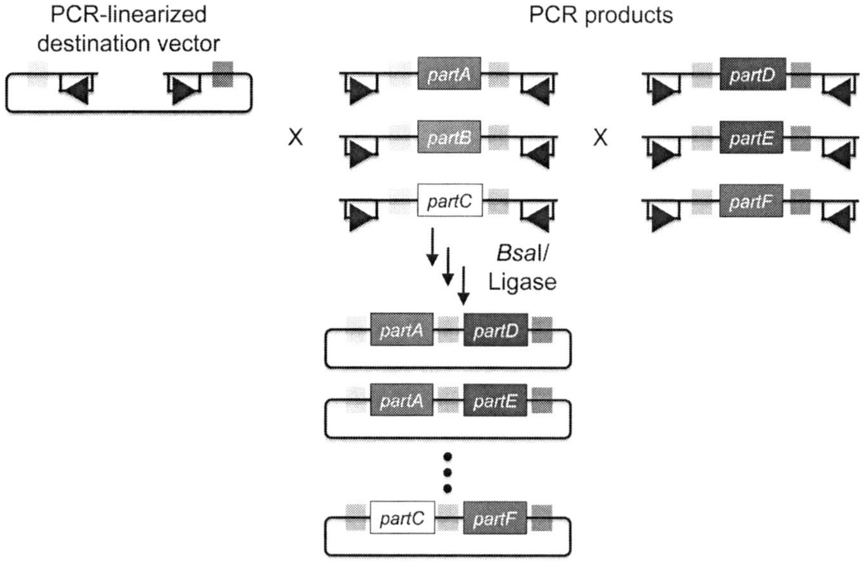

Fig. 14.12 A variant of Golden Gate combinatorial library assembly

Golden Gate assembly is a particularly good choice for constructing combinatorial libraries. As shown in Fig. 14.12, every part in each combinatorial bin (the linearized destination vector is the first bin; *partA*, *partB* and *partC* the second; and *partD*, *partE* and *partF* the third) is flanked by the same two 4 bp overhang sequences. Any two parts in a bin are completely interchangeable with respect to Golden Gate assembly, and only a single pair of oligos is required for each part across the entire assembly. Since the same 4 bp overhang sequences are used throughout a combinatorial bin, it is optimal to place the overhangs in sequence regions that are identical across all of the DNA fragments in the bin. If there are no 4 bp stretches of sequence identity at the termini of the bin's sequence fragments, the combinatorial Golden Gate assembly will result in scars (between 1 and 4 bp in length). Even so, this is far superior to BioBrick assembly that always results in 6 bp scar sequences, and very preferable to SLIC/Gibson/CPEC assembly which, while potentially scarless if all sequences have about 15 bp of sequence identity at their termini, will either result in longer scar sequences, or require many more oligos per combinatorial part to achieve a comparable scar length.

Golden Gate Limitations and Obstacles

Perhaps the most significant limitation of the Golden Gate method is that it is less sequence–independent than SLIC/Gibson/CPEC, in the sense that, like BioBrick assembly, the selected type IIs recognition site (e.g. *Bsa*I) should be absent from the internal portions of all of the DNA fragments to be assembled. However, like BioBrick assembly, once the modifications are made to remove these sites, they never have to be remade. In addition, since the overhangs are only 4 bp in length, and we would like at least 1 and preferably 2 bp to be different between each and every overhang in an assembly reaction, it may not be possible to find a set of overhangs that are compatible with each other that allows for a single multi-part assembly step, especially if the number of fragments to assemble together becomes large (greater than about ten fragments), or if the %GC content of the fragment termini is highly skewed to one extreme or the other. While rarely necessary in practice, in these circumstances, it is possible to do a hierarchical Golden Gate assembly.

Conclusion

In contrast with the traditional multiple cloning site approach, standardized DNA assembly methodologies, including BioBrick, SLIC, Gibson, CPEC and Golden Gate, enable and facilitate part re-use, assembly design automation, and high-throughput physical assembly protocols (putatively utilizing liquid-handling robotics). The utility of these standardized methods is becoming increasingly compelling as molecular biology tasks progress from single gene (over-expression) to the implementation of entire metabolic pathways and behavioral genetic circuitry, and as the emergence of parts characterization data further motivates the repeated use of well-specified parts.

The standardized DNA assembly methodologies described in this chapter (see Table 14.1) are complementary; no single method is universally preferable to the others. For example, while in vivo BioBrick assembly does not require PCR, purification, or intensive sequence validation steps, a different technique must be used to package the set of DNA fragments into BioBrick format in the first place. The SLIC/Gibson/CPEC methodologies provide largely sequence–independent scar-less multi-part DNA assembly (unachievable with BioBricks), but Golden Gate is often a better choice for combinatorial library generation.

The list of standardized methodologies presented here is far from exhaustive. Other comparable approaches, such as uracil-specific excision reagent (USER) assembly [2], offer their own sets of advantages and drawbacks. It is also important to point out that, although standardized, some methodologies, including multi-site Gateway cloning [14], which is currently limited to 4 non-vector backbone parts and results in large scar sequences, are not flexible enough for many modern synthetic biology applications, including metabolic engineering and genetic logic systems, and as such have not been discussed here.

The extent of standardization differs for each DNA assembly methodology. While the BioBrick, SLIC, Gibson, CPEC and Golden Gate methods all utilize standardized sets of reagents and protocols, BioBrick assembly additionally constrains the physical composition of the parts to be assembled. Such physical standardization may bolster the reproducibility and robustness of the assembly process, but at the same time can limit sequence design flexibility. Although not strictly necessary, it is

Table 14.1 Comparison of the presented DNA assembly method variants[a,b]

Method	Reagents
BioBrick[a]	*Eco*RI, *Bgl*II, *Bam*HI, *Xho*I, ligase
SLIC	T4 DNA polymerase
Gibson	T5 DNA exonuclease, Phusion DNA polymerase, ligase
CPEC	Phusion DNA polymerase
Golden Gate[b]	*Bsa*I, high-concentration ligase

Method	Pros
BioBrick[a]	No PCR required; very cost-effective (*in vivo*); physical composition standardization yields high reproducibility and robustness.
SLIC	Sequence and ligase independent; scar-less multi-part assembly; inexpensive reagents.
Gibson	Sequence independent; scar-less multi-part assembly; less sensitive to secondary structure at termini than SLIC; possibly higher efficiencies than SLIC.
CPEC	Sequence and ligase independent; scar-less multi-part assembly; less sensitive to secondary structure at termini than Gibson; small assembly pieces allowed.
Golden Gate[b]	Scar-less multi-part assembly; excels at combinatorial assembly.

Method	Cons
BioBrick[a]	No internal BioBrick restriction sites allowed; 6 bp obligate scar sequences; only binary assembly possible; parts to assemble must already be in BioBrick format.
SLIC	PCR-dependent; two-step assembly protocol; sensitive to stable secondary structure at assembly piece termini.
Gibson	PCR-dependent; expensive reagent costs; minimum assembly piece size 250 bp.
CPEC	PCR-dependent; termini mis-priming possible throughout assembly pieces; direct sequence repeats problematic.
Golden Gate[b]	PCR-dependent; no internal *Bsa*I restriction sites allowed; high-concentration ligase expensive.

[a]BglBrick variant
[b]PCR-dependent variant without physical composition standardization

similarly possible to standardize the physical composition of the flanking homology sequence regions for SLIC/Gibson/CPEC assembly and the overhang sequences for Golden Gate assembly. Going forward, as increasing numbers of flanking homology and overhang sequences are validated (utilizing laboratory information management systems), and as design rules are honed (with design specification tools such as Eugene [3], it is very likely that highly reproducible and robust assembly processes will be achievable without physical composition pre-requisites. Besides physical composition, an additional layer of standardization can be applied to the SLIC,

Gibson, CPEC and Golden Gate methods through the use of DNA assembly design automation tools, such as the j5 software package currently in development at JBEI, which consistently design assembly protocols, greatly facilitating the aggregation of multiple independent cloning projects in to the same work flow, and even into the same sets of 96-well plates.

As the cost of direct DNA synthesis decreases, forgoing in-house DNA assembly and outsourcing to synthesis companies will become increasingly time and cost-effective. For the forseeable future, though, it is likely that the construction of large combinatorial libraries, especially when coupled with a functional screen or selection that identifies a small subset of candidates of interest for sequence identification and validation, will be accomplished with variations of the standardized DNA assembly methods presented here, with the underlying composite parts putatively derived from direct DNA synthesis.

Glossary

cloning site A restriction site within a vector into which a particular sequence of interest may be inserted, or cloned.

direct DNA synthesis Directed chemical, or chemo-enzymatic, synthesis of (doubled-stranded) DNA with a specified sequence.

DNA assembly The process of constructing a contiguous piece of DNA from a set of composite parts.

DNA hairpin A type of DNA secondary structure in which the strand immediately (with only a loop of a few bases) folds back and anneals to itself.

DNA ligase An enzyme that catalyzes the formation of a covalent bond between two adjacent terminal DNA bases.

DNA oligo A short (generally less than 100 bp) single-stranded DNA fragment (oligomer).

DNA part A DNA sequence fragment that is often associated with a particular biological function, such as a promoter, a coding sequence, a terminator, etc.

DNA polymerase An enzyme that extends the 3′ terminus of a DNA strand (or DNA oligo) using the strand opposite as template.

DNA scar sequence An obligate (non-researcher specifiable) sequence that is introduced into a DNA fragment as a by-product of the DNA assembly process.

DNA secondary structure The physical three-dimensional structure (or two-dimensional projection there of) resulting from a single-stranded DNA fragment annealing to itself.

endonuclease An enzyme that cuts (makes (staggered or blunt) double-stranded breaks) a DNA fragment within (or adjacent to) specific recognized sub-sequence(s).

exonuclease An enzyme that digests a DNA fragment from its termini inward.

expression vector A circular piece of DNA, or plasmid, that can be introduced, or transformed, into a host organism such as *E. coli*, that, in addition to (replicative and) selective (e.g. antibiotic resistance) functionalities, may additionally contain sequence that encodes particular genes of interest and enables the expression of these genes within the host organism.

hierarchical DNA assembly Assembling multiple DNA parts together by means of a multi-level assembly tree (as shown in Fig. 14.4).

overhang sequence The sequence of the $5'$ or $3'$ single-stranded DNA protrusion extending from a non-blunt terminus of a DNA fragment.

PCR-derived mutation A DNA sequence mutation (insertion, deletion, point mutation) introduced into a PCR product as a consequence of imperfect DNA polymerase fidelity during the amplification process.

PCR product A DNA fragment resulting from polymerase chain reaction (PCR), which utilizes DNA oligos and DNA polymerase to amplify a sub-sequence of a DNA template.

restriction enzyme See *endonuclease*.

restriction site A site within a DNA fragment that will be cut by a restriction enzyme.

(silent) DNA point mutation A single base pair mutation in a DNA fragment; considered 'silent' if the mutation falls within a protein coding sequence and does not affect the resulting translated protein sequence.

type IIs endonuclease An endonuclease that cuts distal to the sub-sequence(s) it recognizes; see for example *Bsa*I in Fig. 14.10.

vector backbone Commonly refers to the portion of a DNA vector sequence that minimally encodes (replicative and) selective functions.

Acknowledgments This work conducted by the Joint BioEnergy Institute was supported by the Office of Science, Office of Biological and Environmental Research, of the U.S. Department of Energy under Contract No. DE-AC02-05CH11231.

References

1. Anderson JC, Dueber JE, Leguia M, Wu GC, Goler JA, Arkin AP, Keasling JD (2010) Bglbricks: a flexible standard for biological part assembly. J Biol Eng 4(1):1
2. Bitinaite J, Rubino M, Varma KH, Schildkraut I, Vaisvila R, Vaiskunaite R (2007) User friendly dna engineering and cloning method by uracil excision. Nucleic Acids Res 35(6):1992–2002

3. Densmore D, Kittleson JT, Bilitchenko L, Liu A, Anderson JC (2010a) Rule based constraints for the construction of genetic devices. Circuits and systems (ISCAS), Proceedings of 2010 IEEE international symposium. Paris, pp 557–560
4. Densmore D, Hsiau TH-C, Kittleson JT, DeLoache W, Batten C, Anderson JC (2010b) Algorithms for automated dna assembly. Nucleic Acids Res 38(8):2607–2616
5. Engler C, Gruetzner R, Kandzia R, Marillonnet S (2009) Golden gate shuffling: a one-pot dna shuffling method based on type iis restriction enzymes. PLoS One 4(5):e5553
6. Engler C, Kandzia R, Marillonnet S (2008) A one pot, one step, precision cloning method with high throughput capability. PLoS One 3(11):e3647
7. Ferrin LJ, Camerini-Otero RD (1991) Selective cleavage of human dna: reca-assisted restriction endonuclease (rare) cleavage. Science 254(5037):1494–1497
8. Gibson DG, Young L, Chuang R-Y, Venter JC, Hutchison 3rd CA, Smith HO (2009) Enzymatic assembly of dna molecules up to several hundred kilobases. Nat Methods 6(5):343–345
9. Ho SN, Hunt HD, Horton RM, Pullen JK, Pease LR (1989) Site-directed mutagenesis by overlap extension using the polymerase chain reaction. Gene 77(1):51–59
10. Knight TF Jr (2002) Darpa biocomp plasmid distribution 1.00 of standard biobrick components. Technical report, Synthetic Biology Working Group, MIT
11. Li MZ, Elledge SJ (2007) Harnessing homologous recombination in vitro to generate recombinant dna via slic. Nat Methods 4(3):251–256
12. Quan J, Tian J (2009) Circular polymerase extension cloning of complex gene libraries and pathways. PLoS One 4(7):e6441
13. Shetty RP, Endy D, Knight TF Jr (2008) Engineering biobrick vectors from biobrick parts. J Biol Eng 2:5
14. Sone T, Yahata K, Sasaki Y, Hotta J, Kishine H, Chesnut JD, Imamoto F (2008) Multi-gene gateway clone design for expression of multiple heterologous genes in living cells: modular construction of multiple cdna expression elements using recombinant cloning. J Biotechnol 136(3–4):113–121

Part IV
Enabling Technologies

Chapter 15
Gene Synthesis – Enabling Technologies for Synthetic Biology

Michael Liss and Ralf Wagner

Abstract Biotechnology has enabled us to render the adaptation of living natural resources from a top-down approach (breeding) to a bottom-up process (designing). Common modern cloning techniques allow for the rearrangement of genetic building blocks, the removal of cross-species boundaries and minor modifications of the DNA sequence itself. The availability of in silico gene optimization and in vitro gene synthesis from synthetic oligonucleotides has ushered a new era by confering independency of natural templates. The fast development of this technology during the last decade has dramatically advanced the availability of this service to a present level that by now outperforms classical cloning techniques in terms of flexibility, speed and costs. The exponential increase of biological sequence database contents and the growing need for genes designed for industrial applications, rather than natural function, further drives this market. The fast-growing demand for synthetic genes is attended by a rapid improvement of techniques to enable their stable and reliable production. Downscaling reaction volumes, massive parallelization and automation are integral parts in this development. With the emerging field of synthetic biology the requirements for gene synthesis expand particularly in terms of synthesis speed and construct size to allow for the construction of pathway operons or even complete viral and bacterial genomes. This challenges the engineering of novel techniques to assemble and manipulate synthetic DNA building blocks to large molecular entities efficiently to provide the necessary tools for tomorrows biotechnology.

Keywords Gene optimization · Gene synthesis · Synthetic genes · Synthetic biology · Protein expression · Biosafety

---

R. Wagner (✉)
Life Technologies Inc. Geneart AG, Im Gewerbepark B35, Regensburg 93059, Germany
and
University of Regensburg, Molecular Microbiology and Gene Therapy,
Franz-Josef-Strauss-Allee 11, Regensburg 93053, Germany
e-mail: ralf.wagner@lifetech.com

H. Koeppl et al. (eds.), *Design and Analysis of Biomolecular Circuits: Engineering Approaches to Systems and Synthetic Biology*, DOI 10.1007/978-1-4419-6766-4_15,

Introduction

The foremost attribute of evolved natural biological systems is an adapted design to survive in particular natural ecological niches, not necessarily matching specific human needs. Nevertheless, for millennia mankind has employed the principle of consecutive selection of random mutations to breed the phenotypes of crops, livestock, pets and even microbes (e.g. brewers yeast) towards features and behaviour that benefit our own survival.

Not long ago, this top-down approach was improved chiefly by accelerating the speed of mutagenesis, e.g. through colchicine or radiation treatment of seeds. However, as recently as the late 1970s the dawn of modern molecular biology allowed systematic genetic manipulation and redesign of novel strains and genetically modified organisms, mainly by removing cross-species boundaries, rearranging natural genetic building blocks and introducing minor modifications into the natural DNA sequence itself. The key prerequisite for this bottom-up development is the universality of the genetic code, since at that time all genetic templates originated from natural sources. In fact, most attempts to generate organisms with novel phenotypes still rely on a trial-and-error approach due to the fact that living systems are extremely complex by nature and far from being fully understood.

This progress in genetic engineering was a big leap, but somewhat unsatisfying, since true construction and genuine design of machines or other man-made items aim to be as flexible, yet as standardized and predictive as possible. A programming language, for example, consists of a defined syntax that is usually very compliant to solving a given problem, highly predictive and suitable for simulation. Although programming software still involves debugging cycles, this takes place on a much more rational basis than is currently the case in biotechnology.

The big goal for synthetic biology is applying such a regime to biological sciences: working with standardized parts, combining these elements according to given syntax rules but in a highly flexible way, and finally, being able to predict the effect of an assembly as precisely as possible. The prime requirement for this task is the actual availability of genetic elements that do not exist in nature. As such, de novo gene synthesis is considered the key enabling technology for synthetic biology.

Several Technological Developments Enabled Gene Synthesis

When the Nobel Prize in physiology or medicine was awarded to Werner Arber, Daniel Nathans, and Hamilton O. Smith for discovering restriction enzymes and their application to molecular genetics in 1978, an editorial comment in *Gene* pointed out that "...The work on restriction nucleases not only permits us easily to construct recombinant DNA molecules and to analyze individual genes but also has led us into the new era of synthetic biology where not only existing genes are described and analyzed but also new gene arrangements can be constructed and evaluated [1]".

This cornerstone in molecular biology gave birth to the success story of genetic engineering that we have witnessed over the last three decades. Other important milestones during this period were certainly the invention of the polymerase chain reaction (PCR) [2], cheap automated production of oligonucleotides, and high throughput dye-deoxy sequencing systems.

The ability to amplify DNA and related molecular protocols dramatically boosted the availability of natural templates otherwise inaccessible in sufficient amounts for genetic manipulation. In conjunction with easy and cheap accessibility to oligonucleotide synthesis, PCR also allowed direct and flexible manipulation of amplified DNA fragments, although usually limited to minor modifications. Introduction of larger mutations and/or rearrangements of DNA fragments remained only possible through consecutive rounds of alterations, in other words time-consuming and expensive. Furthermore, the switch from radioactive Sanger DNA sequencing to automated fluorescence-based cycling techniques significantly accelerated molecular cloning and facilitated easy examination of each intermediate step. High throughput sequencing also led to the exponential growth of available sequence information in publicly available databases. With a doubling rate of ~18 months, this in turn motivated the development of sophisticated algorithms and web applications to manage and use this vast amount of data (Fig. 15.1).

By the mid-1990s, the records of DNA and protein sequences, structural data, protein interaction networks, expression profiles, etc., became comprehensive enough to substitute for real-life experiments. Today, it is difficult to perform BLAST analysis of a sequence that has not been previously identified, in addition to finding numerous related sequences from many different species, alive or extinct. Moreover, modern next generation high throughput sequencing of complete genomes or even metagenomes predominantly store the data electronically on hard drives, rather than in tangible genomic or cDNA libraries.

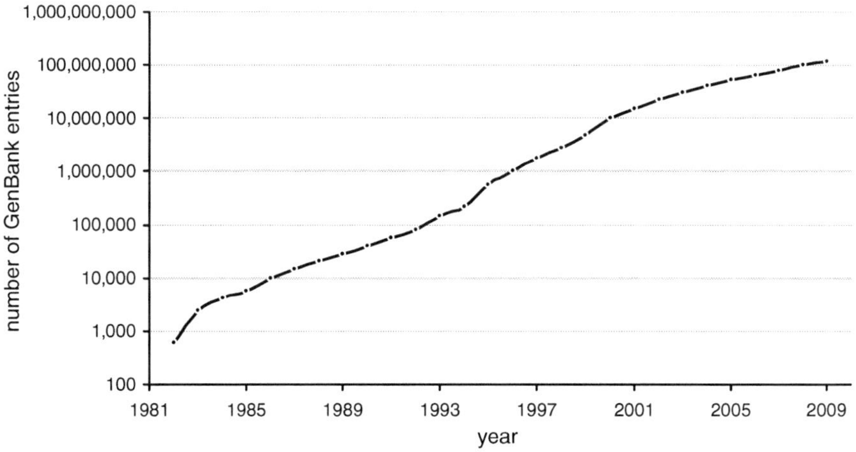

Fig. 15.1 Exponential increase in gene sequence entries in the public GenBank database (ftp://ftp.ncbi.nih.gov/genbank/gbrel.txt)

Ideally, this could free the experimenter from genetic source material, which is often difficult or impossible to obtain. Still, there remains the problem of the fundamental difference between written or electronic sequence data and its physical counterpart preserved in a string of DNA nucleotides. A machine capable of quickly converting a copy/pasted input sequence into a cloned DNA molecule was needed. Consequently, it was primarily the massive expansion of accessible data that created a demand, and therefore a market for such services.

Growing knowledge about biological systems and sequences also provided impetus from a different direction. Although the genetic code, the "language" of life, is universal, the specific requirements or "dialects" of different species can be quite unique, and also increasingly better understood with the availability of data (Fig. 15.4). If a coding sequence is to be manufactured synthetically, why not adapt the frequency of alternative codons towards the preferences of the projected host; and in addition, adjust the GC content, avoid RNA secondary structures, exclude restriction sites, and so on. This not only necessitates data-to-molecule conversion, but involves redesigning the in silico sequence itself. Computational optimization of a protein-coding gene, however, is not trivial. Since 61 codons code for 20 amino acids, the average number of possible alternative codons per protein position is $61/20 \approx 3$. Even a small protein of only 100 amino acids can thus be translated by $3^{100} \approx 5 \times 10^{47}$ different codon combinations. For multiparameter optimization, taking all the above-mentioned constraints into account, it is impossible to test all possible reading frames for an optimal match. The development of novel algorithms together with sufficient computational power was necessary to cope with this challenge [19].

Historical Overview of Gene Synthesis Milestones

The first example of de novo synthesis of a DNA sequence was demonstrated by Khorana and co-workers in 1970 [3]. In an effort taking several years they assembled a 77 bp gene encoding yeast alanine transfer RNA using short oligonucleotides obtained through organic chemistry methods. Five years later, Koester et al. synthesized the first protein-encoding gene (angiotensin II) [4] comprising 33 bp, and in 1977 scientists from Genentech and academic partners not only achieved the first example of recombinant expression of a human protein (somatostatin) in *E. coli*, they also did so without using a natural gene [5]. At that time, 9 years before PCR, it was easier to design the 14 amino acid long somatostatin gene rationally and synthesize it with organic chemistry methods rather than cloning it using a natural template. While in those days gene synthesis was still restricted by the limited availability of synthetic oligonucleotides, the development of automated oligo synthesizers and the subsequent decline in prices of related services motivated the emergence of novel gene synthesis methods. In 1981, Edge et al. used a T4 DNA ligase approach to assemble enzymatically phosphorylated oligonucleotides into a 514 bp synthetic gene encoding human leukocyte interferon [6]. The isolation of heat stable ligases by Barany and Gelfand in 1991 [7], and the introduction of the ligase chain reaction (LCR) made this strategy quite robust and reproducible, since many repeated ligation cycles can be performed at high temperatures, providing stringent hybridization conditions [8]. However, this protocol depends on phosphorylated oligos, full coverage of oligonucleotides for both DNA strands, and a limited possibility to design unique sticky ends providing enough hybridization specificity for controlled assembly. Nonetheless, this technique was the first to be commercially successful, although still labour intensive, time consuming and expensive.

Indeed, it was the invention of the polymerase chain reaction by Kary B. Mullis in 1985 [2] that made de novo gene synthesis accessible to the broad market. After PCR was introduced into genetic engineering, several PCR-based oligonucleotide assembly methods emerged based on one or more primer extension steps with subsequent amplification. Their application crossed the 1,000 bp size barrier in 1990 with the synthesis of a 2.1 kbp fully synthetic plasmid by Young and colleagues [9], and in 1995 when Stemmer et al. used 132 oligonucleotides in a single primer extension reaction of overlapping complementary oligonucleotides with subsequent PCR amplification to construct a 2.7 kbp sized vector in one step [10]. Since then, ever larger synthetic DNA molecules have been constructed, although usually put together from smaller de novo synthesized 1–2 kbp modules by classical ligation and/or recombination. The first example of in vitro construction of a complete viral genome was in 2002 when Cello et al. reported the synthesis of an infectious ~7.5 kbp poliovirus cDNA [11]. Only 2 years later, Kodumal et al. described the assembly of a contiguous 32 kbp polyketide synthase gene cluster [12], and demonstrated the functionality of the operon by successfully expressing recombinant polyketide synthase in *E. coli* and confirming its enzymatic activity. In 2006, the same laboratory synthesized a redesigned polyketide synthase gene cluster, this time taking the codon usage of *E. coli* into account. The resulting operon expressed significantly more protein than the wildtype cluster, even requiring subsequent down-regulation by promoter attenuation in order to balance relative recombinant and wildtype protein levels for optimal polyketide synthesis [13]. The current pinnacle of this advance is the compilation of an entirely synthetic bacterial genome. The group around J. Craig Venter designed, synthesized and assembled the 1.08 Mbp *Mycoplasma mycoides* JCVI-syn1.0 genome starting from digitized genome sequence information. Synthetic building blocks of ~1 kbp were first assembled from oligonucleotides and then recombined into ~10 kbp fragments in yeast. In a next step these were likewise recombined into ~100 kbp intermediates, then into the complete bacterial genome, which was further transplanted into recipient *Mycoplasma capricolum* cells. This resulted in the first self-replicating organism derived from a fully synthetic genome [14] (Fig. 15.2).

Development of Commercial Services Providing Synthetic Genes

In the setting of the brief technological overview above, it is evident that reports of de novo fabrication of long DNA molecules without requiring a natural template created an awareness of new technological opportunities and visions.

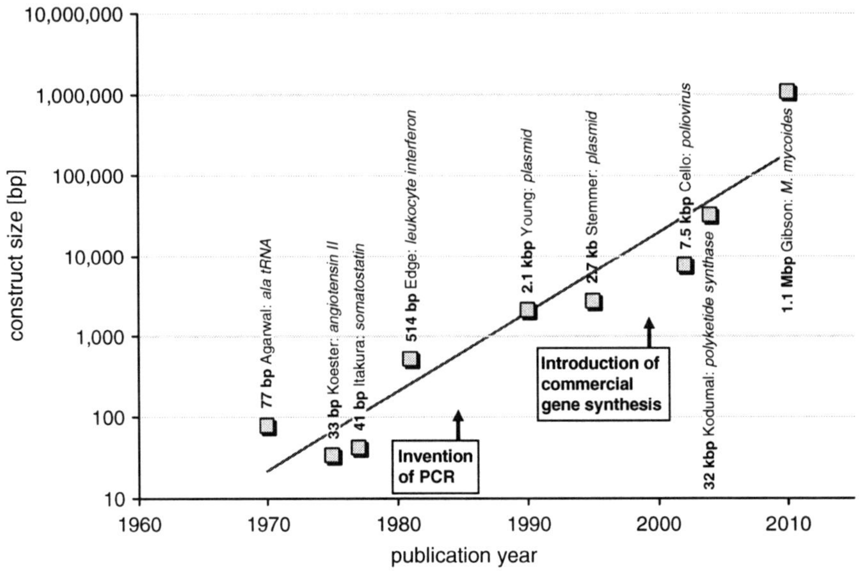

Fig. 15.2 Milestone timeline of longest so-far published synthetic DNA construct

Provided that an accessible source for synthetic genes and comprehensive databases of DNA sequences are available, these are some of the visions:

1. Working with hazardous biological organisms or tissues to isolate genetic material is obsolete and related projects can be operated solely under biosafety level one conditions.
2. Also the availability of non-hazardous DNA sources, such as genomic or cDNA libraries is dispensable.
3. Unwanted sequence variations present in many DNA samples are irrelevant and do not need to be verified or corrected.
4. Novel non-natural gene constructs can be conceived without taking the technical and temporal limitations of classical molecular biology into account.
5. The design of artificial genes can consider special requirements of the host organism to incorporate criteria such as high expression rate, genetic stability, methylation status, etc.

Taken together, de novo gene synthesis has the potential to massively influence safety, availability, reliability, throughput, flexibility, and last but not least, total project costs. Its application not only completely changes the way in which scientists think when designing cloning strategies, but also allows outsourcing of related experimental steps in order to concentrate on less trivial scientific operations. One could electronically access DNA sequences through databases, design constructs on the computer that specifically fit given requirements, with no compromise between utility and production – and then push a button so that the molecule is synthesized and shipped within a matter of days.

These promises were so tempting that consequently around the year 2000 several new startup companies appeared on the market offering such services. By then, most technologies were based on ligation of phosphorylated double-stranded oligonucleotides and protocols for their assembly involved many manual steps. Costs for oligonucleotides were still in the range of 50 ct per base and demand for synthetic genes in the market was still limited. Consequently, the business started out with a relative high price for artificial genes amounting to US$12 per bp or US$10,000–20,000 for an average sized gene. Therefore, at the dawn of commercial gene synthesis, the application of synthetic genes in scientific projects involved careful preparation and budgeting and was still far from wide-spread. It was actually a tempting alternative for researchers to accomplish the synthesis themselves. However, during the following 10 years the price for such services rapidly declined exponentially with a half-life period of ~24 months. Today (July 2010), gene synthesis costs are about $^1/_{30}$th of their original figure and have reached a level that is highly competitive with any alternative cloning method.

What was the reason for this remarkable price drop and what was the reaction of the market? Well, when around the turn of the century the demand for synthetic genes became evident, no less than 30 companies started offering this service. Some were young startups, specializing in gene synthesis; others were established enterprises incorporating it into their existing portfolio. This rapidly led to challenging competition between the providers, not only at the level of product prices but also in service coverage, quality, capacity and delivery time. In the beginning, however, pricing was the deciding factor for the customer and providers started to undercut each other. The falling market price for synthetic genes forced them to drive technological and administrative developments towards being cost-effective in a tight market and coping with an exponential increase in demand. Hence, the number of profitable gene synthesis companies decreased rapidly down to the current status of only a few major providers.

Additionally, the plummeting prices had another notable impact on the market size itself. For the first 6 years the growing demand for synthetic genes was predominantly driven by increasing awareness of their applicability through word-of-mouth recommendation and associated publications, yet it was somehow limited by the relatively high costs. Then, during 2007 the price per base pair dropped below the US$1.00 threshold – and triggered a massive burst in gene synthesis orders, confirming the previous reluctance due to pricing (Fig. 15.3).

Since nowadays related costs are no longer the vital or limiting factor for deciding to work with synthetic genes, providers concentrate more on total synthesis capacity and the reduction and reliability of delivery time. With the ever growing market and the beginning of the era of synthetic biology, the business model for gene synthesis companies changed from a high-priced low-quantity niche market provider to a high throughput supplier of a common research reagent.

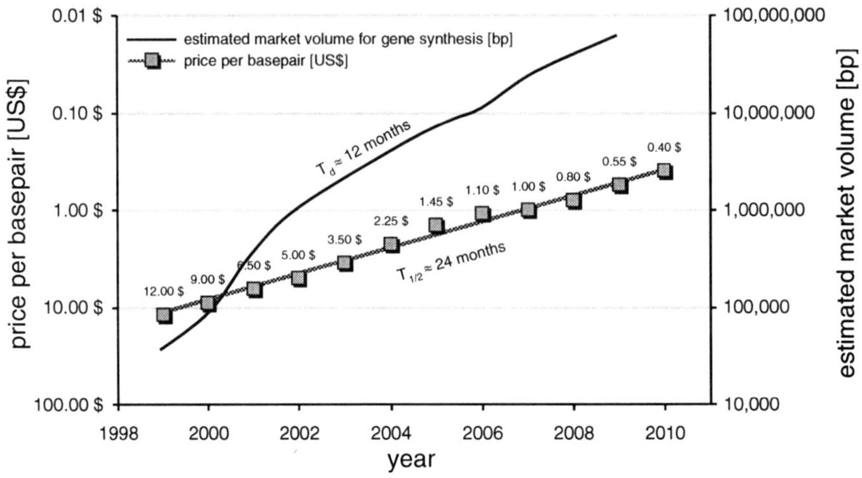

Fig. 15.3 Historical overview of the price development for gene synthesis (half-life ≈24 months) in relationship to the global demand for synthetic genes (doubling time ≈12 months)

Technological Background of de novo Gene Synthesis

The first step in gene synthesis is actually determining and specifying the sequence itself. Given the flexibility to synthesize any conceivable string of nucleotides, it is logical to alter a natural gene to ensure its best performance in the required application or experiment.

The most commonly employed modification of protein-coding genes is adapting codon usage. With the rapidly growing size of natural sequence databases, numerous sequenced genes are listed for many species – up to fully sequenced genomes of the most studied organisms. This information is also compiled into a register, reflecting the relative frequency of alternative codons in each organism, which is deposited in the Codon Usage Database (http://www.kazusa.or.jp/codon/).

Figure 15.4 illustrates part of the codon usage distribution for a representative prokaryote, fungus, plant and mammal. The preferred codons for each organism differ significantly for each amino acid. Different strategies and algorithms have been developed to best adapt a coding gene to the codon usage of the host organism. One possibility is distributing alternative codons within an optimized gene to mirror the codon frequencies of the overall target species. For example, all proline codons in a gene optimized for yeast should occur roughly in the following distribution: CCG : CCA : CCT : CCC = 12% : 42% : 31% : 15%. It is obvious that this also comprises the inclusion of rare codons, albeit only at some randomly chosen positions. Another approach is to copy the positional frequencies of the alternative codons in the wildtype gene into the frequencies of the host. For instance, a proline at position 42 within a human protein is coded by CCA, which is the third best codon for proline in humans. Optimization for yeast expression applies the 3rd best proline codon in

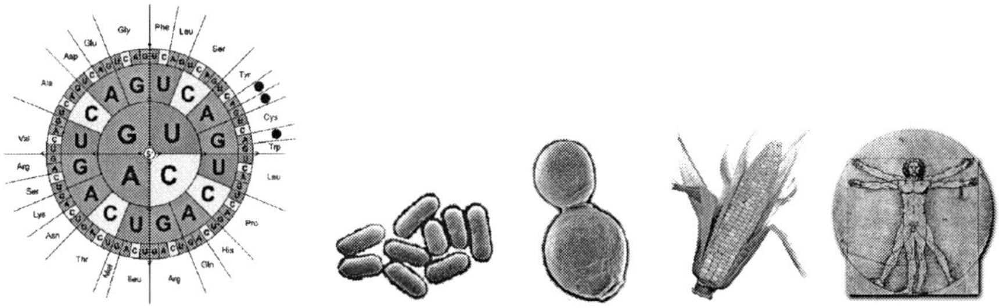

amino acid	codon	*E. coli*	*Yeast*	*Corn*	*Human*
Pro	CCG	**51%**	12%	**28%**	11%
	CCA	19%	**42%**	**25%**	**28%**
	CCT	17%	**31%**	23%	**29%**
	CCC	13%	15%	24%	**32%**
Arg	CGC	**37%**	6%	**25%**	18%
	CGT	**36%**	14%	10%	8%
	CGG	11%	4%	16%	20%
	CGA	7%	7%	8%	11%
	AGA	5%	**48%**	**26%**	21%
	AGG	3%	21%	15%	21%
Leu	CTG	**49%**	11%	**28%**	**40%**
	TTA	13%	**28%**	6%	8%
	TTG	13%	**29%**	14%	13%
	CTT	11%	13%	17%	13%
	CTC	10%	6%	**27%**	20%
	CTA	4%	14%	8%	7%
% GC		52%	40%	55%	52%

Fig. 15.4 Part of a codon usage table from *Escherichia coli, Saccharomyces cerevisiae, Zea mays* and *Homo sapiens* for the amino acids proline, arginine and leucine. The numbers represent the overall frequency of a particular codon amongst all codons for a given amino acid in that species. Frequencies <25% are in *bold* and <10% are in *grey*. The *bottom line* indicates the GC content of all coding genes in that organism

yeast, which is CCC. This system tends to assume that a protein-coding gene is already optimized for high expression in its natural organism. However, in an evolved composite context with a focus on regulation rather than maximum productivity this is rarely the case.

The most common optimization strategy to date is completely avoiding rare codons, and aiming for maximum saturation with the most frequent ones. It has been demonstrated that in *E. coli* these codons correlate with the most abundant tRNA pools [15–17] and that the relative tRNA levels do not change with expression or cellular growth. Thus, the prevalent codons still access the largest available tRNA pools feeding the translational machinery [18].

Codon choice, however, is not the only parameter when contemplating a well-designed gene. Other variables to consider are adjusting GC content, and avoiding direct and reverse repeats, restriction sites, ribosomal entry sites, cryptic splice motifs, polyadenylation signals, sequences controlling mRNA half-life, RNA secondary structures, etc.

Together, this results in multiparameter optimization, requiring sophisticated algorithms and significant computational speed [19].

When it comes down to actually synthesizing the designed gene, the initial building blocks must be produced chemically without an available template. This is still a significant cost factor in gene synthesis and different strategies have been conceived to reduce these expenses.

An appealing idea is to build the complete sequence from an existing collection of oligonucleotides covering every possible nucleotide combination – rather like writing an article via copy/paste from a collection of all possible words instead of typing single letters. That way the oligonucleotides only need to be synthesized once and can be reused. The automated sequential assembly of 6-mers from a set of all $4^6 = 4{,}096$ possible combinations has been described [20], but it remains challenging to repeatedly use and store such a collection of molecules without facing degradation problems.

Another promising, though not yet fully developed option to reduce organic chemistry costs is the application of array-based oligonucleotide synthesis methods – either using microfluidic [21] or photo-programmable chips [22] – and combining this with classical PCR-based gene synthesis techniques. Although consumption of chemicals can be dramatically reduced, after release into solution the oligonucleotides are only present in femtomolar concentrations, which is too low to allow bimolecular priming necessary for de novo gene construction. Therefore, the oligonucleotides released from the chip need to be re-amplified by PCR using universal primers, followed by endonuclease treatment to remove the universal priming region, and finally purification. These extensive post-synthesis treatments, in addition to the setup costs of the oligonucleotide chip, add to the overall costs of this new and visionary gene synthesis method [22].

Another issue with chip-based oligonucleotide synthesis is the poor accuracy of the molecules produced. The large number of sequential chemical reactions on the elongated chain, together with the inherent imperfection of each step lead to an increasing probability of incorporating a mutation within the molecule. Usually these are single nucleotide deletions, insertions or depurinations that occur with a frequency of 0.1–0.5% (0.1% = every 1,000th nucleotide has an error, or 1 out of 50 20-mers carries a mutation). When assembling many oligonucleotides into a longer contiguous molecule, the statistical clustering of mutations within a synthetic gene increases exponentially. Length of the gene and error rate of the oligonucleotides both have a dramatic effect on the final sequence accuracy. Figure 15.5 demonstrates that a synthetic gene of 1,000 bp made from oligonucleotides with an error rate of 0.1% will have a total accuracy of $(100\%–0.1\%)^{1{,}000} = 37\%$, while an error rate of 0.3% decreases the final accuracy down to 5%. In the first case one out of three sequenced clones contains the accurate sequence, in the second case one has to sequence 20 clones to find a correct one.

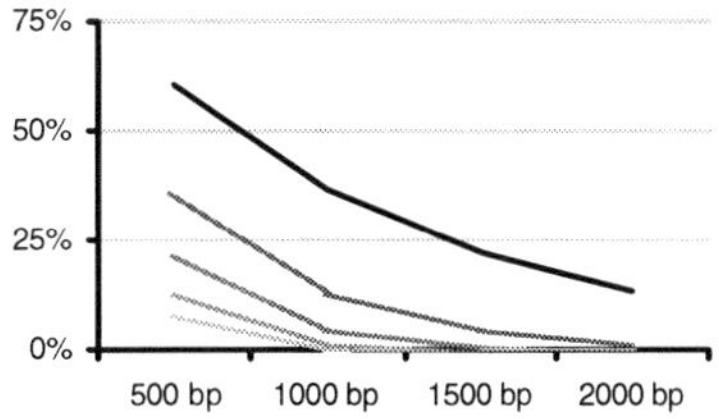

oligonucleotide error rate	gene length			
	500 bp	1000 bp	1500 bp	2000 bp
0.1%	60.638%	36.770%	22.296%	13.520%
0.2%	36.751%	13.506%	4.964%	1.824%
0.3%	22.263%	4.956%	1.103%	0.246%
0.4%	13.479%	1.817%	0.245%	0.033%
0.5%	8.157%	0.665%	0.054%	0.004%

fraction of correct full-length molecules

Fig. 15.5 Effect of oligo-synthesis error rate and gene length on total synthesis accuracy

Different techniques can be applied to reduce remaining errors in the initial gene synthesis product before ligation and transformation. After denaturation and reannealing it is highly unlikely that two synthesized complementary DNA strands with matching mutations in complementary positions will find each other. Only strands with the correct sequence can form perfect duplex DNA – all erroneous molecules will result in dsDNA carrying mismatches. These can serve as entry points for digestion by different enzyme systems or affinity purification by mismatch binding proteins [8, 23, 24]. So far, these procedures are not fully efficient either, and do not guarantee 100% sequence accuracy. They also involve additional production steps and can quickly reduce the total quantity of the DNA product to the extent that additional amplification is required before ligation is possible.

It is obvious that oligonucleotides of the highest possible grade must be used for gene synthesis to reduce downstream efforts and expenses. Compared to other reactions mentioned above, classical phosphoamidite-based solid phase oligonucleotide synthesis still provides unsurpassed quality. Unfortunately, for reasons of cost and delivery speed, most oligo providers have not designed their synthesis protocols to avoid as many errors as possible. However, the protocols can be optimized to eliminate mutations to a higher extent, while concurrently allowing for miniaturization to reduce chemical consumption, for example by conversion into DNA synthesizers in 96-well and even 384-well formats. It is likely that this technique will remain the method of choice over the next few years to provide the raw material for synthetic genes.

The next step in gene synthesis is assembling the oligonucleotides to yield longer contiguous sequences. As discussed above, the maximal final length of these constructs must be considered carefully, in order to limit the likelihood of errors in the product as well as the number of transformants to screen. Currently, the most cost-effective size of these synthetic building blocks is between 1 and 2 kbp. The assembly process is basically a multiplex primer extension reaction, taking place under controlled temperature cycling conditions (Fig. 15.6). In the first cycling round, overlapping primers anneal to each other and are filled in by polymerase to form short double strands. These can again anneal to each other in the subsequent cycle and are extended to fragments bridging four oligonucleotides. This progression continues until fragments arise containing the complete length of the intended product. Once achieved, terminal primers, present in excess, take effect and amplify the full-length product exponentially.

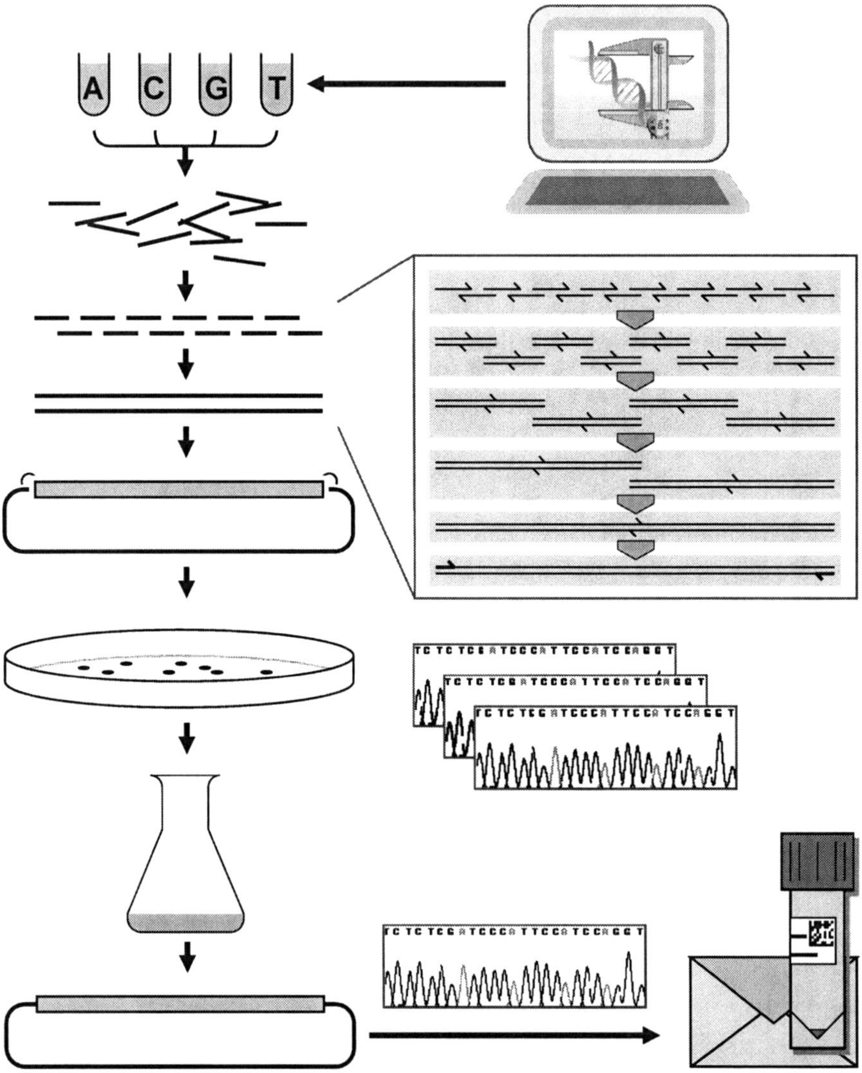

Fig. 15.6 Schematic overview of gene synthesis workflow. In silico designed sequence data are converted into a set of oligonucleotides by automated organic chemistry. These are stepwise assembled, elongated and amplified into a full-length fragment (see box), which is then ligated into a cloning vector. After transformation, *E. coli* colonies are screened for error-free insert sequences and a correct colony is cultivated for plasmid isolation. After a final sequence verification of the plasmid preparation it is ready for shipping or further assembly into larger constructs

In a next step, this PCR product is ligated into a minimal cloning vector using classical restriction endonuclease techniques. After transformation into *E. coli* and bacterial cultivation, some colonies are selected for plasmid preparation and the accuracy of the synthesized DNA construct is verified by sequencing. Altogether, conditions for mass production are chosen to have a >95% chance of picking at least one correct fragment with a single screen. This sets a limitation on the total size of the initial product, since longer molecules accumulate mutations exponentially – accordingly resulting in increased necessary screening efforts for correct fragments. Therefore, in order to compile synthetic gene constructs exceeding 1–2 kbp, they are reassembled from the sequence-verified first building blocks. Since assembling DNA elements is the nuts and bolts of biotechnology, several techniques exist to do so efficiently, although not all of them are equally apt for sequence-independent gene synthesis.

The straightforward approach for DNA fragment linkage is classical manipulation with restriction enzymes and ligase [25]. This method, however, is quite inflexible in terms of junction sequence design and involves well-known problems regarding availability and uniqueness of appropriate restriction sites. Type II class S restriction sites can eliminate scar sequences at the boundaries. These enzymes can produce sticky ends outside their recognition sequence, while the nucleotides of the adjacent cohesive stretch can be chosen freely, representing a common part of the intended compound product for ligation [26]. Designing this common part to have a length of ~20 bp allows flexible and specific attachment of two or more DNA fragments by fusion PCR, but is limited to moderate overall size and inherits an additional source of sequence errors [27]. The DISEC-TRISEC and LIC-POR methods employ the exonuclease activity of Klenow or T4 DNA polymerase to generate compatible single stranded overhangs, which are then combined with or without ligase, respectively [28, 29]. In vitro recombination extends this technology by annealing the overhangs under more stringent conditions at elevated temperatures and then filling and closing gaps with a heat stable polymerase and ligase. This already allows for efficient assembly of molecules in the range of 100–200 kbp [30]. To access even larger fragment sizes, recent protocols took advantage of the recombination efficiency of yeast, enabling the assembly of complete bacterial genomes of sizes exceeding 1 Mbp [14] (Table 15.1).

The complete process of gene synthesis – from sequence submission to shipping the final plasmid – is a process involving many different disciplines: sales, bioinformatics, organic chemistry, molecular biology, export and logistics must all play hand in hand to shift the entire workflow from small-scale to an industrial high throughput operation. A laboratory information management system (LIMS) is essential to track every intermediate in the multi-step production when dealing with hundreds and thousands of syntheses in parallel. Equally, an increasing degree of automation is mandatory to avoid exponential growth in production volume necessitating an equivalent increase in manpower. Pipetting robots communicate flawlessly with a LIMS network and *vice versa*. Some steps are also simply not manageable by humans anymore, such as the move from 96 to 384-well plates, or the decrease of reaction volumes below 1 μL. It is the interplay between LIMS, automation and miniaturization that creates the prerequisites necessary for a smooth and robust production platform enabling cheap and fast production of synthetic genes.

Table 15.1 Overview of available techniques for DNA fragment assembly

Method	Description	Accessible size/Sequence independent?	References
Type II class P restriction endonucleases	Conventional cleavage with restriction endonucleases and ligation of cohesive palindromic ends	∼25 kbp No	Maniatis [25]
Type II class S restriction endonucleases	Usage of restriction enzymes cutting outside their recognition sequence and generation of non-palindromic cohesive ends	∼25 kbp Yes	Padgett [26]
Fusion PCR	Short sequence overlaps allow dsDNA to act as megaprimers in PCR reaction and elongate to full-length constructs	∼10 kbp Yes	Mullinax [27]
DISEC-TRISEC	Concerted action of Klenow and T4 DNA polymerase produce compatible cohesive ends for ligation	∼25 kbp Not fully	Dietmaier [28]
LIC-POR	Ligation-independent assembly of compatible cohesive ends produced by T4 DNA polymerase	∼25 kbp Not fully	Aslanidis [29]
In vitro recombination	Compatible ssDNA overhangs generated by 3′ exonuclease are annealed at high temperatures, filled in by Taq DNA polymerase and linked by Taq ligase	∼150 kbp Yes	Gibson [32]
Transformation associated recombination	Homologous recombination of dsDNA in yeast	∼1.1 Mbp Yes	Gibson [14]

Fields of Application for Synthetic Genes

The first reports of genes constructed from synthetic oligonucleotides were primarily motivated by the relative complexity of attaining these molecules using alternative, perhaps not yet developed molecular techniques [4, 5]. The ensuing rapid development of genetic manipulation, in particular the invention of PCR, later offered much faster access to natural genetic material. Thus, for some years the potential of synthetic genes fell into oblivion, until the coverage of sequence databases and the limited flexibility and performance of natural genes stimulated a new need for synthetic genes.

Availability and Safety

Today, the conversion of electronic sequence data into actual bioactive molecules is a vital tool in biotechnology. In many cases, the natural source material for possibly isolating genes is simply not available, or it is too labour-intensive to conduct the necessary steps required to attain a full-length gene. Biosafety may also be an issue for choosing artificial genes, since working with isolated genes removed from the context of the complete organism is classified as level 1 (no risk) in most cases. An example where both conditions were relevant was the recent outbreak of the 2009 influenza H1N1 pandemic. The immediate public release of the heamagglutinin and neuraminidase sequences from Mexican patient isolates enabled researchers and vaccine developing companies to obtain the corresponding genes within a matter of days, without necessary shipment of potentially infectious material nor the need for biosafety precautions. Another protective measure of synthetic genes using alternative codons is their decreased ability to recombine with otherwise homologous wildtype sequences, which may be an issue with viral sequences or human oncogenes.

Origin and Reliability

Particularly industrial projects require most steps in research and production to be well documented and certified for regulatory and intellectual property reasons. This also includes the audit trail of the origin of research reagents. It can sometimes be challenging to retrace a gene's laboratory history, or it may derive from sources or collections that do not meet regulatory demands. The source of a physical gene manufactured by an ISO certified provider circumvents this problem and is a straightforward strategy for gapless documentation. It also assures the full sequence fidelity according to project design requirements, since according to experience, many constructs derived from in-house, public and commercial gene collections are not identical to the documented sequence.

Expression Efficiency

To date, most experiments in biotechnology include the recombinant expression of proteins, either to change the host's phenotype or to directly obtain and purify the overproduced polypeptide. As described before, the dissimilar genetic and biochemical setup of different species usually causes non-optimal transcription, processing, stability and translation of the extrinsic gene or mRNA. Employing multiparameter optimization allows adaptation of a coding sequence to the requirements of the host so that it performs like a native gene. Moreover, since most natural genes have not evolved for maximum expression, optimization can introduce this feature. With an overall effect on protein production yields ranging from a 10% increase to obtaining high expression of a previously undetectable gene product, optimization not only improves cross-species performance but also autologous expression, for example the production of human genes in mammalian cells [33].

Protein Performance

It is not only the genes that are in sub-optimal shape for technological and industrial purposes but also their products. Ever more recombinant proteins are being employed in healthcare, the chemical and food industry, agriculture and everyday household products. Here, they must perform under conditions that are substantially different from their previous natural environment. Viral antigens for immunization ought to be highly immunogenic, humanized antibodies for cancer therapy must recognize distinct cellular targets, enzymes in laundry detergents have to perform under the harsh conditions of a washing machine, to name just a few. Proteins need to be engineered in order to be of commercial use. However, rational computation and prediction of necessary alterations is extremely difficult, and in most cases unachievable, since we still lack sufficient knowledge to deduce the three-dimensional protein structure from its amino acid sequence. It is common practice to involve methods of directed evolution here – the generation and selection of many protein variants. While earlier methods to produce gene collections or gene libraries for this purpose involved tedious targeted, or random mutagenesis, gene synthesis provides access to these collections much faster and on a more rational basis. During gene fabrication the use of oligonucleotides carrying controlled impurities (degenerations) at defined positions allows the production of libraries that result in proteins where only the relevant amino acids are prone to substitutions. This narrows down the desired fuzziness of the variants to the areas of interest and dramatically increases the success rate of protein improvement through directed evolution.

Cost, Capacity and Speed

The considerable decline in prices for synthetic genes has today created a source of biological DNA sequences that economically outcompetes the classic genetic engineering methods. Molecular cloning steps, necessary in many projects as groundwork, can be outsourced and internal resources focused on genuine research goals. Relocating the manual DNA manipulation to an automated industrial manufacturing process also dramatically increases the processable unit size – more genes can be obtained in a shorter time – a vital necessity in the competitive domains of commercial and scientific biotechnology.

Flexibility of Design: Artificial Genes, Operons and Genomes

The freedom to access any imaginable DNA sequence not only allows the modification and adaptation of naturally occurring molecules, but also enables the manifestation of some very new visions of synthetic biology [31]. A major goal within this field is to design and construct new metabolic pathways within a producer cell. This must address three major obstacles. First, for a stable and efficient series of reactions, the enzymes involved must be expressed in a highly concerted manner. Very much like other engineering technologies, this demands the availability of standardized regulatory parts and elements. Ideally, promoters, ribosome binding sites, terminators, DNA-binding proteins, corresponding protein landing sites, etc., should be available with various well-characterized potencies and specificities. Together with sophisticated computer-aided design and simulation tools, these elements ought to be combined a priori to compile novel pathways. Secondly, fast and efficient congregation of new gene clusters or operons requires the simultaneous assembly of such parts in a robust, yet flexible way. Classical restriction sites do not allow for arbitrary combination of multiple elements simultaneously. Novel in vitro recombination technologies in conjunction with artificial modular junction sites can offer solutions in this direction.

Thirdly, establishing an extrinsic (foreign) biochemical pathway within a living cell must always be perceived in the context of the entire metabolism. The availability of only one diffusion space does not allow efficient spatial separation of distinct reaction steps, and participating intermediates can always interfere with both the projected pathway and total cell fitness. Therefore, one aim is to construct simplified "chassis" strains with genomes reduced to the lowest number of genes necessary for cellular survival and growth [32]. Here again, knocking out dispensable genes one by one by conventional methods is likely to be a highly tedious strategy. More likely, the in vitro synthesis of complete genomes, designed from scratch, will provide a much faster and more flexible way to bring these organisms to life. The technical feasibility of this approach was just recently demonstrated by the synthetic construction of a complete *Mycoplasma mycoides* genome comprising 1.08 Mbp and its transplantation into and reprogramming of *Mycoplasma capricolum* [14]. It is reasonable to assume that this cornerstone will drive further developments towards modular gene construction kits in conjunction with compatible host strains, allowing for true engineering strategies in biological sciences.

References

1. Szybalski W, Skalka A (1978) Nobel prizes and restriction enzymes. Gene 4:181–182
2. Saiki RK, Scharf S, Faloona F, Mullis KB, Horn GT, Erlich HA, Arnheim N (1985) Enzymatic amplification of beta-globin genomic sequences and restriction site analysis for diagnosis of sickle cell anemia. Science 230(4732):1350–1354
3. Agarwal KL, Buchi H, Caruthers MH, Gupta N, Khorana HG, Kleppe K, Kumar A, Ohtsuka E, Rajbhandary UL, Van de Sande JH, Sgaramella V, Weber H, Yamada T (1970) Total synthesis of the gene for an alanine transfer ribonucleic acid from yeast. Nature 227:27–34

4. Koster H, Blocker H, Frank R, Geussenhainer S, Kaiser W (1975) Total synthesis of a structural gene for the human peptide hormone angiotensin II. Hoppe Seylers Z Physiol Chem 356:1585–1593
5. Itakura K, Hirose T, Crea R, Riggs AD, Heyneker HL, Bolivar F, Boyer HW (1977) Expression in *Escherichia coli* of a chemically synthesized gene for the hormone somatostatin. Science 198:1056–1063
6. Edge MD, Green AR, Heathcliffe GR, Meacock PA, Schuch W, Scanlon DB, Atkinson TC, Newton CR, Markham AF (1981) Total synthesis of a human leukocyte interferon gene. Nature 292:756–762
7. Barany F, Gelfand DH (1991) Cloning, overexpression and nucleotide sequence of a thermostable DNA ligase-encoding gene. Gene 109:1–11
8. Young L, Dong Q (2004) Two-step total gene synthesis method. Nucleic Acids Res 32:e59
9. Mandecki W, Hayden MA, Shallcross MA, Stotland E (1990) A totally synthetic plasmid for general cloning, gene expression and mutagenesis in *Escherichia coli*. Gene 94:103–107
10. Stemmer WP, Crameri A, Ha KD, Brennan TM, Heyneker HL (1995) Single-step assembly of a gene and entire plasmid from large numbers of oligodeoxyribonucleotides. Gene 164:49–53
11. Cello J, Paul AV, Wimmer E (2002) Chemical synthesis of poliovirus cDNA: generation of infectious virus in the absence of natural template. Science 297:1016–1018
12. Kodumal SJ, Patel KG, Reid R, Menzella HG, Welch M, Santi DV (2004) Total synthesis of long DNA sequences: synthesis of a contiguous 32-kb polyketide synthase gene cluster. Proc Natl Acad Sci USA 101:15573–15578
13. Menzella HG, Reisinger SJ, Welch M, Kealey JT, Kennedy J, Reid R, Tran CQ, Santi DV (2006) Redesign, synthesis and functional expression of the 6-deoxyerythronolide B polyketide synthase gene cluster. J Ind Microbiol Biotechnol 33:22–28
14. Gibson DG, Glass JI, Lartigue C, Noskov VN, Chuang RY, Algire MA, Benders GA, Montague MG, Ma L, Moodie MM, Merryman C, Vashee S, Krishnakumar R, Assad-Garcia N, Andrews-Pfannkoch C, Denisova EA, Young L, Qi ZQ, Segall-Shapiro TH, Calvey CH, Parmar PP, Hutchison CA 3rd, Smith HO, Venter JC (2010) Creation of a Bacterial Cell Controlled by a Chemically Synthesized Genome. Science 329:52–56
15. Ikemura T (1981) Correlation between the abundance of Escherichia coli transfer RNAs and the occurrence of the respective codons in its protein genes: a proposal for a synonymous codon choice that is optimal for the *E. coli* translational system. J Mol Biol 151:389–409
16. Ikemura T (1985) Codon usage and tRNA content in unicellular and multicellular organisms. Mol Biol Evol 2:13–34
17. Ikemura T (1982) Correlation between the abundance of yeast transfer RNAs and the occurrence of the respective codons in protein genes. Differences in synonymous codon choice patterns of yeast and *Escherichia coli* with reference to the abundance of isoaccepting transfer RNAs. J Mol Biol 158:573–597
18. Emilsson V, Naslund AK, Kurland CG (1993) Growth-rate-dependent accumulation of twelve tRNA species in Escherichia coli. J Mol Biol 230:483–491
19. Raab D, Graf M, Notka F, Schoedl T, Wagner R (2010) The GeneOptimizer Algorithm: Using a sliding window approach to cope with the vast sequence space in multiparameter DNA sequence optimization. Syst Synth Biol 4:215–225
20. Van den Brulle J, Fischer M, Langmann T, Horn G, Waldmann T, Arnold S, Fuhrmann M, Schatz O, O'Connell T, O'Connell D, Auckenthaler A, Schwer H (2008) A novel solid phase technology for high-throughput gene synthesis. Biotechniques Sep;45(3):340–343
21. Zhou X, Cai S, Hong A, You Q, Yu P, Sheng N, Srivannavit O, Muranjan S, Rouillard JM, Xia Y, Zhang X, Xiang Q, Ganesh R, Zhu Q, Matejko A, Gulari E, Gao X (2004) Microfluidic PicoArray synthesis of oligodeoxynucleotides and simultaneous assembling of multiple DNA sequences. Nucleic Acids Res 32(18):5409–5417. Print 2004
22. Tian J, Gong H, Sheng N, Zhou X, Gulari E, Gao X, Church G (2004) Accurate multiplex gene synthesis from programmable DNA microchips. Nature 432:1050–1054
23. Greger B, Kemper B (1998) An apyrimidinic site kinks DNA and triggers incision by endonuclease VII of phage T4. Nucleic Acids Res 26(19):4432–4438

24. Smith J, Modrich P (1997) Removal of polymerase-produced mutant sequences from PCR products. Proc Natl Acad Sci USA 94(13):6847–6850
25. Maniatis T, Fritsch EF, Sambrook J (1982) Molecular cloning: a laboratory manual. Cold Spring Harbor laboratory, Cold Spring Harbor, NY
26. Padgett KA, Sorge JA (1996) Creating seamless junctions independent of restriction sites in PCR cloning. Gene 168(1):31–35
27. Mullinax RL, Gross EA, Hay BN, Amberg JR, Kubitz MM, Sorge JA (1992) Expression of a heterodimeric Fab antibody protein in one cloning step. Biotechniques 12(6):864–869
28. Dietmaier W, Fabry S, Schmitt R (1993) DISEC-TRISEC: di- and trinucleotide-sticky-end cloning of PCR-amplified DNA. Nucleic Acids Res 21(15):3603–3604
29. Aslanidis C, de Jong PJ (1990) Ligation-independent cloning of PCR products (LIC-PCR). Nucleic Acids Res 18(20):6069–6074
30. Gibson DG, Benders GA, Axelrod KC, Zaveri J, Algire MA, Moodie M, Montague MG, Venter JC, Smith HO, Hutchison CA 3rd (2008) One-step assembly in yeast of 25 overlapping DNA fragments to form a complete synthetic Mycoplasma genitalium genome. Proc Natl Acad Sci USA 105(51):20404–20409. Epub 2008 Dec 10
31. Heinemann M, Panke S (2006) Synthetic biology – putting engineering into biology. Bioinformatics 22(22):2790–2799. Epub 2006 Sep 5. Review
32. Gibson DG, Benders GA, Andrews-Pfannkoch C, Denisova EA, Baden-Tillson H, Zaveri J, Stockwell TB, Brownley A, Thomas DW, Algire MA, Merryman C, Young L, Noskov VN, Glass JI, Venter JC, Hutchison CA 3rd, Smith HO (2008) Complete chemical synthesis, assembly, and cloning of a Mycoplasma genitalium genome. Science 319(5867):1215–1220
33. Fath S, Bauer AP, Liss M, Spriestersbach A, Maertens B, Hahn P, Ludwig C, Schaefer F, Graf M, Wagner R (2011) Multiparameter RNA and codon optimization: a standardized tool to assess and enhance autologous mammalian gene expression. PloS ONE 6:e17596

Chapter 16
On the Construction of Minimal Cell Models in Synthetic Biology and Origins of Life Studies

Pasquale Stano and Pier Luigi Luisi

Abstract In this chapter we describe the concept of minimal living cells, defined as synthetic or semi-synthetic cells having the minimal and sufficient number of components to be endowed with the main biological properties of living cells. The construction of minimal cells starting from isolated compounds is an issue in synthetic biology, origins of life studies, and biotechnology. We start by discussing the different concepts underlining the three above-mentioned fields, by comparing the different viewpoints and highlighting common perspectives. We focus on the first two approaches, firstly describing our recent investigation on the construction of semi-synthetic minimal cells (developed in the Synthcells project), based on the use of liposomes as cell models. A short review of most relevant studies in the field is also given. The emphasis is then shifted to more basic biophysical aspects that emerged from these studies and that can significantly contribute to the understanding of the origins of primitive cells. In particular, we report the unexpected finding of spontaneous self-concentration of proteins and other solutes inside lipid vesicles. This recent discovery gives rise to a several theoretical and experimental implications that are shortly discussed. As a conclusion, we comment on the state-of-the-art in the field, next developments, and future challenges, and highlight how this research may contribute to improve our understanding of life.

Keywords Autopoiesis · Minimal living cells · Autopoiesis · Self-reproduction · Semi-synthetic cells · Liposomes · Origins of life · Protocells · Fatty acids

Why Minimal Cells?

In recent years we have been involved in a novel aspect of chemical-biological investigation, namely the laboratory construction of cell models. Such investigation is relevant in different avenues of research, from the new emerging discipline of

P.L. Luisi (✉)
Biology Department, University of Roma Tre; V.le G. Marconi 446, 00146 Rome, Italy
e-mail: luisi@mat.ethz.ch

H. Koeppl et al. (eds.), *Design and Analysis of Biomolecular Circuits: Engineering Approaches to Systems and Synthetic Biology*, DOI 10.1007/978-1-4419-6766-4_16,

synthetic biology (SB), to biotechnology, to more basic science as in the case of origins of life studies. To date, several research groups in US, Europe and Japan are developing and studying liposome-based cell models following an approach that was experimentally pioneered by one of us (PLL) in the early 1990s at the Swiss Federal Institute of Technology (Zurich). A parallel input to modern developments was also given by David Deamer (University of California Santa Cruz), who also focused his research on fatty acid vesicles as primitive cell models.

The current experimental approach to the construction of "synthetic" cells is based, as anticipated, on the use of lipid vesicles (liposomes) as microscopic self-boundary systems capable of hosting simple or complex biomacromolecular systems. In such way, the structural and functional analogy of liposome-based cells with natural cells is very evident. However, concepts, experimental approaches and historical developments of such studies vary with the different perspectives and contexts.

From the viewpoint of origins of life studies, that are at the roots of early studies, liposome-based cell model are physical model of primitive cells. They are used to gain knowledge about the transition from non-living, non organized matter to the first living systems, which are necessarily confined in cellular or cell-like structures. In this context, "minimal" cells are those cells or cell models that contain the minimal number of molecules in order to display living-like properties. The need of specifying that such number should be minimal comes from the fact that realistics model of primitive cells cannot be so complex as modern (evolved) cells. To be recalled here is the fact that the simplest living cells on our planet contain at least 500 genes, and a total of several thousand of molecular components. But precisely this high complexity elicits the question, whether all this complexity is essential for life, or whether instead the main biological functions can be accomplished with much simpler systems. This question comes also from the consideration that early cells, the ones which started our life on Earth, must conceivably be much simpler than the modern cells. In this way, then, minimal cells should be simpler than current living cells. The construction of a minimal cell would experimentally show *a possible route* for the emergence of living systems from non-living molecules, taking advantage of simple physico-chemical factors such as self-assembly, self-organization, and self-confinement. We will see later how the theory of autopoiesis helps in codifying the desired pattern in a rigorous way.

From the viewpoint of biotechnology and SB, synthetic cells can be seen as (1) a biotech tool or (2) a product of modern and sophisticated bioengineering. In the simpler version, synthetic cell-like systems (not necessarily living systems) are complex molecular systems organized within a membrane boundary that can be used for a variety of application, such as molecular screening, bioreactors, biosensors, and ultimately as advanced drug delivery systems [85]. A visionary perspectives – also described by Le Duc and coworkers in a recent review [118], foresees the use of synthetic cells, endowed by synthetic genetic circuitry inside, as targetable vehicles for in vivo nanomedicine. A short description of this view will be given below. On the other hand, also for the quickly growing SB, the concept of minimal cells is very germane [28,29,42]. From the viewpoint of classical SB (bioengineering-inspired), the construction of minimal cells might mean the minimization and/or reduction approach of already-existing cells at the aim of removing unessential part, so that the resulting still-living cells can be manipulated for useful applications. Such approach can be developed only thanks to recent technical and theoretical advancements (often integrating systems biology), but the companion concept of genome reduction and minimal genome has been largely discussed from the viewpoint of comparative genomics. The work of Craig Venter and coworkers on the assembly of a synthetic genome and its transplantation [33] can be seen as an example of synthetic cells in the SB paradigm.

Very recenty, however, one of the authors (PLL) has coined the term "chemical SB" for describing another SB approach to synthetic cells, which remarkably differs from the standard SB way [58]. By chemical SB we mean the research on synthetic biological systems (from molecules to whole cells) that shares with SB the constructive and sometimes engineering approach, but does not aim to optimize systems by genetic manipulation for achieving a pre-determined performance. In chemical SB we aim to construct synthetic systems, e.g., synthetic or semi-synthetic minimal cells, at the aim of understanding why a certain kind of system is successful, and another is not. We therefore aim to achieve a kind of biological learning by constructing models, i.e. a constructive approached opposed to the analytical (dissecting) one. We believe that in this way we can see some underlying principles of the emergence of cellular life. This approach actually complements the classical approach of origins of life studies (bottom-up pathway from simple low molecular-weight molecules to macromolecules to molecular systems) because many of the key transitions in molecular evolution have been not clarified yet or give rise to unanswered questions (the reader interested in open questions in origins of life can find two recent overviews in Stano and Luisi [89, 99]). The philosophical implications of this approach are discussed in a recent article on the epistemic basis of synthetic biology [59].

In this chapter we will shortly review the most relevant experimental approach to the construction of semi-synthetic minimal cells, which is one of our currently developed research program (e.g. in the FP6 Synthcells project), starting from early studies on liposomes to most modern systems based on transcription-translation inside liposomes, to important biophysical effects at the basis of cell assembly.

Autopoiesis and Minimal Life

Autopoiesis (from the Greek, self-production) is a theory developed by Maturana and Varela [109] dealing with the question "what is life?", and attempts to define, beyond the diversity of all living organisms, a common denominator that allows for the discrimination of the living from the non-living. Despite its simplicity and logical robustness, autopoiesis is not a familiar concept in mainstream biological sciences. This is partly due to the fact that autopoiesis theory is not centered on DNA, RNA and on replication, and makes only a minimal use of the term "information".

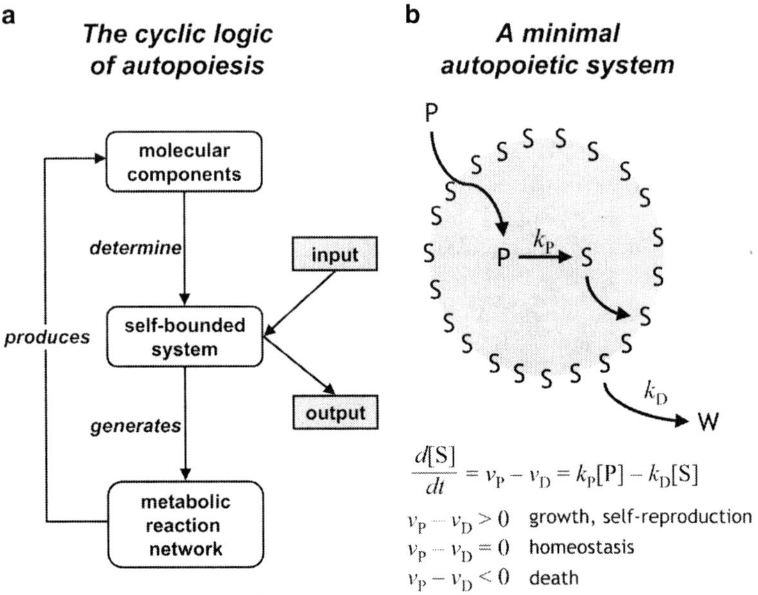

Fig. 16.1 Autopoiesis and minimal autopoietic systems. (**a**) The circular logic of autopoiesis as minimal life. The process of living is seen as cyclic, one in which the internally produced molecular components assemble into the self-bounded functional structure, which generates the microenvironment reaction (metabolic) network, which then produces the molecular components... and so on. The system exchange energy and matter with the external environment. (**b**) A minimal autopoietic system is constituted by a self-bounded system, which can uptake a precursor P from the environment, transform it by one or more reaction(s) into the boundary element S, which can also undergo a degradative process to W. Depending on the relative rates of these processes, the autopoietic system can grow, stay in a homeostatic state, or die (Reproduced from Stano and Luisi [100]. With permission from The Royal Society of Chemistry)

Autopoiesis is a systems theory dealing with the logic of cellular life, identifying the main activity of the cell as the maintenance of its own dynamic and structural organization. This occurs despite the large number of transformations taking place inside its boundary and involving all elementary components of the cell (enzymes, metabolites, RNA, etc.). Self–maintenance is carried out by a constant regeneration (from the inside) of all components (boundary molecules included), that are also continuously transformed and/or disposed of. This is possible thanks to a network of processes that produces all components, that in turn generate the processes producing such components, and so on (Fig. 16.1a) [56]. The outcome is a kind of circular organization. This dynamics is sustained thanks to the external supply of chemicals or energy, being the autopoietic cell a thermodynamically open system. Notice, however, that the internal set of transformation can be maintained and sustained only if the dual entities of process and components are consistently and recursively linked to each other, a feature typical of an operational closure [56, 109].

Can autopoiesis be used as a theoretical framework for the development of experimental research? The answer is yes, and the first examples of "autopoietic chemical systems" have been given by Luisi in the early 1990s [62] by using self-assembling supramolecular systems as micelles [3], reverse micelles [2] and vesicles [111]. We will discuss the latter case in details in the next paragraphs, but let see how and at what extent is it possible to design and construct autopoietic (molecular) systems in the laboratory. By autopoietic we mean in this case, that the synthesis of new elements takes place from within the boundary of the original system. This concept is very clear in the case of aqueous micelles of fatty acids, which are taking up the water-insoluble precursor-like fatty acid esters [3]. The term autopoietic has been used by our group also in the case of self-reproduction of vesicles. In this case the chemistry leading to self-reproduction takes place on the bilayer of the vesicle [111], but this can still be considered chemistry taking place within the boundary. Self-reproduction is only one mode of the activity o fan autopoietic system, and Fig. 16.1b shows a simplified cartoon representing the complete activity of a minimal autopoietic system.

An autopoietic system sustains itself by transforming external components (here indicated as P) into the elements (S) of the autopoietic system, which self-organize into the autopoietic unit. The transformation, here indicated by a simple process (P becoming S) occurs within the system, i.e. within the self-generating boundary that separates and distinguishes the system from the environment. In other words, working out of equilibrium, an autopoietic cell continuously uptakes from the environment the precursor P that is internally transformed into the building block S. Eventually S is destroyed to give a waste product W. In more general terms (by imaging a more complex autopoietic cell), all structural components are continuously renewed by two concurrent anabolic and catabolic processes; but despite this continual turnover of components, the whole organization of the autopoietic unit does not change. Thanks to this mechanism, it is possible to imagine (and possibly construct in the laboratory) simple or complex molecular systems that are based on an autopoietic mechanims. The difference will be in the nature of P, S and of the P-to-S transformation. The building block(s) S can be a single molecule or a set of molecules or macromolecules. The P-to-S transformation can be a single reaction or a set of multiple reactions, like a minimal metabolism. What is important is that the peculiar self-producing dynamics is maintained. From the viewpoint of minimal life, we are therefore interested in finding, first conceptually and then experimentally, the set of (bio)chemical transformations that give rise to an autopoietic system, looking for its minimal complexity. This would correspond to a minimal autopoietic system, possibly living (for a discussion on the definition of life and autopoiesis, see [8,54,56]). Looking at the transformation rates depicted in Fig. 16.1b, it is easy to see that an autopoietic cell grows when the rate of production of S overcome the rate of its disappereance. When the opposite is true, the autopoietic cell dies; and only when the two rates are the same, the autopoietic cell enter into a stationary state that keeps it unchanged (homeostasis). Of the three different outcomes, the first one is particularly interesting, because the growth can be followed by a division of a large autopoietic unit into two or more daughter units, so realizing a self-reproduction of autopoietic cells, i.e., the way to the proliferation of cellular structures from a parent one. Several experimental studies have been devoted to growth and self-reproduction, whereas there is only one report on autopoietic homeostasis [117]. We will shortly comment the results with "naked" lipid vesicles (the interested reader can refer to a recently published review [100]), whereas a more detailed discussion will be focused on liposome-based minimal cells.

Minimal Cells: Concepts and Experimental Approaches

The starting point of our discussion will be the realization of liposome-based cell models and the transition from such simple systems to minimal biological cells.

Self-reproduction of Liposomes

Liposomes, or lipid vesicles, are microscopic water-filled hollow structures (Fig. 16.2) formed spontaneously by self-assembly of lipids or of other molecules

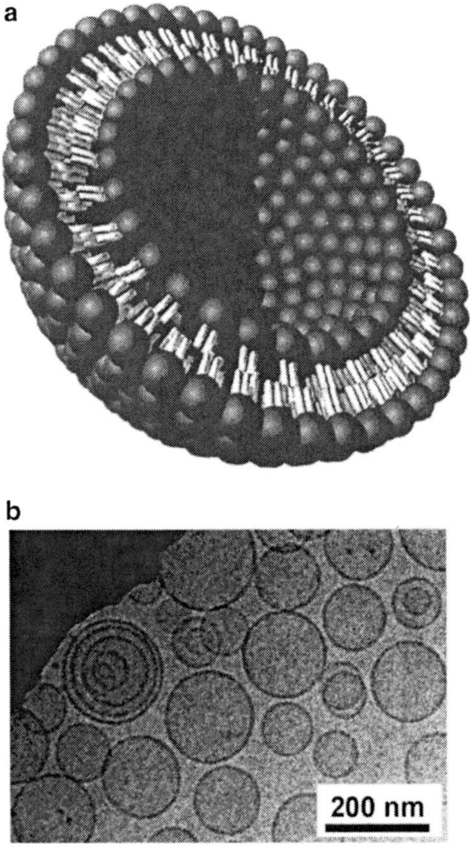

Fig. 16.2 Model of the cross section of a (quite small) spherical unilamellar liposome (**a**), and liposomes as seen by cryo-transmission electronmicroscopy (**b**)

(fatty acids, cationic surfactants, block-copolymers). Since the time of their discovery by Alec Bangham [5] it was immediately clear that they could have been used as cell model, initially to study biological membranes, and later on as more complex cell models. The hydration of lipids in aqueous solution brings about the formation of liposomes in short time, and it has been demonstrated that a variety of molecules (small and large molecular weight; polar or ionic), initially present in the aqueous phase, can be entrapped inside liposomes. Hydrophobic molecules are also entrapped by solubilization in the membrane. Today, together with the classical medical application of liposomes (essentially as drug delivery systems) their use as cell model is well recognized. In particular, there is a large consensus on the fact that in a certain stage of prebiotic evolution liposomes have played a key role for the emergence of cellular life.

This shared view is based on two important pillars, namely: (1) the discovery that very simple molecules as fatty acids (found in meteorites [51] and possibly synthesized abiotically [24, 69, 90] spontaneously form stable vesicles [32, 37]); and (2) the discovery of the spontaneous self-reproduction of fatty acid vesicles at the expenses of a precursor [9, 53, 86, 111].

Fatty acid vesicles, e.g., "oleate" vesicles, form vesicle in a pH range from ca. 7.5 to 9.5. At this pH, the carboxylic group of oleic acid is partially transformed into carboxylate (approximatively 50% when pH $=$ 8.5) so that the oleic acid/oleate mixture is structurally competent for self-assembly in the lamellar phase, giving rise to membranes and therefore vesicles. At lower and higher pH values oleic acid is present in form of oil droplets and oleate micells, respectively. It has been shown that oleic acid vesicles may grow and divide at the expenses of externally added precursors, namely oleic anhydride or oleate micelles (reviewed in [100]). In this way, they establish an autopoietic self-reproduction mechanism that closely resembles the general scheme shown in Fig. 16.1b.

In the first series of experiments, oleic anhydride has been used as a precursor of fatty acid vesicles [111]. In particular, when oleic anhydride (a water insoluble compound that forms a separate oily phase) is added to an aqueous solution at pH 8.5, only a very small amount of anhydride is converted to oleate, namely the tiny soluble fraction and possibly the molecules at the anhydride/water interface. In contrary, when preformed oleate vesicles are present in the solution, the hydrolysis of the oily anhydride proceeds much faster. In both cases a sigmoidally-shape kinetic profile indicate that the underlying mechanism is autocatalytic. Following mechanistic investigations, essentially based on kinetic data and light scattering experiments, it has been shown that oleate vesicles uptake oleic anhydride in their membrane, where it is hydrolized to form two new oleate molecules, so that the vesicle surface growth. Such growth brings about new vesicles by division of the parent one, most probably because of the physical instability of the enlarged vesicles. The fact that the overal vesicle number increases means that more and more oleic anhydride molecules are then taken up by the newly formed vesicles, and the product of such dynamics is that the number of vesicles further increases, in autocatalytic fashion. This mechanism has an autopoietic signature, but its specific feature comes from the fact that vesicle growth is followed by vesicle division (this is not implicit in autopoiesis), and it can be seen as an outcome of vesicle physics. Taking advantage of such mechanism, it was shown that it is indeed possible to design and realize vesicle-based cell models by enclosing a biochemical reaction such as the poly(A) synthesis from ADP catalyzed by polynucleotide phosphorilase [112], or the RNA replication catalyzed by the Qbeta replicase [78] into self-reproducing liposomes. More recent studies have used ferritin-containing vesicles to address the question of mechanism and solute redistribution in self-reproducing vesicles [6, 7].

One of the limits of the anhydride-based approach is that the reaction occurs in a biphasic system, making a real-time monitoring of oleate vesicle transformation difficult, and hindering mechanistic investigation. The second approach to autopoietic vesicle self-reproduction is based instead on oleate micelles [9, 15, 20, 53, 66, 86, 88, 102, 119]. As described before, oleate form vesicles or micelles depending on the pH, i.e., depending on the protonation state of the carboxylate group. This means that it is possible to transform oleate micelles into oleate vesicles thanks to a pH change. Such transformation is analogous to the oleic anhydride to oleate vesicle one, but has the advantage of occuring in one phase (the aqueous phase), and of being subject of spectroscopic observations. Spectroturbidimetry, dynamic light scattering, fluorescence measurements, stopped-flow techniques, laser-scanning microscopy and electronmicroscopy have been used to investigate in great detail such phenomenon, that is analogous to the growth and transformation of oleate vesicles fed by oleic anhydride. Also in this case, oleate molecules are taken up by pre-existing vesicles, whose surface increase brings about to a growth and a division into new vesicles. The progression of vesicle number is also in this case autocatalytic and sigmoidal kinetic profiles are obtained. Although there is still not full agreement on the molecular detail of such process (for critical reviews, see [65, 100]), a remarkable effect has been described, namely the "conservation" of vesicle size in different generation of vesicles obtained by self-reproduction. This effect has been named "matrix effect" to emphasize how the size of pre-existing vesicles (the matrix) affects the size of newly formed one [9, 53, 86]. Freeze-fracture electronmicroscopy studies evidence the presence of possible self-reproduction intermediates, the "twin-vesicles" (Fig. 16.3) [102] that may account for a possible pseudo-symmetric division mechanism.

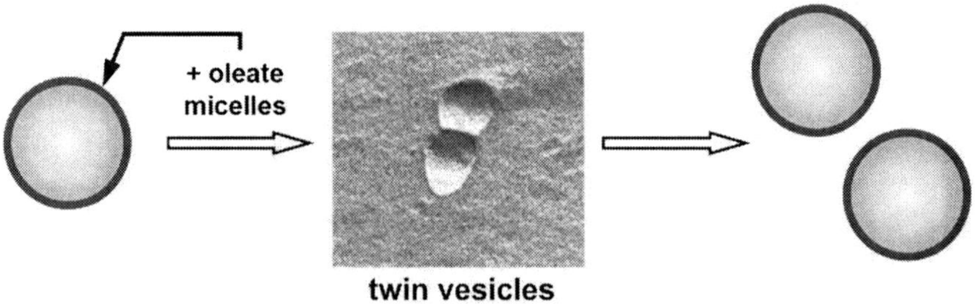

Fig. 16.3 Freeze–fracture electron microscopy of oleate vesicles (taken after 40 s from the addition of oleate micelles) reveals that "twin vesicles" can be the actual intermediates of the self-reproduction. Twin vesicles are not present at the end of the reaction (Reproduced from Stano and Luisi [100]. With permission of the Royal Society of Chemistry)

In recent years the number of investigation on fatty acid vesicles increased considerably, and more information is now available on their membrane permeability, stability in the presence of magnesium, growth and other transformations. Reviews of this work (from Szostak's laboratory) have been recently published [10, 67, 93].

In conclusion, the self-reproduction of "naked" vesicles and a rich landscape of vesicle behavior are well established facts, that can be seen as the one of the fundaments of more complex studies on vesicles as cellular model (the other fundament is provided by the realization of simple enzymatic reactions inside vesicles, see next section). Clearly, due to the key importance of self-reproduction in living systems, the fact that it occurs in very simple vesicles readily prompted the next question, namely: is it possible to build a vesicle-based cell model that self-reproduce thanks to a minimal internal chemical "metabolism"? Such question is equivalent to say (see Fig. 16.1b): is it possible to build a minimal autopoietic cell capable of self-reproduction?

Vesicle-Based Minimal Synthetic Cells: From the Origins of Life to Synthetic Biology

In the context of synthetic cells, the concept of autopoietic self-reproduction can therefore be exploited to design and build vesicle-based systems that share with living systems some of their essential properties. Before starting the discussion, it must be recalled that also other surfactant-based self-assembling compartments, such as micelles and reverse micelles, are capable of self-reproduction, and actually their behavior was discovered before the case of vesicles [2, 3, 62]. Clearly, from the viewpoint of biological analogy, both from the side of origins of life and synthetic biology the use of such structures – more distant from the cellular structure than lipid vesicles – is not very appealing. However this cannot be true if one is interested in making a purely synthetic molecular system without the need of being bound to cell-like architecture. This is the case of a research program called "Los Alamos Bug" that aims to construct micelle-based minimal living particles by combining micelles self-reproduction, PNA as genetic material, and a minimal catalytic mechanism based on transition metal catalists [87].

Coming back to vesicle-based systems, constructing a minimal cells means to design – according to the requirement of the study – a kind of minimal metabolism that should be implemented within the compartment, in order to mimic as much as possible the autopoietic mechanism of Fig. 16.1. One of the first attempts dates back to 1992, when Luisi and coworkers [92] designed a lipid-producing enzymatic system inside lipsomes (see below).

In general terms, to design such kind of systems (not only limited to the lipid production, however, but aimed to build an autopoietic cell) we can find at least two possibilities which may conceptually differ in the question they would like to answer to, but essentially follow experimentally similar methodologies (Fig. 16.4a). The

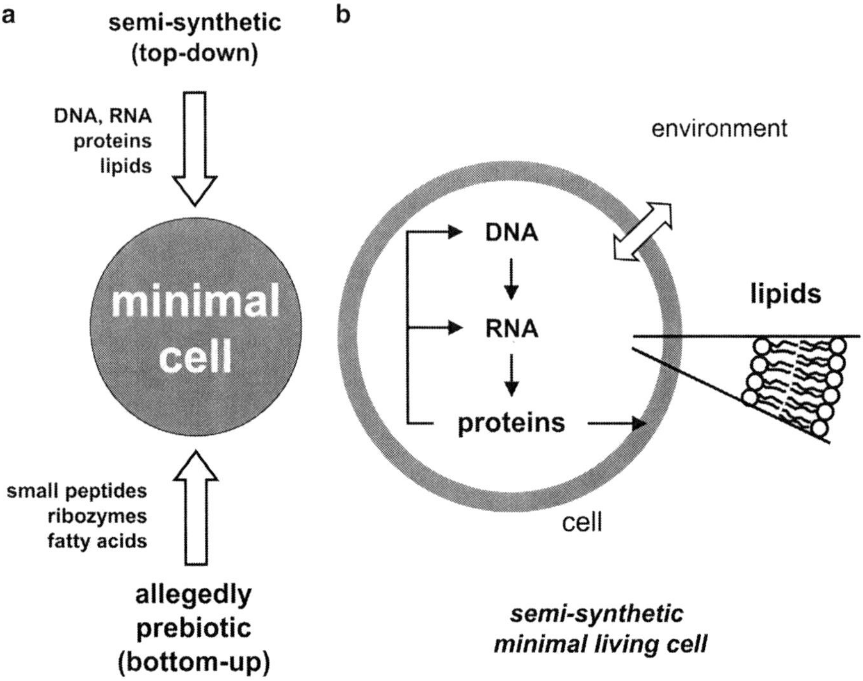

Fig. 16.4 Comparison between semi-synthetic and prebiotic approach to minimal cell (**a**). The notion of semi-synthetic minimal cell based on the entrapment of DNA, RNA, and proteins inside lipid vesicles (**b**)

two main routes are sometimes known as "bottom-up" and "top-down" approaches, although such nomenclature needs some specification (and possibly be revised accordingly). Modern living cells – the only autopoietic living structures we know – derive from millions of years of evolution. Minimal cells, on the other hand, are structures composed by the minimal and sufficient number of components. We have to find a connection between these two aspects thanks to two different perspectives: primitiveness and minimization.

The first route, sometimes known in literature as "bottom-up" [55, 63] is characterized by the research of a plausible path that starts with simple molecules and ends with living cells. It is therefore a typical origin of life scenario, where prebiotic plausibility has as dominant role. In particular one has to look for chemical compounds that may give rise to an autopoietic mechanism and that are allegedly formed in prebiotic conditions. Here the schools of metabolism-first or RNA-first scenarios foresee, respectively, the emergence of minimal cells based on the encapsulation of primitive metabolism based on small molecule catalysis, primitive enzymes, transition metal ions, etc. (for a general discussion on the origins of early cells see [73, 94]); or the formation of RNA-based cells containing only ribozymes capable of self-replicantion, peptide ligase or peptide transferase activities, and lipid synthesis [104]. Both views have pros and cons, but both lack extensive experimental investigations. This is clearly due to the difficulty in defining prebiotically plausible compounds, intermediates, and conditions like pH, temperature, redox regime, and so on. To date, this route to minimal cell has not been very successful, but it is in principle the preferred way for constructing realistic chemical models of primitive cells.

Investigation on minimal living cells based on well-known biochemical species is the second route, sometimes known in literarture as "top-down" [63]. Clearly, primitiveness in strict sense cannot be associated to this route, but it can be of great interest when the resulting minimal cells are viewed as simplified cell models endowed by a minimal set of functions, with a minimal complexity. From the experimental viewpoint, the "top-down" approach (also known as semi-synthetic one) is already quite advanced, due to the availability of compounds to be used in constructing such simplified cells (DNA, RNA, ribosomes, enzymes, etc.). It consists in the insertion of the minimal number of present-day biomolecules inside liposomes, aiming at obtaining a minimal living cells (Fig. 16.4b). The "minimal genome" specify the minimal number of genes required for a minimal cell. One of the latest version of the minimal genome – based on comparative genomics – is due to the group of Moya [34], consisting in 206 genes, classified as follows: 16 genes for DNA processing, 106 genes for RNA processing and protein synthesis, 15 genes for protein processing, 5 genes for cell processing, 56 genes for basic metabolism, 8 poorly characterized genes. A functional genomic analysis has provided the number 151 for the number of genes for a minimal cell [28]. The hypothetical organism characterized by the minimal genome is simple when compared with natural living microorganisms, but still represents a challenge for laboratory construction.

A possibly alternative (and third) route to minimal cells could be instead instead quite different from the first two, we may call this as fully synthetic route and foresees the use of metal catalysts entrapped inside polymersomes (liposomes made by block copolymers), or by other fully synthetic or hybrid combinations.

Relevance in Basic Science, Biotechnology, Drug Delivery

Before starting the more technical discussion on the state of the art of minimal cell research, it is important to emphasize their multi-facets relevance in diverse fields.

Clearly, the construction of a synthetic or semi-synthetic minimal cell is per se a fundamental scientific enterprise, regardless possible applications in biotechnology or medicine. In fact, the first scientific question that a minimal cell address is on the capability of demonstrate that life is indeed an emergent property of a chemical system composed by elements that are non-living. This view is probably accepted by the majority of scientists, but has not been experimentally demonstrated yet. Constructing minimal life in the lab does not mean "playing God". It is firstly a way to deepen our knowledge on what is life and what are the minimal phisico-chemical and organizational features that allow the emergence of it as a system property. Being a property of the whole (higher hierarhical level), it stems from the properties of the single chemical components at the lower hierarhical level, but there is no need of a central control unit, as evident in the autopoietic mechanism (Fig. 16.1). It is possible to get such knowledge by the synthetic approach, which nicely complement the analytical (dissecting) one, typical of many biological investigation of the twentieth century. In the context of origin of life, it is also clear that the synthetic approach to the understanding of the origin of cellular life is the only one that can be follow experimentally, due to the fact that we do not have primitive cells to analyze. Therefore, keeping in mind two main constrains, namely the prebiotic plausibility of chemical structures and a kind of continuity between primitive and modern cells, the research on primitive cell models might actually fill some of the gaps in the scientific scenario of origin of life hypothesis. The last aspect inherent to basic science is related to the field of complex systems, which are often investigated by computer simulation, and the need of experimental approaches would enrich significantly such studies.

Biotechnological applications of synthetic cells (intended here as produced by a SB approach) can be sketched only approximately, because the potential applications are so wide that it is difficult to foresee the scenarios. These goes from fabrication of fine chemicals for pharmaceutical industry to hydrogen production, from polymer synthesis to bioremediation, to biosensing. The recent report of Craig Venter and colleagues on the assembly of a synthetic cell from a synthetic chromosome and a natural cytoplasm [33] of a recipient cell has created interests and fear in the society. Actually, despite the tremendous impact of synthetic biology on the media, the real challenge is now to produce something useful [49].

The third field of interest of synthetic cells (possibly not alive) is in pharmacology and medical diagnostics. In recent years, considerable progress has been made toward using liposomes as drug delivery vehicles [50]. From the viewpoint of synthetic cell, however, one can imagine more sophisticated systems, aimed, for instance, to protein delivery or to the use of minimal cell as nanofactories for local production of drugs. Due to their low bioavailability, many naturally occurring proteins can not be used in their native form in diseases caused by insufficient amounts or inactive variants of those proteins. The strategy of delivering proteins to biological compartments using carriers represents the most promising approach to improve protein bioavailability [4]. On the other hand, it is very fashinating the idea of creating a minimal cell that can reach the site of desease, establish a two-way communication with natural cells and act consequently, providing the optimal response to the requirements (Fig. 16.5).

This visionary scenario implies that a synthetic minimal cell should be endowed of internal genetic/metabolic circuitry that are able to send/receive chemical signal and capable of being activated consequently [118]. These performance are not yet achieved experimentally but it seems that the semi-synthetic approach can indeed be developed in such direction.

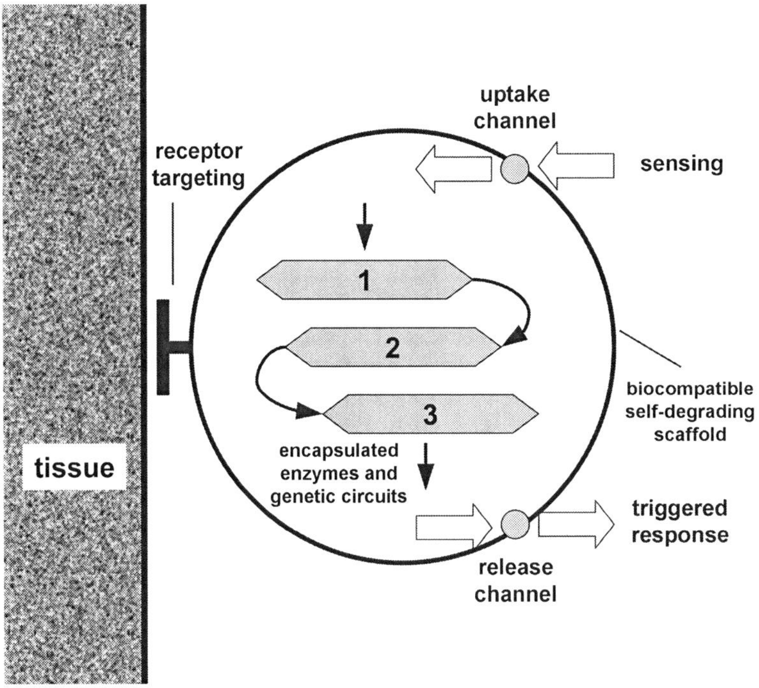

Fig. 16.5 Minimal semi-synthetic cell for nanomedical applications, as proposed by LeDuc and coworkers. Redrawn after [118]

The State of the Art of Minimal Cell Studies

A road map for the construction of semi-synthetic cells consists in sequential steps of increasing complexity, starting from very simple biochemical reactions inside liposomes, to more complex design, up to reach the desired (minimal) complexity of a semi-synthetic living cell. We have classically distinguished the milestones in this research as: (1) simple enzymatic reactions inside liposomes, (2) transcription/translation reactions; (3) self-reproduction of the core and of the shell components. We will shortly review some of the most relevant advancements. Notice, however, that milestone 3 is not yet reached whereas milestones 1 and 2 can be considered as almost standard achievements. Further discussion can be found in Luisi [18, 57, 63, 100].

Enzyme Reactions Inside Liposomes

The realization of simple enzyme reactions inside liposomes can be seen as the first milestone toward the construction of semi-synthetic cells. It is important to mention, however, that the research on confined enzymes is not only relevant to the construction of semi-synthetic cells. One of the (still open) question is whether – and at what extent – enzyme reactivity might change inside small compartments. This can shed light on unexpected behavior of confinement and compartmentation of metabolic-like reactions. Experiments with one enzyme inside liposomes can be considered a standard achievement, although particular enzymes could have a problematic manipulation procedure, due to sensitiveness to standard liposome formation methods, to their specific properties (for instance, membrane enzymes), or to biochemical incompatibility with lipids forming the liposome membrane. Liposomology has a variety of techniques that allow the formation of liposomes in quite different way, so that it can be said that is virtually possible to entrap all kind of enzymes inside liposomes. The use of enzyme-containing liposomes generally aims to study the properties of the enzymes in confined space, or as indirect to study the membrane permeability of a certain solute, or to develop a simple "bioreactor". In many cases, the substrate is added externally and it penetrates into the liposomes by diffusion, by using a channel (e.g. porins [36, 107, 115], hemolysin [76]), or by doping the liposome membrane with sublytic concentration of detergents [80, 105]. Once inside, the product of the reaction diffuses away or stays entrapped inside. It is not possible to list here the very huge amount of work done with entrapped enzymes inside liposomes (including: alkaline phosphatase, amylase, asparaginase, chymotrypsin, elastase, galactosidase, lysozyme, pepsin, perossidase, glucose oxidase, glucose-6-phosphate-dehydrogenase, hexokinase, glucuronidase, phosphotriesterase, superoxide dismutase, tyrosinase, urease, carbonic anhydrase, luciferase, lipase, etc.). A recent review has been published by Walde and Ichikawa [110].

Multi-enzyme Reactions Inside Liposomes

In contrary to the simpler cases, less work has been done on performing multi-enzymatic reactions inside liposomes. Clearly, this is a relevant issue to be investigated because the cellular metabolism is indeed a multi-enzymatic reaction network catalyzed by enzymes. Therefore, investigating the properties of simpler networks is of great value for origin of life studies (recalling the concepts of minimal and primitive metabolism, and of the spontaneous onset of metabolic and autocatalytic cycles), as well as in synthetic biology (in order to develop bioreactors for diverse biotechnological applications or in drug delivery).

Although not very well known, the work initiated by Thomas M.S. Chang about 40 years ago on encapsulation of cell homogeneate inside 50 mm cellulose or nylon artificial microcompartments can be considered as the first attempt to construct multi-enzymatic vesicle-based systems. Aiming to develop what he defined "enzyme therapy" (i.e. the administration of enzymes in order to counterbalance physiological deficiency in the blood), Chang designed several systems containing cyclic enzymatic routes. For example, it is described the preparation of a four-enzymes system (urease, glutamate dehydrogenase, glucose dehydrogenase, glutamate-pyruvate transaminase) that converts urea, pyruvate and glucose to alanine and glucuronate (glutamate and ketoglutarate are intermediates) [14].

Modern research has been also carried out, although in limited way, by encapsulating enzymes in the vesicle's water core or by chemical binding on the membrane. Again, our report of 1991 on the co-entrapment of four enzymes inside lecithin vesicles aimed at synthesizing lecitin from within must be recalled here [92]. This work also started the modern attempts to develop semi-synthetic minimal cells. As it will be mentioned later, the aim of this work was the in vitro biosynthesis of phosphatidylcholine from glycerol-3-phosphate, acyl-CoAs and CDP-choline by exploiting a four-enzymes path reconstituted inside liposomes.

Looking at recent literature, the preferred coupled system is composed by glucose oxidase and horseradish peroxidase inside liposomes [39,44] or polymersomes [25, 108], possibly with the involvement of a third enzyme (a lipase). The other system that has received attention is the couple bacterhodopsin/ATP synthase (both membrane proteins) that have been reconstituted in liposomes at the aim of producing ATP after irradiation [19, 30, 84]. Interestingly, this system shows how physical energy (photons) is transformed into chemical concentration energy (proton gradient), which is finally converted into chemical bond energy (ATP molecules).

It is clear that, in contrary to DNA or RNA based systems that will be introduced in the next section, no much attention has been given to the realization of metabolic cycles inside liposomes. The field of multi-enzymic reactions inside lipid vesicles, being not well investigated, can indeed help in understanding the establishment of confined biochemical networks and their physico-chemical properties. We argue that next generation of enzyme-containing vesicles might well involve multi-enzymic routes and cycles.

DNA Synthesis, RNA Replication, Protein Synthesis

We have now reached the main corpus of experiments on minimal cell construction, namely the synthesis of biopolymers (DNA, RNA, proteins) inside lipid vesicles. These advancements can be defined as the state of the art. In our previous review [63] we listed about 15 papers that describe DNA synthesis (via PCR inside liposomes), RNA synthesis, and protein synthesis by coupled transcription-translation. To date, there are more than 30 of such reports, and the technical sophistication increased correspondingly. This impressive acceleration (+100% in 4 years) is certainly due to the fact that more scientist find the construction of synthetic cells very attractive and challenging. Moreover, the popularity of SB certainly affected positively on the spreading of this research. It is not the aim of this chapter describing in details all published results; a compilation of records can be find elsewere [18, 63, 101].

Here we would like to comment on the general strategy related to such studies and their relevance for the construction of minimal living cells by commenting few selected examples. Emphasis will be given to the connection between such systems and autopoietic mechanims that should underlie the dynamics of a minimal living cell.

Two Selected Cases of Pioneering Research

Historically, the first reaction of this type (synthesis of biopolymers) carried out inside lipid vesicles was reported independently by Luisi and coworkers [112] and by Deamer and coworkers [13]. It consisted in the polymerization of ADP to give poly(A) (a ribonucleic acid, although with the trivial sequence A-A-A-...) catalyzed by the enzyme polynucleotide phosphorilase (PNP). The reaction was called "the Oparin's reaction revisited" [112] by referring to the analogous reaction carried out decades before by Oparin inside coacervates, the early cell models based on coagulation of macromolecules into microparticles [83]. The reaction has some relevance because it is a way to synthesize RNA from activated nucleotides without a template. Here, coacervates were substituted by lipid vesicles, in particular oleate vesicles [112] or DMPC vesicles [13]. In the first case, it was possible to carry out poly(A) synthesis inside liposomes and a simultaneous liposome self-reproduction, by feeding vesicles with oleic anhydride (see section "Self-reproduction of Liposomes"). In other words, during the occurrence of poly(A) synthesis, the vesicles grow thanks to anhydride uptake and hydrolysis, divide and give new vesicles. Quite probably (but it was not demonstrated) the new vesicles could still support poly(A) synthesis, at least until PNP is present in each vesicle [112]. In this way, it was possible to run simultaneously liposome internal reactions and the liposome shell growth.

The results of a similar approach were published 1 year later by Oberholzer et al. [82], who carried out RNA synthesis inside oleate vesicles. RNA synthesis was achieved by means of a RNA-dependent RNA polymerase (Qbeta replicase; [38,70], at the expenses of nucleotides triphosphates and on a RNA template (co-entrapped). Also in this case, the internal RNA synthesis was carried out simultaneously to external oleic anhydride addition, so that the well-known mechanism of uptake-hydrolysis and vesicle growth/division could take place. Again, in principle the reaction could occur also in second- and third-generation vesicles if Qbeta replicase was still contained inside such vesicles.

It is important to remark the similarity of these two first examples (Fig. 16.6).

In both an internal chemical reaction produced a ribonucleic acid (RNA) thanks to enzyme catalysis (PNP or Qbeta replicase). In both, the vesicle self-reproduction was triggered by external oleic anhydride addition. Despite the success of observing the desired behavior, such design was not totally satisfactory for two reasons: (1) it would be advisable to synthesize lipids from an internalized reaction, as requested by autopoietic scheme (Fig. 16.1b); (2) the internal reaction proceeds only if the catalyst (PNP or Qbeta replicase) is maintained after generations, but this is clearly impossible because their number is constant and they do not undergo reproduction. In other words, some of the new vesicles will be not capable of supporting internal RNA synthesis because of the "dilution" of the catalyst among newly generated vesicles. We have called this problem as "death by dilution".

How to face these two limitations? The first one can be approached by inserting the lipid-synthesizing enzymes inside liposomes, and produce lipids at the expenses of some available precursors. Early attempts have already focused on the enzymatic production of lecithin in lecithin liposomes [92]. The metabolic

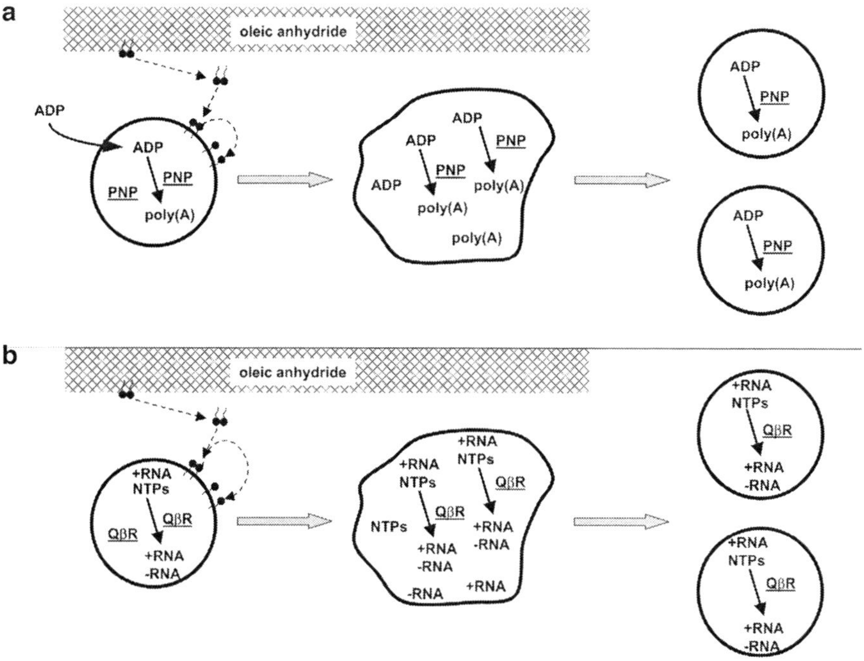

Fig. 16.6 Synthesis of: (**a**) poly(A) [112]; (**b**) and RNA [82] in self-reproducing vesicles. See text for details

pathway chosen for carry out this synthesis was the so-called salvage pathway, which converts glycerol-3-phosphate to phosphatidic acid, to diacylglycerol and, finally, to phosphatidylcholine. The four enzymes needed to accomplish these reactions were simultaneously inserted into liposomes by the detergent depletion method, and the synthesis of new phosphatidylcholine (10% yield) was followed by radioactive labelling, but it was not possible to clearly demonstrate the liposome self-reproduction. More recently, we have studied again this system by expressing the first two enzymes of the path inside liposomes (namely, glycerol-3-phosphate acyltransferase G3PAT and lysophosphatidic acid acyltransferase LPPAT), starting from the corresponding genes (Fig. 16.7) [47]. After careful optimization of the conditions for liposome encapsulation, protein synthesis and protein/lipid interactions, the two proteins were expressed in functional form. Despite the great success of synthesizing for the first time an active membrane-enzyme inside liposomes (G3PAT), the low lipid production yields hindered any morphological change observation.

It is therefore evident that one of the key step for assembling a minimal living cell is the synthesis of the vesicle shell. Such synthesis should be carried out by elements of the minimal cell itself, as indicated by autopoiesis. From the chemical viewpoint, "membrane synthesis" means "lipid synthesis", or – in more general terms – synthesis of water-insoluble compounds typically characterized by long aliphatic chains. The lipid salvage pathway uses, as source of long acyl chains,

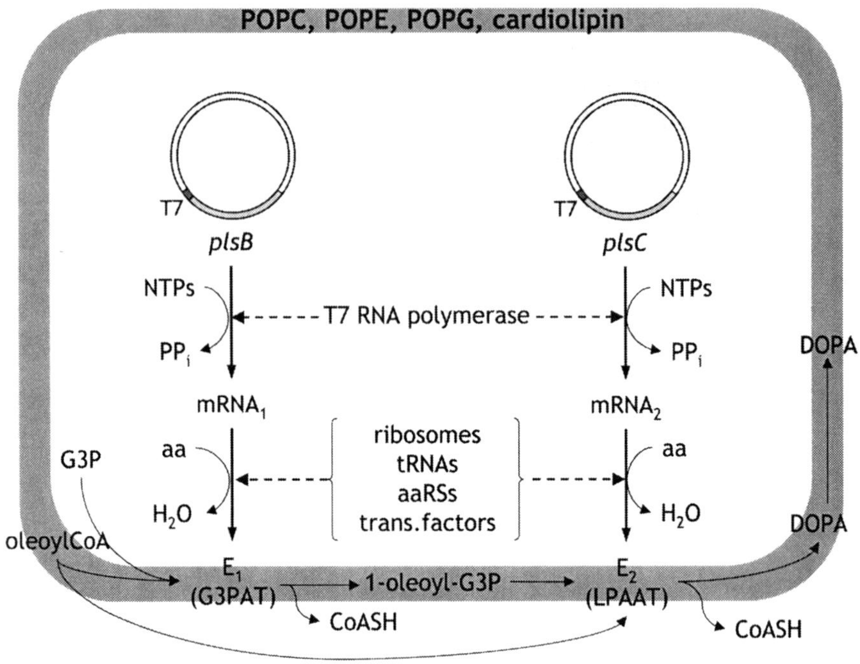

Fig. 16.7 Semi-synthetic minimal cell designed to produce dioleolyphosphatidic acid (DOPA) starting from glycerol-3-phosphate (G3P) and oleoyl-coenzyme A (oleoyl-CoA). See text for details (Reproduced from Kuruma et al. [47]. With permission of Elsevier)

oleoyl- or palmitoyl-coenzyme A, which can be added externally in forms of micelles. But would it be possible to design a total synthesis of fatty acids? Looking at the modern prokaryote fatty acid synthase (FAS) system (around 20 proteins), it is evident that the complexity of multi-enzyme system goes beyond our current capability. Mammalian FAS (540 kDa), despite its very complex architecture, has been recently tested for fatty acid synthesis inside liposomes [74], giving a low yield of palmitate (0.1% with respect to the amount of already present lipids).

The second limitation of the PNP/Qbeta replicase systems, namely the need of a reproduction of the whole set of catalysts entrapped inside the cell, can be solved only by designing a set of core reactions that are overall autocatalytic, i.e., where all components of the network are produced from within, as suggested by the autopoietic theory. From a conceptual viewpoint, this could be done by any molecular system that perform such transformation chains. In practice, however, we simply do not know any other chemical system different than cellular metabolism. The simplest thing to do is therefore the use of a minimal genetic/metabolic network that is still capable of self-producing all its molecular elements. We notice here that the shift from a minimal version of modern cell reactions to simpler systems (possibly of prebiotic relevance), namely those involving small molecule catalysis, ribozyme catalysis or metal complex catalysis, would be very interesting, especially if a logical thread can join such primitive systems and the cellular metabolism.

Clearly, this corresponds to the above-mentioned bottom-up approach (see section "Vesicle-Based Minimal Synthetic Cells: From the Origins of Life to Synthetic Biology"). As already emphasized, we simply miss the "tools" for constructing a minimal cell by using primitive catalysts (we do not even know what they are), except the nature of the liposome membrane, which very probably consisted of fatty acids mixed with other hydrophobic or amphiphilic molecules. Let us therefore see how nucleic acids and proteins can be used to build a minimal cell. In section "What Next", we will see a possible way to reduce the molecular complexity associated with the use of modern (evolved) molecules.

Working for the Production of Nucleic Acids and Proteins Inside Liposomes

The main ingredients for achieving the reproduction of nucleic acids and proteins are the enzymes that duplicate DNA and transcribe it into RNA and the ribosomal system for protein synthesis. In principle, if a minimal cell produces the minimal set of enzymes and RNAs to carry out these transformation it can reproduce all its internal components, as requested by the autopoietic dynamics. Clearly, this should be accompained by the synthesis of the cell membrane.

To date, there has been several efforts to synthesize functional proteins inside liposomes. The first report was due to Oberholzer et al. [81] who performed the ribosomal synthesis of poly(Phe) inside liposomes by using a poly(U) chain as messenger RNA. Shortly after, the group of Yomo [116] showed the synthesis of the green fluorescent protein (GFP); next, several other groups have reported on this system [41,76,77,79]. Remarkably, the work of Noireaux and Libchaber [76] involved the simultaneous synthesis of GFP and alpha-hemolysin, another water soluble peptide (33.2 kDa), which forms a membrane pore by self-assembly into eptamers, so that the internal liposomal synthesis of GFP could be sustained for 4 days, fed by externally added precursors (amino acids, ATP, ...). In the report of Ishikawa et al. [41], a second water-soluble protein was also synthesized, namely the T7RNA polymerase, that contributed to produce GFP by a well designed two-step genetic circuit (the *t7rna* polymerase gene, under SP6 promoter is firstly transcribed by SP6RNA polymerase, producing the T7RNA polymerase, which – in turn – transcribe the *gfp* gene under T7 promoter, to give finaly GFP). It can be said that although only a handful of functional water-soluble proteins have been actually synthesized inside liposomes (GFP, T7RNA polymerase, alpha-hemolysin, beta-glucuronidase [40], beta-galactosidase [45, 103] and Qbeta-replicase [45]) there is a common agreement on the possibility of producing water-soluble proteins inside liposomes. This is remarkable because the synthesis of protein is the way for achieving the minimal number of functions for minimal cells.

More complex is the case of membrane proteins. The only available study, carried out in our laboratory [47] shows the difficulties involved, especially due to the requirement of correct insertion of the protein in the membrane and the need of well defined lipid microenvironment around the protein in order to be active. Very recently, it has been described the cell-free production (on the external surface of liposomes) of the F0 subcomplex of F0/F1-ATP synthase [48].

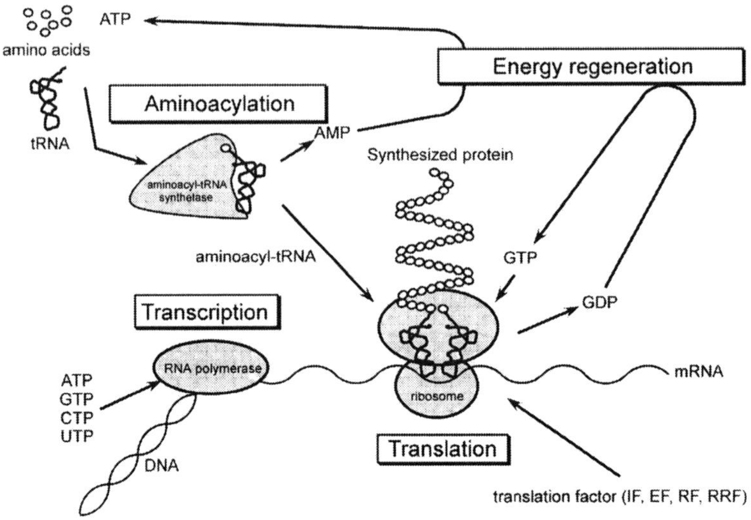

Fig. 16.8 Reactions occurring in the PURE system (Reproduced from Shimizu et al. [96]. With permission of Elsevier)

Protein synthesis inside liposomes is achieved by coentrapping all components of the transcription/translation machinery, from RNA polymerase to ribosomes, to tRNAs, to aminoacyl-tRNA-synthases, and an energy recycling system. This mixture can be derived by cell extracts (e.g., from *E. coli*, from yeast, from rabbit reticulocyte, from insect muscle) or created artificially by mixing purified components of known composition. The latter transcription/translation kit is now commercially available with the tradename of PURE system® [95, 96]. We have emphasized the great importance of using the PURE system for minimal cell contruction [75] because it fulfills the requirement of using the minimal number of components, in contrary to the not well-defined (black-box) composition of cell extracts. Research dated after 2006 is typically done (when possible) by incorporating the PURE system inside liposomes (Fig. 16.8).

PURE system is composed by 36 purified enzymes, ribosomes, a set of t-RNAs and other low molecular weight compounds. We have estimated that the PURE system contains ca. 80 different macromolecules [98], which become ca. 130 when the ribosomal proteins are explicitally taken into account. It follows that in first approximation the reproduction of the transcription/translation minimal machinery should involve the production of ca. 130 different macromolecules (proteins and tRNAs/rRNAs).

From the viewpoint of DNA and RNAs synthesis inside liposomes, less work has been done, possibly because these productions are seen as more feasible and less critical than protein synthesis. However, the control of DNA replication as well as the synthesis of functional ribosomes and tRNA might not be trivial. The first experiment was again coming by the Luisi's group in Zurich [78], consisting in the accomplishment of PCR inside conventional liposomes. DNA strands were produced from a template, dNTPs and DNA polymerase by the classical thermal method, without damaging the phospholipid vesicle structure. A more recent work from the group of Szostak [68] shows that also fatty acid vesicles can be used at high temperature, demonstrating the enzyme-free oligomerization of activated nucleotides (phosphorimidazolides) that diffuse passively into the vesicles and react on an oligo-dC template in the presence of a primer. Shohda and Sugawara [97] demonstrated that DNA polymerase Klenow fragments from *E. coli*, DNA template and dNTPSs react inside giant vesicles to give new DNA strands.

From the viewpoint of RNA synthesis inside vesicles, we can distinguish among messenger RNA and transfer/ribosomal RNAs. The several report on protein production clearly demonstrated that DNA transcription occurs readly. The direct demonstration was given several years ago by several groups [27,71,106] (who also studied the passive diffusion of externally added NTPs into vesicles [71]). These reports are based on DNA transcription without any kind of regulation mechanism. A very recent report by the team of the New England Biolabs [1] describes the in vitro genetic reconstruction of bacterial transcription initiation by coupled synthesis and detection of RNA polymerase holoenzyme. This could be useful to control the transcription of genes in vitro, by a totally self-produced molecular machinery.

RNA, however, has a key relevance in origins of life scenario, in the well-known hypothesis of RNA world [35, 113]. According to this hypothesis, RNA was the first functional biopolymer, capable of carrying genetic information and performing catalysis (by a "ribozyme"). In particular, it should be able to replicate itself. Research on self-replicating RNA has been a classical subject in the field, where the recent report of Joyce and coworkers [52] on the self-sustained replication of 80-nucleotides long ribozymes remarkably show the latest efforts to find a "replicant" RNA molecule. In the past years, a hypothetical scenario with ribozymes entrapped inside liposomes has been put forward [104]. This RNA-based minimal cell should contain at least two ribozymes: a replicase and a lipid synthase. The replicase replicates itself and the lipid synthase; whereas the lipid synthase produces lipids. These reactions occur at the expenses of a rich medium containing all the required building blocks. In such way, all the components of the cell are produced from within, i.e., the system should display an autopoietic organization.

The experimental efforts to construct a RNA-based minimal cells focus on RNA replication inside vesicles. This is implemented by using Qbeta replicase, an RNA-dependent RNA polymerase. This enzyme has been classically used in in vitro RNA evolution experiments [38,70]. The goal is to produce alternatively sense and anti-sense RNA strands from a template and NTPs. As mentioned in section "**Two Selected Cases of Pioneering Research**", the Qbeta replicase system was firstly used by Oberholzer et al. [82] to obtain the RNA production inside self-reproducing oleate vesicles. A most recent report, due to the Yomo's group [45] is based on a RNA strand that codes for Qbeta replicase. Coentrapped with the ribosomal machinery inside liposomes, the RNA strand acts a template for the in situ synthesis of Qbeta replicase, that in turn synthesize the antisense RNA strand. The overall reaction is an autocatalytic cycle producing RNA and Qbeta replicase. In principle, such system could be coupled with vesicle self-reproduction to obtain a more complex core-and-shell self-reproducing system, but also in this case one of the catalysts (the ribosome needed for the Qbeta replicase synthesis) is not produced, so that a fully autopoietic scheme cannot be developed.

The need of a in vitro synthesis of ribosome is therefore one of the key step for the minimal cell project. Prokaryote ribosomes self-assemble from isolated 58 components (55 proteins and 3 RNAs), but until now no one succeeded to produce functional ribosomes starting from the corresponding 58 genetic sequences. In this respect, a first attempt to fill the gap has been provided by Jewett and Church (work presented as poster at the Fourth Synthetic Biology conference, Hong Kong 2008), who looked for in vitro (cell-extract based) conditions for *E. coli* ribosome assembly starting from rRNA synthesis.

Toward Self-reproduction of Core and Shell Components

We can now discuss the central theme of the minimal cell construction, namely the achievement of a simultaneous (coupled) reproduction of core-and-shell cell components. The distinction between "core" components and "shell" components is only a useful way to dividing the problem into two parts, but, as requested by the autopoietic organization, we should look at autpoietic systems as a unity rather than separate parts.

Here an important remark must be done. In agreement with autopoiesis, the essential feature for a living system is the self-production of all its components, which are continuously synthesized and degraded, so that an homeostatic state is reached. We may call this dynamics as a self- maintenance one. When we speak of self-reproduction, however, we generally means an increase of the number of molecular species in autopoietic systems so that after growth, they can divide to give rise to new units. This is clearly possible if anabolic (constructive) routes overcome catabolic (destructive) ones, and if the division follows the growth phase. Self-reproduction is important for biology because it lets a population of living systems numerically grow, but – strictly speaking – such transformation is not necessary in autopoietic systems.

According to the semi-synthetic approach, a minimal cell is constructed from genes, RNAs, proteins and lipids. In order to self-reproduce the whole cell, it is needed a molecular mechanism for DNA duplication, RNA synthesis, protein synthesis, and lipid synthesis (Fig. 16.9).

We have seen in the previous sections the relevant advancements done in these directions. In particular, it is given as granted the synthesis of messenger RNA and the synthesis of water soluble proteins, whereas there are still obstacles in the synthesis of ribosomal RNA, transfer RNA and membrane proteins. The replication of long DNA strands with a minimal set of enzymes and protein factors has not been achieved yet, despite the success of copying short DNA sequences.

Vesicle growth by external addition of lipid precursors (oleic anhydride or oleate vesicles) is a feasible route, and it has been used to self-reproduce vesicles which were simultaneously synthesizing poly(A) or RNA inside [82, 112], respectively),

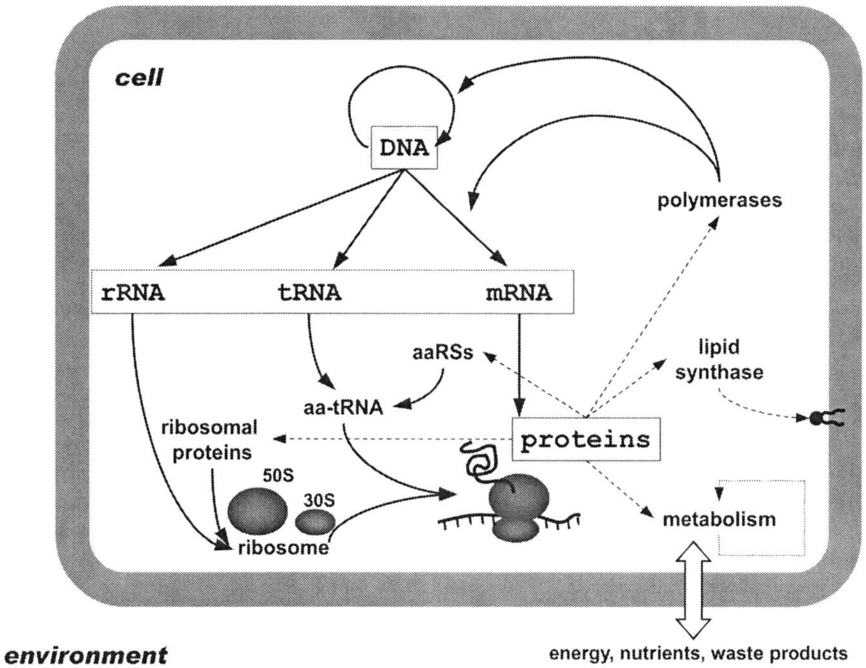

Fig. 16.9 A schematic view of a semi-synthetic minimal cell that is be able to synthesize DNA, RNA, proteins and lipids

but there are no recent reports on the self-reproduction of vesicles expressing a protein inside, for example. An attempt to divide a GFP-synthesizing water-in-oil compartment has been given by Fiordemondo and Stano [26]. Phospholipid synthesis by entrapped enzymes or in situ synthesized enzymes has been reported by Schmidli et al. [92] and Kuruma et al. [47], respectively, but in both cases the lipid production was not sufficient to observe the growth-division process.

In other words, there are still several open questions and technical gaps to be solved before the experimental goal of a self-reproducing minimal cell can be reached. Moreover, several aspects are involved for such desired behavior: (1) the redistribution of water- and membrane-solutes from parent vesicle to daughter ones, in order to be sure than the second generation is still capable of sustain the whole mechanims; (2) the synchronicity of the events, in order to avoid that, e.g., a faster shell reproduction results in production of new cells missing not-yet-replicated internal components; (3) the need of a good (balanced) surface-to-volume growth ratio, in order to avoid osmotic crisis or hamper the division mechanism.

What Next?

In addition to the well established goal of achieving a self-reproducing minimal cell, there are other two relevant directions that can be followed in next years.

The first one is oriented to a further reduction of semi-synthetic minimal cells toward "primitiveness". Luisi et al. [63, 64] have already discussed the hypothetical strategy for further reducing the apparent minimal complexity of semi-synthetic cells. We may recall here that according to the concept of minimal genome (see for example, [34]), about 200 genes are needed for a minimal cell construction. We have also recalled that the number of different macromolecules (RNAs and proteins, including the 55 ribosomal proteins) contained in the PURE system is about 130 (the "missing" 70 genes – to get the minimal genome of 200 genes – codify for metabolic enzymes). One possible way to redesign a minimal cell is not limited to number reduction, but also to qualitative changes. For example, we can imagine enzymes that have lower specificity and can process (at lower efficiency) diverse substrates, e.g. a single polymerase that synthesize DNA and RNA. Another possibility is a reduction of the ribosomal proteins, from 55 to a lower number, or possibly eliminating all proteins in favour of simpler peptides (preliminary results in our laboratory show that simple polycationic peptides like polyarginine condense rRNA in ribosomal-like particles); since it is known that the peptidyl-trasferase activity of ribosomes is due to rRNA, it can be reasonable to observe a minimal peptidyl-transferase activity in extremely reduced ribosome [22, 60]. It has been estimated that the very small number of about 50 genes would support – in well established conditions and at very low efficiency – cellular life [64]. Clearly, such extremely reduced minimal cell would strongly interest the origin of life community.

The second direction for future developments, instead, is oriented to shift the classical approach focused on the synthetic cell as isolated object to the concept of synthetic cell population and of the interaction among cells. Here there are several intriguing possibilities, such as: (1) exploiting the fusion between minimal, perhaps "limping cells" [63] as a way to exchange/redistribute functional molecules so that the resultant population has higher chance to develop a fully functional cell [12]; (2) design and perform dedicated investigation of phenomena occurring at the population-level, like competition [16, 17] and selection [72] among minimal cells; (3) studying explicity the diversity of minimal cells within a population – derived from stochastic events – and exploit it to design selection/competition experiments [11,91,114]; (4) move from single-cell scenario a multi-cells one (cell clusters) (this is currently under investigation in our laboratory, [11]); (5) focus on communication between synthetic cells or between synthetic and natural cells [21, 31].

Biophysical Aspects of Minimal Cell Construction

One of the open questions in biology, with clear relevance also in origin of life, is about the minimal physical size of cells. Several speculations have been published on the subject [46], based on calculations and on the controversial existence of nanobacteria [43]. We have approached the problem from the experimental viewpoint by creating a minimal system for protein production in 200 nm (diameter) lipid vesicles. The argument of using protein synthesis as a paradigm of cellular metabolism holds because – as we have seen before – more than one half of the minimal genome actually serves to decodify the DNA sequences into proteins. To date, expressing a protein inside liposomes does not correspond to entrap ca. 130 genes corresponding to the compounds necessary for transcription/translation, but only one gene of interest, the remaining macromolecular compounds are given in form of RNAs and proteins (these are ca. 80 different macromolecules, not 130, because 55 ribosomal proteins and 3 rRNAs count as a single object).

We have then prepared small liposomes in the presence of ca. 80 different macromolecular compounds, each present at a concentration around 0.1–1 μM, including the *gfp* gene. External GFP synthesis was avoided by the addition, after liposome formation, of RNase, and then internal fluorescence was monitored in real-time, showing that GFP was actually synthesized inside 200 nm (diam.) vesicles. The yield, when normalized for the very small entrapped volume was surprisingly six times higher than the corresponding reaction in bulk water. The most interesting conclusion, however, was that accurate statistical analysis suggest that the probability of co-entrapping about 80 different macromolecules (each in single copy) inside 200 nm (diam.) vesicles is vanishing small, namely 10 to the −27th power. Moreover, this value further decreases if multiple copies of each compounds are considered. We concluded that in order to justify the experimental observation, a strong deviation from Poisson statistics is expected, which will therefore bring about a kind of super-concentration of solutes inside liposomes should occurs in the moment of liposome formation [98].

Recently we have investigated in greater details the physical mechanism of liposome formation at the aim of verify the above mentioned hypothesis. It is interesting to remark that the construction of minimal cell models, as in this case, may indeed reveal interesting biophysical phenomena. We believe that this facet of the synthetic or constructive approach (i.e., gaining knowledge on biological systems by constructing them) is an additional value that cannot be found in classical analytical approaches.

In particular, we have entrapped ferritin inside spontaneously formed liposomes. Ferritin is a water soluble protein containing a core of about 4,500 iron atoms – and therefore visible by electronmicroscopy. After liposome formation, a large number of liposomes (ca. 8,000) have been visualized by cryo-transmission electromicroscopy and analyzed in terms of size, lamellarity and ferritin content. The expected behavior was a ferritin occupancy distribution that follows the Poisson distribution around the average value given by the expected ferritin concentration. In contrary, we have found that the large majority of vesicles were empty and only few of them were filled with a very high amount of ferritin (Fig. 16.10). Detailed numerical analysis suggests that the ferritin occupancy distribution follows a power law rather than the Poisson pattern [61]. In addition to ferririn, recent experiments have shown that the entrapment of ribosomes follow a similar behavior (Souza et al., submitted).

These results gives a new vista on the mechanism of liposome formation and more in general on the biophysics of the origin of the metabolism. One question in the origin of life is indeed the timing at which membrane compartments came

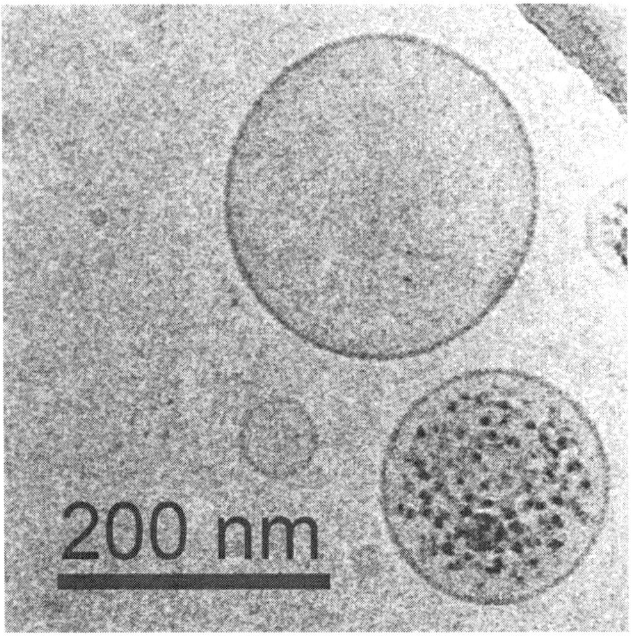

Fig. 16.10 Cryo-transmission electronmicrographs showing empty and ferritin-filled vesicles

into the picture as host for the first forms of metabolism. If we assume proteins and nucleic acid first, then it is difficult to conceive how all macromolecular components would have been entrapped at a later time into a single compartment. On the other hand, the hypothesis that metabolism originated from inside the compartment, meets the difficulty that we would then have to conceive semi-permeable, sophisticated membranes in prebiotic times, which does not appear plausible. With the present study we believe to be able to offer a partial solution to this riddle, opening at the same time a new vista on the principles of entrapment of solute in vesicles. In other words, thanks to their spontaneous formation, functional compartments (rich of functional solutes) might have been selected for further steps along the pathway from inanimate to living matter.

Concluding Remarks

The great interest shown by the synthetic biology community for minimal cell research [23] witnesses that the central idea of constructing synthetic cells is one of the main question in biology, helps the understanding of what is life, and promises interesting applications in modern biotechnology and medicine.

Acknowledgements This work has been funded by the SYNTHCELLS project (Approaches to the Bioengineering of Synthetic Minimal Cells, EU FP6 Grant #FP6 043359); by the Human Frontiers Science Program (RGP0033/2007-C); by the Italian Space Agency (Grant Nr. I/015/07/0); and by the Italian PRIN2008 program (Grant Nr. 2008FY7RJ4). It is also developed within the COST Systems Chemistry CM0703 Action.

Abbreviations

Acyl-CoAs	Acyl-coenzymes A (oleoyl-CoA, palmitoyl-CoA)
ADP	Adenosindiphosphate
CDP-choline	Cytidinediphosphocholine
dNTPs	Deoxynucleotide triphosphates
DMPC	Dimyristoylphosphatidylcholine
NTPs	Nucleotide triphosphates
Poly(A)	Poly(adenylic acid)
PNP	Polynucleotide phosphorilase

References

1. Asahara H, Chong S (2010) In vitro genetic reconstruction of bacterial transcription initiation by coupled synthesis and detection of RNA polymerase holoenzyme. Nucleic Acids Res doi:10.1093/nar/gkq377
2. Bachmann PA, Walde P, Luisi PL, Lang J (1990) Self-replicating reverse micelles and chemical autopoiesis. J Am Chem Soc 112:8200–8201
3. Bachmann PA, Luisi PL, Lang J (1992) Autocatalytic selfreplicating micelles as models for prebiotic structures. Nature 357:57–59
4. Balasubramanian V, Onaca O, Enea R, Hughes DW, Palivan CG (2010) Protein delivery: from conventional drug delivery carriers to polymeric nanoreactors. Expert Opin Drug Deliv 7:63–78
5. Bangham AD, Horne RW (1964) Negative staining of phospholipids and their structural modification by surface-active agents as observed in the electron microscope. J Mol Biol 8:660–668
6. Berclaz N, Bloechliger E, Mueller M, Luisi PL (2001b) Matrix effect of vesicle formation as investigated by cryotransmission electron microscopy. J Phys Chem B 105:1065–1071
7. Berclaz N, Mueller M, Walde P, Luisi PL (2001a) Growth and transformation of vesicles studied by ferritin labeling and cryotransmission electron microscopy. J Phys Chem B 105:1056–1064
8. Bitbol M, Luisi PL (2004) Autopoiesis with or without cognition: defining life at its edge. J R Soc Interface 1:99–107
9. Bloechliger E, Blocher M, Walde P, Luisi PL (1998) Matrix effect in the size distribution of fatty acid vesicles. J Phys Chem 102:10383–10390
10. Budin I, Szostak JW (2010) Expanding roles for diverse physical phenomena during the origin of life. Annu Rev Biophys 39:245–263
11. Carrara P (2010) Studies on giant vesicles as minimal cell models. PhD thesis, in Italian, University of RomaTre, Rome

12. Caschera F, Stano P, Luisi PL (2010) Reactivity and fusion between cationic vesicles and fatty acid anionic vesicles. J Colloid Interface Sci 345:561–565
13. Chakrabarti AC, Breaker RR, Joyce GF, Deamer DW (1994) Production of RNA by a polymerase protein encapsulated within phospholipid vesicles. J Mol Evol 39:555–559
14. Chang MTS (1987) Recycling of NAD(P) by multienzyme systems immobilized by microencapsulation in artificial cells. Method Enzymol 136:67–82
15. Chen IA, Szostak JW (2004a) A kinetic study of the growth of fatty acid vesicles. Biophys J 87:988–998
16. Chen I, Szostak JW (2004b) The emergence of competition between model protocells. Science 305:1474–1476
17. Cheng Z, Luisi PL (2003) Coexistence and mutual competition of vesicles with different size distributions. J Phys Chem B 107:10940–10945
18. Chiarabelli C, Stano P, Luisi PL (2009) Chemical approaches to synthetic biology. Curr Opin Biotechnol 20:492–497
19. Choi HJ, Montemagno CD (2007) Light-driven hybrid bioreactor based on protein-incorporated polymer vesicles. IEEE Trans Nanotechnol 6:171–176
20. Chungcharoenwattana S, Ueno M (2004) Size control of mixed egg yolk phosphatidylcholine EggPC/oleate vesicles. Chem Pharm Bull 52:1058–1062
21. Cronin L, Krasnogor N, Davis BG, Alexander C, Robertson N, Steinke JH, Schroeder SL, Khlobystov AN, Cooper G, Gardner PM, Siepmann P, Whitaker BJ, Marsh D (2006) The imitation game – a computational chemical approach to recognizing life. Nat Biotechnol 24:1203–1206
22. D'Aguanno E (2009) Studies on prebiotic models of ribosomes. Graduate thesis, in Italian, University of RomaTre, Rome
23. De Lorenzo V, Danchin A (2008) Synthetic biology: discovering new worlds and new words. EMBO Rep 9(9):822–827
24. Deamer DW, Oro J (1980) Role of lipids in prebiotic structures. Biosystems 12:167–175
25. Delaittre G, Reynhout IC, Cornelissen JJLM, Nolte RJM (2009) Cascade reactions in an all-enzyme nanoreactor. Chem Eur J 15:12600–12603
26. Fiordemondo D, Stano P (2007) Lecithin-based water-in-oil compartments as dividing bioreactors. Chembiochem 8:1965–1973
27. Fischer A, Franco A, Oberholzer T (2002) Giant vesicles as microreactors for enzymatic mRNA synthesis. Chembiochem 3:409–417
28. Forster AC, Church GM (2006) Towards synthesis of a minimal cell. Mol Syst Biol 2:45
29. Forster AC, Church GM (2007) Synthetic biology projects in vitro. Genome Res 17:1–6
30. Freisleben HJ, Zwicker K, Jezek P, John G, Bettin-Bogutzki A, Ring K, Nawroth T (1995) Reconstitution of bacteriorhodopsin and ATP synthase from Micrococcus luteus into liposomes of the purified main tetraether lipid from Thermoplasma acidophilum: proton conductance and light-driven ATP synthesis. Chem Phys Lipids 78:137–147
31. Gardner PM, Winzer K, Davis BG (2009) Sugar synthesis in a protocellular model leads to a cell signalling response in bacteria. Nat Chem 1:377–383
32. Gebicki JM, Hicks M (1973) Ufasomes are stable particles surrounded by unsaturated fatty acid membranes. Nature 243:232–234
33. Gibson DG et al (2010) Creation of a bacterial cell controlled by a chemically synthesized genome. Science 329:52–56
34. Gil R, Silva FJ, Pereto J, Moya A (2004) Determination of the core of a minimal bacteria gene set. Microbiol Mol Biol Rev 68:518–537
35. Gilbert W (1986) The RNA world. Nature 319:618–618
36. Graff A, Winterhalter M, Meier W (2001) Nanoreactors from polymer-stabilized liposomes. Langmuir 17:919–923
37. Hargreaves WR, Deamer D (1978) Liposomes from ionic, single-chain amphiphiles. Biochemistry 17:3759–3768
38. Haruna I, Nozu K, Ohtaka Y, Spiegelman S (1963) An RNA replicase induced by and selected for a viral RNA. Isolation and properties. Proc Natl Acad Sci USA 50:905–911

39. Hill KJ, Kaszuba M, Creeth JE, Jones MN (1997) Reactive liposomes encapsulating a glucose oxidase-peroxidase system with antibacterial activity. Biochim Biophys Acta 1326:37–46
40. Hosoda K, Sunami T, Kazuta Y, Matsuura T, Suzuki H, Yomo T (2008) Quantitative study of the structure of multilamellar giant liposomes as a container of protein synthesis reaction. Langmuir 24:13540–13548
41. Ishikawa K, Sato K, Shima Y, Urabe I, Yomo T (2004) Expression of a cascading genetic network within liposomes. FEBS Lett 576:387–390
42. Jewett MC, Forster AC (2010) Update on designing and building minimal cells. Curr Opin Biotechnol doi:10.1016/j.copbio.2010.06.008
43. Kajander EO, Ciftcioglu N (1998) Nanobacteria: an alternative mechanism for pathogenic intra- and extracellular calcification and stone formation. Proc Natl Acad Sci USA 95:8274–8279
44. Kaszuba M, Jones MN (1999) Hydrogen peroxide production from reactive liposomes encapsulating enzymes. Biochim Biophys Acta 1419:221–228
45. Kita H, Matsuura T, Sunami T, Hosoda K, Ichihashi N, Tsukada K, Urabe I, Yomo T (2008) Replication of genetic information with self-encoded replicase in liposomes. Chembiochem 9:2403–2410
46. Knoll A, Osborn MJ (eds) (1999) Size limits of very small microorganisms. National Academic Press, Washington, DC
47. Kuruma Y, Stano P, Ueda T, Luisi PL (2009) A synthetic biology approach to the construction of membrane proteins in semisynthetic minimal cells. Biochim Biophys Acta 1788:567–574
48. Kuruma Y, Suzuki T, Ueda T (2010) Production of multi-subunit complexes on liposome through an *E. coli* cell-free expression system. Methods Mol Biol 607:161–171
49. Kwok R (2010) Five hard truths for synthetic biology. Nature 463:288–290
50. Lasic DD, Papahadjopoulos D (eds) (1998) Medical applications of liposomes. Elsevier, Amsterdam
51. Lawless JG, Yuen GU (1979) Quantitation of monocarboxylic acids in the Murchison carbonaceous meteorite. Nature 282:431–454
52. Lincoln TA, Joyce GF (2009) Self-sustained replication of an RNA enzyme. Science 323:1229–1232
53. Lonchin S, Luisi PL, Walde P, Robinson BH (1999) A matrix effect in mixed phospholipid/fatty acid vesicle formation. J Phys Chem B 103:10910–10916
54. Luisi PL (1998) About various definitions of life. Orig Life Evol Biosph 28:613–622
55. Luisi PL (2002) Toward the engineering of minimal living cells. Anat Rec 268:208–214
56. Luisi PL (2003) Autopoiesis: a review and a reappraisal. Naturwissenschaften 90:49–59
57. Luisi PL (2006) The emergence of life: from chemical origin to synthetic biology. Cambridge University Press, Cambridge
58. Luisi PL (2007) Chemical aspects of synthetic biology. Chem Biodivers 4:603–621
59. Luisi PL (2010a) Epistemology notes on synthetic biology. In: Chiarabelli C, Luisi PL (eds) Chemical synthetic biology. Wiley, New York, pp 343–362
60. Luisi PL (2010b) On the minimal cell. Orig Life Evol Biosph 40:464–466
61. Luisi PL, Allegretti M, Souza T, Steineger F, Fahr A, Stano P (2010) Spontaneous protein crowding in liposomes: A new vista for the origin of cellular metabolism. ChemBiochem 11:1989–1992
62. Luisi PL, Varela FJ (1990) Self-replicating micelles – a chemical version of minimal autopoietic systems. Orig Life Evol Biosph 19:633–643
63. Luisi PL, Ferri F, Stano P (2006) Approaches to semi-synthetic minimal cells: a review. Naturwiss 93:1–13
64. Luisi PL, Oberholzer T, Lazcano A (2002) The notion of a DNA minimal cell: a general discourse and some guidelines for an experimental approach. Helv Chim Acta 85:1759–1777
65. Luisi PL, Souza T, Stano P (2008) Vesicle behavior: in search of explanations. J Phys Chem B 112:14655–14664
66. Luisi PL, Stano P, Rasi S, Mavelli F (2004) A possible route to prebiotic vesicle reproduction. Artif Life 10:297–308

67. Mansy SS, Szostak JW (2009) Reconstructing the emergence of cellular life through the synthesis of model protocells. Cold Spring Harb Symp Quant Biol LXXIV:1–8
68. Mansy S, Schrum JP, Krishnamurthy M, Tobé S, Treco DA, Szostak JW (2008) Template directed synthesis of a genetic polymer in a model protocell. Nature 454:122–125
69. McCollom TM, Ritter G, Simoneit BRT (1999) Lipid synthesis under hydrothermal conditions by Fischer-Tropsch-type reactions. Orig Life Evol Biosph 29:153–166
70. Mills DR, Peterson RL, Spiegelman S (1967) An extracellular Darwinian experiment with a self-duplicating nucleic acid molecule. Proc Natl Acad Sci USA 58:217–224
71. Monnard PA, Deamer DW (2002) Membrane self-assembly processes: steps toward the first cellular life. Anat Rec 268:196–207
72. Morigaki K, Dallavalle S, Walde P, Colonna S, Luisi PL (1997) Autopoietic self-reproduction of chiral fatty acid vesicles. J Am Chem Soc 119:292–301
73. Morowitz H (1992) Beginnings of cellular life. Metabolism recapitulates biogenesis. Yale University Press, New Haven, CT
74. Murtas G (2009) Internal lipid synthesis and vesicle growth as a step toward self-reproduction of the minimal cell. Syst Synth Biol. doi: 10.1007/s11693-009-9048-1
75. Murtas G, Kuruma Y, Bianchini P, Diaspro A, Luisi PL (2007) Protein synthesis in liposomes with a minimal set of enzymes. Biochem Biophys Res Commun 363:12–17
76. Noireaux V, Libchaber A (2004) A vesicle bioreactor as a step toward an artificial cell assembly. Proc Natl Acad Sci USA 101:17669–17674
77. Nomura SM, Tsumoto K, Hamada T, Akiyoshi K, Nakatani Y, Yoshikawa K (2003) Gene expression within cell-sized lipid vesicles. Chembiochem 4:1172–1175
78. Oberholzer T, Albrizio M, Luisi PL (1995a) Polymerase chain reaction in liposomes. Chem Biol 2:677–682
79. Oberholzer T, Luisi PL (2002) The use of liposomes for constructing cell models. J Biol Phys 28:733–744
80. Oberholzer T, Meyer E, Amato I, Lustig A, Monnard PA (1999) Enzymatic reactions in liposomes using the detergent-induced liposome loading method. Biochim Biophys Acta 1416:57–68
81. Oberholzer T, Nierhaus KH, Luisi PL (1999) Protein expression in liposomes. Biochem Biophys Res Commun 261:238–241
82. Oberholzer T, Wick R, Luisi PL, Biebricher CK (1995) Enzymatic RNA replication in self-reproducing vesicles: an approach to a minimal cell. Biochem Biophys Res Commun 207:250–257
83. Oparin AI, Serebrowskaya KB, Auerman TL (1961) Biokhimiya 26:499–504
84. Pitard B, Richard P, Dunarach M, Girault G, Rigaiud JL (1996) ATP synthesis by the F0F1 ATP synthase from Thermophilic bacillus PS3 reconstituted into liposomes with bacteriorhodopsin 1. Factors defining the optimal reconstitution of ATP synthases with bacteriorhodopsin. Eur J Biochem 235:769–778
85. Pohorille A, Deamer D (2002) Artificial cells: prospects for biotechnology. Trends Biotechnol 20:123–128
86. Rasi S, Mavelli F, Luisi PL (2003) Cooperative micelle binding and matrix effect in oleate vesicle formation. J Phys Chem B 107:14068–14076
87. Rasmussen S, Chen L, Deamer D, Krakauer DC, Packard NH, Stadler PF, Bedau MA (2004) Evolutionary transitions from nonliving to living matter. Science 303:963–965
88. Rogerson ML, Robinson BH, Bucak S, Walde P (2006) Kinetic studies of the interaction of fatty acids with phosphatidylcholine vesicles (liposomes). Colloid Surface B 48:24–34
89. Ruiz-Mirazo K, Luisi PL (2010) Open questions on the origins of life: introduction to the special issue. Orig Life Evol Biosph 40:353–355
90. Rushdi AI, Simoneit BRT (2001) Lipid formation by aqueous Fischer-Tropsch-type synthesis over a temperature range of 100 to 400 °C. Orig Life Evol Biosph 31:103–118
91. Saito H, Kato Y, Le Berre M, Yamada A, Inoue T, Yoshikawa K, Baigl D (2009) Time-resolved tracking of a minimum gene expression system reconstituted in giant liposomes. Chembiochem 10:1640–1643

92. Schmidli PK, Schurtenberger P, Luisi PL (1991) Liposome mediated enzymatic synthesis of phosphatidylcholine as an approach to self-replicating liposomes. J Am Chem Soc 113: 8127–8130
93. Schrum JP, Zhu TF, Szostak JW (2010) The origins of cellular life. Cold Spring Harb Perspect Biol. doi: 10.1101/cshperspect.a002212
94. Shapiro R (2007) A simpler origin for life. Sci Am 296:46–53
95. Shimizu Y, Inoue A, Tomari Y, Suzuki T, Yokogawa T, Nishikawa K, Ueda T (2001) Cell free translation reconstituted with purified components. Nat. Biotechnol 19:751–755
96. Shimizu Y, Kanamori T, Ueda T (2005) Protein synthesis by pure translation systems. Methods 36:299–304
97. Shohda K, Sugawara T (2006) DNA polymerization on the inner surface of a giant liposome for synthesizing an artificial cell model. Soft Matter 2:402–408
98. Souza T, Stano P, Luisi PL (2009) The minimal size of liposome-based model cells brings about a remarkably enhanced biological activity. Chembiochem 10:1056–1063
99. Stano P, Luisi PL (2007) Basic questions about the origins of life: proceedings of the erice international school of complexity (fourth course). Orig Life Evol Biosph 37:303–307
100. Stano P, Luisi PL (2010a) Achievements and open questions in the self-reproduction of vesicles and synthetic minimal cells. Chem Commun 46:3639–3653
101. Stano P, Luisi PL (2010b) Reactions in liposomes. In: Brinker UH, Mieusset JL (eds) Molecular encapsulation: organic reactions in constrained systems. Wiley, Chichester, pp 455–491
102. Stano P, Wehrli E, Luisi PL (2006) Insights on the oleate vesicles self-reproduction. J Phys Condens Matter 18:S2231–S2238
103. Sunami T, Hosoda K, Suzuki H, Matsuura T, Yomo T (2010) Cellular compartment model for exploring the effect of the lipidic membrane on the kinetics of encapsulated biochemical reactions. Langmuir 26:8544–8551
104. Szostak JW, Bartel DP, Luisi PL (2001) Synthesizing life. Nature 409:387–390
105. Treyer M, Walde P, Oberholzer T (2002) Permeability enhancement of lipid vesicles to nucleotides by use of sodium cholate: basic studies and application to an enzyme-catalyzed reaction occurring inside the vesicles. Langmuir 18:1043–1050
106. Tsumoto K, Nomura SM, Nakatani Y, Yoshikawa K (2001) Giant liposome as a biochemical reactor: transcription of DNA and transportation by laser tweezers. Langmuir 17:7225–7228
107. Vamvakaki V, Fournier D, Chaniotakis NA (2005) Fluorescence detection of enzymatic activity within a liposome based nano-biosensor. Biosens Bioelectron 21:384–388
108. Van Dongen SFM, Nallani M, Cornelissen JJLM, Nolte RJM, van Hest JCM (2009) A three-enzyme cascade reaction through positional assembly of enzymes in a polymersome nanoreactor. Chem Eur J 15:1107–1114
109. Varela F, Maturana HR, Uribe RB (1974) Autopoiesis: the organization of living system, its characterization and a model. Biosystems 5:187–196
110. Walde P, Ichikawa S (2001) Enzymes inside vesicles: preparation, reactivity and applications. Biomol Eng 18:143–177
111. Walde P, Wick R, Fresta M, Mangone A, Luisi PL (1994a) Autopoietic self-reproduction of fatty acid vesicles. J Am Chem Soc 116:11649–11654
112. Walde P, Goto A, Monnard PA, Wessicken M, Luisi PL (1994b) Oparin's reactions revisited: enzymatic synthesis of poly(adenylic acid) in micelles and self-reproducing vesicles. J Am Chem Soc 116:7541–7544
113. Woese CR (1979) A proposal concerning the origin of life on the planet Earth. J Mol Evol 13:95–101
114. Yamaji K, Kanai T, Nomura SM, Akiyoshi K, Negishi M, Chen Y, Atomi H, Yoshikawa K, Imanaka T (2009) Protein synthesis in giant liposomes using the in vitro translation system of Thermococcus kodakaraensis. IEEE Trans Nanobiosci 8:325–331
115. Yoshimoto M, Wang S, Fukunaga K, Fournier D, Walde P, Kuboi R, Nakao K (2005) Novel immobilized liposomal glucose oxidase system using the channel protein OmpF and catalase. Biotechnol Bioeng 90:231–238
116. Yu W, Sato K, Wakabayashi M, Nakatshi T, Ko-Mitamura EP, Shima Y, Urabe I, Yomo T (2001) Synthesis of functional protein in liposome. J Biosci Bioeng 92:590–593

117. Zepik HH, Bloechliger E, Luisi PL (2001) A chemical model of homeostasis. Angew Chem Int Ed 40:199–202
118. Zhang Y, Ruder WC, LeDuc PR (2008) Artificial cells: building bioinspired systems using small-scale biology. Trends Biotechnol 26:14–20
119. Zhu TF, Szostak JW (2009) Coupled growth and division of model protocell membranes. J Am Chem Soc 131:5705–5713

Chapter 17
Fluorescent-Based Quantitative Measurements of Signal Transduction in Single Cells

Serge Pelet and Matthias Peter

Abstract Budding yeast (*Saccharomyces cerevisiae*) has been widely used as a model system to study fundamental biological processes. Genetic and biochemical approaches have allowed in the last decades to uncover the key components involved in many signaling pathways. Generally, most techniques measure the average behavior of a population of cells, and thus miss important cell-to-cell variations. With the recent progress in fluorescent proteins, new avenues have been opened to quantitatively study the dynamics of signaling in single living cells. In this chapter, we describe several techniques based on fluorescence measurements to quantify the activation of biological pathways. Flow cytometry allows for rapid quantification of the total fluorescence of a large number of single cells. In contrast, microscopy offers a lower throughput but allows to follow with a high temporal resolution the localization of proteins at sub-cellular resolution. Finally, advanced functional imaging techniques such as FRET and FCS offer the possibility to directly visualize the formation of protein complexes or to quantify the activity of proteins in vivo. Together these techniques present powerful new approaches to study cellular signaling and will greatly increase our understanding of the regulation of signaling networks in budding yeast and beyond.

Keywords Cellular signaling · Fluorescent proteins · Microscopy · Flow cytometry · FRET · FCS

Introduction

Cells have engineered elaborate capacities to sense their environment in order to detect nutrient sources, hormones or stress. These extra-cellular cues are often sensed by membrane-associated receptors, which transduce this information inside the cell via signaling cascades. These highly interconnected biochemical pathways

S. Pelet (✉)
Institute of Biochemistry, Department of Biology, Schaffmattstr 18, CH 8093-Zürich, ETH Zürich
e-mail: serge.pelet@bc.biol.ethz.ch

H. Koeppl et al. (eds.), *Design and Analysis of Biomolecular Circuits: Engineering Approaches to Systems and Synthetic Biology*, DOI 10.1007/978-1-4419-6766-4_17,

can integrate multiple inputs to deliver the appropriate response such as proliferation, differentiation, adaptation or apoptosis. These various cellular fates are generally controlled by the activation of key signaling molecules and/or the activation of a specific transcriptional expression program.

To better understand these signaling networks and to build predictive mathematical models, it is essential to gather quantitative measurements at different steps along the information transduction cascade. For many decades, biologists have optimized standard biological methods such as western blotting and mass spectrometry to deliver more accurate and quantitative estimations of protein levels and modifications. However, their requirements for large quantities of material generally limit these techniques to the measurements of the average response of a population to a stimulus.

Single Cell Analysis

Since the seminal work from Ferrel and Machleder performed on single maturing oocytes [1], it has been realized that population averaged measurements can prevent the discovery of more complex regulatory mechanisms. As schematically represented in Fig. 17.1, the output of a pathway can increase gradually with stimulus or display an all-or-none output pattern. However, due to stochastic differences

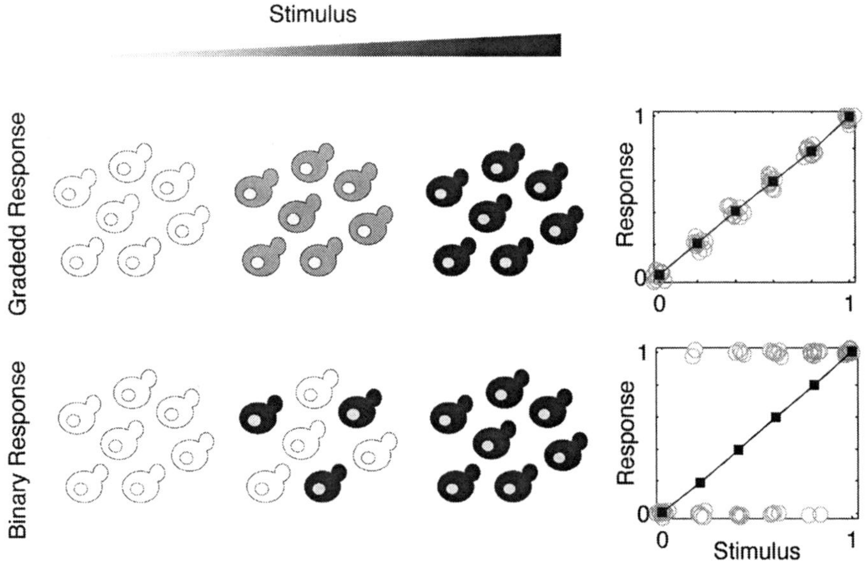

Fig. 17.1 Schematic difference between a graded and binary response to an increasing stimulus. The graphs show how different single cell behaviours (○) can lead to an identical population measurement (■)

between individual cells, the population averaged measurements can be identical in both situations. Only measurements performed at the single cell level can reveal this fundamental difference in signal transduction.

In the case of the maturing oocyte [1], the averaged population response displays an apparent Hill coefficient of one, which is typical for a graded response. Fitting of the single cell measurements results in a Hill coefficient of 42, which describe a highly switch-like activation. The discovery of this ultra-sensitive response allowed the authors to postulate the presence of a positive feedforward loop, acting at the transcriptional level. Inhibition of the protein production by addition of cycloheximide renders the output of the signaling cascade more graded.

This example of the ability of single cell measurements to uncover hidden regulatory mechanisms is being applied to a wide variety of signal transduction pathways [2, 3]. Signaling cascades often include such feedback or feedforward loops to exquisitely control the output of the pathway or generate ultra-sensitivity of the cellular response. Moreover the dynamics of signal transduction can be best studied at the single cell level since many informative behaviors, such as oscillations or activity bursts, could be averaged out in population measurements [4, 5].

Cellular Noise

As mentioned in the previous section, the different outcomes measured at the output of a signaling cascade can arise from small stochastic differences between individual cells [6]. Therefore single cell experiments have been designed to study the source of cellular noise. Using a set of yellow and cyan fluorescent reporters, which are under the control of the same promoter, Elowitz and co-workers [7] defined the concept of intrinsic and extrinsic noise.

As sketched in Fig. 17.2, if all cells were well-stirred chemical reactors with an identical number of internal components, one would expect them to produce the same amount of cyan and yellow fluorescent proteins. However, it is well known that each cell is different from the others with a unique composition of proteins (for instance: number of polymerases and ribosomes), even within a clonal population [7, 8]. Under these conditions, one would expect each individual cell to have various abilities to express proteins. However, in a given cell, the cyan and yellow fluorescent proteins should be expressed at the same level. This cell-to-cell variation is defined as the extrinsic noise. There is an additional component of noise that will uncouple the cyan and yellow expression levels. Due to a limited number of components, the expression of each individual gene can become more stochastic. If there are for instance only a small number of transcription factors that can promote the expression of the cyan and yellow reporters, the final amount of protein produced will depend on the stochastic binding event of the transcription factor to either promoter and will therefore not be identical for both genes. The noise, which results in the variation of the cyan and yellow proteins levels in the same cell, is defined as the intrinsic noise.

The ability to isolate the origin of the cellular noise can help understand the processes that control the signaling dynamics of a given systems. As an example,

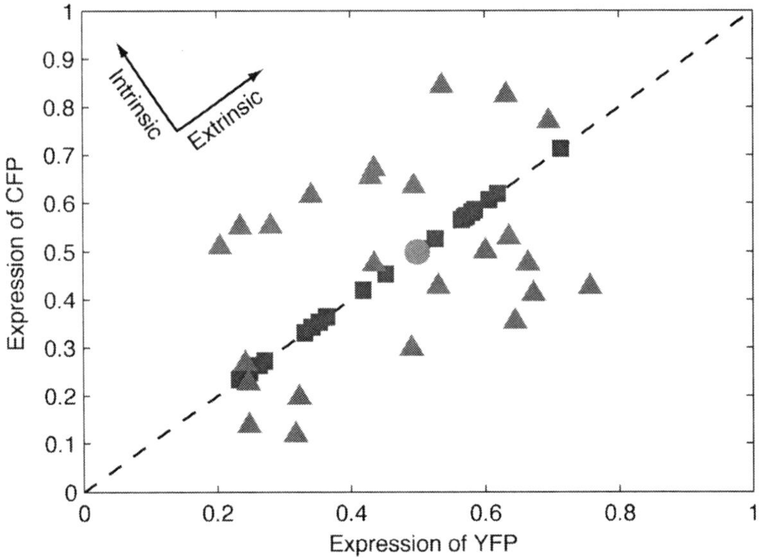

Fig. 17.2 Representation of intrisic and extrinsic noise in a biological system. In absence of noise, the expression of yellow (YFP) and cyan (CFP) reporter proteins under the control of the same promoter would be identical in all cells (•). In the sole presence of extrinsic noise, the expression of YFP and CFP would be highly correlated (■). In the presence of stochastic variations in each cell, YFP and CFP can be expressed at different levels in the same cell (▲)

Colman-Lerner et al. [9] found a high level of correlation between the expression of a reporter specific for a signaling cascade and the expression of a constitutively expressed gene. This implies that most of the noise in expression comes from the ability of the individual cells to produce proteins (defined as expression capacity). Importantly, this demonstrates that the transduction of the signal occurs in this pathway with a high fidelity although the number of signaling molecules varies from cell to cell.

Single Cell Measurements

In the study from Ferrell and Machleder [1], the large size of the oocyte allowed them to perform biochemical experiments directly on individual cells. This is unfortunately not the case for the majority of the commonly used biological model systems where proteins are found at too low abundance for biochemical investigations. Although recent advances in mass spectroscopy have allowed to reach single cell detection levels for some proteins or metabolites [10, 11], this technique is not yet generally used to analyze signal transduction pathways but might add, in the future, a different approach to analyze the content of single cells. For this chapter, we will focus our discussion on the experimental methods that currently offer single cell resolution which are optical methods such as microscopy and flow cytometry.

Table 17.1 Differences between flow cytometry and microscopy

	Microscopy	Flow cytometry
Number of cells per measurement	10^2	10^5
Data analysis	Complex	Simple
Time-lapse	Yes	No
Measurement	Sub-cellular resolution	Total cellular intensity

Different imaging modalities can be used to visualize cells in a microscope. Absorption or interference methods generate a contrast in the image. However, their sensitivity is limited by the fact that one measures a relative change in the intensity of the transmitted light. On the opposite, fluorescence microscopy offers a much higher sensitivity, since one measures the apparition of a signal over a dark background. Detectors such as photo-multiplier tubes or electron-multiplying-CCD have single photon sensitivities. Within the linearity range of the detector, the fluorescence signal will scale linearly with the amount of fluorophore present in the sample. The possibility to image single proteins in living cells has been demonstrated recently [12, 13]. Therefore fluorescent microscopy has evolved as the most commonly used imaging modality with high specificity and unmatched sensitivity.

Flow cytometry has been specifically developed to analyze single cells in solution. While the first instruments were based on impedance or absorption measurements, fluorescence rapidly emerged as the preferred detection method, thus enabling the development of fluorescent probes used for both microscopy and flow cytometry.

Although both techniques allow the quantification of fluorescent signal emerging from cells, the resulting measurements have different characteristics (Table 17.1). The typical number of cells measured by microscopy is on the order of 100 per image, while flow cytometry can easily measure 10,000 cells, thereby generating statistically more significant measurements. Microscopy with time-lapse imaging and subcellular resolutions can offer more insight in the sample at the expense of a more complex data analysis. Therefore these two techniques provide different and complementary measurements.

Fluorescent Proteins

For many decades chemical dyes were the only contrast method used for fluorescent detection. A large variety of compounds have been synthesized specifically to tag organelles or report on ionic changes in cells. Fluorescently-tagged antibodies have also been extensively used to target fluorescent probes to specific proteins. Unfortunately, these dyes are often toxic for the cell and use of antibody staining requires fixation of the specimen, thereby excluding live cell imaging. The discovery of the green fluorescent protein (GFP) [14, 15] has opened a new area in the field of fluorescent microscopy. It became possible to genetically encode a fluorescent tag for

Table 17.2 Commonly used fluorescent proteins

Protein	$\lambda_{Excitation}$	$\lambda_{Emission}$	Other variants
ECFP	433	475	Cerulean, CyPet, mTFP
EGFP	488	507	Emerald, Azami green
EYFP	514	527	Venus, mCitrine, YPet
mCherry	587	610	mRFP, dsRed, tdTomato

any protein by extending its sequence with the one from GFP. It therefore became feasible to visualize, in living cells, the location and dynamic behavior of any protein [16]. As shown in Table 17.2, a large variety of fluorescent proteins have been engineered [17]. From the original GFP, a number of spectral variants have been created by mutations of only a few residues, such as the cyan and yellow fluorescent proteins. Generation of bright and monomeric red proteins has been more challenging, but many variants are now also available. Codon optimized variants for different organisms have been generated to allow for better expression of the proteins.

There are however a few drawbacks associated with the use of fluorescent proteins. Firstly, as with any other protein tags, the addition of the fluorescent protein (~240 amino acids) to the native sequence of a protein can sometimes impair its function and thereby lead to a mis-localization of the protein. Secondly, the fluorescence of the protein can be affected by its environment in the cell. Although the chromophore is shielded by the β-barrel structure of the protein, cations often lead to changes in the brightness of the proteins. This can also be taken advantage of as, for instance, a pH-sensitive fluorescent protein has been generated [18]. The photo-stability of the fluorophore can also be an issue, which can lead to degradation of the signal after extended illuminations. Researchers have also put that drawback to use by studying the recovery of a fluorescence signal after photo-bleaching (FRAP) [19, 20]. This method has allowed to uncover very fast protein dynamics and the possible applications of this technique have been extended by the engineering of photo-activable (PA-GFP) [21] or photo-switchable (Kaede) [22] fluorescent proteins.

Finally, the slow maturation time of GFP can result in some experimental artifacts. The fluorescent protein is expressed and folded rapidly, but the apparition of the fluorescence is delayed by roughly 30–45 min due to a slow oxidation reaction which needs to take place at the core of the protein to form the chromophore [14]. Despite these few technical issues, fluorescent proteins have revolutionized the bioimaging field in the last two decades. Engineering of newer, brighter and more stable version will allow to image less abundant proteins for extended periods of time.

Budding Yeast As a Model Organism

Saccharomyces cerevisiae is a small single cell, eukaryotic organism. It is characterized by an asymmetric cell division, which coined its common name: budding yeast. The daughter cell grows as a bud from the mother cell and separates after nuclear division and cytokinesis. Its genome has been sequenced more than 10 years ago [23] and to date approximately 75% of the open reading frames have been assigned a function. The biochemical tools developed in yeast to manipulate its genome offer a great advantage over other model systems. It is possible to modify genes at their endogenous genomic locus and therefore express proteins tagged with GFP at their physiological level. Since over-expression of proteins often perturbs the dynamics of signal transduction, retaining the native conditions of a signaling cascade avoids artifacts.

As a unicellular fungus, budding yeast shares many properties common to both plants and animals. Due to the evolutionary conservation of cellular pathways, it has been widely used as model system to study cell cycle regulation, metabolism and signal transduction. Mechanisms used in yeast to transduce information from the exterior to the interior of the cells are also present in higher eukaryotes. As an example, the mitogen activated protein kinase (MAPK) pathway, which consists of a module of three kinases that activate each other is conserved from yeasts to mammals [24]. Misregulation of MAPK signal transduction in mammalians cells has been implicated in multiple diseases such as cancer and inflammation. These pathways are therefore considered as potential drug targets. Although they serve different functions, the general architecture of MAPK pathways is conserved. Therefore considerable work has been invested into understanding the regulation of these pathways in the simplified setting offered by yeast.

Possibly the best understood signal transduction cascade is the MAPK mating pathway in budding yeast [25,26]. Decades of thorough genetic and biochemical experiments have allowed to identify the components and architecture of this signaling pathway. The challenge is now to understand how these proteins work together to generate a faithful and robust output in response to the stimulus. Quantitative single cell measurements have been widely employed to achieve this task. Below, we will illustrate each technique presented by an example of their application to the yeast mating pathway.

Haploid yeast (mat-a or mat-α) sense mating pheromones (α- or a-factor) secreted by a cell of opposite mating type. As depicted in Fig. 17.3, the pheromone binds to a G-protein coupled receptor (Ste2p) which triggers dissociation of the trimeric G-protein. The free G$\beta\gamma$-subunit (Ste4p/Ste18p) recruits the scaffold protein Ste5p to the plasma membrane. This scaffold protein has binding sites for all three kinases of the MAPK cascades: Fus3p, the MAP kinase; Ste7p, the MAP kinase kinase; Ste11p, the MAP kinase kinase kinase. This latter protein is the most up-stream components of the cascade and resides pre-activated at the plasma membrane. Recruitment of Ste5p to the membrane brings Ste7p in close proximity of Ste11p, and results in the activation of the pathway by phosphorylation of Ste7p which, in turn, phosphorylates Fus3p [27]. Activated Fus3p orchestrates the mating response by phosphorylating multiple targets, among them Dig1p, which represses transcription of pheromone response genes by inhibiting the transcription factor Ste12p. All these events will result in the arrest of the cell cycle in G1 and the extension of a polarized mating projection towards the source of pheromone.

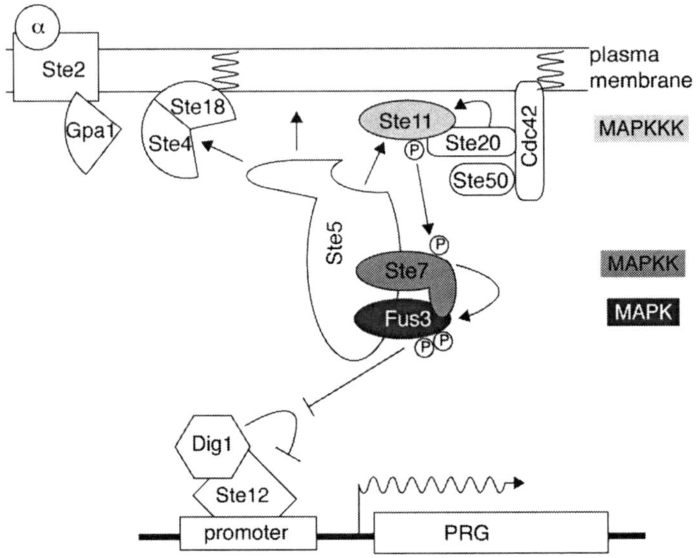

Fig. 17.3 Schematic drawing of the pheromone signalling pathway. Relocation of the scaffold protein Ste5p to the plasma membrane results in the activation of the MAPK Fus3p, which can in turn promote expression of pheromone responsive genes (PRG)

Flow Cytometry

Fluorescence-based flow cytometry has been developed in the late 1960s. While it has mostly been used with immuno-fluorescent staining and DNA marker dyes, with the advent of fluorescent proteins, it has become a tool of choice for signal transduction studies in yeast. Its ability to provide quantitative measurement for a large number of cells within a relatively short amount of time allows to access single cell measurements with a high statistical significance [28].

Instrumentation

To observe individual cells in suspension, the solution is aspirated at a low flow rate in the flow cytometer. A fast flow of so-called sheath fluid surrounds the medium to generate a thin core of sample (Fig. 17.4). This hydrodynamic focusing technique allows to precisely position the flow of medium in the laser beam and present the cells one by one in front of an excitation source. Every time a cell crosses the laser path it will lead to scattering of the excitation light. The forward scattered light is detected by blocking the main laser beam and detecting the amount of light that passes around the obscuration bar. The amount of forward scattered light detected scales with the volume of the cell. The light scattered at a 90° (Side-Scattering) is

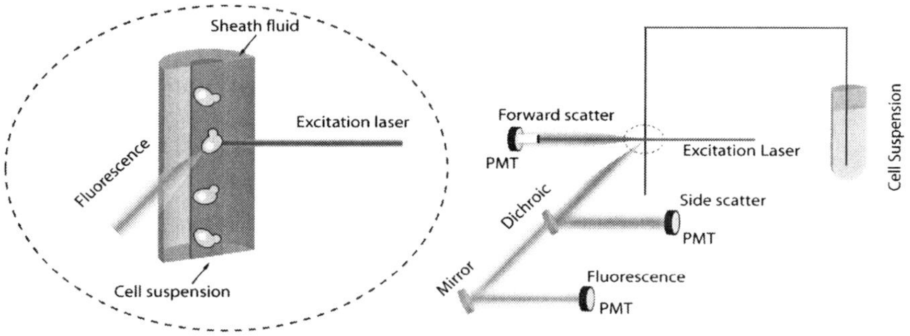

Fig. 17.4 Schematic drawing of the flow cytometer detection system. Hydrodynamic focusing with the sheath fluid forces cells to pass one by one in front of the excitation laser. The resulting fluorescence and side-scatters are separated by appropriate optics and detected by photo-multiplier tubes (PMT). The forward scatter is detected by blocking the excitation beam

also collected and provides an estimate of the granularity of the cell. Using the same excitation source, multiple fluorescence signals can be detected using appropriate filters to select the portion of spectrum of interest. The light is detected by photomultiplier tubes, which have a high sensitivity and large dynamic range. For every scattering event, the fluorescence intensity is measured in all channels, therefore allowing to obtain for each individual cell a measurement of forward and side scatter along with multiple fluorescence measurements at various wavelengths. The rate of event detection is in the order of 1000 cells per seconds, thus enabling to obtain in a short amount of time statically significant data sets.

Typical flow cytometers are equipped with a laser at 488 nm and multiple fluorescence detection channels around 530, 580 and 660 nm optimized for the detection of common synthetic dyes. Except for the GFP and YFP variants, which can both be excited at 488 nm, this configuration is unfortunately not ideal for detecting combinations of fluorescent proteins. More complex set-up offer however multiple laser lines and a dozen of fluorescent detection channels which can be selected to measure CFP or RFP variants.

Applications

Flow cytometry is mostly used in conjunction with antibodies to label specific proteins. This technique is extremely successful in the analysis of the mammalian immune system to identify specific cell types or probe for the expression of surface receptors [29, 30]. It has also been exploited to quantify cell cycle profiles. Using DNA intercalating fluorophores, it is straightforward to follow doubling of the DNA content thereby monitoring the progression of cells through the cell cycle [31].

In yeast, since it is possible to tag a protein with GFP at its genomic locus, it becomes feasible to precisely quantify its expression level in single cells and study the variability of its abundance in a population. Given that the expression of a protein is mainly controlled by the promoter region upstream of the ORF (~1,000 base pairs before the start codon), synthetic constructs can be generated based on a promoter driving the expression of a fluorescent protein. The promoter can be selected among the proteins that are known to be transcriptionally induced by a specific pathway [32]. Flow cytometric measurements can then be used to quantify the expression of this fluorescent reporter as function of stimulus to obtain a dose-response curve or the temporal evolution of the expression of this construct.

As an example, we show in Fig. 17.5a, the expression of a fluorescent reporter driven by the promoter of the pheromone-responsive gene (*FIG1*). Each sample is induced with 1 μM of α-factor and cycloheximide is added at different time points to block protein expression. After 1–2 h of incubation to allow for protein maturation, 10,000 cells are measured for each sample. The sample taken before induction displays a basal level of fluorescence. This signal comes from the auto-fluorescence of the yeast and is certainly a limiting factor for the sensitivity of this method. Two hours of pheromone treatment leads to a 20-fold increase in fluorescence intensity of the cells. Interestingly, 30 min after induction only a fraction of cells expressed the fluorescent reporter. This is due to the fact that only cells in the G1 phase of the cell cycle can induce the pathway [33, 34]. The other cells will first need to complete cell division before becoming signaling competent (as shown by microscopy in Fig. 17.5b). This explains the large extrinsic noise in the cellular response observed in the flow cytometry measurements.

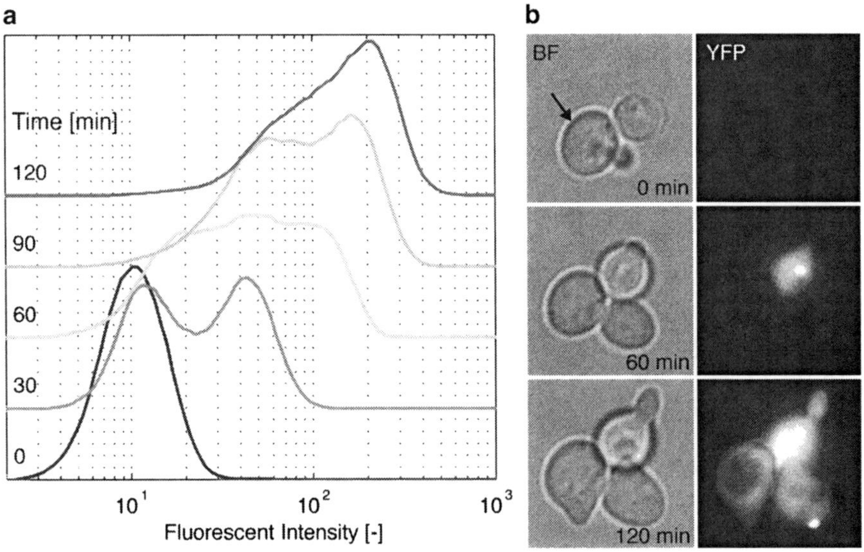

Fig. 17.5 Temporal evolution of the expression of a fluorescent reporter (*pFIG1*-YFP) upon activation of the mating pathway. **(a)** Flow cytometry. Cells were treated with cycloheximide and fluorescence was quantified after full maturation of the protein. **(b)** Live-cell microscopy. The budding cell (*arrow*) clearly show a delayed expression of the fluorescent reporter

One shortcoming of the flow cytometric measurements is that every cell can only be measured once. Therefore one cannot follow the fate of a given cell after pathway induction. This is somewhat compensated by the fact that the number of cells quantified in each measurement is so large that the sampling represents the behavior of all cells in the population. Another strategy, which can be used in certain situation, is the ability of the device to sort cells based on a set of measured parameters. For instance, it is possible to separate cells expressing and not-expressing a fluorescent reporter upon activation of the pathway. These cells can be re-cultured and stimulated a second time to verify if their behavior is inherited [35].

Microscopy

Microscopy has always been a tool of choice for biologists to study individual cell morphologies and phenotypes. With the recent advances in electronic detectors and image analysis software, this method is becoming highly quantitative and therefore can provide valuable data for mathematical modeling of biochemical pathways.

Instrumentation

Quantification of signaling pathways requires the measurements of the output by recording changes in the fluorescent signal. Many technical improvements have increased the sensitivity and reliability of microscopes in the last years. But it is probably equally important to properly control the stimulus applied to the cells. Given that microscopes offer the possibility to monitor cells over long periods of time, it also provides the opportunity to control the stimulus during the course of the experiment. Therefore researchers have employed flow channels and microfluidic devices to improve their understanding of the dynamics of signaling pathways.

Microscopy Samples and Microfluidic Devices

To perform live imaging, cells are typically attached to the bottom of a well slide. The media present on top of the cells allows to sustain growth for many hours. If required, the microscope can be enclosed in an incubation chamber with temperature, humidity and CO_2 control to provide the best conditions for cell growth. Stimulation of a signaling pathway can be achieved by adding soluble chemicals, such as sugars, drugs, pheromone or stress agents to the medium in the well. The cellular response induced by these compounds can then be observed microscopically in real time.

A limitation from this set-up is the inability to remove compounds from the medium. Therefore, for more refined experiments, flow chambers with coverslip

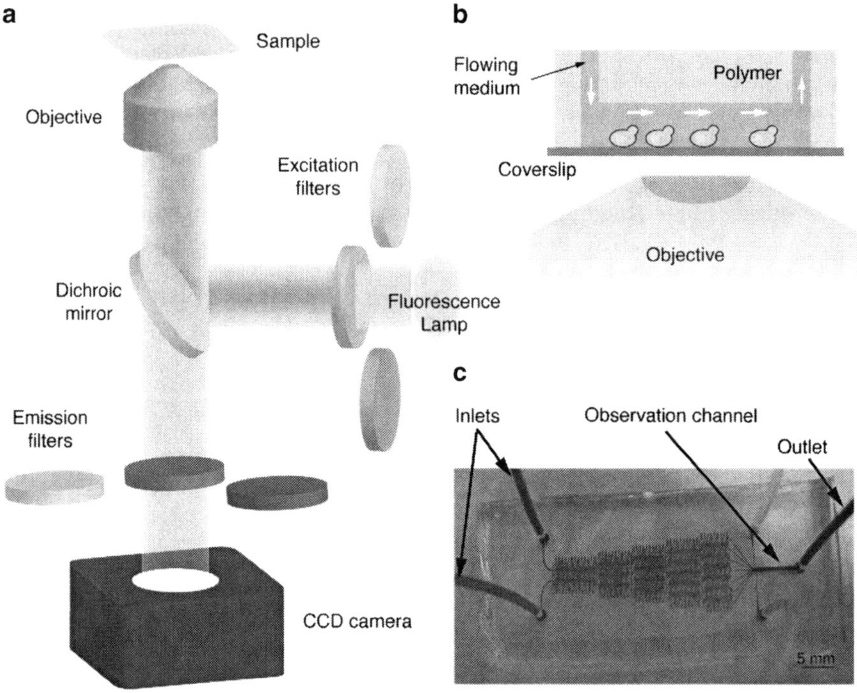

Fig. 17.6 **(a)** Light path in an epi-fluorescent microscope. The wide band emission of the fluorescence lamp is filtered and reflected by the dichroic mirror. The light is focused by the objective on the specimen. The emitted fluorescence is collected by the objective and is transmitted by the dichroic mirror. It is filtered by the emission filter before an image is recorded by the CCD camera. **(b)** In a flow channel, cells are attached to the coverslip and the medium flows in the channels molded in the polymer. **(c)** Picture of a gradient generating micro-fluidic device. The two inlet flows are mixed to generate a smooth concentration ramp in the observation channel

bottoms have been developed (Fig. 17.6b). This allows to keep a constant input of fresh medium and thereby offers the possibility to rapidly change the content of the medium allowing both addition and removal of stimuli to the cells.

A particular type of flow chambers are microfluidic devices, which are made of 10–100 μm channels molded in a polymer [36, 37]. As shown in Fig. 17.6c, these devices are much more than a simple small dimension flow channel because they integrate all the element of the flow control and mixing within the chip. The small dimensions of the channels prevent turbulent mixing of flows and thereby allow the generation of complex temporal stimulation patterns such as pulses or ramps [38, 39]. Due to the small volume present in these devices, the media switching times can be in the order of a few seconds. Using diffusion between two different media, it is also possible to generate stable concentration gradients for example to study the establishment of oriented cell polarity [40]. Another benefit of the small dimensions of microfluidic devices is the possibility to keep all cells in the same focal plane for multiple generations by designing chambers with a height of only a few microns [41, 42].

Microscope Light-Path

The light source in most epi-fluorescence microscopes is a mercury or metal-halide lamp. These lamps have a very broad emission covering the UV and visible part of the spectrum. An excitation filter is necessary to select the band of wavelength used for the excitation of the sample. Since the lamp is continuously emitting light it is also necessary to place a fast shutter in front of the lamp to precisely control the illumination time of the sample. High power light-emitting diodes (LED) are a new alternative for fluorescence imaging with the great advantage of being switched on electronically within a few microseconds. LEDs have well-defined emission spectra of 10–30 nm width, thus four to six units have to be coupled together to excite a wide range of fluorophores.

The filtered excitation light is reflected by a dichroic mirror and sent to the objective to excite the sample. The fluorescent light is emitted isotropically and only a small fraction is collected by the objective. It will pass through the dichroic mirror and is then filtered by an emission filter before being detected by the CCD camera (Fig. 17.6a). The exposure time of the sample to the excitation light determines the brightness of the image. This parameter has to be selected carefully to obtain a good signal-to-noise ratio in the image while avoiding saturation of the high intensity pixels. For time-lapse imaging, it is also crucial to take into account the bleaching of the sample to avoid a decrease in the image quality at later time points.

The choice of the objective is always a trade-off between the size of the field of view and the optical resolution. For yeast cells, a 40× objective is usually ideal to quantify overall cellular fluorescence and allows to observe about 100 cells simultaneously. To observe cellular organelles, a higher magnification is necessary (60× or 100×) at the expense of fewer cells being imaged. To increase the statistics of the measurements, it is possible to record images at different locations in the sample. Microscopes are often equipped with highly reliable motorized XY-stages, which allow to repeatedly record multiple fields of view. The trade-off lies now between the number of stage positions to acquire and the dynamics of the process to be quantified. For very fast signaling events, this is clearly a limiting factor (1 min time interval can allow imaging of approx. ten positions).

Image Analysis

While flow cytometric assays directly provide a measure of the total cellular intensity, microscopy provides an image, which needs to be analyzed to extract the desired information. This is a complex task, which requires advanced image analysis algorithms. As shown in Fig. 17.7, a microscopy image is in fact a 3D intensity map of pixel intensities. The thresholding process consists in finding the optimal intensity value to discriminate between background and object pixels. To identify the contour of individual cells, a water-shedding algorithm is often applied, which is based on the shape or intensity information from the segmented objects. Multiple fluorescent tags can be used to identify organelles and generate secondary objects within the cell. Once the final object shape has been obtained, it becomes

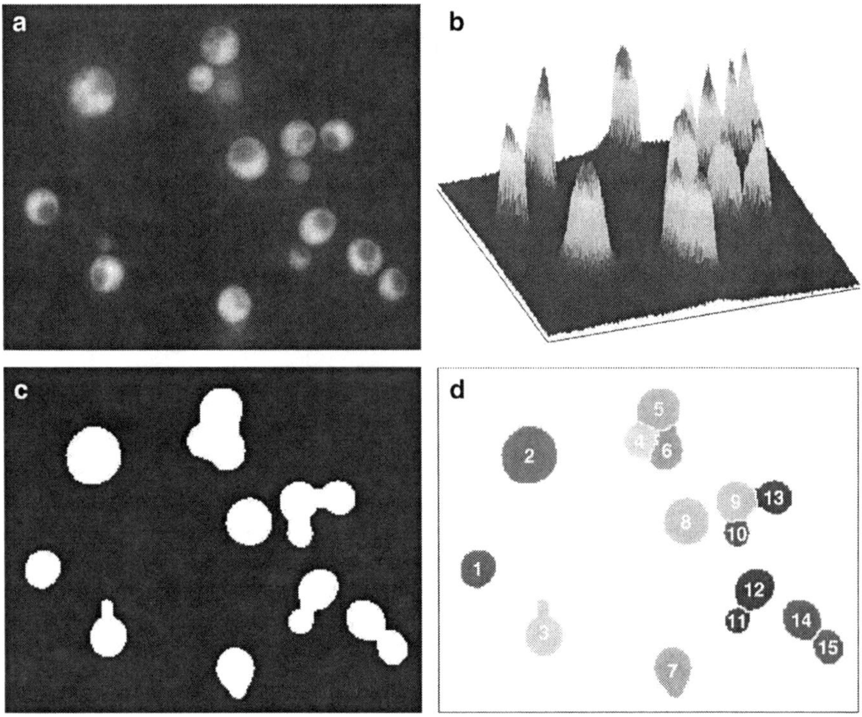

Fig. 17.7 The original image (**a**) can be represented as 3D surface, where each pixel of the image has a given intensity (**b**). An intensity threshold is placed on the image to distinguish background and object pixels (**c**). The thresholded image is split between individual cells with a watershedding algorithm and each cell is labeled (**d**)

relatively straightforward to extract the desired information. Geometrical measurements such as area, diameter or eccentricity can be extracted from the shape of the object. By quantifying the intensity in each pixel of the object it is possible to obtain many more features such as the mean, maximum or total fluorescent intensity of the object.

Multiple software packages have been developed to analyze microscopy images [43, 44]. In addition, ImageJ (http://rsb.info.nih.gov/ij/) and the image processing toolbox from Matlab (Mathworks) offer many low level routines that can be combined to generate a specific analysis workflow combining segmentation, objects recognition and feature measurements.

Applications

Using the same expression reporter described for the flow cytometry assays, it is possible to follow in real time the apparition of the fluorescence signal in a cell after inducing the signaling pathway. Unfortunately, due to the slow maturation of

Table 17.3 Examples of relocation assays for signal transduction quantification

Relocation to	Stimulus	Protein
	General stress	Msn2p [49]
Nucleus	Osmotic shock	Hog1p [47]
	G1 cell cycle stage	Whi5p [45]
Plasma membrane	Mating pheromone	Ste5p [48]
Vacuolar membrane	Glucose levels	Vma5p [50]
Vacuole	Autophagy	Atg8p [51]
Polarity landmark	α-factor gradient	Cdc24p [52]

the protein, a roughly 30 min. delay is observed between the expression of the protein and the apparition of the fluorescence (Fig. 17.5b). The other issue with these types of reporters is the fact that the fluorescent proteins have a long lifetime in the cell. Observing the shut-down of a transcriptional process is therefore problematic because the cells remain fluorescent for many hours after protein production has been stopped. To circumvent this problem, destabilized fluorescent protein have been generated which have a short lifetime [45].

For more dynamic measurements of signaling cascade activation, it is sometimes possible to use a GFP-tagged protein that changes location upon stimulation of the cell. These types of experiments cannot be performed by flow cytomery because there is no net change of fluorescence. The sub-cellular resolution from the microscope is required to detect these events. Table 17.3 lists a few examples of relocation assays that can be used to quantify the activity of different signal transduction cascades.

One commonly used assay is the relocation of a protein to the nucleus of the cell. Many transcription factors or transcriptional regulators can display large variations in their nuclear concentration upon activation [46]. Figure 17.8a illustrates the nuclear relocation of the Hog1p MAP kinase. Within a few minutes after osmotic stress, the change in fluorescence localization is apparent. This nuclear accumulation of the MAPK is closely linked to its activity state [47] and therefore serves as a read-out for the activity of this pathway in single cells [38, 39].

In the mating pathway, activation of the signaling cascade can be assessed by the membrane relocation of the scaffolding protein Ste5p. This assay was used to demonstrate the presence of a negative feedback loop from the MAPK, which tunes down the activation of the pathway after induction [48]. As shown in Fig. 17.8b, Ste5p relocation can already be detected a few minutes after addition of pheromone to the cells. This clearly demonstrates the high temporal resolution, which can be obtained by these assays in contrast to the slow dynamics measured for fluorescent protein expression, which are further hindered by the delay due to maturation of the fluorophore.

One general consideration when using reporters based on full-length proteins or endogenous promoters is that they often integrate multiple signals from various pathways. As an example, the Msn2p nuclear relocation is sensitive to a wide range of stresses such as starvation, osmotic shock or light-induced stress. The response

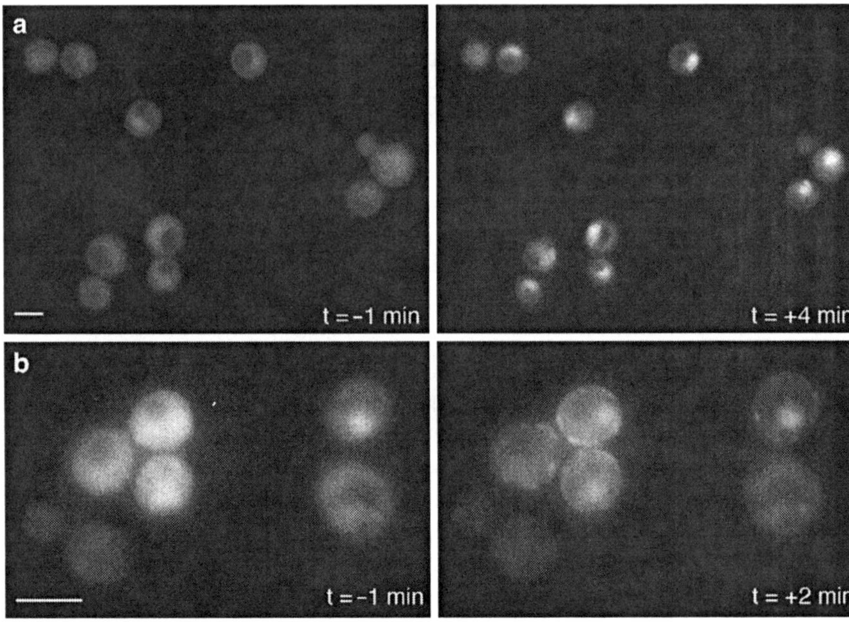

Fig. 17.8 Quantification of signaling by relocation of active proteins. (**a**) The MAPK Hog1p tagged with mCherry accumulates in the nucleus upon activation of the osmo-stress pathway with NaCl. (**b**) Relocation of the scaffod protein Ste5p to the plasma membrane upon treatment with mating pheromone (scale bar 5 μm)

to the stimulus can therefore be strongly influenced by the experimental conditions. In the case of Msn2p, after careful analysis of its phosphorylation pattern, a shorter version of the protein has been constructed to report specifically to glucose starvation and not to other stresses [50]. Engineering of synthetic reporter based on short regulated region of signaling proteins might offer more specificity and sensitivity to the assay.

Functional Microscopy

Activation of a signaling cascade triggers the assembly (or disassembly) of protein complexes and the change in activity of proteins by modifications such as phosphorylation, acetylation or ubiquitination. Relocation and expression assays report on the integration of multiple elements of the signaling cascade. Functional imaging techniques such as Förster resonance energy transfer (FRET) and fluorescence correlation spectroscopy (FCS) techniques can monitor one single interaction or modification in a signaling cascade.

Förster Resonance Energy Transfer

Förster Resonance Energy Transfer is defined as the transfer of excitation energy of a donor to an acceptor fluorophore via dipole-dipole interaction. Due the nature of this interaction, the efficiency of the transfer (E) decreases with the sixth power of the distance between the two dyes R [53, 54].

$$E = 1 - \frac{I_F}{I_D} = \frac{R_0^6}{R_0^6 + R^6}$$

Where I_F and I_D are respectively the fluorescent intensities of the donor in presence and absence of FRET. The Förster distance R_0 corresponds to the separation where 50% of the donor fluorescence is converted into the energy transfer process.

$$R_0^6 = \frac{9000 \ln(10)\, \kappa^2 Q_D}{128\pi^5 N_A n^4} J$$

This parameter depends on the index of refraction of the medium, n; Avogadro's number, N_A; the fluorescence quantum yield of the donor molecule, Q_D; the orientation factor between the two dipoles, κ and the overlap integral, J. This integral scales with the overlap of the emission spectrum of the donor and the absorption spectrum of the acceptor. For a well-matched fluorophore pair, R_0 is on the order of 50–60 Å [55, 56].

Due to the high sensitivity of the FRET process with distance it is possible to probe the association of two proteins *in vivo* by tagging them with a set of fluorescent protein (CFP and YFP or GFP and mCherry). Indeed, unless these two proteins belong to the same complex, it is extremely unlikely for them to come by chance close enough to generate a FRET signal (Fig. 17.9a).

This technique has been used to study the activation of the mating pathway at two different points along the signal transduction cascade. A loss of FRET was observed upon dissociation of the hetero-trimeric G-protein in α-factor treated cells [48, 57]. Moreover, dissociation of the repressor Dig1p from the transcription factor Ste12p has also been probed by FRET. Since this dissociation is triggered by phosphorylation of Ste12p and Dig1p by Fus3p, the loss of FRET reports faithfully on the nuclear activity of the MAPK [48].

The FRET efficiency is commonly quantified by measuring the acceptor sensitization (i.e. the additional fluorescence generated by the FRET process in the acceptor channel upon donor excitation). For FRET measurements, it is necessary to acquire at least three images for the donor, the acceptor and the emission of the acceptor upon donor excitation [58].

Although appealing, the application of this technique is challenging. Indeed, the size of the interacting proteins and of the fluorescent tags can lead to a large distance and unfavorable orientation between the chromophores, which can result in an undetectable FRET signal. Moreover, due to the variation in the level of expression of the two proteins, it is hard to decouple the change in FRET efficiency from the

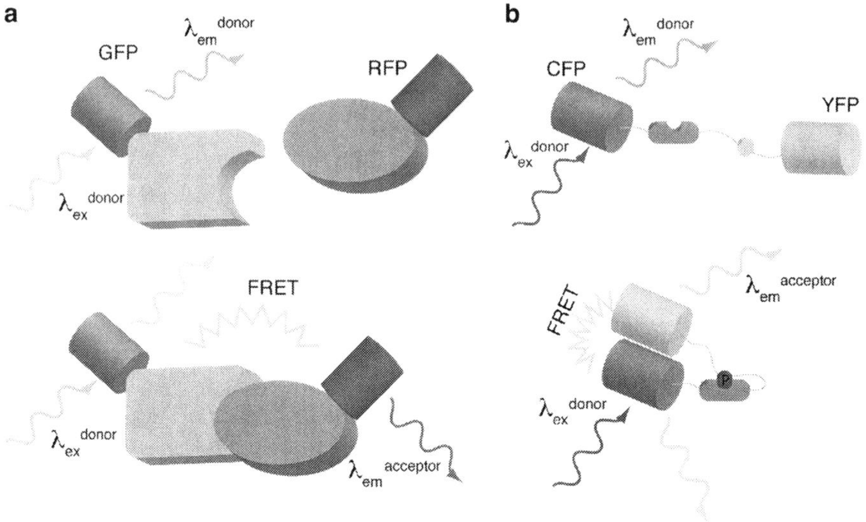

Fig. 17.9 Two different strategies to make use of the FRET process *in vivo* to quantify signal transduction. **(a)** Protein-protein interraction: upon binding of the two fluorescently-tagged proteins, the two flluorescent protein variants come in close proximity and energy transfer can occur. **(b)** FRET sensor: a synthetic construct made of two fluorescent proteins separated by a functional linker which will undergo a large conformation change upon modification or ligand binding

changes in the intensities of the donor and acceptors. Finally, the FRET efficiency measures a combination of the amount of energy transfer and the number of complexes formed. The same efficiency value can be measured from a sample with a high transfer efficiency and only few complexes present or a sample with low energy transfer but with all donors engaged in complexes.

Only the measurement of the fluorescence lifetime of the donor can decouple the efficiency of the FRET process, the amount of complexes formed and the concentration of donor present [59].

$$E = 1 - \frac{\tau_F}{\tau_D}$$

The deactivation of the donor excited-state by FRET shortens the lifetime of the excited state of the dye (τ_F) compared to a non-fretting donor (τ_D). This technique requires a complex acquisition set-up to record the lifetime of the fluorescence in every pixel of the image [60, 61]. This method was used to detect a gradient of MAPK activity arising from the tip of the mating projection. The authors quantified by fluorescence lifetime imaging the FRET between a GFP-tagged Fus3p and a Cy3-labeled antibody specific for the active form of the MAPK [62].

Another way to use Förster resonance energy transfer for signaling studies is to develop FRET sensors. These sensors consist of pair of fluorescent proteins linked by a short peptide. Due to the high sensitivity of the FRET efficiency to chromophore distance and orientation, a change in the conformation of the linking peptide can lead to a strong change in the FRET signal (Fig. 17.9b). This strategy has been used to sense intracellular metabolites by engineering a binding site in the linker region [63, 64]. It is also possible to probe kinase activity by combining in the FRET sensor a phospho-acceptor site for a specific kinase and a sequence binding to this phophorylated residue [65, 66]. This can offer a direct read-out for the activity of a protein upon stimulation of the pathway. The great advantage of this technique is that the two chromophores are linked together and therefore are always expressed at the same level. Many of the problems involved in quantification of the FRET process for protein interactions are thus circumvented. The FRET efficiency can be simply calculated from the ratio between two images acquired with donor excitation and emission filters for the donor or the acceptor.

Fluorescence Correlation Spectroscopy

In contrast to FRET, Fluorescence correlation spectroscopy (FCS) offers the possibility to study protein complex formation quantitatively and with a high sensitivity independently of distance or orientation [67–69]. This technique relies on the measurement of the fluctuations in the fluorescence signal generated by single molecules moving through the confocal volume of a microscope. This volume is on the order of 10^{-14} L and contains on average ten molecules for a concentration of approx. 10 nM. If one of these molecules leaves the observation volume, a drop of 10% in the fluorescent signal should be detected. Based on Poisson statistics, if the average number of molecule is 10, the standard deviation is the square root of 10, and we therefore expect a variation in the signal of roughly 30% over time. The rate at which those fluctuations happen is directly related to the diffusion coefficient of the molecule in the medium and therefore the residence time of each molecule in the confocal volume (Fig. 17.10).

To analyze these fluctuations an auto-correlation of the fluorescence signal as function of time is calculated. The auto-correlation curve $G(\tau)$ will obey to the following equation for a simple diffusion model:

$$G(\tau) = \frac{G_0}{\langle N \rangle} \frac{1}{1 + \frac{\tau}{\tau_D}} \frac{1}{\sqrt{1 + \frac{r_0^2}{Z^2} \frac{\tau}{\tau_D}}}$$

The brightness factor G_0 as well as the lateral r_0 and axial Z dimensions of the confocal volume are parameters that have to be determined experimentally with a reference sample. Interestingly, the amplitude of the correlation function scales with the inverse of the average number of molecules in the volume $\langle N \rangle$. This follows naturally from the fact that relative changes in intensity, when one molecule leaves the confocal volume will be smaller as the concentration increases. Therefore the sensitivity of FCS is maximized at low concentrations. The other parameter which can be extracted from the fitting of the auto-correlation function is the diffusion time of the dye (τ_D). This parameter could be used to measure the binding

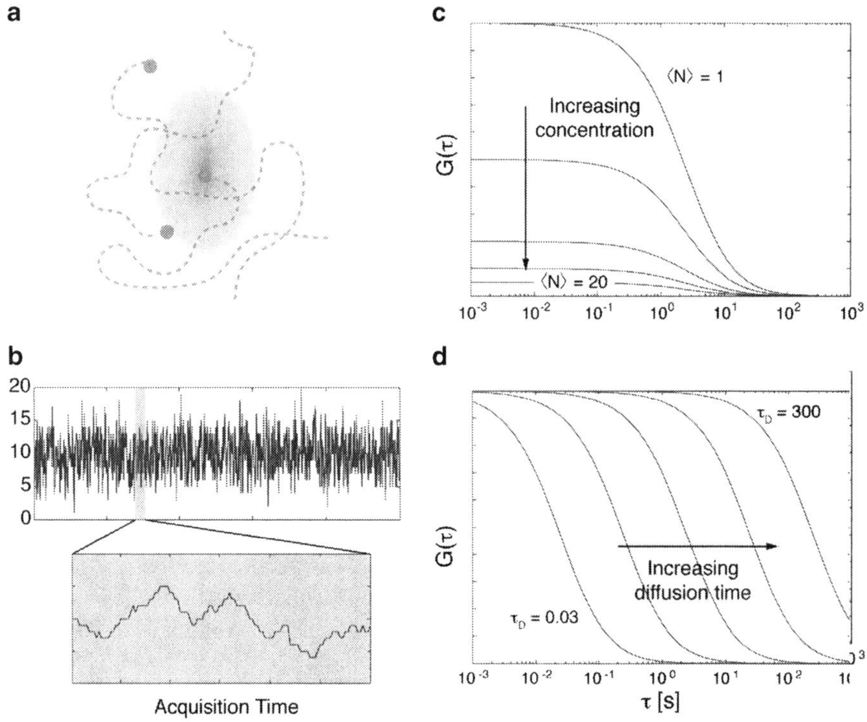

Fig. 17.10 Random diffusion of fluorescent particles through the confocal volume of a microscope (**a**) generates a fluctuating fluorescence signal (**b**). Autocorrelation of these fluctuations allows the determination of the concentration (**c**) and the diffusion time (**d**) of the fluorescently-tagged particle

of a fluorescently-tagged protein to a complex. Unfortunately, since the diffusion constant scales approximately with the cubic root of the mass, the change in size between the free and complex-bound protein needs to be very large to be able to quantify it. It is however possible to probe homo-dimer formation due to a change in brightness of the particles [70].

Studying the interaction of two proteins can be achieved by tagging both of them with different fluorescent protein variants (typically GFP and mCherry). Under these conditions, one can record the fluctuations in both color channels in parallel and calculate a cross-correlation curve [71,72]. If the two proteins of interest are present in the same complex, they will generate similar fluctuations in both detection channels, which will result in an increased cross-correlation signal. This technique allows to obtain the absolute concentration of the proteins along with the ratio of proteins engaged in the complex and therefore enables the characterization of the affinity constant of the two proteins. This method has been applied successfully *in vitro*, but experimental artifacts such as autofluorescence have slowed down its adoption for live cell assays.

Two groups investigated complex formation in the mating pathway by FCS [62, 73]. Maeder et al. studied the pair-wise interactions between Ste11p, Ste7p, Ste5p and Fus3p in the cytoplasm of α-factor treated and untreated cells. They quantified the cytoplasmic concentration of each proteins and by cross-correlation measurements the concentration of the complexes between the two proteins. Their measurements lead them to conclude that there is no change in the cytoplasmic complex formation upon α-factor treatment for any of the complexes investigated. In contrast to this, Slaughter et al. [73] performing similar measurements found a regulated interaction for Fus3p and Ste7p upon α-factor stimulation. They could not detect an interaction between the two proteins in cycling cells but measured the formation of a complex upon α-factor stimulation. Both studies however agree on the fact that most of the change in protein interactions happen at the cell membrane and particularly at the tip of the mating projection were these proteins accumulate. Unfortunately, it is difficult to probe this region by FCS since the proteins will have an increased residence time due to their involvement in membrane bound complexes and therefore will not give rise to fluctuations.

Outlook

Flow cytometry and quantitative microscopy significantly improved our understanding of signal transduction in single cells. While flow cytometry allows to rapidly quantify the total fluorescence of a large number of cells, microscopy allows to follow in real time the dynamics of signal transduction in the same cell by monitoring the relocation or expression of reporter proteins. Using FRET or FCS, it becomes possible to track directly the changes in activity or binding affinity of key signaling molecules.

The pace of the technical development of microscopy and flow cytometry will keep increasing in the coming years and will blend the boundary between these two complementary techniques. The cytometer, which offers excellent statistics, delivers poor information content for each cell. A newer generation of instruments provide the capability of imaging the cells in the flow [74, 75]. This allows to extract more complex features from each cell and will for instance allow to study a nuclear relocation event with high temporal resolution and improved statistics.

On the other side, genome-wide screens have been performed on microscopes, which resulted in the development of high-throughput acquisition and image analysis techniques [76]. These screens, which involved the measurements of millions of cells, were mostly performed for endpoint measurements. Signaling studies with their special needs for high temporal resolution are less prone to such approaches, but a combination of microfluidics and microscopy could solve this issue [77].

The major challenge will remain the development of sensitive and high fidelity sensors for the measurements of signal transduction. We have presented in this chapter a number of tools used to monitor the activation of signaling cascades such as expression reporters, relocation assays or FRET sensors. It is now possible to combine several of these sensors in the same cell to correlate the activity of the pathway at multiple levels in single cells.

Acknowledgements We would like to thank Reinhard Dechant for critical reading of the manuscript. Work in the laboratory of M.P. is supported by Unicellsys, SPMD, the SystemsX.ch initiative (YeastX and LiverX projects), the ETHZ and the Swiss National Science Foundation (SNF).

References

1. Ferrell JE, Machleder EM (1998) The biochemical basis of an all-or-none cell fate switch in Xenopus oocytes. Science 280:895–898
2. Muzzey D, van Oudenaarden A (2009) Quantitative time-lapse fluorescence microscopy in single cells. Annu Rev Cell Dev Biol 25:301–327
3. Santos SDM, Verveer PJ, Bastiaens PIH (2007) Growth factor-induced MAPK network topology shapes Erk response determining PC-12 cell fate. Nat Cell Biol 9:324–330
4. Cai L, Dalal CK, Elowitz MB (2008) Frequency-modulated nuclear localization bursts coordinate gene regulation. Nature 455:485–490
5. Garmendia-Torres C, Goldbeter A, Jacquet M (2007) Nucleocytoplasmic oscillations of the yeast transcription factor Msn2: evidence for periodic PKA activation. Curr Biol 17:1044–1049
6. Kaern M, Elston TC, Blake WJ, Collins JJ (2005) Stochasticity in gene expression: from theories to phenotypes. Nat Rev Genet 6:451–464
7. Elowitz MB, Levine AJ, Siggia ED, Swain PS (2002) Stochastic gene expression in a single cell. Science 297:1183–1186
8. Taniguchi Y et al (2010) Quantifying *E. coli* proteome and transcriptome with single-molecule sensitivity in single cells. Science 329:533–538
9. Colman-Lerner A et al (2005) Regulated cell-to-cell variation in a cell-fate decision system. Nature 437:699–706
10. Amantonico A, Oh JY, Sobek J, Heinemann M, Zenobi R (2008) Mass spectrometric method for analyzing metabolites in yeast with single cell sensitivity. Angew Chem Int Ed Engl 47:5382–5385
11. Monroe E, Jurchen J, Rubakhin S, Sweedler J (2007) Single-cell measurements with mass spectrometry. In: Xu XN (ed) New frontiers in ultrasensitive bioanalysis: advanced analytical chemistry applications in nanobiotechnology, single molecule detection, and single cell analysis. John Wiley & Sons, Inc., New York, p 269
12. Elf J, Li G-W, Xie XS (2007) Probing transcription factor dynamics at the single-molecule level in a living cell. Science 316:1191–1194
13. Xie XS, Choi PJ, Li G-W, Lee NK, Lia G (2008) Single-molecule approach to molecular biology in living bacterial cells. Annu Rev Biophy 37:417–444
14. Tsien RY (1998) The green fluorescent protein. Annu Rev Biochem 67:509–544
15. Zimmer M (2002) Green fluorescent protein (GFP): applications, structure, and related photophysical behavior. Chem Rev 102:759–781
16. Huh W-K et al (2003) Global analysis of protein localization in budding yeast. Nature 425:686–691
17. Shaner NC, Steinbach PA, Tsien RY (2005) A guide to choosing fluorescent proteins. Nat Methods 2:905–909
18. Miesenböck G, De Angelis DA, Rothman JE (1998) Visualizing secretion and synaptic transmission with pH-sensitive green fluorescent proteins. Nature 394:192–195
19. van Drogen F, Peter M (2004) Revealing protein dynamics by photobleaching techniques. Methods Mol Biol 284:287–306

20. van Drogen F, Stucke VM, Jorritsma G, Peter M (2001) MAP kinase dynamics in response to pheromones in budding yeast. Nat Cell Biol 3:1051–1059
21. Patterson GH, Lippincott-Schwartz J (2004) Selective photolabeling of proteins using photoactivatable GFP. Methods 32:445–450
22. Ando R, Hama H, Yamamoto-Hino M, Mizuno H, Miyawaki A (2002) An optical marker based on the UV-induced green-to-red photoconversion of a fluorescent protein. Proc Natl Acad Sci USA 99:12651–12656
23. Goffeau A et al (1996) Life with 6000 genes. Science 274:546, 563–547
24. Chen RE, Thorner J (2007) Function and regulation in MAPK signaling pathways: lessons learned from the yeast Saccharomyces cerevisiae. Biochim Biophys Acta 1773:1311–1340
25. Bardwell L (2005) A walk-through of the yeast mating pheromone response pathway. Peptides 26:339–350
26. Rudolf F, Pelet S, Peter M (2007) Regulation of MAPK signaling in yeast. Top Curr Genet 20:187–204
27. Lamson R, Takahashi S, Winters M, Pryciak PM (2006) Dual role for membrane localization in yeast MAP kinase cascade activation and its contribution to signaling fidelity. Curr Biol 16:618–623
28. Nolan JP, Yang L (2007) The flow of cytometry into systems biology. Brief Funct Genomic Proteomic 6:81–90
29. Drouet M, Lees O (1993) Clinical applications of flow cytometry in hematology and immunology. Biol Cell 78:73–78
30. Shapiro HM (1983) Multistation multiparameter flow cytometry: a critical review and rationale. Cytometry 3:227–243
31. Rieseberg M, Kasper C, Reardon KF, Scheper T (2001) Flow cytometry in biotechnology. Appl Microbiol Biotechnol 56:350–360
32. Gasch AP et al (2000) Genomic expression programs in the response of yeast cells to environmental changes. Mol Biol Cell 11:4241–4257
33. Oehlen LJ, Cross FR (1994) G1 cyclins CLN1 and CLN2 repress the mating factor response pathway at start in the yeast cell cycle. Genes Dev 8:1058–1070
34. Strickfaden SC et al (2007) A mechanism for cell-cycle regulation of MAP kinase signaling in a yeast differentiation pathway. Cell 128:519–531
35. Acar M, Becskei A, van Oudenaarden A (2005) Enhancement of cellular memory by reducing stochastic transitions. Nature 435:228–232
36. Whitesides GM (2006) The origins and the future of microfluidics. Nature 442:368–373
37. Bennett MR, Hasty J (2009) Microfluidic devices for measuring gene network dynamics in single cells. Nat Rev Genet 10:628–638
38. Hersen P, McClean MN, Mahadevan L, Ramanathan S (2008) Signal processing by the HOG MAP kinase pathway. Proc Natl Acad Sci USA 105:7165–7170
39. Muzzey D, Gómez-Uribe CA, Mettetal JT, van Oudenaarden A (2009) A systems-level analysis of perfect adaptation in yeast osmoregulation. Cell 138:160–171
40. Paliwal S et al (2007) MAPK-mediated bimodal gene expression and adaptive gradient sensing in yeast. Nature 446:46–51
41. Charvin G, Cross FR, Siggia ED (2008) A microfluidic device for temporally controlled gene expression and long-term fluorescent imaging in unperturbed dividing yeast cells. PLoS ONE 3:e1468
42. Lee PJ, Helman NC, Lim WA, Hung PJ (2008) A microfluidic system for dynamic yeast cell imaging. Biotechniques 44:91–95
43. Carpenter AE et al (2006) CellProfiler: image analysis software for identifying and quantifying cell phenotypes. Genome Biol 7:R100
44. Gordon A et al (2007) Single-cell quantification of molecules and rates using open-source microscope-based cytometry. Nat Methods 4:175–181
45. Bean JM, Siggia ED, Cross FR (2006) Coherence and timing of cell cycle start examined at single-cell resolution. Mol Cell 21:3–14
46. Cyert MS (2001) Regulation of nuclear localization during signaling. J Biol Chem 276:20805–20808

47. Reiser V, Ruis H, Ammerer G (1999) Kinase activity-dependent nuclear export opposes stress-induced nuclear accumulation and retention of Hog1 mitogen-activated protein kinase in the budding yeast *Saccharomyces cerevisiae*. Mol Biol Cell 10:1147–1161
48. Yu RC et al (2008) Negative feedback that improves information transmission in yeast signalling. Nature 456:755–761
49. Görner W et al (1998) Nuclear localization of the C2H2 zinc finger protein Msn2p is regulated by stress and protein kinase A activity. Genes Dev 12:586–597
50. Dechant R et al (2010) Cytosolic pH is a second messenger for glucose and regulates the PKA pathway through V-ATPase. EMBO J 29:2515–2526
51. Kraft C, Deplazes A, Sohrmann M, Peter M (2008) Mature ribosomes are selectively degraded upon starvation by an autophagy pathway requiring the Ubp3p/Bre5p ubiquitin protease. Nat Cell Biol 10:602–610
52. Shimada Y, Wiget P, Gulli M-P, Bi E, Peter M (2004) The nucleotide exchange factor Cdc24p may be regulated by auto-inhibition. EMBO J 23:1051–1062
53. Förster T (1948) Zwischenmolekulare Energiewanderung und Fluoreszenz. Ann Phys 2:55–75
54. Jares-Erijman EA, Jovin TM (2003) FRET imaging. Nat Biotechnol 21:1387–1395
55. Patterson GH, Piston DW, Barisas BG (2000) Förster distances between green fluorescent protein pairs. Anal Biochem 284:438–440
56. Wu P, Brand L (1994) Resonance energy transfer: methods and applications. Anal Biochem 218:1–13
57. Yi T-M, Kitano H, Simon MI (2003) A quantitative characterization of the yeast heterotrimeric G protein cycle. Proc Natl Acad Sci USA 100:10764–10769
58. Berney C, Danuser G (2003) FRET or no FRET: a quantitative study. Biophys J 84:3992–4010
59. Pelet S, Previte MJR, So PTC (2006) Comparing the quantification of Forster resonance energy transfer measurement accuracies based on intensity, spectral, and lifetime imaging. J Biomed Opt 11:34017
60. Becker W et al (2004) Fluorescence lifetime imaging by time-correlated single-photon counting. Microsc Res Tech 63:58–66
61. Verveer PJ, Squire A, Bastiaens PIH (2001) Frequency-domain fluorescence lifetime imaging microscopy: a window on the biochemical landscape of the cell. In: Periasamy A (ed) Methods in cellular imaging. Oxford University Press, New York, pp 273–294
62. Maeder CI et al (2007) Spatial regulation of Fus3 MAP kinase activity through a reaction-diffusion mechanism in yeast pheromone signalling. Nat Cell Biol 9:1319–1326
63. Fehr M, Ehrhardt DW, Lalonde S, Frommer WB (2004) Minimally invasive dynamic imaging of ions and metabolites in living cells. Curr Opin Plant Biol 7:345–351
64. Fehr M, Frommer WB, Lalonde S (2002) Visualization of maltose uptake in living yeast cells by fluorescent nanosensors. Proc Natl Acad Sci USA 99:9846–9851
65. Ting AY, Kain KH, Klemke RL, Tsien RY (2001) Genetically encoded fluorescent reporters of protein tyrosine kinase activities in living cells. Proc Natl Acad Sci USA 98:15003–15008
66. Harvey CD et al (2008) A genetically encoded fluorescent sensor of ERK activity. Proc Natl Acad Sci USA 105:19264–19269
67. Magde D, Elson E, Webb W (1972) Thermodynamic fluctuations in a reacting system – measurement by fluorescence correlation spectroscopy. Phys Rev Lett 29:705–708
68. Schwille P, Haupts U, Maiti S, Webb WW (1999) Molecular dynamics in living cells observed by fluorescence correlation spectroscopy with one- and two-photon excitation. Biophys J 77:2251–2265
69. Schwille P (2001) Fluorescence correlation spectroscopy and its potential for intracellular applications. Cell Biochem Biophys 34:383–408
70. Chen Y, Müller JD, So PT, Gratton E (1999) The photon counting histogram in fluorescence fluctuation spectroscopy. Biophys J 77:553–567
71. Hwang LC, Wohland T (2007) Recent advances in fluorescence cross-correlation spectroscopy. Cell Biochem Biophys 49:1–13
72. Schwille P, Meyer-Almes FJ, Rigler R (1997) Dual-color fluorescence cross-correlation spectroscopy for multicomponent diffusional analysis in solution. Biophys J 72:1878–1886

73. Slaughter BD, Schwartz JW, Li R (2007) Mapping dynamic protein interactions in MAP kinase signaling using live-cell fluorescence fluctuation spectroscopy and imaging. Proc Natl Acad Sci USA 104:20320–20325
74. George TC et al (2004) Distinguishing modes of cell death using the ImageStream multispectral imaging flow cytometer. Cytometry A 59:237–245
75. George TC et al (2006) Quantitative measurement of nuclear translocation events using similarity analysis of multispectral cellular images obtained in flow. J Immunol Methods 311:117–129
76. Pepperkok R, Ellenberg J (2006) High-throughput fluorescence microscopy for systems biology. Nat Rev Mol Cell Biol 7:690–696
77. Taylor R et al (2009) Dynamic analysis of MAPK signaling using a high-throughput microfluidic single-cell imaging platform. Proc Natl Acad Sci USA

Index

H. Koeppl et al. (eds.), *Design and Analysis of Biomolecular Circuits: Engineering Approaches to Systems and Synthetic Biology*, DOI 10.1007/978-1-4419-6766-4,

D